Advances in Plant Breeding Strategies

Volume 8

Series Editors

Jameel M. Al-Khayri, Department of Agricultural Biotechnology, King Faisal University, Al-Hassa, Saudi Arabia

Shri Mohan Jain, Department of Agricultural Sciences, University of Helsinki, Helsinki, Finland

This book series is intended to provide a platform for plant scientists to present the latest advances in both innovative and traditional methodologies applied in plant breeding for the development of new crop varieties under different environmental conditions leading to sustainable food production, food and nutrition security, food supply chain, germplasm conservation and utilization. This series is an excellent reference source for plant breeders, geneticists, biotechnologists, and molecular biologists using molecular tools, bioinformatics, artificial intelligence, and other emerging technologies. The subject matter of all volumes will relate to postgraduate students specializing in agriculture, biotechnology, and molecular breeding as well as policymakers, seed companies, researchers, and professionals interested in plant genetic improvement.

Jameel M. Al-Khayri · Shri Mohan Jain ·
Suprasanna Penna
Editors

Biodiversity and Genetic Improvement of Herbs and Spices

Springer

Editors
Jameel M. Al-Khayri
Department of Agricultural Biotechnology
King Faisal University
Al-Ahsa, Saudi Arabia

Shri Mohan Jain
Department of Agricultural Sciences
University of Helsinki
Helsinki, Finland

Suprasanna Penna
Amity Center for Nuclear Biotechnology
Amity Institute of Biotechnology
Amity University of Maharashtra
Mumbai, Maharashtra, India

ISSN 3004-8737 ISSN 3004-8745 (electronic)
Advances in Plant Breeding Strategies
ISBN 978-3-031-81837-0 ISBN 978-3-031-81838-7 (eBook)
https://doi.org/10.1007/978-3-031-81838-7

This Springer imprint is published by the registered company Springer Nature Switzerland AG
The registered company address is: Gewerbestrasse 11, 6330 Cham, Switzerland

Preface of the Series

Contemporary plant breeders rely on conventional and emerging methodologies to provide sustainable food production worldwide. However, the human population is growing at an alarming rate, especially in developing countries, and food availability to feed new mouths could gradually become a serious problem. Agriculture production is negatively impacted by environmental pollution, rapid industrialization, water scarcity, soil erosion of fertile topsoil, limited areas of potentially arable land, lack of improvement of local plant types, erosion of genetic diversity, and dependence on just a few major crop species for the world's food supplies. Breeding programs of these crops have been very much dependent on the ready availability of genetic variation, whether spontaneous or induced. Plant breeders and geneticists are under constant pressure to sustain food production by using innovative breeding strategies and enhancing minor crops that are well adapted to the marginal land and source of nutrition as well as abiotic and biotic stresses. An inherent limitation in traditional breeding is the introgression of one or few genes in a cultivar which is carried out via backcrossing for several generations requiring extended time.

With the development of new molecular tools, molecular marker-assisted breeding helped rapid introgression of transgene and reduction of linkage drag. The continued development and adaptation of plant biotechnology approaches including genetic engineering, genome editing, molecular markers, genomics, proteomics, bioinformatics, artificial intelligence, and nanotechnology have established new tools for the creation, analysis, and manipulation of genetic variation for the development of improved cultivars to achieve agricultural sustainability and food security. A multidisciplinary approach combining traditional plant breeding, mutation breeding, plant biotechnology, and molecular biology would be strategically ideal for developing improved crop varieties to feed the world, provide natural pharmaceutical sources, and supply fiber and ornamental crops for industry, esthetic and environmental suppose. However, breeders are challenged by the constraints of environmental stress, the limitation of arable land expansion, and erosion of genetic resources particularly under the impact of global climate change.

This book series is intended to provide a platform for plant scientists to present the latest advances in both innovative and traditional methodologies applied in plant breeding for the development of new crop varieties under different environmental conditions leading to sustainable food production, food and nutrition security, food supply chain, germplasm conservation, and plant sustainable utilization. This series is an excellent reference source for plant breeders, geneticists, biotechnologists, and molecular biologists using molecular tools, bioinformatics, artificial intelligence, and other emerging technologies. The subject matter relates to postgraduate students specializing in agriculture, biotechnology, and molecular breeding as well as policymakers, seed companies, researchers, and professionals interested in plant genetic improvement.

Al-Ahsa, Saudi Arabia — Jameel M. Al-Khayri
Helsinki, Finland — Shri Mohan Jain

Preface

Research and development in crop plants has always been on the forefront of breeding programmes aiming at high productivity. Herbs and spices have been used since prehistoric times to enrich foods as preservatives, flavor agents and, in healthcare. There have been instances of their usage throughout the world, in different countries. The World Health Organization has estimated that 70–80% of the world's population depends on herbal sources in their healthcare. A great deal of research focus is now given to improve the herbs and spices for improvement, phytochemistry and plant health.

This volume 10: *Biodiversity and Genetic Improvement of Herbs and Spices* presents the breeding objectives, genetic resources, molecular markers, biotechnological advances to aid plant improvement programs. The book has 17 chapters covering important herbs and spices: Bay leaf, Cinnamon, Coriander, Dill, Lemongrass, Long pepper, Nutmeg, Onion, Oregano, Parsley, Pepper, Saffron, Sage, Sesame, Thyme, Common Turmeric, Scentless Turmeric. Chapters, written by experts, begin with an introduction, detailed discussion of the topic accompanied by high-quality color images, and conclude with future study directions. Chapters comprehensively review the contemporary literature on the subject and reflect the experiences of the authors. Chapters conclude with future research perspective and a comprehensive list of pertinent references.

The book is an excellent reference source for plant biologists, breeders, biotechnologists and geneticists engaged in the improvement of Herbs and Spices. It is suitable for both advanced undergraduate postgraduate students and researchers specializing in agriculture, biotechnology, and molecular breeding as well as entrepreneurs and policymakers concerned with analysis improvement of plant varieties.

The editors would like to sincerely thank the authors from different countries for their efforts and scientific contributions and, to Springer team for the constant support during the publication of this book.

Al-Ahsa, Saudi Arabia Jameel M. Al-Khayri
Helsinki, Finland Shri Mohan Jain
Mumbai, India Suprasanna Penna

Acknowledgment The editors extend their appreciation to the support of the Deanship of Scientific Research, Vice Presidency for Graduate Studies and Scientific Research, King Faisal University, Saudi Arabia (Grant No. KFU251077).

Contents

Editors and Contributors

About the Editors

Prof. Jameel M. Al-Khayri is affiliated with the Department of Agricultural Biotechnology, King Faisal University, Saudi Arabia. He received B.S. in Biology in 1984 from the University of Toledo, M.S. in Agronomy in 1988, and Ph.D. in Plant Science in 1991 from the University of Arkansas. He is the National Correspondent of the International Association of Plant Tissue Culture and Biotechnology. Al-Khayri has authored 130 research articles, 70 chapters and edited several journal special issues. In addition, he edited 30 reference books on plant biotechnology, genetic resources, breeding, genomics and nanotechnology. In addition to teaching, advising and research, he held administrative responsibilities as the Assistant Director of Date Palm Research Center, Head of Department of Plant Biotechnology and Vice Dean for Development and Quality Assurance. Al-Khayri served as a Member of Majlis Ash-Shura (Saudi Legislative Council) for the 2009–2012 term. He is interested in agricultural innovations and sustainability.

Prof. Shri Mohan Jain, (Adjunct) is currently associated with the Department of Agricultural Sciences, University of Helsinki, Helsinki, Finland. Received M.Phil. 1973 and Ph.D., 1978, Jawaharlal Nehru University, New Delhi, India. He served as a Technical Officer, Plant Breeding, and Genetics, International Atomic Energy Agency (IAEA), Vienna, Austria, 1999–2005 when the agency was awarded Nobel Peace Prize, 2005. He is on the Editorial Board of *Euphytica*, *In Vitro*, *Propagation of Ornamental Plants*; a reviewer in plant cell and organ culture, *Plant Cell Reports*, and *Plants*. His publications are more than 225 peer-reviewed journals, book chapters, and conference proceedings, and edited 72 books; invited speaker and acted as a Chairperson in several international conferences worldwide. He is a former consultant to the IAEA, the European Union, The Government of Grenada, the Iranian Private Company, and the Egyptian Government.

Prof. Suprasanna Penna, Ph.D. (Genetics) is the Director of Amity Institute of Biotechnology, Amity Center for Nuclear Biotechnology, Amity University of Maharashtra, Mumbai, India. He was formerly with Bhabha Atomic Research Centre, Mumbai, as the Head of Nuclear Agriculture and Biotechnology Division. He is also an IAEA Expert consultant. His significant contributions include plant biotechnology, mutation breeding, stress tolerance, and biopolymers for enhancing plant productivity in plants. His research on plant abiotic stress tolerance in crop plants and halophytes led to elucidating mechanisms of salinity tolerance. He is on the editorial board of several international journals and has published over 400 publications. He has edited several books on plant mutation breeding, salinity tolerance, plant genetic diversity, plant biotechnology and plant-metal interactions and was the Guest Editor of special issues of International journals.

Contributors

Sawsan Abd-Ellatif Department of Bioprocess Development, Genetic Engineering and Biotechnology Research Institute (GEBRI), City of Scientific Research and Technology Applications (SRTA-City), Borg EL-Arab, Alexandria, Egypt

Mohamed A. Abdein Seeds Development Department, El-Nada Misr Scientific Research and Development Projects, Mansoura, Egypt

Jameel M. Al-Khayri Department of Agricultural Biotechnology, College of Agriculture and Food Sciences, King Faisal University, Hofuf, Al-Ahsa, Saudi Arabia

Suliman M. S. Alghanem Department of Biology, College of Science, Qassim University, Buraydah, Saudi Arabia

Haifa A. S. Alhaithloul Department of Biology, College of Science, Jouf University, Sakaka, Saudi Arabia

Abhishek Anand The Neotia University, Kolkata, West Bengal, India

Adithya Kodagu Anand Department of Studies in Biotechnology, School of Sciences, University of Mysore, Mysore, Karnataka, India

Chole Pranjali Bajrang Department of Life Sciences, School of Sciences, Christ University, Bengaluru, Karnataka, India

Akshatha Banadka Department of Life Sciences, School of Sciences, Christ University, Bengaluru, Karnataka, India;
Department of Biotechnology and Genetics, School of Sciences, JAIN (Deemed to be University), Bengaluru, Karnataka, India

Anastasia Boutsika Institute of Plant Breeding and Genetic Resources, Hellenic Agricultural Organization—Dimitra (ELGO-Dimitra), Thermi, Thessaloniki, Greece

Vahdettin Ciftci Department of Field Crops, Faculty of Agriculture, Bolu Abant Izzet Baysal University, Bolu, Türkiye

Maha M. Elshamy Department of Botany, Faculty of Science, Mansoura University, Mansoura, Egypt

Jaya Gangwar Department of Life Sciences, Christ University, Bengaluru, Karnataka, India

Ioannis Ganopoulos Institute of Plant Breeding and Genetic Resources, Hellenic Agricultural Organization—Dimitra (ELGO-Dimitra), Thermi, Thessaloniki, Greece

Zsuzsanna György Group of Horticultural Plant Genetics, Department of Plant Biotechnology, Institute of Genetics and Biotechnology, Hungarian University of Agriculture and Life Sciences, Budapest, Hungary

Amira A. Ibrahim Department of Botany and Microbiology, Faculty of Science, Arish University, Al-Arish, Egypt

Juhi Puthukulangara Jaison Department of Life Sciences, Christ University, Bengaluru, Karnataka, India

Kadanthottu Sebastian Joseph Department of Life Sciences, Christ University, Bengaluru, Karnataka, India

Kalliopi I. Kadoglidou Institute of Plant Breeding and Genetic Resources, Hellenic Agricultural Organization—Dimitra (ELGO-Dimitra), Thermi, Thessaloniki, Greece

Md Kamaruzzaman The Neotia University, Kolkata, West Bengal, India

Sushil K. Kothari The Neotia University, Kolkata, West Bengal, India

Muhittin Kulak Department of Herbal and Animal Production, Vocational School of Technical Sciences, Iğdır University, Iğdır, Türkiye

Sanjeet Kumar College of Agriculture, Acharya Narendra Deva University of Agriculture & Technology (ANDUA&T), Ayodhya, India

Nattika Kumsee School of Agricultural Technology, King Mongkut's Institute of Technology (KMITL), Ladkrabang, Bangkok, Thailand

B. T. Manjunath Department of Life Sciences, School of Sciences, CHRIST (Deemed to be University), Bengaluru, Karnataka, India

Ifigeneia Mellidou Institute of Plant Breeding and Genetic Resources, Hellenic Agricultural Organization—Dimitra (ELGO-Dimitra), Thermi, Thessaloniki, Greece

Mohamed F. Mohamed Department of Vegetable Crops, Faculty of Agriculture, Assiut University, Assiut, Egypt

Praveen Nagella Department of Life Sciences, School of Sciences, Christ University, Bengaluru, Karnataka, India

Ponnam Naresh Division of Vegetable Crops, ICAR-Indian Institute of Horticultural Research (ICAR-IIHR), Bengaluru, India

Sarita Pandey The Neotia University, Kolkata, West Bengal, India

Madhavan Parthasarathy Department of Life Sciences, School of Sciences, Christ University, Bengaluru, India

Zsuzsanna Pluhár Department of Medicinal and Aromatic Plants, Institute of Horticultural Science, Hungarian University of Agriculture and Life Sciences, Budapest, Hungary

Roshni Pydi Division of Vegetable Crops, ICAR-Indian Institute of Horticultural Research (ICAR-IIHR), Bengaluru, India

Nalumakkal Mini Raj College of Agriculture, Vellanikkara, Kerala Agricultural University, Thrissur, India

K. R. Rakesh Rao Department of Studies in Biotechnology, School of Sciences, University of Mysore, Mysore, Karnataka, India

Rashmi Ramakrishnan Department of Life Sciences, School of Sciences, Christ University, Bengaluru, India

El-Sayed S. Abdel Razik Department of Livestock Research, Arid Lands Cultivation Research Institute (ALCRI), City of Scientific Research and Technology Applications (SRTA-City), Borg EL-Arab, Alexandria, Egypt

Sneha Sabu Friedrich schiller University Jena, Jena, Germany;
Erfurt Research Centre for Horticultural Crops (FGK), University of Applied Sciences, Erfurt, Erfurt, Germany

Khaled F. M. Salem Department of Biology, College of Science and Humanitarian Studies, Shaqra University, Qwaieah, Saudi Arabia;
Department of Plant Biotechnology, Genetic Engineering and Biotechnology Research Institute (GEBRI), University of Sadat City, Sadat City, Egypt

Gholam-Reza Sharifi-Sirchi Department of Agricultural Biotechnology, Faculty of Agriculture, Shahid Bahonar University of Kerman, Kerman, Iran

Sibel Turan Sirke Department of Agricultural Biotechnology, Faculty of Agriculture, Erciyes University, Kayseri, Türkiye

Wudali Narasimha Sudheer Department of Life Sciences, School of Sciences, Christ University, Bengaluru, Karnataka, India

Penna Suprasanna Amity Institute of Biotechnology, Amity University Maharashtra, Mumbai, India

Patcharaporn Suwor School of Agricultural Technology, King Mongkut's Institute of Technology (KMITL), Ladkrabang, Bangkok, Thailand

Azza A. Tawfik The Faculty of Agriculture, Assiut University, Assiut, Egypt

V. L. Vasantha Department of Life Sciences, School of Sciences, CHRIST (Deemed to be University), Bengaluru, Karnataka, India

Hasirumane Chinnappagowda Vikram Department of Plantation Crops and Spices, College of Agriculture, Vellanikkara, Kerala Agricultural University, Thrissur, India

Pallavi Vyas Amity Institute of Biotechnology, Amity University Maharashtra, Mumbai, India

Mai M. Wahba Department of Botany, Faculty of Science, Mansoura University, Mansoura, Egypt

Aliki Xanthopoulou Institute of Plant Breeding and Genetic Resources, Hellenic Agricultural Organization—Dimitra (ELGO-Dimitra), Thermi, Thessaloniki, Greece

Abdurrahim Yilmaz Department of Field Crops, Faculty of Agriculture, Bolu Abant Izzet Baysal University, Bolu, Türkiye

Chapter 1
Bay Leaf (*Laurus nobilis* L.) Breeding

Abdurrahim Yilmaz, Sibel Turan Sirke, Vahdettin Ciftci, and Muhittin Kulak

Abstract Bay (*Laurus nobilis* L.), one of the most well-known and effective fragrant and medicinal plants in Turkey, has a wide range of applications, including tincture, tea and medicine, aromatherapy, and the pharmaceutical industry because of its distinctive components. It is mainly widespread in Turkey, but also in countries such as Greece, Algeria, Morocco, France, Belgium, Mexico, Spain, Portugal, Libya, Romania, Albania, Syria, Crimea and the Canary Islands. Being collected from the wild and not cultivated, a standard variety of the plant has not been developed and introduced hitherto. Bay leaves are used as a fragrant flavoring and spice in foods such as fish, meat, stew, soups, casseroles, puddings, and in vinegar and some other beverages. The plant has not been handled for classical breeding studies. Instead, the plant has been widely monitored for its chemical content and composition and assayed for its biological activities, especially antioxidants and antimicrobials. However, it is very important to reveal the existing wealth of laurel genetic resources through studies and to use them with a focus on developing new varieties. In this regard, there are some genetic diversity and molecular characterization studies conducted on laurel genotypes. Concerning its potential market, Turkey meets approximately 90% of the world's bay market. For this reason, the plant deserves to be investigated and evaluated more in Turkey. Herewith in the chapter, we addressed the plant's

A. Yilmaz (✉) · V. Ciftci
Department of Field Crops, Faculty of Agriculture, Bolu Abant Izzet Baysal University, 14030 Bolu, Türkiye
e-mail: abdurrahimyilmaz@ibu.edu.tr

V. Ciftci
e-mail: vahdettinciftci@ibu.edu.tr

S. T. Sirke
Department of Agricultural Biotechnology, Faculty of Agriculture, Erciyes University, 38030 Kayseri, Türkiye
e-mail: sibelturan@erciyes.edu.tr

M. Kulak
Department of Herbal and Animal Production, Vocational School of Technical Sciences, Iğdır University, Iğdır, Türkiye
e-mail: muhittin.kulak@igdir.edu.tr

J. M. Al-Khayri et al. (eds.), *Biodiversity and Genetic Improvement of Herbs and Spices*, Advances in Plant Breeding Strategies 8, https://doi.org/10.1007/978-3-031-81838-7_1

economic and medicinal values as well as phenomics and genomics in detail. This information will benefit scientists concerned about the plant's breeding and future.

Keywords Bay · Laurel · Essential oil · Aromatic · Soap · Aromatherapy · Perfumery · Antioxidant · Medicinal plant · Turkey

1.1 Introduction

The *Lauraceae* family, as a woody plant family, is represented by 50 genera, including 2500–3000 species, being spread in the subtropical and tropical regions of the world (Carpenter et al. 2010). The family species has great economic significance for their uses as spices and essential oils (Oliveira-Filho et al. 2015). Of those species, Bay (*Laurus nobilis* L.) has been not only an aromatic plant but also a cleansing agent for thousands of years (Dobroslavic et al. 2022; Paparella et al. 2022). Leaf essential oils are widely evaluated in the perfumery industry (perfume, soap, and body lotions) and the pharmaceutical industry. Furthermore, the plants have a wide array of uses, viz. tincture, tea, and aromatherapy. Also, the relevant plant is used as a food preservative in Turkey owing to its antimicrobial and antifungal properties (Yilmaz and Ciftci 2021a).

On the global export scale, Turkey is the primary producer of the bay, with export to 64 countries. The annual production amount is considered to be between 7000 and 7500 tons (Paparella et al. 2022). According to the SCOPUS database (Fig. 1.1), also, Turkey is in the first order in terms of published documents, with 192 papers. Then, Turkey is followed by Italy (181), the United States (111), Spain (69), Tunisia (51), Iran (45), Brazil (43), France (37), Germany (37), and Greece (36). Since the bay plant has not been cultivated, it is generally collected from nature. For this reason, bay areas are increasingly being destroyed because of excessive collection and bad-agricultural practices (Yilmaz and Ciftci 2021a). Excessive cuttings cause a waste of genetic resources (Ayanoğlu et al. 2018). Since the genetic resources of the bay cannot be evaluated sufficiently in studies on breeding, no developed, registered, and introduced a variety of bay is present. In addition, the production amount is not at a sufficient level against the increasing demands. For this reason, it is essential to reveal the richness of bay genetic resources with the studies to be carried out and to use with a focus on developing new varieties (Yilmaz and Ciftci 2021a).

1.1.1 Name

It is called as bay, laurel, sweet bay, bay laurel (English); laurel (Spanish); laurier sauce, laurier (French); lorbeerbaum, lorbeer (German); ye gui (Chinese); gekke iju (Japanese); lauro, alloro (Italian); habbul gar, rend gar (Arabic); laur brogorodry

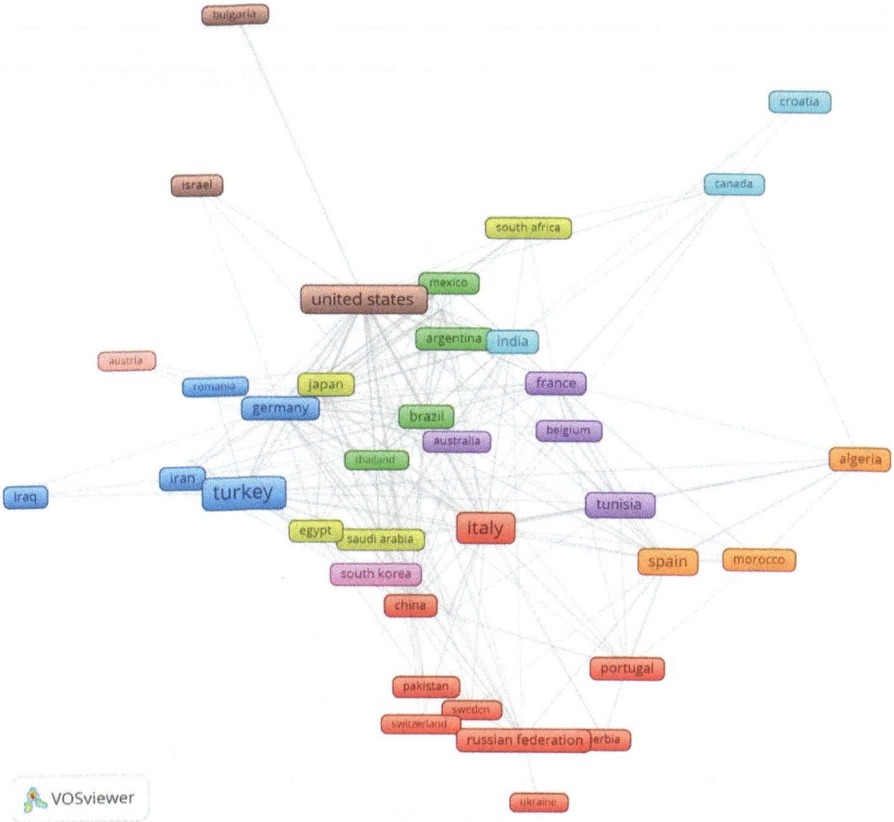

Fig. 1.1 Countries retrieved from the documents of *Laurus nobilis*. According to the SCOPUS search on August 31, 2022, 1119 documents were recorded using the "*Laurus nobilis*" TITLE-ABS-KEY section. The retrieved documents were subjected to visualization analysis using VOSviewer (https://www.vosviewer.com/). Accordingly, 101 countries were obtained, but after setting criteria as a minimum five number of documents, 46 countries met the threshold and were used for analysis

(Russian); lager (Swedish); laakerinlehti, laarkeripuu (Finnish) and wacurnzyn szlachetny (Polish) (Yilmaz 2020).

1.1.2 *Origin and Distribution*

Its popularity dates back to ancient Rome and Greece and is considered a sign of victory and a symbol of peace in sports and military competitions. At that time, bay branches were crossed to form crowns placed on the winners' heads in competitions. This expressed high respect, recognition, and honor. The poets who won the bay wreath were "graduated poets," and thus, those who deserved the tremendous royal

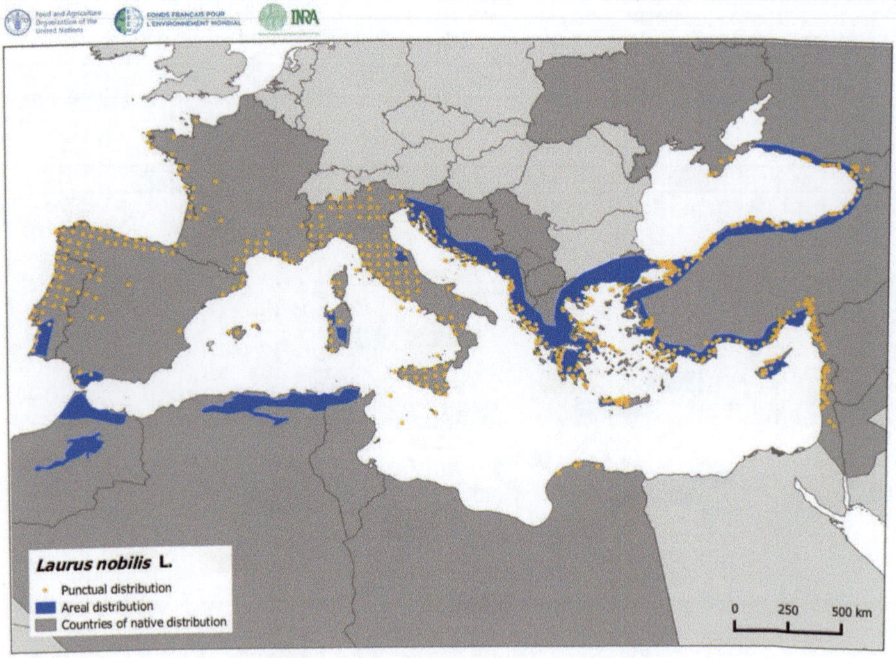

Fig. 1.2 Geographic distribution of bay in the Mediterranean basin. *Source*: FAO

praise were called "nobilis" (Anzano et al. 2022). With all this mythological history, it is seen that the origin of the bay plant is Mediterranean countries in some sources, and Asia Minor (Anatolia) and the Balkans in some sources. The bay plant, which has been seen on the entire coastline of the Mediterranean since ancient times, spreads in countries such as Turkey, Greece, Algeria, Morocco, France, Belgium, Mexico, Spain, Portugal, and the Canary Islands. Other distribution areas are Libya, Romania, Albania, Syria, Crimea, and all other Western Mediterranean Basin (Fig. 1.2; Yilmaz and Ciftci 2021a).

1.1.3 Phenomics

Gene discovery and its introduction to farm conditions are significant challenges under ever-changing environmental conditions. In this regard, a holistic approach in plant biology requires scaling up from the leaf to the whole plant, from the gene to the biochemical process and crop. Plant phenomenology is a critical predictor of estimating the relevant traits/measurements (Furbank 2009). Bay is bush-type evergreen and can grow up to 2.15 m in length Davis (1982) (Fig. 1.3). Its aromatic scented leaves are alternate, narrow, and oblong-lanceolate (Conforti et al. 2006). The leaves are elliptical-lanceolate, wavy margins, leathery, dark green, 3–4 cm wide, and

5–8 cm long (Afifi et al. 1997). This plant, which is not overly branched, reaches a height of approximately 20–30 ft (Said and Hussein 2014). Its leaves have a balmy, fragrant, sweet, and aromatic solid odor (Shokoohinia et al. 2014). The plant flowers in April and matures in September (Sari et al. 2006). The flowers of the bay are four-lobed; the male plant has 8–12 stamens, while the female has 2–4 staminodes. The female flowers are pale yellow with a single pistil with an ovary and a superior ovary containing four staminodes 4–5 mm long. Male flowers have 8–14 staminodes, are 6–7 mm long, and are pale yellow. Female plants can be differentiated from males based on their flower production, and the opening of a flower co-occurs (8–9 am) on both female and male plants. Flower production on a single branch is higher in male plants, but the lifespan of female plants is more prolonged than male plants (Pacini et al. 2014). Bay anthers are opened with a valve mechanism specific to a small number of families, including the *Lauraceae* family, close during high humidity and rain. (Hufford 1996).

Fig. 1.3 Bay plant view from Turkey (Nadeem et al. 2018a)

Franchi et al. (2011) divided plants into four primary groups based on dehydration: orthodox pollen and recalcitrant pollens, recalcitrant pollen and orthodox seeds, orthodox pollen and seeds, and recalcitrant pollen and seeds. Bay belongs to a group with both seeds and recalcitrant pollen, as it requires 30% water content for both seed and pollen dispersal. It has been reported that the viability of bay pollen is less than one day (Pacini et al. 2014), similar to persistent pollen (Nepi et al. 2001). The fruit ripening in the bay takes place in September–October, and fruit drop begins if it is not harvested on time. The fruit is black, oval, or ellipsoid shaped and 10–15 mm in size (Conforti et al. 2006; Sari et al. 2006).

1.1.4 Economic Importance and Health Benefits

Owing to their aromatic properties, bay leaves are highly interested in economics in Turkey, France, Spain, Italy, Portugal, Mexico, Morocco, and Algeria (Karik et al. 2019; Yilmaz and Ciftci 2021a). In Turkey, bay leaves are used as spices and aromatic flavorings for fish, meat, soups, stews, pudding, vinegar, and beverages. It's an essential and fixed oil used for food, cosmetic and medicinal purposes (Yilmaz and Ciftci 2021a). On the other hand, it is seen that early and late harvests affect the quality of the plants negatively (Bak et al. 2021). Therefore, harvesting at the right time is essential for high-quality production.

The consumption of native species increases daily due to its beneficial effects on health (Muradoğlu et al. 2021). Bay leaves are widely used in folk medicine to treat gastrointestinal, rheumatism, and urinary problems (Yilmaz and Ciftci 2021b). Bay leaves have antioxidant, antifungal, antimicrobial, antibacterial, anti-inflammatory, anticonvulsant, antihypertensive, antiseptic, analgesic, diaphoretic, curing diabetes, and relieving stomach disorders properties (Afifi et al. 1997; Sayyah et al. 2002, 2003; Nayak et al. 2006; Ramos et al. 2012; Yilmaz et al. 2013; Bayar et al. 2018; Sıriken et al. 2018; Mssillou et al. 2020; Stefanova et al. 2020; Guedouari and Nabiev 2021; Mohammed et al. 2021; Bouadid et al. 2023; Khalil et al. 2024). Information about the health benefits of the bay compiled from various studies, is given in Table 1.1.

After a basic search on *L. nobilis*, we analyzed the documents using visualization analysis. Most of the research have been addressed essential oil and its antioxidant and antibacterial activities. Figure 1.4 confirms and supports the limited uses of the plants. However, the other reports also utter that the bay plant is suitable for fatigue, indigestion, insomnia, menstrual irregularities, and rheumatic diseases (Baytop 1984; Acartürk 1997; Duke 1997; Özhatay et al. 1997; Sayyah et al. 2002; Sayyah et al. 2003; Simic et al. 2003; Rodilla et al. 2008). It is also used as a natural dye in the cosmetics, pharmaceutical, and food industries for its anthocyanin content in fruit (Ramadan et al. 2019). Especially in recent years, due to the great interest in medicinal plants with antioxidant and antimicrobial properties (Yilmaz et al. 2021a), the value of the bay plant in the health sector is increasing daily.

Table 1.1 Health benefits of bay according to some academic research

Health benefits	References
Antibacterial	(Ramos et al. 2012; Sırıken et al. 2018; Tural et al. 2019; Stefanova et al. 2020; Tomar et al. 2020)
Anticonvulsant	(Sayyah et al. 2002)
Antidiarrheal	(Qnais et al. 2012)
Antifungal	(Bayar et al. 2018; Mssillou et al. 2020)
Antihypertensive	(Bouadid et al. 2023)
Anti-Inflammatory	(Sayyah et al. 2003; Guedouari and Nabiev 2021; Khalil et al. 2024)
Antimicrobial	(Yilmaz et al. 2013; Vilela et al. 2016)
Antioxidant	(Conforti et al. 2006; Ramos et al. 2012; Yilmaz et al. 2013; Mssillou et al. 2020)
Analgesic	(Sayyah et al. 2003)
Curing diabetes	(Mohammed et al. 2021)
Gastroprotective	(Afifi et al. 1997; Speroni et al. 2011)
Woung healing	(Nayak et al. 2006; Semiz and Deveci 2023)

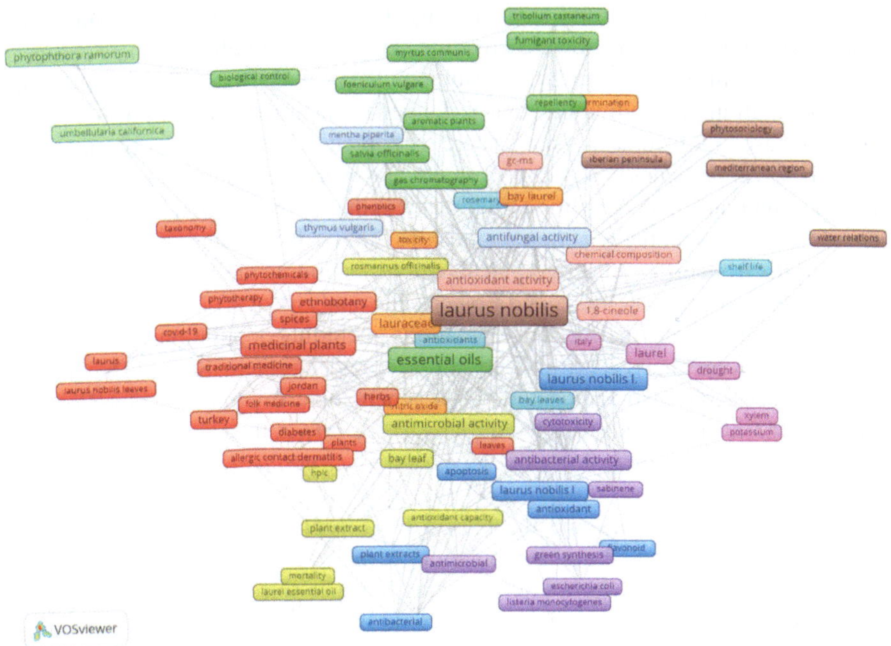

Fig. 1.4 Keywords retrieved from the documents of *Laurus nobilis*. According to the SCOPUS search on August 31, 2022, 1119 documents were recorded using the "*Laurus nobilis*" TITLE-ABS-KEY section. The retrieved documents were subjected to visualization analysis using VOSviewer (https://www.vosviewer.com/). Accordingly, 3173 keywords were obtained, but after setting criteria as a minimum of five occurrences, 110 keywords met the threshold and were used for analysis

1.2 Germplasm Biodiversity and Conservation

Germplasm is the gene pool for properties diversity and deciphers significant properties in plant improvement (Yilmaz et al. 2021b). Heterogeneous structure and the larger population size of germplasm confine its easy availability and usage for different breeding programs (Yilmaz et al. 2022).

1.2.1 Germplasm Diversity

The rapid expansion of the global population presents significant environmental challenges, driving substantial changes in modern agriculture (Yilmaz et al. 2023; Yilmaz and Yilmaz 2025). To ensure economic stability, agricultural production systems must undergo urgent transformation (Yilmaz et al. 2024). Germplasm resources are essential tools for assessing the diversity of agricultural traits and playing critical roles in crop improvement (Eren et al. 2022). In this context, germplasm diversity plays a crucial role in adapting plants to changing environmental conditions and ensuring long-term agricultural sustainability. Genetic diversity assessment is one of the most significant steps toward breeding stages since it delivers a new resource of variation (Yilmaz and Ciftci 2021b; Bulunuz Palaz et al. 2022). Genetic diversity among populations and individuals can be defined using biochemical, morphological, and DNA-based molecular markers. There were relatively few reports of molecular marker-based approaches to bay improvement; not even a genetic map is available for this important plant. The Amplified fragment length polymorphism (AFLP) markers were utilized to reveal the genetic relationships among 14 populations of *L. azorica* and *L. nobilis* from different geographical regions, including the Madeira and Canary Islands, Iberian Peninsula, Italy, and France. The analysis of AFLP revealed a high genetic diversity among the Iberian populations, including populations from the rest of the populations analyzed from Italy, France, and Northern Spain. The Iberian Peninsula accessions also demonstrated low genetic diversity to accessions from the Madeira and Canary Islands, defined initially as *L. azorica*, then to samples from the Mediterranean area, morphologically identified as *L. nobilis* (Arroyo-García et al. 2001).

Three *Lauraceae* genera and the plastid DNA (cpDNA) sequence (trnD-trnT, trnK-matK) analysis of 57 Laurus populations revealed monophyly for Laurus with high sequence similarity within Laurus. Three lineages were obtained: a first clade containing the Aegean region, the second clade in Eastern lineage, corresponding to the near East populations and Turkey, and lastly, a western clade that grouped all populations from the western and central Mediterranean regions and Macaronesia (Rodríguez-Sánchez et al. 2009). The inter-simple sequence repeats (ISSRs) were also used for genetic analysis of Laurus species, including eight taxa (representing three tribes, four genera, and seven species) grown in Egypt (Mohamed et al. 2016) and *L. nobilis* populations distributed in the Aegean region (Sevindik 2019). The ISSR markers were sufficient to resolve the relationship within *Lauraceae* (Mohamed et al. 2016), while the polymorphism rate for these markers is slightly low among *L. nobilis* populations in the Aegean/Turkey region (Sevindik 2019).

High-resolution markers can enable the recognition of many genotypes that cannot be detected with conventional markers. Simple sequence repeats/Microsatellites (SSRs) are sequences with cascading one to six nucleotide motifs that can be used to characterize individuals with higher resolution levels. These highly polymorphic, multi-allelic, and codominant markers are ideal for population and diversity studies (Paparella et al. 2022). Genetic analysis of sixty-six Mediterranean bay trees separated *L. nobilis* into two main gene pools, one from the Mediterranean (Turkey) and the other from eastern western (Algeria, Tunisia, and France) (Marzouki et al. 2009). The same authors also reported that *L. nobilis* has a higher genetic differentiation than the other angiosperms. Moreover, Arroyo et al. (2010) characterized a total of 63 genotypes containing 37 from Spain (*L. nobilis*) and 26 from Macaronesian islands (*L. azorica*) with 20 newly designated polymorphic SSR markers. One hundred ninety-six alleles belonged to the *L. nobilis* species, and 222 alleles belonged to the *L. azorica* species. Recently, 95 bay genotypes selected from the flora of Hatay province for their superior characteristics were characterized using six SSR markers. Eighty-two alleles were obtained with a mean of 16.4 of five polymorphic loci (Bulut et al. 2018).

It aimed to examine the genetic diversity of Turkish bay germplasm, which has a greater variety, unlike former small-scale studies recently. Turkish bay germplasm' (94 genotypes from the Black Sea, Aegean, Marmara, and Mediterranean regions) genetic diversity was identified using 13 iPBS, 10 SCoT, and 16 ISSR markers (Karik et al. 2019; Yilmaz 2020; Yilmaz and Ciftci 2021b). A total of 373, 195, and 227 clean and scorable bands were identified with these markers, respectively (Fig. 1.5). The map representing the genotypes is given in Fig. 1.6. The results of these studies put forward sources for contributing to the conservation of genetic diversity, advanced breeding techniques, and managing genetic resources for the breeders.

1.2.2 Genetic Characterization

DNA advancement based on different molecular markers and molecular breeding offered many new strategies to breeders to resolve the problems encountered during conventional breeding. Genetic characterization analysis studies have been one of the most effective assays for grouping many organisms, including plants, plant pathogens, and other organisms (Coşkan et al. 2022; Morca et al. 2022). In recent years, molecular tools have been used to elucidate genetic relationships between different varieties in aromatic plants, some aspects of genetic diversity, and comparisons between molecular marker analysis and plant chemical compositions. However, in the case of the bay, utilization of these technologies is at a very early stage, and in the future, more concentrated efforts are needed in this direction. (Paparella et al. 2022).

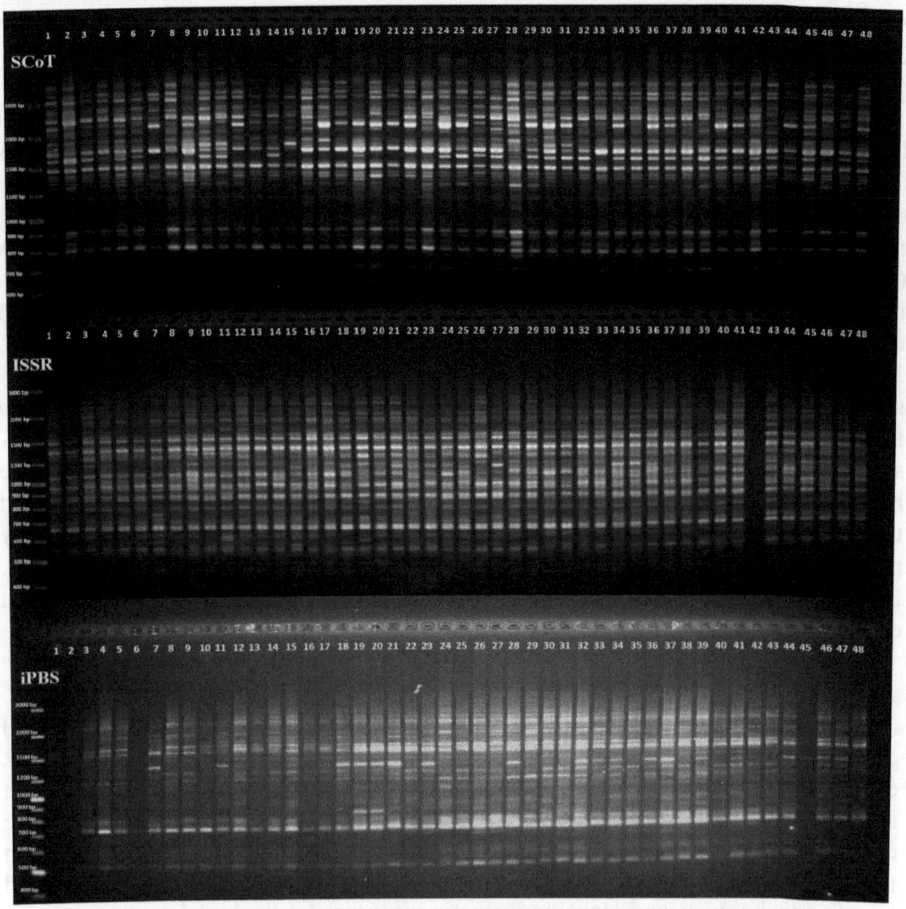

Fig. 1.5 The gel images from SCoT, ISSR, and iPBS Retrotransposon analysis represent 48 bay genotypes in Turkey (Karik et al. 2019; Yilmaz 2020; Yilmaz and Ciftci 2021b)

1.2.3 Genetic Resources Conservation Approaches

Biologists, geneticists, plant breeders, and educators have access to different fields, laboratory facilities, and levels of genetic information. However, there is a general need for a wide range of plant genetic resources, particularly novel germplasm (materials with a genetic variation or existing breeding populations not currently available), disease-free, viable, and acceptable legal conditions (Volk et al. 2021). Genetic resources included in plant breeding programs serve as sources of new genes. Researchers have made significant attempts to preserve many crops' genetic resources. However, little work has been done to conserve dioecious plants' genetic resources comprehensively. *L. nobilis* is mainly distributed in Mediterranean countries such as Turkey, France, Spain, and Italy. Researchers in many countries have

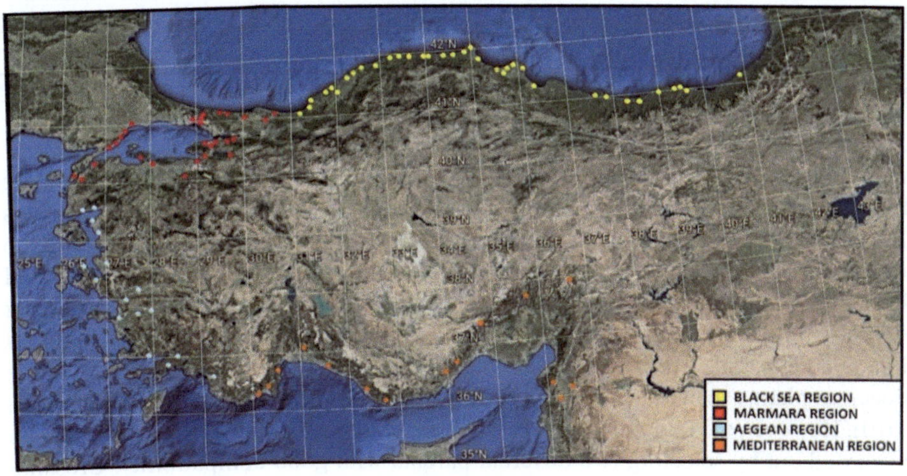

Fig. 1.6 The map represents 94 bay genotypes in Turkey from SCoT and ISSR analysis (Yilmaz and Ciftci 2021b)

made little effort to conserve the genetic resources of this plant. In order to improve the breeding programs of this plant, nobilis specimens from different climates should be collected and preserved. Currently, the seeds of approximately 600 *L. nobilis* genotypes from the Mediterranean, Marmara, Black Sea, and Aegean regions of Turkey were collected in the Aegean Agricultural Research Institute of the Ministry of Agriculture of the Republic of Turkey. The nursery plot was also created in this institution. Each genotype is labeled and preserved for further metabolomic and genomic analysis. In addition, a bay garden was established with the germination of the collected seeds (Nadeem et al. 2018a). The seeds of these specimens are preserved in İzmir Aegean Agricultural Research Institute to enable interested researchers worldwide to breed this plant.

1.2.4 Cytogenetics

Different levels of tetraploids were defined in Laurus, and the most common karyotype was determined as $2n = 4x = 48$ (Ehrendorfer et al. 1968; Arroyo et al. 2010). The genome size of the bay has not been investigated so far.

1.3 Molecular Breeding

1.3.1 Molecular Marker-Assisted Breeding

Marker-assisted selection (MAS) is the selection of desired traits to manage genetic developments in plants whose chromosome regions are determined by DNA markers (Fujino et al. 2019). MAS has changed the ways of plant breeding by exceeding the limits of conventional breeding. This method requires markers that can be obtained through genome-wide association mapping (GWAS) or QTL mapping. QTL mapping is biparental mapping in which two parents are crossed to develop mapping populations (Nadeem et al. 2018b). QTL maps have been generated for various traits of many dioecious plants, such as *Ginkgo biloba* (Liu et al. 2017) and date palm (Billotte et al. 2010). However, QTL mapping of the bay leaf is not available yet. These studies can be carried out by selecting two different parents with good compatibility. Crossing these parents will result in the F1 generation (Nadeem et al. 2018a). Double haploids, F2, inbreeding, and backcross are used for annual population mapping. But the bay is a perennial plant, and these mapping populations need a very long time to develop. Thus, the F1 generation can be used as a mapping population in the bay. QTL maps have been found to be very effective in plant breeding studies. However, it has some disadvantages, such as less allelic diversity and less recombination. Mapping population growth is a significant disadvantage of this technique for dioecious plants such as bay. GWAS has emerged as a more helpful tool for breeding annual and perennial plants that can overcome these limitations.

1.3.2 Bioinformatics

Different bioinformatic approaches such as sequencing and population studies are commonly used to explain better the current data (Çelik et al. 2022). The bioinformatics of bay has not been developed as it has in many other agriculture crops. Still, several general data banks can be followed forward and can use for this plant genome. Important among them are Genome Net (https://www.genome.jp/db/pcidb/kna_species/3114) and Powo Science (https://powo.science.kew.org/taxon/465049-1) which provide genomic data and NCBI (http://www.ncbi.nlm.nih.gov/) can be generally scanned for all bioinformatics purposes.

1.4 Tissue Culture Applications

Plant tissue culture techniques, which are widely used in forestry, agriculture, and plant breeding, include applied biotechnological methods such as haploid plant production, somoclonal variation, in vitro germplasm conservation, in vitro selection, somatic cell hybridization, gene transfer, in vitro cloning of plants, secondary metabolite production, synthetic seed production, virus elimination, and mass propagation via micropropagation. These methods have been successfully applied to protect plant species with dormant or recalcitrant seeds, providing tremendous advantages over traditional protection methods such as cutting propagation (Oseni et al. 2018; Soydemir and Yilmaz 2022). These methods have been successfully applied to protect plant species with dormant or recalcitrant seeds, providing tremendous advantages over traditional protection methods such as cutting propagation (Oseni et al. 2018). Tissue culture techniques used in dioecious plants like bay are promising tools, aiming to overcome the difficulties of traditional cultivation methods (Al Gabbiesh et al. 2015). Direct shoot regeneration and callus-derived somatic embryogenesis are the two main tissue culture techniques. Somatic embryogenesis in other Laurus species has been successful, while all attempts to induce shoot induction from callus have been unsuccessful to date (Chen and Chang 2009). A zygotic embryo formation process results in the proliferation of tissues (Nadeem et al. 2018b). Al Gabbiesh et al. (2015) suggested a protocol for bay and provided a 62.5% success rate for callus regeneration. Souayah et al. (2002) reported an array of in vitro and conventional techniques, highlighting that successful reproduction depends on the culture, sampling date, and type of cuttings. As reported for quite a number of crop and non-crop species, in vitro methods are considered to be precious to overcoming the proliferation problem of the bay (Chourfi et al. 2014). In this context, Chourfi et al. (2014) investigated the germination of the bay through micro-cuts and seeds as a propagation tool and tested the acclimatization of these plants. To our best knowledge and survey, no reports on the isolation of bay protoplasts are available hitherto. Few studies have been conducted with this technique in other dioecious plants because of rapid protoplast death. Establishing a protocol from bay protoplast for callus formation is necessary. A successful callus was obtained from date palm protoplast by aiming to transfer date palm genes related to important traits (Chabane et al. 2007). Since bay plants are reputed for their essential oil components and their uses in a wide range of industries, tissue culture techniques might be employed for the relevant plant. In this regard, using the micropropagation method, high amounts of production can be achieved in a very short time by taking a small explant from the desired bay plant.

1.5 Mutation Breeding

The exploitation of genetic diversity is significant for obtaining a high yield, resistance to disease and pests, and developing new cultivars (Nadeem et al. 2018a; Güler et al. 2021). Tools such as tissue culture-derived somaclonal variation, T-DNA splice mutagenesis, and chemical mutagens have been developed to induce genetic variability (Bairu et al. 2011; Amosova et al. 2019; Rajan and Singh 2021). Ethyl methane sulfonate (EMS) has been evaluated as a chemical mutagen in many crops, including bay, according to a Jain (2012) protocol for date palm trees. The study noted that superior mutants should be selected since they might perform well during regeneration. One of the common physical mutagens is gamma radiation. Due to the success obtained for many species, the relevant test of mutagens is also needed. Mutation breeding has gained popularity among breeders to produce plants with suitable traits through induced or spontaneous mutation techniques. Both techniques in various dioecious plants have been successfully applied (Jain 2012). It is essential to evaluate genetic variations in any crop to meet reproductive needs. In vitro selection with somaclonal variations has been a promising tool in breeding plants, especially for abiotic and biotic stresses. Somaclonal variations refer to changes that occur during regenerating plants, cell cultures, and their progeny. Somatic cells can be applied as sources of variation that can be evaluated in the breeding of bay to produce new individuals with identified essential characteristics.

1.6 Hybridization

Considerable efforts have been made to examine the reproductive mechanism of angiosperms, ranging from dioecism to self-incompatibility and from forced crossing to obligate self-reproduction (Cardoso et al. 2018; Tebele et al. 2021). The sexual system is the central part that affects the evolutionary continuity, lineage, polyploidy, and first spreading of many species (Nadeem et al. 2018b). Bay is an evergreen, perennial, dioecious, and small tree. Difficulties in seed germination, dioecious nature, and sexual reproduction are the fundamental ingredients hindering the breeding programs of the bay (Souayah et al. 2002). Hybridization methods are essential in plant breeding, especially in dioecious plants. Natural hybridization is possible in the bay plant, but the probability of success is low due to the low rates seen in other dioecious plants, such as the date palm. So far, successful plant hybrids have been produced in artificial hybridization, and successful studies have been carried out on various dioecious plants such as date palm (Gros-Balthazard 2013), *Ocotea catharinensis* Mez (Santa-Catarina et al. 2004), and common persimmon (Choi et al. 2003). On the other hand, there are disadvantages, such as the inability to control genetic mutations and a decrease in genetic diversity in artificial hybridization.

Diversity declines through hybridization, extinction of species or populations, destruction of reproductive barriers, and a coalition of different earlier evolutionary lineages (Buerkle et al. 2003). Spontaneous hybridization can be used if both females and males begin to flower simultaneously, and pollination by insects or wind can occur quickly (Gros-Balthazard 2013). Autumn is a season-best technically suited for propagation by cuttings. However, both methods (artificial hybridization and spontaneous) were reported to provide low growth recovery (Laurent 2007). Low growth recovery is a massive problem for plant reproduction. Significant seed loss by birds and difficulties in pollination results in small seed yields in the plant (Al Gabbiesh et al. 2015). Such conventional methods could not meet the homogeneity levels in sufficient quantities; moreover, they could be done for a long time.

1.7 Conclusion and Prospects

Despite the excellent medicinal importance of bay, there is a considerable gap or lack of its genetic makeup, and reproductive aspects once compared to other crop and non-crop species. The genomics of bays collected from different regions needs to be investigated further. Since many studies have been done on their metabolomics, there is a need for studies on increasing the concentrations of essential compounds, especially in their content, with modern breeding programs such as genomic selection and genome-wide association mapping. To better understand this plant, mechanisms need to investigate sex differentiation, polyploidy level, and chromosome number and to determine the whole genome sequence. Although there are many reports on the importance of this plant species, it is impossible to obtain enough information about the plant regarding breeding with little conservation effort or conservation of genetic resources. In this context, it is necessary to characterize existing germplasm collections to identify and classify accessions and to create a catalog of information for germplasm management and different breeding programs. In recent years, molecular tools have been used to explain genetic relationships between different cultivars, some aspects of genetic diversity, and molecular marker relationships of plant chemical compositions in various medicinal and aromatic plants. However, understanding the molecular basis of flavor production is no easy task. Multiple aspects of plant secondary metabolism are still unknown at the molecular level. Finally, the routes from genomics to proteomics are not illustrated. Addressing these questions requires concentrated breeding efforts and multidisciplinary approaches, including genomics, transcriptomics, proteomics, bioinformatics, gene editing and genetic transformation which might open new perspectives in future bay improvement programs.

Appendix

List of major Institutes and Departments engaged in research on bay (*Laurus nobilis* L.)

1. Aegean Agricultural Research Institute, Cumhuriyet District, Çanakkale Asfaltı Street. No: 138, 35660 Menemen/Izmir, Turkey, e-mail: etae@tarim.gov.tr
2. Sivas University of Science and Technology, Faculty of Agricultural Sciences and Technology, Yenişehir District, Kardesler Street, No:7/1, 58000, Sivas, Turkey, e-mail: bilgi@sivas.edu.tr
3. Bolu Abant Izzet Baysal University, Faculty of Agriculture, Karaköy District, 14030, Bolu Turkey, e-mail: zdf@ibu.edu.tr

References

Acartürk R (1997) Şifalı Bitkiler Flora ve Sağlığımız Ovak Yayınları Yayın No: 1, Ankara

Afifi FU, Khalil E, Tamimi SO, Disi A (1997) Evaluation of the gastroprotective Effect of *Laurus nobilis* seeds on ethanol-induced gastric ulcer in rats. J Ethnopharmacol 58(1):9–14

Al Gabbiesh AH, Ghabeish M, Kleinwächter M, Selmar D (2015) Plant regeneration through somatic embryogenesis from calli derived from leaf bases of *Laurus nobilis* L. (Lauraceae). Plant Tissue Cult Biotechnol 24(2):213–221

Amosova AV, Zoshchuk SA, Volovik VT, Shirokova AV, Horuzhiy NE, Mozgova GV, Yurkevich OY, Artyukhova LVA, Samatadze TE, Muravenko OV (2019) Phenotypic biochemical and genomic variability in generations of the rapeseed (*Brassica napus* L.) mutant lines obtained via chemical mutagenesis. PLoS ONE 14(8):e0221699

Anzano A, de Falco B, Grauso L, Mottı R, Lanzotti V (2022) Laurel *Laurus nobilis* L: a review of its botany traditional uses phytochemistry and pharmacology. Phytochem Rev 1–51

Arroyo JM, Rigueiro C, Rodríguez R, Hampe A, Valido A, Rodríguez-Sánchez F Jordano P (2010) Isolation and characterization of 20 microsatellite loci for laurel species (Laurus Lauraceae). AM J BoT 97:e26–e30

Arroyo-García R, Martínez-Zapater JM, Prieto JF, Álvarez-Arbesú R (2001) AFLP evaluation of genetic similarity among laurel populations (*Laurus* L.). Euphytica 122:155–164

Ayanoğlu F, Kaya DA, Bahadirlı NP (2018) Effects of planting density and harvesting time on leaf and essential oil yield of bay laurel (*Laurus nobilis* L.) cultured in shrub form. In International conference on advanced materials and systems (ICAMS). The National Research Development Institute for Textiles and Leather-INCDTP, pp 289–294

Bairu MW, Aremu AO, Van Staden J (2011) Somaclonal variation in plants: causes and detection methods. Plant Growth Regul 63(2):147–173

Bak T, Güler E, Karadeniz T (2021) Harvest time alters fruit quality traits of barberries (*Berberis vulgaris*). In: Pakyürek M (ed) Recent headways in pomology Adiyaman, IKSAD, Turkey, pp 355–375. https://iksadyayinevi.com/home/recent-headways-in-pomology/

Bayar Y, Onaran A, Yilar M, Gul F (2018) Determination of the essential oil composition and the antifungal activities of bilberry (*Vaccinium myrtillus* L.) and bay laurel (*Laurus nobilis* L.). J Essent Oil Bear Plants 21(2):548–555

Baytop T (1984) Türkiye'de Bitkiler ile Tedavi İÜ, Eczacılık Fakültesi Yayınları No: 40, Istanbul, p 520

Billotte N, Jourjon MF, Marseillac N, Berger A, Florı A, Asmady H, Adon B, Singh R, Nouy B, Potier F, Cheah SC (2010) QTL detection by multi-parent linkage mapping in oil palm (*Elaeis guineensis* JacQ). Theor Appl Genet 120(8):1673–1687

Bouadid I, Amssayef A, Eddouks M (2023) Study of the antihypertensive effect of *Laurus nobilis* in rats. Cardiovasc Hematol Agents Med Chem (For Curr Med Chem Cardiovasc Hematol Agents) 21(1):42–54

Buerkle CA, Wolf DE, Rieseberg LH (2003) The origin and extinction of species through hybridization. In: Population viability in plants. Springer, Berlin, pp 117–141

Bulunuz Palaz E, Demirel F, Adalı S, Demirel S, Yilmaz A (2022) Genetic relationships of salep orchid species and gene flow among *Serapias vomeracea* × *Anacamptis morio* hybrids. Plant Biotechnol Rep 1–13

Bulut MÇ, Özmen CY, Ergül A, Ayanoğlu F (2018) Genetic characterization of bay laurel (*Laurus nobilis* L.) populations using microsatellite markers and flow cytometry. Mustafa Kemal Üniv Ziraat Fakültesi Dergisi 23(2):242–253

Cardoso JCF, Viana ML, Matias R, Furtado MT, Caetano APDS, Consolaro H, Brito VLGD (2018) Towards a unified terminology for angiosperm reproductive systems. Acta Bot Brasilica 32:329–348

Carpenter RJ, Truswell EM, Harris WK (2010) Lauraceae fossils from a volcanic Palaeocene oceanic island Ninetyeast Ridge Indian Ocean: ancient long-distance dispersal? J Biogeogr 37(7):1202–1213

Çelik A, Santosa AI, Gibbs AJ, Ertunç F (2022) Prunus necrotic ringspot virus in Turkey: an immigrant population. Adv Virol 167(2):553–562

Chabane D, Assani A, Bouguedoura N, Haïcour R, Ducreux G (2007) Induction of callus formation from difficile date palm protoplasts by means of nurse cultur. CR Biol 330(5):392–401

Chen YC, Chang C (2009) Plant regeneration through somatic embryogenesis from young leaves of *Cinnamomum kanehirae* Hayata. Taiwan J for Sci 24:117–125

Choi YA, Tao R, Yonemori K, Sugiura A (2003) Genomic in situ hybridization between persimmon (*Diospyros kaki*) and several wild species of Diospyros. J Jpn Soc Hort Sci 72(5):385–388

Chourfi A, Alaoui T, Echchgadda G (2014) In vitro propagation of the bay laurel (*Laurus nobilis* L.) in Moroccn. South Asian J Exp Biol 96–103

Conforti F, Statti G, Uzunov D, Menichini F (2006) Comparative chemical composition and antioxidant activities of wild and cultivated *Laurus nobilis* L. leaves and *Foeniculum vulgare* subsp. piperitum (Ucria) coutinho seeds. Biol Pharm Bull 29(10):2056–2064

Coşkan S, Morca AF, Akbaş B, Çelik A, Santosa AI (2022) Comprehensive surveillance and population study on plum pox virus in Ankara Province of Turkey. J Plant Dis Protect 1–11

Davis PH (1982) Flora of Turkey, vol 7. Edinburg University Press, Edinburg, p 947

Dobroslavic E, Repajic M, Dragovic-Uzelac V, Elez Garofulic I (2022) Isolation of *Laurus nobilis* leaf polyphenols: a review on current techniques and future perspectives. Foods 11:235

Duke JA (1997) The green pharmacy: new discoveries in herbal remedies for common diseases and conditions from the world's foremost authority on healing herbs. Rodale Press, New York, NY, p 501

Ehrendorfer F, Krendl F, Habeler E, Sauer W (1968) Chromosome numbers and evolution in primitive angiosperms. Taxon 17:337–353

Eren B, Keskin B, Demirel F, Demirel S, Türkoğlu A, Yilmaz A, Haliloğlu K (2022) Assessment of genetic diversity and population structure in local alfalfa genotypes using iPBS molecular markers. Genet Resour Crop Evol 1–12

Franchi GG, Piotto B, Nepi M, Baskin CC, Baskin JM, Pacini E (2011) Pollen and seed desiccation tolerance in relation to degree of developmental arrest dispersal and survival. J Exp Bot 62(15):5267–5281

Fujino K, Hirayama Y, Kaji R (2019) Marker-assisted selection in rice breeding programs in Hokkaido. Breed Sci 69(3):383–392

Furbank RT (2009) Plant phenomics: from gene to form and function. Funct Plant Biol 36(10):5–6

Gros-Balthazard M (2013) Hybridization in the genus Phoenix: a review. Emir J Food Agric 25(11):831

Guedouari R, Nabiev M (2021) Anti-inflammatory activity of different extracts from *Laurus nobilis* growing in Algeria. Algerian J Environ Sci Technol 7(4)

Güler E, Bak T, Karadeniz T, Muradoğlu F (2021) Relationships of fruit characteristics of rosehips (*Rosa canina* L.) grown in Bolu city center. J Inst Sci Technol 11(2):831–838

Hufford LA (1996) The origin and early evolution of angiosperm stamens. In: The anther: form function and phylogeny. Cambridge University Press, Cambridge, pp 58–91

Jain SM (2012) In vitro mutagenesis for improving date palm (*Phoenix dactylifera* L.). Emir J Food Agric 24(5):400

Karik Ü, Nadeem MA, Habyarimana E, Ercişli S, Yildiz M, Yılmaz A, Yang SH, Chung G, Baloch FS (2019) Exploring the genetic diversity and population structure of Turkish laurel germplasm by the iPBS-retrotransposon marker system. Agronomy 9(10):647

Khalil NA, ALFaris NA, ALTamimi JZ Mohamed Ahmed IA (2024) Anti-inflammatory effects of bay laurel (*Laurus nobilis* L.) towards the gut microbiome in dextran sodium sulfate induced colitis animal models. Food Sci Nutr. https://doi.org/10.1002/fsn3.3946

Laurent B (2007) Le grand livre des plantes aromatiques, vol 108. Rustica, Paris

Liu H, Cao F, Yin T, Chen Y (2017) A highly dense genetic map for Ginkgo biloba constructed using sequence-based marker. Front Plant Sci 8:1041

Marzouki H, Nasri N, Jouaud B, Bonnet C, Khaldi A, Bouzid S, Fady B (2009) Population genetic structure of *Laurus nobilis* L. inferred from transferred nuclear microsatellites. Silvae Genet 58:270–276

Mohamed ASH, Ahmed W, Rabia SS, Mourad MM (2016) Implications of morphology and molecular criteria in taxonomy of lauraceae juss. Egypt J Exp Biol (BoT) 12:45–52

Mohammed RR, Omer AK, Yener Z, Uyar A, Ahmed AK (2021) Biomedical effects of *Laurus nobilis* L. leaf extract on vital organs in streptozotocin-induced diabetic rats: experimental research. Ann Med Surg 61:188–197

Morca AF, Çelik A, Coşkan S, Santosa AI, Akbaş B (2022) Population analysis on tomato spotted wilt virus isolates inducing various symptoms on tomato pepper and *Chenopodium album* in Turkey. Physiol Mol Plant Pathol 118:101786

Mssillou I, Agour A, El Ghouizi A, Hamamouch N, Lyoussı B, Derwich E (2020) Chemical composition antioxidant activity and antifungal effects of essential oil from *Laurus nobilis* L. flowers growing in Moroccn. J Food Qual

Muradoğlu F, Gürsoy S, Güler E (2021) Multivariate analysis revealed the morphological variability among crataegus species. Yuzuncu Yıl Univ J Agric Sci 31(4):961–972

Nadeem MA, Nawaz MA, Shahid MQ, Doğan Y, Comertpay G, Yıldız M, Hatipoğlu R, Ahmad F, Alsaleh A, Labhane N, Özkan H, Chung G, Baloch FS (2018b) DNA molecular markers in plant breeding: current status and recent advancements in genomic selection and genome editing. Biotechnol Biotechnol Equip 32(2):261–285

Nadeem MA, Aasim M, Kırıcı S, Karık Ü, Nawaz MA, Yılmaz A, Maral H, Khawar KM, Baloch FS (2018) Laurel (*Laurus nobilis* L.): a less-known medicinal plant to the world with diffusion genomics phenomics and metabolomics for genetic improvement. Biotechnol Approach Med Aromat Plants 631–653

Nayak S, Nalabothu P, Sandiford S, Bhogadi V, Adogwa A (2006) Evaluation of wound healing activity of *Allamanda cathartica* L. and *Laurus nobilis* L. extracts on rats. BMC Complement Altern Med 6:1–6

Nepi M, Franchı GG, Padnı E (2001) Pollen hydration status at dispersal: cytophysiological features and strategies. Protoplasma 216(3–4):171

Oliveira-Filho AA, Fernandes HM, Assis TJ (2015) Lauraceae's family: a brief review of cardiovascular effects. Int J Pharmacogn Phytochem Res 7:22–26

Oseni OM, Pande V, Nailwal TK (2018) A review on plant tissue culture a technique for propagation and conservation of endangered plant species. Int J Curr Microbiol App Sci 7(7):3778–3786

Özhatay N, Koyuncu M, Atay S, Byfıeld A (1997) Türkiye'nin Doğal Tıbbi Bitkilerinin Ticareti Hakkında Bir Çalışma. Doğal Hayatı Koruma Derneği, İstanbul, p 121. ISBN: 975-96081-9-7

Pacini E, Sciannandrone N, Nepi M (2014) Floral biology of the dioecious species *Laurus nobilis* L. (Lauraceae) Flora-Morphol Distrib Funct Ecol Plants 209(3):153–163

Paparella A, Nawade B, Shaltiel-Harpaz L, Ibdah M (2022) A review of the botany volatile composition biochemical and molecular aspects and traditional uses *of Laurus nobilis*. Plants 11(9):1209

Qnais EY, Abdulla FA, Kaddumi EG, Abdalla SS (2012) Antidiarrheal activity of *Laurus nobilis* L. leaf extract in rats. J Med Food 15(1):51–57

Rajan RP, Singh G (2021) A review on application of somaclonal variation in important horticulture crops. Plant Cell Biotechnol Mol Biol 161–175
Ramadan AZ, Mandil HAL, Anadanı L (2019) Effect of ripening time of fruits in the chemical composition (%) of fatty acids in syrian laurel oil. Int J Curr Res Chem Pharm Sci 6(7):27–32
Ramos C, Teixeira B, Batista I, Matos O, Serrano C, Neng NR et al (2012) Antioxidant and antibacterial activity of essential oil and extracts of bay laurel *Laurus nobilis* Linnaeus (Lauraceae) from Portugal. Nat Prod Res 26(6):518–529
Rodilla JM, Tinoco MT, Morais JC, Gimenez C, Cabrera R, Benito DM, Castillo L, Gonzalez-Coloma A (2008) *Laurus novocanariensis* essential oil: seasonal variation and valorization. Biochem Syst Ecol 36:167–176
Rodríguez-Sánchez F, Guzmán B, Valido A, Vargas P, Arroyo J (2009) Late Neogene history of the laurel tree (*Laurus* L., Lauraceae) based on phylogeographical analyses of Mediterranean and Macaronesian populations. J Biogeogr 36:1270–1281
Said CM, Hussein K (2014) Determination of the chemical and genetic differences of laurus collected from three different geographic and climatic areas in Lebanon. Eur Sci J ESJ 10(10)
Santa-Catarina C, Hanai LR, Dornelas MC, Viana AM, Floh EI (2004) SERK gene homolog expression polyamines and amino acids associated with somatic embryogenic competence of *Ocotea catharinensis* Mez. (Lauraceae). Plant Cell Tissue and Organ Culture 79(1):53–61
Sari AO, Oguz B, Bilgic A (2006) Breaking seed dormancy of laurel (*Laurus nobilis* L.) New For 31:403–408
Sayyah M, Valizadeh J, Kamalinejad M (2002) Anticonvulsant activity of the leaf essential oil of *Laurus nobilis*against pentylenetetrazole-and maximal electroshock-induced seizures. Phytomedicine 9:212–216
Sayyah M, Saroukhani G, Peirovı A, Kamalinejad M (2003) Analgesic and anti-inflammatory activity of the leaf essential oil of *Laurus nobilis* Linn. Phytother Res 17:733–736
Semiz N, Deveci MZY (2023) Effects of common centaury (*Centaurium erythraea*) oil and laurel (*Laurus nobilis*) seed oil on full-thickness excisional skin wound healing in rats. Ankara Üniv Vet Fak Dergisi 1–10
Sevindik E (2019) Molecular genetic diversity of some *Laurus nobilis* L. (Lauraceae) populations grown in the Aegean region/Turkey. AdV Food ScI 41:28–31
Shokoohinia Y, Yegdaneh A, Amin G, Ghannadi A (2014) Seasonal variations of *Laurus nobilis* L. leaves volatile oil components in Isfahan Iran. Res J Pharmacogn 1(3):1–6
Simic M, Kundakovic T, Kovacevic N (2003) Preliminary assay on the antioxidative activity of *Laurus nobilis* extracts. Fitoterapia 74(6):613–616
Sırıken B, Yavuz C, Güler A (2018) Antibacterial Activity of *Laurus nobilis*: a review of literature. Med Sci Discov 5(11):374–379
Souayah N, Khouja ML, Khaldi A, Rejeb MN, Bouzid S (2002) Breeding improvement of *Laurus nobilis* L. by conventional and in vitro propagation techniques. J Herbs Spices Med Plants 9:101–105
Soydemir HE, Yılmaz A (2022) Bitki doku kültüründe antibiyotik kullanımı. In: Modern Tarım Uygulamaları. IKSAD Publishing House pp 279–289
Speroni E, Cervellati R, Dall'Acqua S, Guerra MC, Greco E, Govoni P, Innocenti G (2011) Gastroprotective effect and antioxidant properties of different *Laurus nobilis* L. leaf extracts. J Med Food 14(5):499–504
Stefanova G, Girova T, Gochev V, Stoyanova M, Petkova Z, Stoyanova A, Zheljazkov VD (2020) Comparative study on the chemical composition of laurel (*Laurus nobilis* L.) leaves from Greece and Georgia and the antibacterial activity of their essential oil. Heliyon 6(12):e05491
Tebele SM, Marks RA, Farrant JM (2021) Two decades of desiccation biology: a systematic review of the best studied angiosperm resurrection plants. Plants 10(12):2784
Tomar O, Akarca G, Gök V, Ramadan MF (2020) Composition and antibacterial effects of Laurel (*Laurus nobilis* L.) leaves essential oil. J Essent Oil Bear Plants 23(2):414–421
Tural S, Durmaz Y, Urçar E, Turhan S (2019) Antibacterial activity of thyme, laurel, rosemary and parsley essential oils against some bacterial fish pathogen. Acta Aquat Turc 15:440–447

Vilela J, Martins D, Monteiro-Silva F, González-Aguilar G, de Almeida JM, Saraiva C (2016) Antimicrobial effect of essential oils of *Laurus nobilis* L. and *Rosmarinus officinallis* L. on shelf-life of minced "Maronesa" beef stored under different packaging conditions. Food Packag Shelf Life 8:71–80

Volk GM, Byrne PF, Coyne CJ, Flint-Garcia S, Reeves PA, Richards C (2021) Integrating genomic and phenomic approaches to support plant genetic resources conservation and use. Plants 10(11):2260

Yilmaz (2020) Molecular characterization of laurel (*Laurus nobilis* L.) genotypes collected from different regions of Turkey. Ph.D thesis. Bolu Abant İzzet Baysal University Institute of Science and Technology

Yilmaz A, Ciftci V (2021a) Status of laurel plant (*Laurus nobilis* L.) in Turkey. Eur J Sci Technol 22:325–330

Yilmaz A, Ciftci V (2021b) Genetic relationships and diversity analysis in Turkish laurel (*Laurus nobilis* L.) germplasm using ISSR and SCoT markers. Mol Biol Rep 48:4537–4547

Yilmaz ES, Timur M, Aslim B (2013) Antimicrobial antioxidant activity of the essential oil of Bay Laurel from Hatay Turkey. J Essent Oil Bear Plants 16(1):108–116

Yilmaz A, Guler E, Soydemir HE, Demirel S, Mollahaliloglu S, Karadeniz T, Ciftci V (2021a) Miracle plant: Black Chokeberry (*Aronia melanocarpa*). MAS J Appl Sci 6:83–94

Yılmaz A, Yeken MZ, Ali F, Barut M, Nadeem MA, Yılmaz H, Naeem M, Hacıoğlu BT, Arslan Y, Kurt C, Aasim M, Baloch FS (2021b) Genomics phenomics and next breeding tools for genetic improvement of safflower (*Carthamus tinctorius* L.). In: Oil crop genomics. Springer, Cham, pp 217–269

Yilmaz A, Yilmaz H, Turan S, Celik A, Nadeem MA, Demirel F et al (2022) Biotechnological advancements in coriander (*Coriandrum sativum* L.). Avrupa Bilim Teknoloji Dergisi 35:203–220

Yilmaz A, Yildirim E, Yilmaz H, Soydemir HE, Güler E, Ciftci V, Yaman M (2023) Use of arbuscular mycorrhizal fungi for boosting antioxidant enzyme metabolism and mitigating saline stress in Sweet Basil (*Ocimum basilicum* L.). Sustainability 15(7):5982

Yilmaz H, Karatas R, Demirel F, Soysal S, Türkoğlu A, Yilmaz A, Ciftci V (2024) Variations in protein, gluten, Zeleny sedimentation and yield of certain wheat (*Triticum aestivum* L.) cultivars under different climatic conditions. Euphytica 220(12):190

Yilmaz H, Yilmaz A (2025) Hidden hunger in the age of abundance: the nutritional pitfalls of modern staple crops. Food Sci Nutr. https://doi.org/10.1002/fsn3.4610

Chapter 2
Advances in Cinnamon (*Cinnamomum verum* L.) Breeding Strategies

Amira A. Ibrahim, Sawsan Abd-Ellatif, El-Sayed S. Abdel Razik, and Khaled F. M. Salem

Abstract Known by its scientific name, *Cinnamomum verum* L., cinnamon is a spice, highly regarded and is utilised in a variety of contexts, including culinary and medical applications. It is vital to cultivate enhanced cinnamon cultivars that display improved qualities, such as larger yield, superior quality, increased disease resistance, and tolerance to a variety of agroclimatic situations. This is because the demand for cinnamon is expanding at an alarming rate. An strategy that is viable for creating improved varieties of cinnamon that have favourable agronomic properties has been made possible by the success that has been made in cinnamon breeding procedures. Cinnamon's productivity, quality, and resilience could be improved by the integration of conventional and molecular breeding approaches, as well as through the efficient utilisation of varied germplasm resources. It is absolutely necessary to devote the necessary resources to cinnamon breeding research in order to effectively meet the ever-increasing demand for this highly sought-after spice on a global scale. The

A. A. Ibrahim (✉)
Department of Botany and Microbiology, Faculty of Science, Arish University, Al-Arish 45511, Egypt
e-mail: amiranasreldeen@sci.aru.edu.eg

S. Abd-Ellatif
Department of Bioprocess Development, Genetic Engineering and Biotechnology Research Institute (GEBRI), City of Scientific Research and Technology Applications (SRTA-City), Alexandria 21934, Egypt
e-mail: sabdellatif@srtacity.sci.eg

E.-S. S. A. Razik
Department of Livestock Research, Arid Lands Cultivation Research Institute (ALCRI), City of Scientific Research and Technology Applications SRTA-City), Alexandria 21934, Egypt
e-mail: eshabaan@srtacity.sci.eg

K. F. M. Salem
Department of Biology, College of Science and Humanitarian Studies, Shaqra University, Qwaieah, Saudi Arabia

Department of Plant Biotechnology, Genetic Engineering and Biotechnology Research Institute (GEBRI), University of Sadat City, Sadat City, Egypt

K. F. M. Salem
e-mail: khaled.salem@gebri.usc.edu.eg; khaledfathi@su.edu.sa

J. M. Al-Khayri et al. (eds.), *Biodiversity and Genetic Improvement of Herbs and Spices*, Advances in Plant Breeding Strategies 8, https://doi.org/10.1007/978-3-031-81838-7_2

objective of this chapter is to provide a complete summary of the latest breakthroughs in breeding procedures that have been utilised for the enhancement of cinnamon. This discussion covers a wide range of topics that are associated with cinnamon breeding, including the investigation of germplasm, the specification of selection criteria, the methods of breeding, and the molecular techniques. In addition, this chapter will shed light on the difficulties that have been experienced in cinnamon breeding research, and it will also offer some insights into the potential opportunities that exist within this particular subject.

Keywords Biodiversity · Biotechnology · Breeding · True Cinnamon · Genetic diversity · True Cinnamon · Genetic diversity · *Cinnamomum verum* Blume

2.1 Introduction

Several tree species in the Lauraceae family are responsible for producing the commercially accessible spice compounds known as cinnamon. There are many kinds of cinnamon, but only around four are grown commercially. *Cinnamomum burmannii* (also known as Korintje or Indonesian Cinnamon or Padang Cassia), *Cinnamomum loureiroi* (also known as Saigon Cinnamon, Vietnamese cassia, or Vietnamese cinnamon), and *Cinnamomum verum* (also known as Ceylon, true, or Mexican Cinnamon) are the types of cinnamon discussed in this study. Commercial cultivation of *C. Verum* takes place on a considerable scale in the islands of Seychelles and Madagascar, and it is native to Sri Lanka, India, Madagascar, Brazil, and the Caribbean (Paranagama et al. 2010; Mohammed 1993). The plant in issue is native to the West Tropical region of Africa and has several practical applications, including consumption (in the form of the bark) and inclusion (via the leaf) in a wide variety of foods and drinks. Bark infusion has a long history of use in traditional medicine to treat several conditions, including pain, arthritis, rheumatism, nasopharyngeal affections, and stomach problems. Diarrhea and dysentery are also treated using the plant's leaves, bark, and roots. Herbs and spices have long been used for their nutritional and traditional medicinal value because they contain many naturally occurring bioactive chemicals linked to a wide range of health benefits. Both cinnamon bark extract and cinnamon powder see widespread use. Currently, over 300 distinct cinnamon species are known to exist. However, *Cinnamomum verum* J. Presl, *Cinnamomum aromaticum* or *Cinnamomum cassia* L. J. Presl, *Cinnamomum burmanni* (Nees and Th. Nees) Nees ex Blume 2012 and *Cinnamomum loureiroi* Nees are the most widely used species for culinary and nutraceutical uses. Physical appearance and chemical composition can be shown to vary among various species. When it comes to phytochemical composition, different genotypes have been shown to have noticeably varied polyphenol and volatile phenol levels (Nabavi et al. 2015).

Furthermore, plant growth conditions and environmental pressures impact secondary metabolite composition (Avula et al. 2015; Ford et al. 2019; Pagliari et al. 2023). There are several chemicals of relevance found only in the bioactive

Cinnamomum accessions. These chemicals include vanillic acid, caffeine, gallic acid, p-coumaric acid, ferulic acid, kaempferol, cinnamic acid, and cinnamaldehyde. Neuroprotective, hepatoprotective, cardioprotective, and gastroprotective effects on human health have been identified in these substances (Khasnavis and Pahan 2012). Most of these compounds have antioxidant properties, increasing the levels of antioxidant enzymes like catalase (CAT), superoxide dismutase (SOD), and glutathione peroxidase (GPx) (Liao et al. 2012; Hemmati et al. 2018). Colitis, arthritis, and diabetes are just some of the inflammatory conditions researched to cinnamon phytocomplexes (Hagenlocher et al. 2017; Hariri and Ghiasvand 2016; Kim and Kim 2017, 2019).

Several food products and dietary supplements have been found to include cinnamon extract and/or its specific bioactive components because of its positive properties. However, it must be recognized that the use of individual molecules is severely confined. This is primarily because, in many cases, the beneficial effect is delivered by the coordinated efforts of multiple molecules. Further contributing to their limited efficacy is that these isolated compounds are not widely distributed systemically and have low bioavailability (Quero et al. 2020). Whole cinnamon phytoextracts obtained using aqueous or hydroalcoholic extractions, which efficiently boost the concentration of different medicinal components, have seen a dramatic uptick in popularity for this same reason in recent years.

In contrast to confectionary goods, functional foods, including yogurt, creams, ice creams, food supplements, and dietetic foods, typically use extracts in their preparation (Liu et al. 2022; Pagliari et al. 2023). Growing *C. burmannii* in Africa has been the subject of a few published accounts. The chemical and pharmaceutical properties and the cultivation techniques, of *Cinnamomum verum* will be the primary focus of this chapter.

2.1.1 Geographical and Distribution of Cinnamon

Experts in the field of wood anatomy have clarified Cinnamomum's wood anatomy. However, the descriptions lack detail and are restricted because they are based on examining a small sample of species. Different species of wood can be identified through anatomical study. In regions where Cinnamomum species are common, this method can be very helpful for gaining insight into the wood's possible uses (Andianto et al. 2015). European explorers showed great interest in spices like cinnamon (*C. verum*) and cassia in the fifteenth and sixteenth centuries. Trade monopolies were created in the region by the Portuguese, who first occupied Sri Lanka in 1536, and then by the Dutch, who took power in 1656. Around 1770, the crop was first farmed in Sri Lanka after being gathered from wild locations (Prota 2023). By 1767, French botanist Jean Baptiste D'Arnault had collected nutmeg and cinnamon in the Lesser Antilles and Venezuela (Zumbroich 2005), proving that these spices had been introduced to the Caribbean by then. Since 1772 it was first brought to the Seychelles, Cinnamon has expanded rapidly, leading to the development of large cinnamon forests by the turn of the twentieth century. The Seychelles now identify cinnamon as the plant with the greatest known spread and, presumably, the greatest population from sea level to the highest peaks. In 1825, people in Java, Indonesia, began growing their food. Despite the promising beginning, it rapidly declined in the years that followed. Several countries have since adopted Ceylon cinnamon, as reported by the Seychelles Ministry of Environment (undated). From its original home in Sri Lanka, cinnamon spread to the northeastern United States, Bangladesh, Borneo, Southeast Brazil, Cambodia, the Caroline Islands, China, the Cook Islands, Fiji, the Gulf of Guinea Islands, India, Java, the Leeward Islands, Mauritius, Myanmar, the Philippines, Samoa, Seychelles, Society Islands, Taiwan, Tanzania, Vietnam, and the Windward Islands (Fig. 2.1).

Dalchini (*Cinnamomum verum*) is a common name for a tall, evergreen tropical shrub that can grow 6–8 m tall. It has smooth, reddish-brown bark and a dense, layered structure. The glabrous, ovate- and lanceolate-shaped leaves of this plant's opposite and subopposite arrangements. The texture of these leaves might be described as hard and leathery. The upper surface of the leaves is brightly colored, while the undersides have a paler tint, and three to five noticeable veins are visible. The petiole is about half an inch long and has a flattened shape. Flowers are clustered in cymes or panicles that may be axillary or subterminal. The fruit is rectangular or ovate and measures around 1.5–2 cm in length. It has a finely apiculate structure and can be either dry or slightly meaty, depending on preparation (Pathak and Himanshu 2021) (Fig. 2.2).

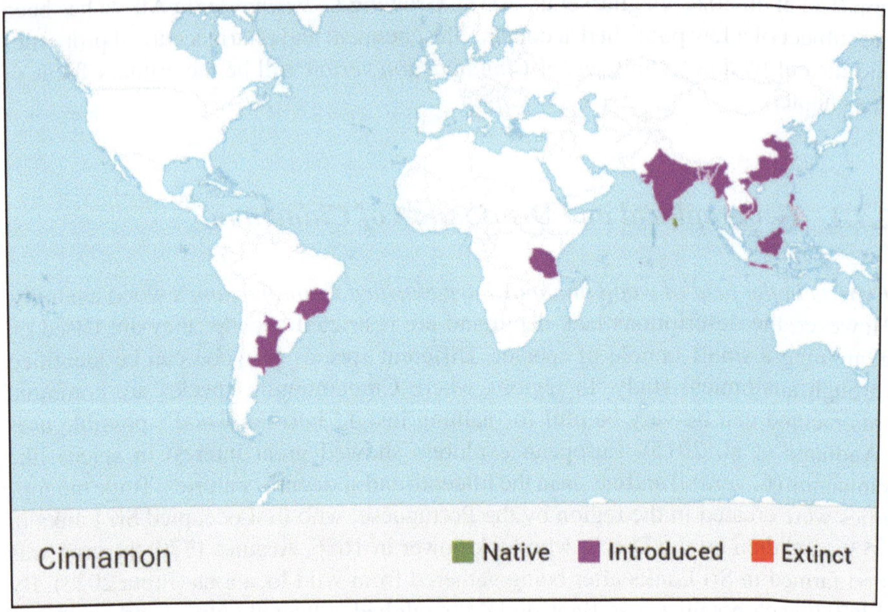

Fig. 2.1 Map of geographical distribution of native, introduced and extinct locations of *Cinnamomum* cultivation

Fig. 2.2 Cinnamon bark **a** *C. verum* J. Presl and **b** *C. cassia* L. *Source* Newerli-Guz and Śmiechowska (2022)

2.1.2 *Origin and Nomenclature*

Cinnamon (*Cinnamomum verum*), commonly known as the authentic Ceylon cinnamon tree, is indigenous to South Asia. However, it was widely transported across tropical regions during the eighteenth century. Birds consume the fruits, facilitating their widespread dispersal, leading to the invasive colonization of various islands in the Indian and Pacific Oceans, with particular emphasis on the Seychelles and Samoa. *C. verum* is recognized as an invasive species in the Seychelles, encompassing Mahe and Praslin Islands and on the island of Mayotte. After its establishment, this tree species dominates in moist, predominantly lowland tropical forests. As a result of competitive interactions, it gradually influences the diversity of native forest species. The dense canopy of this tree species obstructs sunlight, suppressing other plant species' growth. Consequently, this leads to forming stands with low species richness, which can extend over expansive areas (Bakewell-Stone 2023).

2.1.3 *History of Cinnamon*

For centuries, people have been using cinnamon for a wide variety of different dishes. Due to its high therapeutic effect, it has been used in Ayurvedic medicine as an antiemetic, antidiarrheal, ant flatulent, and stimulant. The Egyptians used it for the process of mummification. In the sixteenth and seventeenth centuries, Portuguese merchants enabled the export of *C. Verum*, a spice grown in Sri Lanka. Cinnamon was first cultivated in Java during the seventeenth century Dutch occupation and was brought to Europe by the East India Company. Sri Lanka has replaced Ceylon as the primary source of cinnamon oils, which are in high demand in the pharmaceutical and food industries. Cinnamon oil from China is also used in the pharmaceutical industry (Pathak and Himanshu 2021).

2.1.4 *Taxonomic Tree of* C. Verum

Domain: Eukaryota
Kingdom: Plantae
Phylum: Spermatophyta
Subphylum: Angiospermae
Class: Dicotyledonae
Order: Laurales
Family: Lauraceae
Genus: Cinnamomum
Species: *Cinnamomum verum* (Bakewell-Stone 2023).

Cinnamon (*C. verum*), a member of the Lauraceae family, was initially described by J. Presl in 1823 (Kew 2023). The etymology of cinnamon can be traced back to the Greek term 'kinnamomon', which translates to "spice". The Greeks adopted the term from the Phoenicians, indicating early trade relations with the Eastern region. Cinnamon has been documented in the Old Testament, Sanskrit, and Greek medicinal texts. A comprehensive survey conducted in northeast India documented a total of 14 Cinnamomum species that are commonly used as the spice known as 'cinnamon' (Baruah and Nath 2007). This result coincides with the United States Department of Agriculture's Agricultural Research Service (USDA-ARS 2009), which also recognized 14 taxonomic groups. Baruah and Nath (2007) present a taxonomic key to help distinguish between species based on a set of evaluated features. However, it has been noted by other sources that the number of species native to the Asia–Pacific region and tropical America is substantially larger, estimated at over 350. About 50 are tropical and subtropical trees and shrubs, most of which are valued for their economic contributions. *C. verum* is where true cinnamon comes from, and the Latin word vera, meaning truth However, other species, most notably *C. cassia*, may also be used to produce cinnamon for commercial purposes. The island of Ceylon, now known as Sri Lanka, is the original home of this species; the term 'Verum' is used to denote this. The given data is the official scientific designation for *C. verum*.

2.2 Botanical Aspects of Cinnamon

Evergreen trees and shrubs between 5 and 15 m tall belong to the Cinnamomum genus. Southeast Asia, China, Australia, and Africa all have their fair share of plants. The leaves are 7–18 cm in length and ovate-oblong in shape; the bark is often used as a spice (Cardoso-Ugarte et al. 2016). Panicle-shaped flowers have a greenish color and a musky scent and are clustered in dense clusters. The berry-like fruit is only around a centimeter in diameter and is purple (Cardoso-Ugarte et al. 2016; Leela 2008). The plant is a typical sight in tropical rainforests, where it thrives on both mountainsides and lowland forests. It grows in both wet and dry conditions but prefers the former. However, they are becoming less widespread in areas with

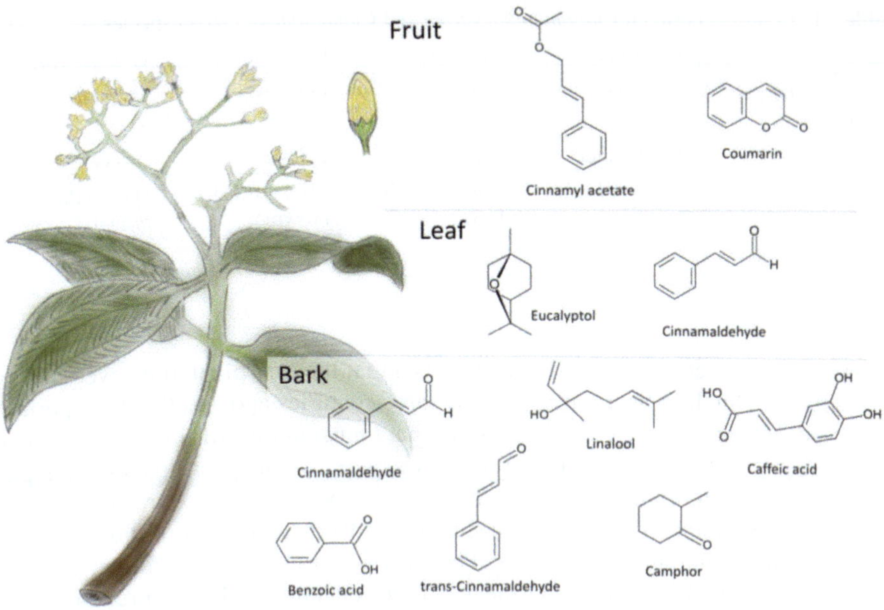

Fig. 2.3 Selected bioactive compounds of cinnamon in different parts of the plant. *Source* Błaszczyk et al. (2021)

seasonal climates (Jantan et al. 2008; Cardoso-Ugarte et al. 2016). Cinnamomum verum, one of the most common types of cinnamon, has been shown to have only 17 mg of coumarin per kilogram (Bansode 2012). The chemical composition of cinnamon is the source of its beneficial effects. Plant extracts from different places have drastically varying chemical compositions. This suggests that they have a wide range of individual traits. The volatile oils extracted from the cinnamon tree's bark, stem, and roots all have almost the same ratio of monoterpene hydrocarbons. Their essential components, however, distinguish them. The bark of the cinnamon tree is the principal source of cinnamon aldehyde. Eugenol is abundant in the oil pressed from the leaves, but camphor predominates in the oil pressed from the root bark (Gruenwald et al. 2010; Ramadan and Ibrahim 2021) (Fig. 2.3).

2.3 Germplasm Resources of Cinnamon Blume

Despite cinnamon's cultural and culinary significance over the centuries, no systematic scientific effort has been to catalog the wide range of genetic variation within the species due to natural selection and human breeding practices. Compared to other tree species like clove and nutmeg, cinnamon has more genetic diversity,'

Table 2.1 Contrasts between *Cinnamomum verum* and *Cinnamomum cassia* species

Taxa	*C. verum*	*C. cassia*	*C. burmannii*	*C. loureiroi*
Country where it originates	Sri Lanka	China	Indonesia	Vietnam
Flavor	Mild sweet	Bitter spicy	Spicy	Sweet spicy
Color	Light reddish brown	Dark reddish brown	Dark reddish brown	Dark reddish brown
Coumarin content (g/kg)	0.017	0.31	2.15	6.97

(Krishnamoorthy et al. 1997). However, given that the first plantations were established by the Dutch in Sri Lanka less than three centuries ago and that long periods elapse between replanting, it is fair to say that cinnamon is still in its early stages of domestication. Cinnamon's importance as a spice crop suggests that organoleptic traits, rather than morphological ones, have been depended on for intraspecific classification in Sri Lanka (Table 2.1). In Sinhalese, distinct varieties of cinnamon are known by their unique slang names. Peni Miris Kurundu means "sweet and hot cinnamon," while Thitta Kurundu means "bitter cinnamon," Peni Kurundu means "sweet cinnamon," Naga Kurundu means "snake cinnamon," Veli Kurundu means "sandy cinnamon," Sevel kurundu means "slimy cinnamon," and Kahata Kurundu means "acrid cinnamon" (Anonymous 1996; Wijesinghe and Pathirana 2000). The chemical composition in different types of cinnamon was illustrated in Table 2.2 (Feltes et al. 2023).

2.4 Physiology and Phenology

The growth of seedling roots exhibits an initial phase of rapid development, characterized by the emergence of a robust taproot and the subsequent formation of multiple lateral roots. This process establishes a moderately deep and extensive root system (Prota 2023). A solitary central stem is typically present in its natural state, although trees can be intentionally coppiced during cultivation. The untrimmed tree exhibits a multitude of branches, frequently characterized by a downward inclination, originating at a low point on the main stem. The growth process occurs in distinct phases, with young leaves initially exhibiting a reddish hue, eventually transitioning to a dark green coloration. According to Orwa et al. (2009), it has been observed that in Sri Lanka, the trees typically undergo flowering in January, followed by fruit ripening approximately six months later. The botanical name for true cinnamon is *C. verum*. The study conducted by Fleischmann et al. (1999) revealed that competition between the endemic palm species *Phoenicophorium borsigianum* and the highly invasive alien species *C. verum* in Seychelles is significantly affected by the

Table 2.2 The volatile components found in the bark and leaf of *C. verum* and *C. cassia* essential oils

Compounds	Content (%)			
	Bark		Leaf	
	C. cassia	*C. verum*	*C. verum*	*C. cassia*
1,3-dimethyl-benzene	0.23	0.15	–	–
Styrene	0.19	0.14	–	–
Benzaldehyde	0.41	0.29	0.05	0.1
Camphene	0.35	0.21	–	–
Acetophenone	0.96	Tr	–	–
β-Pinene	0.15	0.44	–	–
Linalool	0.68	–	–	–
Camphor	0.97	0.53	–	–
Benzene propanal	0.64	0.53	–	–
Borneol	0.19	0.12	–	–
Cis-cinnamaldehyde	1.95	2.29	–	–
Trans-cinnamaldehyde	77.21	74.49	16.25	30.65
Eugenol	0.21	7.29	79.75	–
Geranyl acetate	0.14	0.12	–	–
Benzene,1-(1,5-dimethyl-4-hexenyl)-4-methyl	0.37	0.15	–	–
Cinnamyl acetate	0.14	0.49	–	–
α-muuroleno	0.47	0.11	–	–
3-Methoxy-1,2-propanediol	–	–	–	29.3
Cinnamyl alcohol	–	–	0.07	0.65
Acetaldehyde	–	–	–	0.47
o-Methoxy cinnamaldehyde	–	–	–	25.39
Coumarin	–	–	0.05	6.36

availability of photosynthetically active radiation. Both species exhibited a negative correlation between seedlings' average leaf area and light levels. *P. borsigianum* exhibited a notably superior response to low levels of gap light compared to *C. verum*. Furthermore, both species demonstrated no additional response when exposed to high levels of direct sunlight. *P. borsigianum* and potentially other endemic palms filter the distribution and abundance of establishing *C. verum* seedlings. According to Fleischmann et al. (1999), there was a significant negative correlation between the mortality of *C. verum* and levels of photosynthetically active radiation. However, no such correlation was observed for *P. borsigianum*. Cinnamon and cinnamon-tree flowers, neither cru imports by country in 2023 were illustrated in Table 2.3.

Table 2.3 Cinnamon and cinnamon-tree flowers, neither cru imports by country in 2023

Reporter	Trade flow	Product code	Product description	Year	Partner	Trade value 1000 USD	Quantity	Quantity unit
United States	Import	90,610	Cinnamon and cinnamon-tree flowers, neither cru	2023	World	91,799.18	15,225,300	Kg
Mexico	Import	90,610	Cinnamon and cinnamon-tree flowers, neither cru	2023	World	48,756.24	4,393,770	Kg
European Union	Import	90,610	Cinnamon and cinnamon-tree flowers, neither cru	2023	World	41,696.96	7,626,520	Kg
Germany	Import	90,610	Cinnamon and cinnamon-tree flowers, neither cru	2023	World	9803.69	1,852,550	Kg
Netherlands	Import	90,610	Cinnamon and cinnamon-tree flowers, neither cru	2023	World	9791.11	1,971,270	Kg
Brazil	Import	90,610	Cinnamon and cinnamon-tree flowers, neither cru	2023	World	9344.67	2,792,590	Kg
Spain	Import	90,610	Cinnamon and cinnamon-tree flowers, neither cru	2023	World	9125.86	1,511,040	Kg
Guatemala	Import	90,610	Cinnamon and cinnamon-tree flowers, neither cru	2023	World	9053.45	825,128	Kg

(continued)

Table 2.3 (continued)

Reporter	Trade flow	Product code	Product description	Year	Partner	Trade value 1000 USD	Quantity	Quantity unit
Egypt, Arab Rep	Import	90,610	Cinnamon and cinnamon-tree flowers, neither cru	2023	World	8780.75	1,679,180	Kg
Japan	Import	90,610	Cinnamon and cinnamon-tree flowers, neither cru	2023	World	7559.62	1,419,590	Kg
France	Import	90,610	Cinnamon and cinnamon-tree flowers, neither cru	2023	World	7474.88	1,263,690	Kg
Malaysia	Import	90,610	Cinnamon and cinnamon-tree flowers, neither cru	2023	World	7312.43	1,822,370	Kg
United Kingdom	Import	90,610	Cinnamon and cinnamon-tree flowers, neither cru	2023	World	6788.93	1,273,820	Kg
Portugal	Import	90,610	Cinnamon and cinnamon-tree flowers, neither cru	2023	World	4214.53	424,310	Kg
Canada	Import	90,610	Cinnamon and cinnamon-tree flowers, neither cru	2023	World	4192.44	753,612	Kg
Turkey	Import	90,610	Cinnamon and cinnamon-tree flowers, neither cru	2023	World	3645.86	1,669,470	Kg

(continued)

Table 2.3 (continued)

Reporter	Trade flow	Product code	Product description	Year	Partner	Trade value 1000 USD	Quantity	Quantity unit
Poland	Import	90,610	Cinnamon and cinnamon-tree flowers, neither cru	2023	World	3454.77	410,685	Kg
China	Import	90,610	Cinnamon and cinnamon-tree flowers, neither cru	2023	World	3422.73	864,167	Kg
Chile	Import	90,610	Cinnamon and cinnamon-tree flowers, neither cru	2023	World	2969.99	340,940	Kg
Nicaragua	Import	90,610	Cinnamon and cinnamon-tree flowers, neither cru	2023	World	2511.75	234,831	Kg
Greece	Import	90,610	Cinnamon and cinnamon-tree flowers, neither cru	2023	World	2204.01	473,070	Kg
Italy	Import	90,610	Cinnamon and cinnamon-tree flowers, neither cru	2023	World	2190.93	381,000	Kg
Bolivia	Import	90,610	Cinnamon and cinnamon-tree flowers, neither cru	2023	World	1854.61	308,013	Kg
Switzerland	Import	90,610	Cinnamon and cinnamon-tree flowers, neither cru	2023	World	1772.09	118,235	Kg

(continued)

Table 2.3 (continued)

Reporter	Trade flow	Product code	Product description	Year	Partner	Trade value 1000 USD	Quantity	Quantity unit
Australia	Import	90,610	Cinnamon and cinnamon-tree flowers, neither cru	2023	World	1670.63	247,609	Kg
Sweden	Import	90,610	Cinnamon and cinnamon-tree flowers, neither cru	2023	World	1392.63	154,304	Kg
South Africa	Import	90,610	Cinnamon and cinnamon-tree flowers, neither cru	2023	World	1344.40	445,755	Kg
Israel	Import	90,610	Cinnamon and cinnamon-tree flowers, neither cru	2023	World	1191.06	222,207	Kg
Panama	Import	90,610	Cinnamon and cinnamon-tree flowers, neither cru	2023	World	1137.05	162,476	Kg
Czech Republic	Import	90,610	Cinnamon and cinnamon-tree flowers, neither cru	2023	World	1030.84	133,465	Kg
Argentina	Import	90,610	Cinnamon and cinnamon-tree flowers, neither cru	2023	World	950.49	86,521	Kg
Estonia	Import	90,610	Cinnamon and cinnamon-tree flowers, neither cru	2023	World	837.36	122,445	Kg

(continued)

Table 2.3 (continued)

Reporter	Trade flow	Product code	Product description	Year	Partner	Trade value 1000 USD	Quantity	Quantity unit
Romania	Import	90,610	Cinnamon and cinnamon-tree flowers, neither cru	2023	World	643.81	70,913	Kg
Mauritius	Import	90,610	Cinnamon and cinnamon-tree flowers, neither cru	2023	World	575.32	128,056	Kg
Denmark	Import	90,610	Cinnamon and cinnamon-tree flowers, neither cru	2023	World	547.95	53,485	Kg
Norway	Import	90,610	Cinnamon and cinnamon-tree flowers, neither cru	2023	World	516.15	70,703	Kg
Ireland	Import	90,610	Cinnamon and cinnamon-tree flowers, neither cru	2023	World	502.05	67,898	Kg
Ukraine	Import	90,610	Cinnamon and cinnamon-tree flowers, neither cru	2023	World	454.83	59,317	Kg
Hungary	Import	90,610	Cinnamon and cinnamon-tree flowers, neither cru	2023	World	395.07	49,106	Kg
Tanzania	Import	90,610	Cinnamon and cinnamon-tree flowers, neither cru	2023	World	380.02	677,406	Kg

(continued)

Table 2.3 (continued)

Reporter	Trade flow	Product code	Product description	Year	Partner	Trade value 1000 USD	Quantity	Quantity unit
Belgium	Import	90,610	Cinnamon and cinnamon-tree flowers, neither cru	2023	World	309.88	44,271	Kg
Maldives	Import	90,610	Cinnamon and cinnamon-tree flowers, neither cru	2023	World	300.96	53,096	Kg
Azerbaijan	Import	90,610	Cinnamon and cinnamon-tree flowers, neither cru	2023	World	285.86	59,736	Kg
Finland	Import	90,610	Cinnamon and cinnamon-tree flowers, neither cru	2023	World	273.51	35,020	Kg
Sri Lanka	Import	90,610	Cinnamon and cinnamon-tree flowers, neither cru	2023	World	241.96	13,482	Kg
Slovak Republic	Import	90,610	Cinnamon and cinnamon-tree flowers, neither cru	2023	World	224.84	21,057	Kg
New Zealand	Import	90,610	Cinnamon and cinnamon-tree flowers, neither cru	2023	World	210.89	27,982	Kg
Bulgaria	Import	90,610	Cinnamon and cinnamon-tree flowers, neither cru	2023	World	208.44	52,804	Kg

(continued)

Table 2.3 (continued)

Reporter	Trade flow	Product code	Product description	Year	Partner	Trade value 1000 USD	Quantity	Quantity unit
Uruguay	Import	90,610	Cinnamon and cinnamon-tree flowers, neither cru	2023	World	207.51	54,406	Kg
Cyprus	Import	90,610	Cinnamon and cinnamon-tree flowers, neither cru	2023	World	147.92	53,107	Kg
Lithuania	Import	90,610	Cinnamon and cinnamon-tree flowers, neither cru	2023	World	143.66	28,116	Kg
Madagascar	Import	90,610	Cinnamon and cinnamon-tree flowers, neither cru	2023	World	140.17	50,869	Kg
Slovenia	Import	90,610	Cinnamon and cinnamon-tree flowers, neither cru	2023	World	136.84	5579	Kg
Hong Kong, China	Import	90,610	Cinnamon and cinnamon-tree flowers, neither cru	2023	World	135.7	24,825	Kg
Brunei	Import	90,610	Cinnamon and cinnamon-tree flowers, neither cru	2023	World	126.13	45,911	Kg
Croatia	Import	90,610	Cinnamon and cinnamon-tree flowers, neither cru	2023	World	111.9	7218	Kg

(continued)

Table 2.3 (continued)

Reporter	Trade flow	Product code	Product description	Year	Partner	Trade value 1000 USD	Quantity	Quantity unit
Guyana	Import	90,610	Cinnamon and cinnamon-tree flowers, neither cru	2023	World	101.5	49,616	Kg
Kazakhstan	Import	90,610	Cinnamon and cinnamon-tree flowers, neither cru	2023	World	95.99	57,272	Kg
Bosnia and Herzegovina	Import	90,610	Cinnamon and cinnamon-tree flowers, neither cru	2023	World	86.34	15,867	Kg
Barbados	Import	90,610	Cinnamon and cinnamon-tree flowers, neither cru	2023	World	70.71	18,112	Kg
Armenia	Import	90,610	Cinnamon and cinnamon-tree flowers, neither cru	2023	World	70.65	18,473	Kg
Aruba	Import	90,610	Cinnamon and cinnamon-tree flowers, neither cru	2023	World	68.54	12,966	Kg
Georgia	Import	90,610	Cinnamon and cinnamon-tree flowers, neither cru	2023	World	63.97	11,113	Kg
Luxembourg	Import	90,610	Cinnamon and cinnamon-tree flowers, neither cru	2023	World	56.14	1963	Kg

(continued)

Table 2.3 (continued)

Reporter	Trade flow	Product code	Product description	Year	Partner	Trade value 1000 USD	Quantity	Quantity unit
Namibia	Import	90,610	Cinnamon and cinnamon-tree flowers, neither cru	2023	World	54.47	28,871	Kg
Iceland	Import	90,610	Cinnamon and cinnamon-tree flowers, neither cru	2023	World	53.47	5233	Kg
Philippines	Import	90,610	Cinnamon and cinnamon-tree flowers, neither cru	2023	World	47.58	9253	Kg
Moldova	Import	90,610	Cinnamon and cinnamon-tree flowers, neither cru	2023	World	40.49	5953	Kg
Malta	Import	90,610	Cinnamon and cinnamon-tree flowers, neither cru	2023	World	37.86	3403	Kg
Botswana	Import	90,610	Cinnamon and cinnamon-tree flowers, neither cru	2023	World	32.34	4670	Kg
Mozambique	Import	90,610	Cinnamon and cinnamon-tree flowers, neither cru	2023	World	29	2547	Kg
Ghana	Import	90,610	Cinnamon and cinnamon-tree flowers, neither cru	2023	World	27.12	10,115	Kg

(continued)

Table 2.3 (continued)

Reporter	Trade flow	Product code	Product description	Year	Partner	Trade value 1000 USD	Quantity	Quantity unit
Serbia, FR(Serbia/ Montenegro)	Import	90,610	Cinnamon and cinnamon-tree flowers, neither cru	2023	World	24.75	2384	Kg
Belize	Import	90,610	Cinnamon and cinnamon-tree flowers, neither cru	2023	World	14.08	68,628	Kg
North Macedonia	Import	90,610	Cinnamon and cinnamon-tree flowers, neither cru	2023	World	12.85	1971	Kg
Antigua and Barbuda	Import	90,610	Cinnamon and cinnamon-tree flowers, neither cru	2023	World	12.71	3458	Kg
Paraguay	Import	90,610	Cinnamon and cinnamon-tree flowers, neither cru	2023	World	10.48	2051	Kg
Suriname	Import	90,610	Cinnamon and cinnamon-tree flowers, neither cru	2023	World	9.39	1716	Kg
Grenada	Import	90,610	Cinnamon and cinnamon-tree flowers, neither cru	2023	World	7.66	4711	Kg
Macao	Import	90,610	Cinnamon and cinnamon-tree flowers, neither cru	2023	World	6.27	1049	Kg

(continued)

Table 2.3 (continued)

Reporter	Trade flow	Product code	Product description	Year	Partner	Trade value 1000 USD	Quantity	Quantity unit
Burkina Faso	Import	90,610	Cinnamon and cinnamon-tree flowers, neither cru	2023	World	4.45	16,200	Kg
Zimbabwe	Import	90,610	Cinnamon and cinnamon-tree flowers, neither cru	2023	World	4.41	1375	Kg
Zambia	Import	90,610	Cinnamon and cinnamon-tree flowers, neither cru	2023	World	3.15	3138	Kg
Kyrgyz Republic	Import	90,610	Cinnamon and cinnamon-tree flowers, neither cru	2023	World	2.31	3945	Kg
Benin	Import	90,610	Cinnamon and cinnamon-tree flowers, neither cru	2023	World	2.26	1580	Kg
Montserrat	Import	90,610	Cinnamon and cinnamon-tree flowers, neither cru	2023	World	1.16	105	Kg
Sao Tome and Principe	Import	90,610	Cinnamon and cinnamon-tree flowers, neither cru	2023	World	1.16	77	Kg
Andorra	Import	90,610	Cinnamon and cinnamon-tree flowers, neither cru	2023	World	1.09	35	Kg

(continued)

Table 2.3 (continued)

Reporter	Trade flow	Product code	Product description	Year	Partner	Trade value 1000 USD	Quantity	Quantity unit
Nigeria	Import	90,610	Cinnamon and cinnamon-tree flowers, neither cru	2023	World	0.78	4000	Kg
Malawi	Import	90,610	Cinnamon and cinnamon-tree flowers, neither cru	2023	World	0.51	282	Kg

2.5 Health Beneficial

Antioxidant molecules play an important part in human life as protectors of bodily integrity. The food industry uses these compounds as additives to prevent spoiling, and they also occur naturally in some foods. The antioxidant compounds in cinnamon are well-known. This seasoning may affect several diseases. Ervina et al. (2019) found that cinnamon bark's polyphenolic and volatile oil components contribute to the infusion's potent antioxidant capabilities. High antioxidant content can be dealt with by using a simple approach to making an extract (Ervina et al. 2016; Yashin et al. 2017). Because only *C. verum* cinnamon has any coumarin, it is typically kept for therapeutic uses. Multiple studies have demonstrated cinnamon's medicinal properties and has been used for centuries (Ribeiro-Santos et al. 2017).

The aging process and the onset of many diseases can be slowed by consuming cinnamon, which is why it is widely acknowledged as a natural source of antioxidants. J. Presl) *Cinnamomum zeylanicum*, *Cinnamomum verum*, and *Cinnamomum cassia* L.) show the most promise. Biologically active compounds, including eugenol, trans-cinnamaldehyde, and linalool, can be found in the bark of this particular species. There has been research into the role of phytonutrients in palm oil stability. They all stymied the production of acids, ketones, aldehydes, and hydrocarbons. Oil's oxidation process can be efficiently slowed by adding antioxidants, reducing the production of harmful oxidative derivatives and free fatty acids (Shahid et al. 2018).

The compounds found in cinnamon have been hypothesized to have protective effects on human health. Cinnamon antioxidants are encapsulated to increase their effectiveness. Adding these compounds to your diet has health benefits since they help lower oxidative stress, a known inflammation trigger (Pisoschi et al. 2021). The importance of antioxidants to human health, especially in preventing metabolic diseases like diabetes, is widely accepted. The yield of cinnamon extracts is highly sensitive to the plant's stage of development. Prabhashini et al. (2019) found that ripe cinnamon leaves are most effective at reducing the oxidative stress that contributes to the development of chronic illnesses. Cinnamaldehyde and its derivatives are the major chemical compounds responsible for the effectiveness of natural medicine. However, the team looked at two novel substances with importance in natural medicine due to their ability to reduce oxidative stress: lignan pinoresinol (PRO) and flavonol (−)-(2R,3R)-5,7-dimethoxy-3′,4′methylenedioxy-flavan-3-ol (MFO) (Li et al. 2019). Cinnamon water extracts primarily p-hydroxybenzoic acid, p-coumaric acid, caffeic acid, ferulic acid, vanillin, and gallic acid, as determined by examining the phenolic profile. However, ascorbic acid can also be found in cinnamon ethanol extracts. Chemicals with several biological effects can inhibit enzymes while also inhibiting Fe^{3+} and Cu^{2+} ion reduction. Li et al. (2019) found that glucosidase, butyrylcholinesterase (BChE) and -amylase were particularly sensitive to the inhibitory effects of ethanol extracts.

Cinnamon could be used as a food preservative. Hyperpigmentation, or the browning effect seen in fruits, vegetables, and mushrooms, can be avoided using (E)-cinnamaldehyde as an inhibitor of tyrosinase activation. There are several ways in which cinnamon components can be used, such as in treating illness, the food industry, medicine, and the cosmetics sector (Georgiev et al. 2013).

Cinnamon may contain tannins, non-volatile bioactive substances derived from phenolic compounds. These compounds are powerful antioxidants with great therapeutic promise for treating various diseases and metabolic problems. Both condensed tannins (proanthocyanidins) and hydrolysable tannins (Ghosh 2015) can be found in cinnamon. The pharmacological effects of tannins are beneficial. The beneficial systemic effects result from tannins being hydrolysed in a microbial environment, generating readily available metabolites (Sallam 2021).

In the past decade, cinnamon's many health benefits have made it a popular topic of study. These qualities include reducing the risk of cardiovascular disease and acting as an anti-microbial, anti-parasitic, anti-oxidant, anti-inflammatory, anti-diabetic, and anti-ulcer agent. In addition to its lipid-lowering and anti-hypertriglyceridemia effects, cinnamon has also been shown to have anticancer properties. Ravindran et al. (2004a, b), Vangalapati et al. (2012), Rao and Gan (2014) and Madan and Kannan (2004). According to Shan et al. (2005), trans-cinnamaldehyde, eugenol, and linalool structure 82.5% of the total composition of the essential oils extracted from the bark of *C. Verum*. Trans-cinnamaldehyde accounts for roughly 49.9–62.8% of the total quantity of bark oil, as reported by Singh et al. (2007) and Simic et al. (2004). Extracts of *C. verum* contain numerous active components, including cinnamaldehyde and eugenol. Numerous research on animals and humans in different parts of the world have linked *C. verum* to numerous positive health effects. Benefits include reduced risk of colon cancer, improved cognitive function, and protection against cardiovascular disease and inflammation (Jayaprakasha and Rao 2011). *C. verum*, known for its wound-healing qualities, was studied using 32 rats. Experimental excision wounds were made on the rats, and they were given *C. verum* ointment topically. Extracts of *C. verum* were found to improve wound healing by encouraging epithelialization, as Farahpour and Habibi (2012) reported. When *C. verum* was given orally to Wister rats with incision wounds, the strength at which the wounds closed increased significantly, as Kamath et al. (2003) reported. The amount of hydroxyproline and the breaking strength of the granulation tissue in the dead space wounds also increased significantly. In a rat trial where the liver injury was produced with CCl4, Kamath et al. (2003) found that *C. Verum* protected the liver. Animal studies have shown no serious side effects on the liver or kidneys, suggesting a wide therapeutic window (Ranasinghe et al. 2012). Rao and Gan (2014) gave *C. verum* to female mice for two weeks to see what would happen. The effects of this treatment were then evaluated by counting the number of nuclei in mouse embryos and analysing the distribution of embryos depending on this metric. Rao and Gan (2014) found that when cinnamon was applied to embryos, the number of nuclei decreased, and the distribution of embryos changed significantly depending on nucleus count. These shifts were related to cinnamaldehyde's anti-proliferative properties. Figures 2.4 and 2.5 depict cinnamon's therapeutic applications.

According to several reports (Thakur et al. 2021; Vasconcelos et al. 2018), this is because of its very hydrophobic character. *Cinnamomum verum* and *C. cyminum* essential oils were tested for their ability to kill bacteria like *Salmonella typhi* and

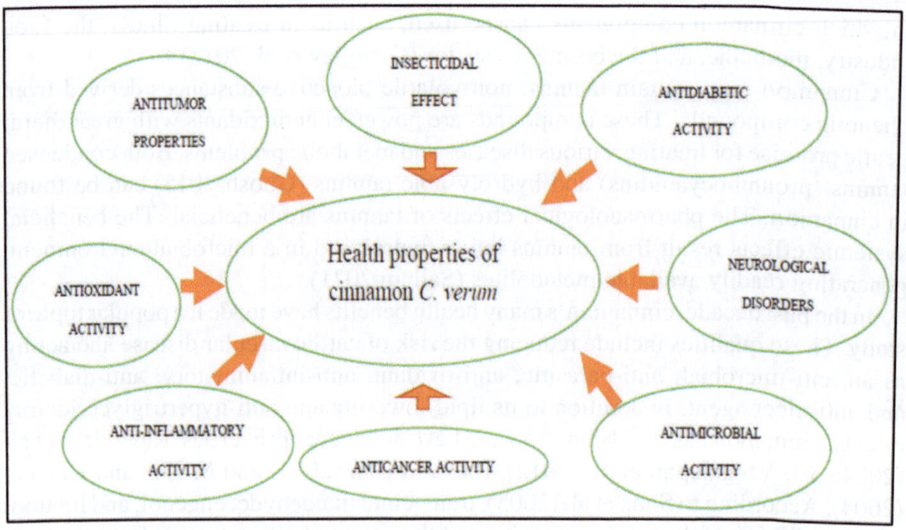

Fig. 2.4 Health properties of cinnamon *C. verum*. *Source* Newerli-Guz and Śmiechowska (2022)

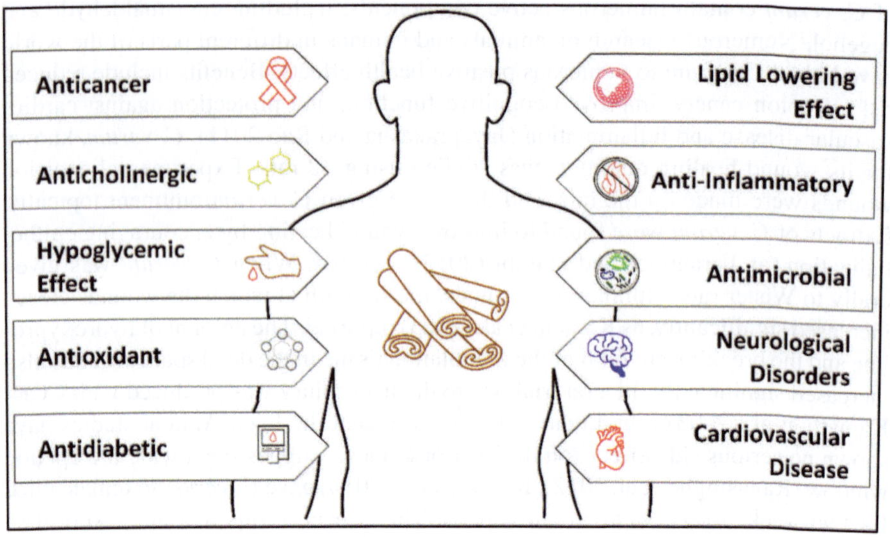

Fig. 2.5 Uses of cinnamon in treating different diseases in humans. *Source* Błaszczyk et al. (2021)

Salmonella paratyphi, as well as *E. coli*, Staphylococcus aureus, *Bacillus licheniformis*, and Pseudomonas fluorescens. The broth technique was used for the analysis. Compared to the other three plant extracts, *C. verum* showed much better antibacterial activity against all tested microorganisms (Wisal 2018; Al-Harbi et al. 2021). Ethyl acetate was found to have potent antibacterial action against *Staphylococcus aureus*,

Escherichia coli, and *Pseudomonas aeruginosa*, as shown in the results. When tested for sensitivity to *Bacillus subtilis*, the petroleum ether extract performed best (Naveed et al. 2013). The effect of 15 essential oil extracts from 15 different plants on 13 different bacterial strains was investigated separately. Abdalla and Abdelgadir (2016) state that cinnamon oil has the strongest antibacterial activity among essential oils. Including *Mycobacterium tuberculosis*, *C. verum* showed antibacterial activity against various Gram-negative and Gram-positive bacteria. The antibacterial activity demonstrated against several pathogens such as *Acinetobacter baumannii*, *Acinetobacter lwoffii*, *Bacillus cereus*, and *Yersinia enterocolitica* is summarized in a recent review by Ranasinghe et al. (2013). The minimum inhibitory concentration (MIC) of *C. verum* extract against the multi-drug resistant *Escherichia coli* AG100 was reported to be 64 g/mL by Voukeng et al. (2012).

2.6 Environmental Impact

Several indigenous palm species in the Seychelles are in danger due to the effective invasion of *C. verum* into mountain mist forests and intermediate forest environments. It has been reported that several species, including *Phoenicophorium borsigianum*, *Deckenia nobilis*, and *Roscheria melanochaetes*, are in jeopardy. *Northia seychellana* and other native trees are also being negatively impacted by this invasive species (Kueffer et al. 2008). *C. verum* has been linked to a decline in native diversity, deterioration of ecosystems, and the loss of habitats or refugia in the Seychelles due to competition, as reported by the Global Invasive Species Database in (2023). Since *C. verum* seeds can travel great distances, the invasive plant will likely keep spreading rapidly across Silhouette Island, Seychelles. However, Gerlach's (2004) study indicated that the natural forest mix in most regions had remained steady over a decade. Invasive plant species are most likely to alter the composition of native plant communities in low-nutrient environments by competing for few resources at ground level. Mature *C. verum* trees were shown to produce a deep topsoil root mat that effectively restricted the growth of seedling trees in a phosphorus-deficient ecology in the Seychelles. Kueffer et al. (2007) found that the increased competition for scarce nutrients was the primary cause of this inhibition. *C. verum* stands preferentially inhibited the establishment of invasive and small-seeded species, demonstrating a below-ground filtering effect on seedling regeneration and impacting secondary forest succession. As Kueffer et al. (2007) noted, this factored significantly towards the rate at which forests recovered. Introducing certain invasive alien plants, such as *C. verum* can speed up ecosystem nutrient cycling. Their litter has a high specific leaf area, increased nutrient concentrations, and lower lignin and polyphenolic levels, all of which contribute to these advantages. This leads to the rapid breakdown of their droppings. Species native to habitats with high nutrient availability are likelier to have the litter qualities mentioned (Kueffer et al. 2008), while species that successfully invade ecosystems with low nutrient availability may not. Even in nutrient-poor soils, pioneer invasive plants like *C. verum* in the granitic Seychelles show increased litter decomposition. As a result, these species can affect nutrient cycling rates. Soil fertility differences between native and introduced trees are small, but they show that the impact of invasive species on nutrient cycling is nuanced and less predictable in nutrient-poor ecosystems. This is especially important when native and non-native species exist because different nutrients may be co-limiting (Kueffer et al. 2008).

2.7 Breeding Strategy and Objectives

Cinnamon farmers have always taken pains to choose the best genotypes from the wild population to grow in their fields, whether on purpose or by accident. Grower-driven mass selection has resulted in today's diversely adapted populations of cinnamon in Sri Lanka and other cinnamon-growing countries. Adopting more refined and targeted breeding strategies is likely to hasten the improvement of desirable traits. Cinnamon breeding efforts can benefit from the insights gathered from more advanced perennial crop breeding programs.

There are substantial outlays of land, labor, time, and infrastructure resources involved in breeding perennial crops. Short-term and long-term breeding strategies must be properly crafted and implemented with the backing of secure and sustainable long-term funding. Partnership between the federal government, provincial agencies, and private grower associations is crucial. The breeding program's parameters should be set according to the needs of the cinnamon trade and the market. The breeders will have a reliable source of money, and the farmers will feel more invested in the project if they are encouraged to make financial contributions. However, as seen in private breeding projects (Abdel-Hamid and Salem 2021; Weerasuriya and Pathirana 2020), it is crucial to ensure that private finance does not compromise the general breeding objectives by unduly favoring frequent variety releases.

Clonal propagation techniques allow for the future mass production of superior plant material, which could help alleviate the economic constraints connected with the larger size and longer generation period of cinnamon plants. However, it is essential to recognize that an inefficient approach has the potential to render all breeding efforts useless. Therefore, it is crucial to carefully organize the breeding program using scientific methods. Pollination control is crucial and should be deliberated thoroughly from the very beginning. The loss of an entire 5-year-to-8-year breeding cycle is possible if such measures aren't implemented. Without a thorough description and documentation of the features demonstrated by elite parental plant types, the utility of program-generated knowledge for future breeders will be limited. Since breeders now have access to more sophisticated molecular tools, they must form a long-term alliance with a lab that specializes in the use of molecular markers (Aboshama et al. 2015; Weerasuriya and Pathirana 2020; Abu-Ellail et al. 2021; Salem et al. 2023a,b).

2.7.1 Cultivated and Wild Gene Pools

Maintaining and applying a diverse gene pool known to have desirable traits is a crucial part of any breeding strategy. The breeding efforts of Sri Lankan scientists are bolstered by their access to domesticated and wild *C. verum* gene pools. Compared to wild populations, the frequency of genes for desirable features is higher in the cultivated crop. Therefore, a convincing argument will be needed if there is a strategy to heavily use the wild gene pool. Screening the parent plants thoroughly to prioritize features like disease and insect pest resistance is essential. Bark and oil output and quality should be prioritized in this screening process. Doing so lets us see if these parents are good candidates for the breeding program (Alsharari et al. 2023; Weerasuriya and Pathirana 2020; Essa et al. 2023).

2.7.2 Unconventional Breeding Methods

No ongoing efforts have been observed to create interspecific hybrids within the Cinnamomum genus. The better quality displayed by *C. verum* suggests that it possesses innate genetic qualities that render gene infusion from similar species unnecessary for improving its characteristics. The same logic can be used to address the problem of pest and disease resistance. Because of their apparent genetic similarity, *C. verum*, *C. tamala*, and several native Sri Lankan species could successfully hybridize. However, the fundamental benefit of such hybridization is not the creation of novel cultivars but rather a deeper understanding of taxonomic relationships (Weerasuriya and Pathirana 2020). Methods for genetic improvement using in vitro experiments; multiple studies have shown that cinnamon can successfully regenerate damaged tissues in culture (Rai and Jagdishchandra 1987; Dayatilake 2000). Possible applications of this technology include somaclonal variation and in vitro mutagenesis for crop development. Additional potential uses for cell and tissue culture methods in cinnamon breeding include in-plant genetic modification, production of polyploidy, and remote hybridization by rescuing embryos.

2.7.3 Induced Mutations

Many efforts have been made to improve perennial fruit species using induced mutations. Furthermore, any helpful mutations within a mutation breeding program can be stabilized using vegetative propagation strategies. So far, (Pathirana 2011; Soliman et al. 2017, 2023). Perennial fruit species have benefited greatly from induced mutations, and any beneficial mutants can be fixed in a mutation breeding programme thanks to vegetative propagation techniques. Efforts to enhance cinnamon have been largely ineffective about radiation's impact on seeds or vegetative organs, and no studies have utilised mutation induction techniques. Thus, before starting mutation breeding programmes in cinnamon, it is necessary to construct growth reduction curves (Pathirana 2011). Induced mutations will be more useful for cinnamon improvement as clonal propagation techniques advance. This method allows for the improvement of a single trait without affecting other characters, as no hybridization steps are involved (Pathirana 2011). Consequently, in order to facilitate future breeding using this technology, it is necessary to establish protocols that account for dose response for various mutagens (Pathirana 2011; Soliman et al. 2017, 2023).

2.8 Genetic Diversity

Analysis of genetic diversity at the molecular level is essential. Numerous studies have used PCR-based markers to assess intraspecific variation (Ibrahim et al. 2023e; Salem and Ibrahim 2024). This study aimed to assess the genetic diversity of *C. zeylanicum* populations collected from the Western Ghats of Karnataka and other states using RAPD markers. Frankel (1974) argues that genetic variety is essential for species survival and should be prioritized in conservation efforts. To secure the long-term survival of *C. zeylanicum.* To effectively conserve and manage the species, assessing its genetic makeup in different regions is necessary. For the investigation of diversity in several red-listed plants, RAPD technology has been used because of its ease of use and rapidity (Li et al. 2002; Fu et al. 2003; Padmalatha and Prasad, 2006a, b; Abd El-Moneim et al. 2021; EL-Mansy et al. 2021; Safhi et al. 2022; ALshamrani et al. 2022; Ibrahim et al. 2023a, b, c, d). This study aimed to use RAPD markers to analyze genetic diversity among the accessions of *C. zeylanicum* to obtain genetic data and a theoretical basis for safeguarding this species. The genetic diversity of 15 different *C. zeylanicum* accessions were therefore analyzed by utilizing RAPD markers. RAPD polymorphism would be a more accurate parameter to measure the pattern of genetic diversity of rare and endangered plants because it reflects variation in the entire genomic DNA. At the same time, RAPD markers are based on the amplification of unknown DNA sequences using single, short, random oligonucleotide primers (Ibrahim et al. 2023a, b, c, d, e; Ibrahim et al. 2024).

Using a molecular technique would be a huge help in correctly identifying species. The genetic diversity of several species of cinnamon grown in Sri Lanka was analyzed by comparing nucleotide sequences from different regions of cpDNA and an ITS region of rDNA. The species studied included *C. verum*, *C. citriodorum*, *C. capparu-coronde*, *C. dubium*, *C. litseafolium*, *C. rivulorum*, *C. sinharajaense*, and *C. camphora.* We looked at the regions of cpDNA that are not part of any genes, called intergenic regions. These regions included the space between trnL and trnF, trnT and trnL, trnH and psbA, and the trnL intron. The trnL intron, trnL-trnF IGS, trnT-trnL IGS, trnH-psbA IGS, and ITS regions of rDNA were mined for cpDNA sequences, which led us to the discovery of several variable sites (Ibrahim et al. 2024). Our results will help shed light on the previously undiscovered genetic diversity of Sri Lankan Cinnamomum species, even if we did not uncover much variation in the cpDNA regions we studied. Since cpDNA variation was low in Sri Lanka, little was learned about the taxonomy of Cinnamomum species. However, it has been shown that the ITS region of rDNA can be used to reliably identify different species (Abeysinghe et al. 2009; Ibrahim et al. 2023e).

2.9 Molecular Genetic Approaches

By shedding light on the evolutionary processes within the genus and the genetic relationships between distinct species, molecular genetic approaches can potentially improve the efficiency of cinnamon breeding (Abd-Ellatif et al. 2022; Soliman et al. 2022). The first step is to create mapping populations with a wide range of cultivars. Next, microsatellite (SSRs) and/or single nucleotide polymorphism (SNP) markers can be developed and used in the generation of high-density (HD) maps. In addition, comparative quantitative trait loci (QTLs) analysis can generate a multi-parent consensus map of cinnamon. In addition, work can be done to create sets of molecular markers that are both practical and easy to use. Quantitative trait loci (QTLs) assessments on growth, blooming, yield, oil content, and quality can be performed with the help of individual populations and HD maps. Comparing the positions of quantitative trait loci (QTLs) across numerous genetic backgrounds is made possible by analyzing several identical features in distinct genetic backgrounds. However, traditional methods of molecular marker production are linked to prohibitive expenses, lengthy timespan requirements, and labor-intensive processes. For crops like cinnamon, where molecular data is scarce, producing markers using high throughput sequencing is more efficient thanks to the decreasing cost of next-generation sequencing (NGS) platforms. In species like cranberries, where molecular genetics data was scarce, the advent of Next-Generation Sequencing (NGS) technology has allowed for the development of Simple Sequence Repeat (SSR) markers (Zalapa et al. 2012; Ibrahim et al. 2023e). New single nucleotide polymorphism (SNP) systems for sequencing-based genotyping have been made possible, in large part, by next-generation sequencing (NGS) technologies (Bhat et al. 2016). The two cinnamon improvement organizations in India and Sri Lanka need to work together and form relationships with molecular biology research facilities to successfully adopt DNA marker-based breeding strategies for cinnamon.

2.10 Propagation and Planting

Propagation can occur through either seed or vegetative methods. Birds have a strong preference for fruits and are capable of spreading their seeds quite easily. Therefore, it is necessary to bag the fruits in order to collect them effectively. The fruit pulp undergoes a process of natural decomposition before the extraction, cleansing, and subsequent drying of the seeds. Seeds tend to experience a rapid decline in viability over time. The germination period for fresh seeds typically ranges from 20 to 25 days. The seeds are typically planted in nurseries or directly in the field. The nursery bed, which should measure 1 m in width, should be filled with properly prepared soil rich in nutrients.

Additionally, it is recommended to provide a light shade for the nursery bed (PROTA 2023). The seeds are planted near each other. After 4 months in the nursery, the seedlings are carefully transferred into individual bags. Following an additional 4–5 months, the seedlings are transplanted from the bags into the field. A cluster of at least five seedlings is consistently planted nearby, forming an undifferentiated clump.

2.10.1 Sexual Propagation

Birds have a strong preference for fruits and are capable of spreading their seeds quite easily. Therefore, it is necessary to bag the fruits to facilitate their collection. The fruit pulp undergoes a process of natural decomposition before the extraction, cleansing, and subsequent drying of the seeds. Seeds demonstrate recalcitrant storage behavior and experience rapid loss of viability. According to a study conducted by Orwa et al. in (2009), it was observed that at ambient temperature, the viability of the substance decreased to 80% after 7 days.

Furthermore, a complete loss of viability was observed after 40 days of storage in polythene bags at room temperature. Seeds are typically planted in a nursery environment, where fresh seeds begin germinating within approximately 20–25 days. After four months, the seedlings undergo the process of transplantation into suitable containers. After approximately 4–5 months, the juvenile plants are transplanted to the field. Seeds may also be directly sown in the field.

2.10.2 Vegetative Propagation

Plants can be artificially multiplied through cuttings, layers, or divisions of established rootstocks. After 12–18 months in a greenhouse, young cuttings with only 2–3 nodes can be transplanted into the field. It is possible to separate old rootstocks. This is accomplished by chopping down mature plants to within 15 cm of the ground and then planting out suitable sections of the rootstock with adherent soil. Harvesting can begin after 1–1.5 years in the field, whereas seedlings require 3 years before they can be harvested. The majority of tree breeding programmes that are in operation for a long time use multiple generations of trees that have been crossed selectively. The best plants can be used as both parents for the next generation of seedlings and as potential new varieties to be released into the market. Methods to significantly enhance genetic gain in a cinnamon breeding programme could involve reducing the duration of the juvenile phase between generations, as this would determine the genetic progress that could be achieved. There may be a minimum number of nodes needed to survive the juvenile phase, as shown by Zimmerman (1972) who found a correlation between plant height and flowering in fruit trees. Consequently, early flowering in cinnamon could be achieved with heavy fertilisation, which promotes strong development, just like in fruit crops. In thick plantations, another method involves training the tree to have only one stem, similar to how chestnut trees are trained (Zimmerman 1971). Breeding efforts in the long run would benefit greatly from agronomic and physiological strategies that reduce the juvenile stage, with a focus on cinnamon.

It is now feasible to produce clonal offspring from selected elite plants thanks to recent successes in vegetative propagation of cinnamon in Sri Lanka (Dayatilake 2000; Samaraweera 2000) and India (Benerjee et al. 1982; Hegde et al. 1989; Ranawere et al. 1995; Thirunavoukkarasu 1996). With the exception of the study conducted by Haldankar et al. (1994) on four specific clones, no findings have been published regarding the field performance of cinnamon material that was grown vegetatively. Cinnamon, in contrast to other plantation crops like tea, requires very little in the way of inputs when cultivated by smallholders (Kularatne 2000). Because of their extensive root system, these plants can withstand harsh weather conditions with little to no help and continue producing for a long time in the field. In addition, there have been no reports of illnesses or pest outbreaks, indicating that the populations have a good resistance to biotic stresses. Sri Lanka has previously had major disease outbreaks in crops that rely on clonal material for production, such as coffee. Hence, even with 24 clonal propagation, it is crucial to release several clones of varied origin to retain cinnamon variety. Converting existing seedling plantings into clonal ones requires meticulous monitoring of disease and pest buildup, as well as comparison of clonal material's performance with improved seedling populations.

2.11 Cultivation

New seeds are planted either in nurseries or in the ground. The nursery bed should contain rich, sandy soil that has been softly shaded and one meter wide. The seeds are scattered in a dense bed. After 4 months in the nursery, clumps of seedlings are moved into bags, and after another 4–5 months, they are transplanted to the field. If you put five or more seedlings in a circle, they will grow into a random clump (Prota 2023).

2.12 Genetic Resources and Breeding

In Sri Lanka, there are various wild and semi-wild types of rice, as well as local cultivars, that have been identified and distinguished by their unique local names. These names are primarily based on the rice's aroma, making them specific to certain geographical locations. Nevertheless, it is noteworthy to mention that there is currently a lack of germplasm collections for *C. verum*, and it is quite surprising that there has been a relatively limited efforts in improvement, given its significant economic value. There is a potential for hybridization with certain species within the genus.

Given that *Cinnamomum* is an open-pollinated species, it would be beneficial to employ selection methods in combination with vegetative propagation of clonal material. The selection process for superior strains is conducted in Sri Lanka and the Seychelles (Prota 2023; Essa et al. 2023). Research by Wijesinghe and Pathirana (2000) has demonstrated that the introduction and dissemination of elite clonal material propagated vegetatively poses a threat to the genetic diversity of this native crop. They suggested setting up cinnamon seed gardens in Sri Lanka's Intermediate and Wet Zones to safeguard the country's varied genetic resources. Sri Lanka's seed-derived plantations, which were started from elite selections by producers and then by the Department of Export Agriculture, provide a rich genetic resource for improving cinnamon. The genomic collections at the NCTRC will benefit from additional sampling from these groups (Alqahtani et al. 2024).

2.13 Conclusions and Prospects

The demand for cinnamon has consistently remained strong, and the future outlook remains promising, as synthetic alternatives have not significantly impacted the market. Cinnamon is gaining significance in local and international markets due to its medicinal and aromatic properties. Prospects have been identified as having potential applications as a food additive, material, and medicinal substance. The primary determinant of consumption is expected to be population growth.

The outlook for essential oils appears to be less promising due to the presence of numerous alternatives. However, there are potential applications in food preservation and ongoing research in this field. Cinnamon essential oil has demonstrated the capacity to effectively impede the proliferation of various bacteria, such as *Staphylococcus aureus*, *Listeria* spp., *Escherichia coli*, and *Salmonella* spp., which are known to contaminate food and contribute to the occurrence of Foodborne Diseases (FBD). The antimicrobial properties exhibited by cinnamon highlight its potential as a viable alternative to the utilization of chemical preservatives in food products. The favorable outcomes and significant commercial viability of employing active film packaging comprised bio-nanocomposite TiO_2 and cinnamon essential oil. The research findings indicated that this packaging method positively impacted the shelf life, quality, and safety of cheese. The provided information is the scientific name for *C. verum*. In this chapter, we have compiled the existing data about the phytochemical composition and pharmacology of *C. verum*. *C. Verum* has exhibited various pharmacological activities, including but not limited to anticancer, antidiabetic, anti-inflammatory, antibacterial, antifungal, antiviral, and antioxidant properties. Nevertheless, it is important to note that comprehensive clinical studies about both species are currently scarce. Consequently, the purpose of this chapter is to present fundamental information that can serve as a catalyst for researchers to undertake further investigations. By doing so, we can enhance our understanding of these species' potential benefits and applications in the treatment and mitigation of various human ailments. Cinnamon is primarily cultivated in various regions across the globe, each with its distinct weather conditions.

Cinnamon and its active constituents, such as cinnamaldehyde, have a range of applications and a recommended dosage limit of 2 g for optimal health benefits. Cinnamon can be incorporated into one's diet to promote overall health and well-being. Cinnamon and its active constituents have been found to positively affect diabetes, obesity, and lipid profile reduction. Cinnamon has traditionally been utilized for various purposes, including enhancing flavor and aroma. It is also utilized to manage various prevalent health conditions such as toothache, oral infection, and halitosis. Excessive consumption of cinnamon can have negative effects on one's health. Therefore, it can be concluded that cinnamon possesses significant utility and benefits for maintaining optimal health. Moreover, it has been observed to be advantageous in preventing various diseases.

References

Abd El-Moneim D, ELsarag EIS, Aloufi S et al (2021) Quinoa (*Chenopodium quinoa* Willd.): genetic diversity according to ISSR and SCoT markers, relative gene expression, and morpho-physiological variation under salinity stress. Plants 10:2802

Abdalla RM, Abdelgadir AE (2016) Antibacterial activity and phytochemical constituents of *Cinnamomum verum* and *Matricaria chamomilla* from Sudan. Bio Bull 2(2):1–5

Abdel-Hamid AME, Salem KFM (2021) Breeding strategies of garden pea (*Pisum sativum* L.). In: Al-Khayri JM, Jain SM, Johnson DV (eds) Advances in plant breeding strategies: vegetable crops, Springer, Cham. https://doi.org/10.1007/978-3-030-66969-0_9

Abd-Ellatif S, Ibrahim AA, Safhi FA et al (2022) Green synthesized of *Thymus vulgaris* chitosan nanoparticles induce relative WRKY-genes expression in *Solanum lycopersicum* against *Fusarium solani*, the causal agent of root rot disease. Plants 11(22):3129. https://doi.org/10.3390/plants11223129

Abeysinghe PD, Wijesinghe KGG, Tachida H, Yoshda T (2009) Molecular characterization of cinnamon (*Cinnamomum verum* Presl) accessions and evaluation of genetic relatedness of cinnamon species in Sri Lanka based on TrnL intron region, intergenic spacers between trnT-trnL, trnL-trnF, trnH-psbA and nuclear ITS. Ress J Agric Biol Sci 5(6):1079–1088

Aboshama HM, Salem KFM, Hassan HMF, Tahoun MH (2015) Estimating heterosis and combining ability for in vivo and in vitro traits using diallel cross on *tomato* (*Solanum lycopersicum* L.). Minufiya J Agric Res 40(3):757–770

Abu-Ellail FFB, Salem KFM, Saleh MM et al (2021) Molecular breeding strategies of beetroot (*Beta vulgaris* ssp. vulgaris var. conditiva Alefeld). In: Al-Khayri JM, Jain SM, Johnson DV (eds) Advances in plant breeding strategies: vegetable crops. Springer, Cham. https://doi.org/10.1007/978-3-030-66965-2_4

Al-Harbi NA, Al Attar NM, Hikal DM, Mohamed SE, Latef AAHA, Ibrahim AA, Abdein MA (2021) Evaluation of insecticidal effects of plants essential oils extracted from Basil, black seeds and lavender against *Sitophilus oryzae*. Plants 10(5): 829. https://doi.org/10.3390/plants10050829

ALshamrani M, Safhi S, Alshaya FA et al (2022) Genetic diversity using biochemical, physiological, karyological and molecular markers of *Sesamum indicum* L. Front Genet 13:1035977

Alsharari SF, Ibrahim AA, Okasha SA (2023) Combining ability for yield, oil content, and physio-biochemical characters of canola (*Brassica napus* L.) under salt stress conditions. SABRAO J Breed Genet 55(4):1003–1024. https://doi.org/10.54910/sabrao2023.55.4.1

Alqahtani MM, Saleh MM, Alwutayd KM, Ibrahim AA et al (2024) Performance and genotypic variability in diverse date palm (*Phoenix dactylifera* L.) cultivars for fruit characteristics. Genet Resour Crop Evol. https://doi.org/10.1007/s10722-023-01719-0

Andianto WI, Waluyo TK, Dungani R et al (2015) Wood anatomical from Indonesian genus cinnamomum (Lauraceae) and their identification key. Asian J Plant Sci 14(1):11–19. https://doi.org/10.3923/ajps.2015.11.19

Anonymous (1996) Cinnamon: cultivation and processing. In: Technical bulletin, vol 5. Department of Export Agriculture, Peradeniya, Sri Lanka

Avula B, Smillie TJ, Wang YH et al (2015) Authentication of true cinnamon (*Cinnamon verum*) utilising direct analysis in real time (DART)-QToF-MS. Food Addit Contam Part A 32:1–8

Bakewell-Stone P (2023) *Cinnamomum verum* (Cinnamon). CABI Compendium. www.cabi.org/fc/datasheet/13573

Bansode VJ (2012) A review on pharmacological activities of *Cinnamomum cassia* Blume. Int J Green Pharm 6(2):102–108

Baruah A, Nath SC (2007) Systematics and diversities of Cinnamomum species used as cinnamon spice in North-East India. J Econ Taxon Bot 31(4):872–887. www.indianperiodical.in/IP/default.aspx

Benerjee DP, Chatterjee BK, Sen S (1982) Air layering of cinnamon (*Cinnamomum zeylanicum* Nees.). South Indian Hort. 30:272–273

Bhat JA, Ali S, Salgotra RK et al (2016) Genomic selection in the era of next-generation sequencing for complex traits in plant breeding. Front Genet 7:221

Błaszczyk N, Rosiak A, Kałużna-Czaplińska J (2021) The potential role of cinnamon in human health. Forests 12:648. https://doi.org/10.3390/f12050648

Cardoso-Ugarte GA, Lopez-Malo A, Sosa-Morales ME (2016) Cinnamon (*Cinnamomum zeylanicum*) essential oils. In: Preedy VR (ed) Essential oils in food preservation, flavor and safety. Academic Press, San Diego, pp 339–347

Dayatilake GA (2000) Possibilities of cloning cinnamon in-vivo and in-vitro. In: Senaratne R, Dayatilake GA (eds) Revitalization of the Cinnamon Industry in Sri Lanka: constraints and remadies. Faculty of Agriculture, University of Ruhuna, Kamburupitiya, Sri Lanka, pp 26–37

EL-Mansy AB, Abd El-Moneim D, ALshamrani SM et al (2021) Genetic diversity analysis of tomato (*Solanum lycopersicum* L.) with morphological, cytological, and molecular markers under heat stress. Hort 7:65

Ervina M, Nawu YE, Esar SY (2016) Comparison of in vitro antioxidant activity of infusion, extract and fractions of Indonesian cinnamon (*Cinnamomum burmannii*) bark. Int Food Res J 23:1346–1350

Essa NM, Ibrahim AA, Soliman MI (2023) Association study between some cultivated species and their wild relatives from Apiaceae, Asteraceae and Brassicaceae families based on molecular and DNA barcoding in Egypt. Genet Resour Crop Evol. https://doi.org/10.1007/s10722-023-01681-x

Farahpour MR, Habibi M (2012) Evaluation of the wound healing activity of an ethanolic extract of *Ceylon cinnamon* in mice. Vet Med 57:53–57

Feltes G, Ballen SC, Steffens J, Paroul N, Steffens C (2023) Differentiating true and false cinnamon: exploring multiple approaches for discrimination. Micromachines 14(10):1819. https://doi.org/10.3390/mi14101819

Fleischmann K (1999) Relations between the invasive *Cinnamomum verum* and the endemic *Phoenicophorium borsigianum* on Mané island, Seychelles. Appl Veg Sci 2(1):37–46

Ford PW, Harmon AD, Tucker AO et al (2019) Cinnamon-differentiation of four species by linking classical botany to an automated chromatographic authentication system. J AOAC Int 102:363–368

Frankel OH (1974) Genetic conservation: an evolutionary responsibility. Genetics 78:53–65

Fu C, Qui Y, Kong H (2003) RAPD analysis for genetic diversity in *Chagium smyrnioides* (Apiaceae), an endangered plant. Bot Bull Acad Sin 44:13–18

Georgiev L, Chochkova M, Totseva I et al (2013) Anti-tyrosinase, antioxidant and antimicrobial activities of hydroxy cinnamoyl amides. Med Chem Res 22:4173–4182

Gerlach J (2004) A 10-year study of changes in forest vegetation on Silhouette island, Seychelles. J Nat Conserv 12(3):149–155. https://doi.org/10.1016/j.jnc.2004.03.002

Ghosh D (2015) Tannins from foods to combat diseases. Int J Pharma Res Rev 4:40–44

Global Invasive Species Database (2023) Species profile: *Cinnamomum verum*. http://www.iucngisd.org/gisd

Gruenwald J, Freder J, Armbruester N (2010) Cinnamon and health. Crit Rev Food Sci Nutr 50:822–834

Hagenlocher Y, Satzinger S, Civelek M et al (2017) Cinnamon reduces inflammatory response in intestinal fibroblasts in vitro and in colitis in vivo leading to decreased fibrosis. Mol Nutr Food Res 61:1601085

Haldankar PM, Nagwekar DD, Patil JL, Gunjate RT (1994) Varietal screening for yield and quality in cinnamon. Indian Cocoa Arecanut Spices J 18:79–81

Hariri M, Ghiasvand R (2016) Cinnamon and chronic diseases. Adv Exp Med Biol 929:1–24

Hegde KR, Sulikeri GS, Hulamani NC (1989) Effect of growth regulators and pre-girdling treatments on rooting of cinnamon (*Cinnamomum verum* Presl.) air layers. South Indian Hort 37:329–332

Hemmati AA, Alboghobeish S, Ahangarpour A (2018) Effects of cinnamic acid on memory deficits and brain oxidative stress in streptozotocin-induced diabetic mice. Korean J Physiol Pharmacol 22:257–267

Ibrahim AA, Abdel Razik ES, Abd-Ellatif S, Ramadan S (2023a) *Nigella sativa* and stressful conditions. In: Latef AAHA (ed) Medicinal plant responses to stressful conditions. CRC Press. https://doi.org/10.1201/9781003242963

Ibrahim AA, Abd-Ellatif S, Abdel Razik ES et al (2023b) Synthesis and applications of cellulose nanomaterials derived from agricultural waste and byproducts. In: Al-Khayri JM, Alnaddaf LM, Jain SM (eds) Nanomaterial interactions with plant cellular mechanisms and macromolecules, Springer, Cham. https://doi.org/10.1007/978-3-031-20878-2_18

Ibrahim AA, Salem KFM, Abdein MA, Ramadan SM (2023c) Advances in summer squash (*Cucurbita pepo* L.) molecular breeding strategies. In: Singh S, Sharma D, Sharma SK, Singh R (eds) Smart plant breeding for vegetable crops in post-genomics era. Springer, Singapore. https://doi.org/10.1007/978-981-19-5367-5_8

Ibrahim AA, Salem KFM, Ramadan S (2023d) *Allium sativum* and stressful conditions. In: Latef AAHA (ed) Medicinal plant responses to stressful conditions. CRC Press. https://doi.org/10.1201/9781003242963

Ibrahim AA, Alwutayd KM, Safhi FA et al (2023e) Characterization and comparative genomic analyses of complete chloroplast genome on *Trema orientalis* L. Genet Resour Crop Evol. https://doi.org/10.1007/s10722-023-01678-6

Ibrahim AA, Abd-Ellatif S, Abdel Razik ESS, Salem KFM (2024) Biodiversity of cereal crops and utilization in food and nutritional security. In: Al-Khayri JM, Jain SM, Penna S (eds) Sustainable utilization and conservation of plant genetic diversity. Sustainable development and biodiversity, vol 35. Springer, Singapore. https://doi.org/10.1007/978-981-99-5245-8_2

Jantan I, Karim B, Santhanam J, Azdina J (2008) Correlation between chemical composition and antifungal activity of the essential oils of eight Cinnamomum species. Pharm Biol 46(6):406–412

Jayaprakasha GK, Rao LJ (2011) Chemistry, biogenesis, and biological activities of *Cinnamomum zeylanicum*. Crit Rev Food Sci Nutr 51:547–562

Kamath JV, Rana AC, Chowdhury AR (2003) Pro-healing effect of *Cinnamomum zeylanicum* bark. Phytother Res 17:970–972

Kew (2023) Plants of the world-online. Royal Botanic Gardens Kew. www.powo.science.kew.org

Khasnavis S, Pahan K (2012) Sodium Benzoate, a metabolite of cinnamon and a food additive, upregulates neuroprotective Parkinson disease protein DJ-1 in astrocytes and neurons. J Neuroimmune Pharmacol 7:424–435

Kim MS, Kim JY (2017) Intestinal anti-inflammatory effects of cinnamon extracts in a co-culture model of intestinal epithelial Caco-2 cells and RAW264.7 macrophages. Appl Biol Chem 60:553–561

Kim MS, Kim JY (2019) Cinnamon subcritical water extract attenuates intestinal inflammation and enhances intestinal tight junction in a Caco-2 and RAW264.7 co-culture model. Food Funct 10:4350–4360

Krishnamoorthy B, Sasikumar B, Rema J et al (1997) Genetic resources of tree spices and their conservation in India. Plant Genet Res Newsl 111:53–58

Kueffer C, Schumacher E, Fleischmann K et al (2007) Strong below-ground competition shapes tree regeneration in invasive *Cinnamomum verum* forests. J Ecol 95(2):273–282. https://doi.org/10.1111/j.1365-2745.2007.01213.x

Kueffer C, Klingler G, Zirfass K et al (2008) Invasive trees show only weak potential to impact nutrient dynamics in phosphorus-poor tropical forests in the Seychelles. Funct Ecol 22(2):359–366. https://doi.org/10.1111/j.1365-2435.2007.01373.x

Kularatne MAR (2000) Present status of cinnamon industry in Sri Lanka. In: Senaratne R, Dayatilake GA (eds) Revitalization of the Cinnamon Industry in Sri Lanka: constraints and remedies. Faculty of Agriculture, University of Ruhuna, Kamburupitiya, Sri Lanka, pp 8–13

Leela J (2008) Cinnamon and cassia. In: Parthasarathy V, Chempakam B, Zachariah T (eds) Chemistry of spices. CABI, USA

Li AL, Li GH, Li YR et al (2019) Lignan and flavonoid support the prevention of cinnamon against oxidative stress-related diseases. Phytomedicine 53:143–153

Li Q, Xu Z, He T (2002) Ex situ genetic conservation of endangered *Vatica guangxiensis* (Dipterocarpaceae) in China. Biol Conserv 106:151–156

Liao JC, Deng JS, Chiu CS et al (2012) Anti-inflammatory activities of *Cinnamomum cassia* constituents in vitro and in vivo. Evid Based Complement Altern Med 2012:429320

Liu W, Chen X, Li H et al (2022) Anti-inflammatory function of plant-derived bioactive peptides: a review. Foods 11:2361

Madan MS, Kannan S (2004) Economics and marketing of cinnamon and cassia—a global view. In: Ravindran PN, Nirmal-Babu K, Shylaja M (eds) Cinnamon and cassia: the genus cinnamomum. CRC Press, Boca Raton, FL, USA

Mohammed I (1993) International trade in non-wood forest products: an overview. FO: Misc/93/11 working paper. Food and Agriculture Organization of the United Nations

Nabavi SF, Di Lorenzo A, Izadi M et al (2015) Antibacterial effects of cinnamon: from farm to food, cosmetic and pharmaceutical industries. Nutrients 7:7729–7748

Naveed R, Hussain I, Tawab A et al (2013) Antimicrobial activity of the bioactive components of essential oils from Pakistani spices against Salmonella and other multi-drug resistant bacteria. BMC Complement Altern Med 13(1):265. https://doi.org/10.1186/1472-6882-13-265

Newerli-Guz J, Śmiechowska M (2022) Health benefits and risks of consuming spices on the example of black pepper and cinnamon. Foods 11:2746. https://doi.org/10.3390/foods11182746

Orwa C, Mutua A, Kindt R et al (2009) Agroforestree database: a tree reference and selection guide. Version 4. World Agroforestry Centre, Nairobi, Kenya. www.worldagroforestry.org/sites/treedbs/treedatabases.asp

Padmalatha K, Prasad MNV (2006a) Interpopulation genetic diversity analysis in *Rauvolfia tetraphylla* L.—an endangered medicinal plant. Plant Biotechnol Rep 8:1–7

Padmalatha K, Prasad MNV (2006b) Optimization of DNA isolation and PCR protocol for RAPD analysis of selected medicinal and aromatic plants of conservation concern from Deccan ecoregion of India. Afr J Biotechnol 5:230–234

Pagliari S, Forcella M, Lonati E et al (2023) Antioxidant and anti-inflammatory effect of cinnamon (*Cinnamomum verum* J. Presl) bark extract after in vitro digestion simulation. Foods 12:452. https://doi.org/10.3390/foods12030452

Paranagama PA, Wimalasena S, Jayatilake GS et al (2010) A comparison of essential oil constituents of bark, leaf root and fruit of cinnamon (*Cinnamomum zeylanicum* Blum), grown in Sri Lanka. J Natl Sci Found Sri 29:147–153

Pathak R, Sharma H (2021) A review on medicinal uses of *Cinnamomum verum* (Cinnamon). J Drug Deliv Ther 11(6-S):161–166. https://doi.org/10.22270/jddt.v11i6-s.5145

Pathirana R (2011) Plant mutation breeding in agriculture. CAB Rev Perspect Agricult Veterin Sci Nutr Nat Resourc 6:1–20

Pisoschi AM, Pop A, Iordache F et al (2021) Oxidative stress mitigation by antioxidants—an overview on their chemistry and influences on health status. Eur J Med Chem 209:112891

Prabhashini W, Mendis K, Premakumara S et al (2019) Antioxidant and glycemic regulatory properties potential of different maturity stages of leaf of Ceylon cinnamon (*Cinnamomum zeylanicum* Blume) in vitro. Evid Based Complement Altern Med

Prota (2023) PROTA4U web database. In: Grubben GJH, Denton OA (eds) Plant resources of tropical Africa, Wageningen, Netherlands. www.prota4u.org

Quero J, Mármol I, Cerrada E et al (2020) Insight into the potential application of polyphenol-rich dietary intervention in degenerative disease management. Food Funct 11:2805–2825

Rai R, Jagdishchandra KS (1987) Clonal propagation of *Cinnamomum zeylanicum* Breyn. by tissue culture. Plant Cell Tiss Org Cult 9:81–88

Ramadan S, Ibrahim AAA (2021) Fruits and vegetables as sources of functional phytochemicals for the prevention and management of obesity, diabetes, and cancer. In: Egbuna C, Hassan S (eds) Dietary phytochemicals. Springer, Cham. https://doi.org/10.1007/978-3-030-72999-8_8

Ranasinghe P, Jayawardana R, Galappaththy P et al (2012) Efficacy and safety of 'true' cinnamon (*Cinnamomum zeylanicum*) as a pharmaceutical agent in diabetes: a systematic review and meta-analysis. Diab Med 29:1480–1492

Ranasinghe P, Pigera S, Premakumara GA et al (2013) Medicinal properties of true cinnamon (*Cinnamomum zeylanicum*): a systematic review. BMC Complement Altern Med 13:275

Ranawere VS, Nawale RN, Khandekar RG, Magdun MB (1995) Effect of season on air layering of cinnamon (*Cinnamomum zeylanicum* Blume). Indian Cocoa Spices Arecanut J 19:81–84

Rao PV, Gan SH (2014) Cinnamon: a multifaceted medicinal plant. Evid Based Complement Alternat Med 2014:642942. https://doi.org/10.1155/2014/642942

Ravindran P, Shylaja M, Nirmal BK, Krishnamoorthy B (2004a) Botany and crop improvement of cinnamon and cassia. In: Ravindran PN, Babu KN (eds) Cinnamon and cassia—the genus cinnamomum. CRC Press, Boca Raton, FL, USA

Ravindran PN, Babu KN, Shylaja M (2004b) Cinnamon and cassia: the genus cinnamomum. CRC Press, Boca Raton, FL

Ribeiro-Santos R, Andrade M, Madella D et al (2017) Revisiting an ancient spice with medicinal purposes: cinnamon. Trends Food Sci Technol 62:154–169

Safhi FA, ALshamrani SM, Jalal AS et al (2022) Genetic characterization of some Saudi Arabia's accessions from *Commiphora gileadensis* using physio-biochemical parameters, molecular markers, DNA barcoding analysis and relative gene expression. Genes 13:2099

Salem KFM et al (2023a) Advances in lettuce (*Lactuca* spp.) molecular breeding strategies. In: Singh S, Sharma D, Sharma SK, Singh R (eds) Smart plant breeding for vegetable crops in post-genomics era. Springer, Singapore. https://doi.org/10.1007/978-981-19-5367-5_11

Salem KFM et al (2023b) Effect of nanomaterials on water and solutes translocation in plants. In: Al-Khayri JM, Alnaddaf LM, Jain SM (eds) Nanomaterial interactions with plant cellular mechanisms and macromolecules and agricultural implications. Springer, Cham. https://doi.org/10.1007/978-3-031-20878-2_2

Salem KFM, Ibrahim AA (2024) Plant biodiversity in the context of food security under climate change. In: Al-Khayri JM, Jain SM, Penna S (eds) Sustainable utilization and conservation of plant genetic diversity. Sustainable development and biodiversity, vol 35. Springer, Singapore. https://doi.org/10.1007/978-981-99-5245-8_1

Sallam E (2021) Effect of gut microbiota biotransformation on dietary tannins and human health implications. Microorganism 9:965

Samaraweera DN (2000) Propagation of cinnamon. In: Senaratne R and Dayatilake GA (eds) Revitalization of the cinnamon industry in Sri Lanka: constraints and remedies. Faculty of Agriculture, University of Ruhuna, Kamburupitiya, Sri Lanka, pp 38–45

Shahid MZ, Saima H, Yasmin A et al (2018) Antioxidant capacity of cinnamon extract for palm oil stability. Lipids Health Dis 1:1–8

Shan B, Cai YZ, Sun M, Corke H (2005) Antioxidant capacity of 26 spice extracts and characterization of their phenolic constituents. J Agric Food Chem 53(20):7749–7759. https://doi.org/10.1021/jf051513y

Simic A, Sokovic MD, Ristic M et al (2004) The chemical composition of some Lauraceae essential oils and their antifungal activities. Phytother Res 18:713–717

Singh G, Maurya S, DeLampasona MP, Catalan CA (2007) A comparison of chemical, antioxidant and antimicrobial studies of cinnamon leaf and bark volatile oils, oleoresins and their constituents. Food Chem Toxicol 45:1650–1661

Soliman MI, Ibrahim AA, El-Metwally MA, Eshak DA (2017) Cytotoxicity effects of biological control and antioxidants using *Vicia faba* chromosomal aberration assay. J Environ Sci Technol 10:230–237. https://doi.org/10.3923/jest.2017.230.237

Soliman SA, Hafez EE, Al-Kolaibe AMG, Abdel Razik E-SS, Abd-Ellatif S, Ibrahim AA, Kabeil SSA, Elshafie HS (2022) Biochemical characterization, antifungal activity, and relative gene expression of two *Mentha* essential oils controlling *Fusarium oxysporum*, the causal agent of *Lycopersicon esculentum* root rot. Plants 11(2):189. https://doi.org/10.3390/plants11020189

Soliman MI, Mohammed NS, EL-Sherbeny G et al (2023) Antibacterial, antioxidant activities, GC-mass characterization, and cyto/genotoxicity effect of green synthesis of silver nanoparticles using latex of *Cynanchum acutum* L. Plants 12:172. https://doi.org/10.3390/plants12010172

Thakur S, Walia B, Chaudhary G (2021) Dalchini (*Cinnamomum zeylanicum*): a versatile spice with significant therapeutic potential: *Cinnamomum zeylanicum*. Inter J Pharm Drug Anal 9(2):126–136

Thirunavoukkarasu M (1996) Induction of rooting in leafy shoot cuttings of *Cinnamomum zeylanicum* Breyn. by IBA. Indian Forester 122:258–262

USDA-ARS (2009) Germplasm resources information network (GRIN). Online database. National Germplasm Resources Laboratory, Beltsville, Maryland, USA. Unpaginated. http://www.ars-grin.gov/cgi-bin/npgs/html/tax_search.pl

Vangalapati M, Satya NS, Prakash DVS, Avanigadda S (2012) A review on pharmacological activities and clinical effects of cinnamon species. Res J Pharm Biol Chem Sci 3(1):653–663

Vasconcelos NG, Croda J, Simionatto S (2018) Antibacterial mechanisms of cinnamon and its constituents. A review. Microb Pathog 120:198–203. https://doi.org/10.1016/j.micpath.2018.04.036

Voukeng IK, Kuete V, Dzoyem JP et al (2012) Antibacterial and antibiotic-potentiation activities of the methanol extract of some Cameroonian spices against Gram-negative multi-drug resistant phenotypes. BMC Res Notes 5:299

Weerasuriya S, Pathirana R (2020) Germplasm resources, crop improvement and biotechnological tools in aid of cinnamon breeding. https://doi.org/10.13140/RG.2.2.30220.54406.

Wijesinghe KGG, Pathirana R (2000) Germplasm screening and varietal development in cinnamon. In: Senaratne R and Dayatilake GA (eds) Revitalization of the Cinnamon Industry in Sri Lanka: constraints and remedies. Faculty of Agriculture, University of Ruhuna, Kamburupitiya, Sri Lanka. pp 15–25

Wisal GA (2018) Antibacterial and antifungal effect of cinnamon. Microbiol Res J Inter 10:1–8. https://doi.org/10.9734/MRJI/2018/41345

Yashin A, Yashin Y, Xia X, Nemzer B (2017) Antioxidant activity of spices and their impact on human health: a review. Antioxidants 6:70

Zalapa JE, Cuevas H, Zhu H et al (2012) Using next-generation sequencing approaches to isolate simple sequence repeat (SSR) loci in the plant sciences. Am J Bot 99:193–208

Zimmerman RH (1971) Flowering in crabapple seedlings. Methods of shortening the juvenile phase. J Am Soc Hort Sci 96:404–411

Zimmerman RH (1972) Juvenility and flowering in woody plants: a review. Hort Sci 7:447–455

Zumbroich TJ (2005) The introduction of nutmeg (*Myristica fragrans* Houtt.) and cinnamon (*Cinnamomum verum* J. Presl) to America. Acta Bot Venezuelica 28(1):155–160

Chapter 3
Revolutionizing *Coriandrum sativum* L. Breeding: Contemporary Approaches and Innovative Strategies

Akshatha Banadka, Chole Pranjali Bajrang, Wudali Narasimha Sudheer, Rashmi Ramakrishnan, Jameel M. Al-Khayri, and Praveen Nagella

Abstract *Coriandrum sativum* L. commonly recognised as cilantro or coriander is one of the oldest spices that has been extensively used in cuisine and as a medicinal plant. It finds extensive application in both traditional medicine and as a seasoning agent in culinary practices. This spice, renowned for its oil-bearing properties, contain abundant linalool and serves as a rich source of plant secondary metabolites. These compounds offer a range of biological benefits, including antibacterial, antifungal, antioxidant, analgesic, anti-inflammatory, and insecticidal properties. Despite its extensive uses in both culinary and medicine, the medicinal herb has been overlooked as a viable agricultural crop. Exploration into traditional breeding methods has yielded the emergence of new varieties. However, traditional breeding has its own limitations and has failed to develop high oil yielding varieties with resistance to the abiotic and biotic stress. However, lately, extensive research has been conducted on modern coriander breeding. These advanced breeding programs not

A. Banadka
Department of Biotechnology and Genetics, School of Sciences, JAIN (Deemed to be University), Bengaluru, Karnataka 560029, India
e-mail: b.akshatha@jainuniversity.ac.in

C. P. Bajrang · W. N. Sudheer · R. Ramakrishnan · P. Nagella (✉)
Department of Life Sciences, School of Sciences, Christ University, Bengaluru, Karnataka 560029, India
e-mail: praveen.n@christuniversity.in

C. P. Bajrang
e-mail: cholepranjali.bajrang@res.christuniversity.in

W. N. Sudheer
e-mail: wudali.sudheer@res.christuniversity.in

R. Ramakrishnan
e-mail: rashmir@res.christuniversity.in

J. M. Al-Khayri (✉)
Department of Agricultural Biotechnology, College of Agriculture and Food Sciences, King Faisal University, Al-Ahsa 31982, Saudi Arabia
e-mail: jkhayri@kfu.edu.sa

J. M. Al-Khayri et al. (eds.), *Biodiversity and Genetic Improvement of Herbs and Spices*, Advances in Plant Breeding Strategies 8, https://doi.org/10.1007/978-3-031-81838-7_3

only help determine the yield and quality of coriander but also allows genetic manipulation in order to develop new plant varieties. In this regard, the current chapter explores various studies with respect to the population structure and genetic diversity of coriander, genome sequencing technology, and molecular markers. The chapter also discusses the biotechnological advancements in coriander breeding and highlights its importance as an effective strategy for enhanced coriander breeding. The chapter establishes a groundwork for enhancing coriander with desirable characteristics using a molecular breeding strategy and provides practical information for coriander cultivators.

Keywords *Coriandrum sativum* · Germplasm · Conservation · Traditional breeding · Modern breeding

3.1 Introduction

Coriandrum sativum L., is the most widely used medicinal herb. It is also identified by other names such as Chinese parsley, cilantro in English, kuzbara in Arab, koriannon in greek, dhanyaka, kusthambari in Sanskrit, and dhania or dhanya in Hindi (Diederichsen and International Plant Genetic Resources Institute 1996). It is a herbaceous plant, classified as both an annual and perennial, capable of attaining heights ranging from 0.20 to 1.40 m. Coriander has a tap root system with an erect ribbed stem. The hollow cylindrical stem has a single branch or multiple branches. The stem appears green and turns yellow or red during the flowering season. The leaves are compound, thin, and brittle. The younger leaves appear rounded and barely slit, while the older leaves are fragile and notched. The spherical fruits, about one centimeter in diameter, when immature, exhibit a green hue, transitioning to a brownish-yellow shade upon reaching maturity (Maroufi et al. 2010).

Cilantro belongs to the Apiaceae family (formerly known as Umbelliferae), alongside other recognized members such as dill and tarragon. It is categorized within the genus Coriandrum, which encompasses only two known species: *Coriandrum tordylium* and *Coriandrum sativum*. The taxonomic classification of coriander given by Pimenov and Leonov (1993), is listed below (Table 3.1). Numerous varieties of this popular herb namely Caribe, Cilantro, Jantar, and Thuringian have evolved over time. The region of origin of coriander is unclear to date. However, it is presumed that the coriander would have originated from the Mediterranean region based on plant characteristics as well as historical findings. According to Vavilov and Dorofeev (1935), the origin of coriander was from regions of Abyssinia, Central Asia, and the Near East. The centers of origin of different varieties of coriander as suggested by Ivanova and Stoletova (1990) are Abyssinia, Central Asia, India, and Northern Africa. Owing to the archeological approach for identification of the place of origin of coriander, the existence of the herb is mentioned in an Egyptian papyrus dated 1550 BC.

Table 3.1 Taxonomic classification of coriander

Kingdom	Plantae
Subkingdom	Viridiplantae
Division	Tracheophyta
Subdivision	Spermatophytina
Class	Magnoliopsida
Order	Apiales
Family	Apiaceae
Genus	Coriandrum
Species	*sativum*

Records of the herb's presence date back to the library of the Assyrian king Ashurbanipal in the seventh century BC (Harten 1974). In Sanskrit literature dating to 5000 BC, coriander is termed as 'kusthumbari' (Prakash 1990). Currently, coriander is cultivated worldwide, with production in countries including France, Argentina, Italy, Mexico, India, Myanmar, the Netherlands, Pakistan, Romania, Turkey, Spain, the UK, and the USA (Khan et al. 2014) (Fig. 3.1). It is recorded that the invaders of the Bronze age introduced coriander to Britain to flavor the barley gruel and the British used coriander to preserve meat by mixing it with vinegar and cumin. Chinese and Israelites used it in their cooking. This herb was then introduced to Mexico and Peru of South America by Spanish conquistadors and became the integral ingredient in most of their dishes (Our Herb Garden 2013).

3.2 Nutritional Composition and Chemical Profile of Coriander

Coriander is a rich source of nutrition composed of 11.49% of proteins, 10.53% of starch, minerals such as calcium (0.14%), iron (0.01%), and phosphorus (0.06%) and vitamins like ascorbic acid (vitamin C), niacin, retinol (vitamin A) and riboflavin (Shahwar et al. 2012). It is reported to contain 0.37% of essential oils (Ramezani et al. 2009). The coriander leaves are known to contain compounds like dodecanal, (E)-2-dodecenal, (E)-2-decenal, (E)-2-tetradecenal, (E)-2-tridecenal, 2-decen-1-ol, (E)-2-hexadecenal, (E)-2-undecenal, pentadecenal, α-pinene and the highest percentage of linalool. The coriander seeds are reported to contain volatile compounds like α-pinene, linanool, camphene, camphor, D-limonene, decanal, geraniol, limonene, and γ-terpinene (Shahwar et al. 2012) (Fig. 3.2). It is known to contain phenolic compounds like β-resorcylic acid, formononetin, gallic acid, gentisic acid, hesperidin, naringenin, rutin, salicylic acid, vanillin, syringic acid, and veratric acid and flavanols such as kaempferol myricetin, and quercetin, and hydroxycinnamic acids like chlorogenic acid, coumaric acids, and ferulic acid (Nagella et al. 2014). The essential oil contains polyphenols such as caffeic acid, caffeoylquinic acid

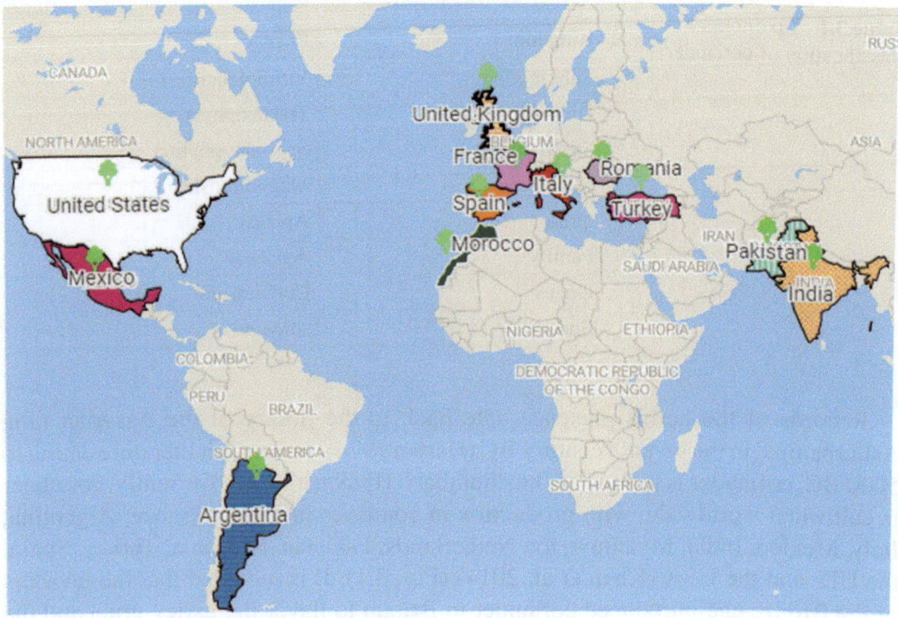

Fig. 3.1 Geographic distribution of coriander

feruloyl quinic acid, and rutin (Kaiser et al. 2013). The roots contain phenolic acids like cinnamic acid, isocoumarin-3-one, p-coumaric acid, 4,4,5,7,8-pentamethyl-3,4-2 1,3,4-tris(trimethylsilyloxy)octade-can-2-amine, and L-valine (Tang et al. 2013).

3.3 Coriander Applications in Culinary, Medicine, and Industry

The annual herb is commonly used in Indian, Thai, South American, and Middle Eastern cuisines. The coriander seeds have a lemony flavor and fresh cilantro has a pungent taste. The coriander seeds are used in brine and pickling. The ground coriander seeds are used as a spice in curry, hearty soups, and stews. The leaves are used in preparation of bread, garnishing, and salads and used for flavoring in drinks and mayonnaise. Whole coriander is used in spice rubs and for blending in rich sauce (Cooking 2022). Besides its use in culinary, it is potentially used in medicine. The therapeutic properties have been reported in Latin and Greek literature (Manniche 1989). The coriander exhibits properties such as anticancer, analgesic, anticonvulsant, antidiabetic, anthelmintic, antimicrobial, antihypertensive, anti-mutagenic, antioxidant, anxiolytic, diuretic, hypnotic activity, and many other properties (Nadeem et al. 2013).

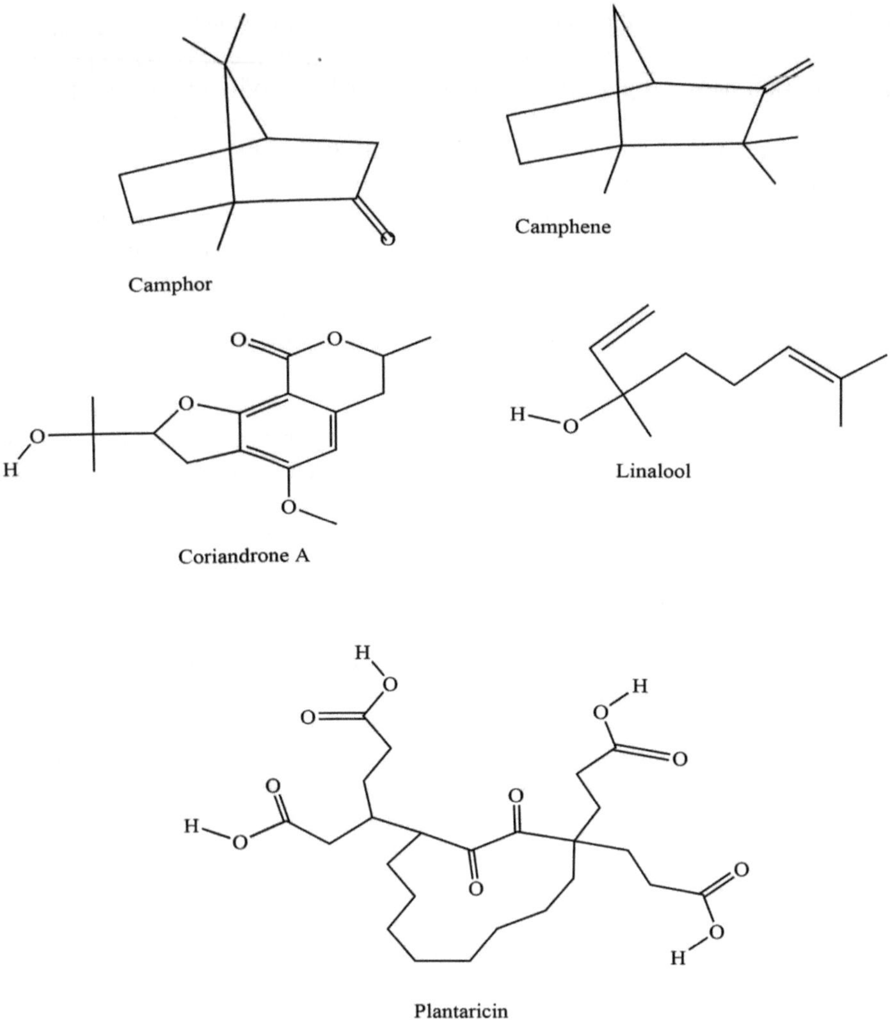

Fig. 3.2 Important phytochemicals present in *Coriandrum sativum* L. *Source* https://pubchem.ncbi.nlm.nih.gov

In a study conducted by Yepez et al. (2002), coriander demonstrated antioxidant properties by effectively scavenging free radicals in DPPH. Additionally, the polyphenolic fraction exhibited antioxidant activities, including catalase, glutathione reductase, glutathione peroxidase, and superoxide dismutase (Hashim et al. 2005). The compounds like α-pinene, camphene, carotenoids, linalool, and limonene are responsible for the antioxidant properties (Guerra et al. 2005; Wei and Shibamoto 2007). The herb displays antimicrobial activity against various pathogenic bacteria and fungi across a broad spectrum. Reports indicate that it possesses antibacterial activity against *Bacillus megaterium*, *Yersinia enterocolitica*, *Staphylococcus aureus*,

Salmonella typhimurium, Pseudomonas aeruginosa, Listeria monocytogenes, Klebsiella pneumoniae, Escherichia coli and *Enterococcus faecalis* (Delaquis et al. 2002; Keskin and Toroglu 2011). The antimicrobial activity can be attributed to the existence of the antimicrobial peptide Plantaricin (2E)-alkenals, alkanals, (2E)-Dodecenal, (2E)—undecenal (Zare-Shehneh et al. 2014). Coriander is used to inhibit the growth of fungi such as *Tricoconis padwickii, Rhizopus stolonifer, Penicillium stoloniferum, Penicillium expansum, Monilia sitophila, Fusarium oxysporum, Drechslera halodes, Curvularia lunata, Bipolaris oryzae, Aspergillus niger*, and *Alternaria alternata* (Darughe et al. 2012; Lalitha et al. 2011). Plantaricin, a peptide, has shown antimicrobial properties against fungi such as A. *niger* and *Penicillium lilacinum and bacteria such as P. aeruginosa, K. pneumoniae*, and *S. aureus* bacteria (Zare-Shehneh et al. 2014).

The coriander seeds and fruits have been reported to exhibit antidiabetic activity in hyperglycemic Meriones shawi rats by stimulating insulin secretion, alleviating insulin resistance, and enhancing glucose metabolism (Aissaoui et al. 2011). In individuals who consumed 2.5 g of ground coriander twice a day for 60 days, the fast blood sugar levels were reduced (Abascal and Yarnell 2012). The coriander extract could effectively reduce the lipid uptake in rats with induced hyperlipidemia when administered at the rate of 1 g/kg (Lal et al. 2004). The hydro-methanolic extract of coriander fruits has reduced myocardial infarction in rats with cardiotoxicity (Patel et al. 2012). Coriander extract has also been shown to exhibit hepatoprotective properties against oxidative stress by its antioxidant activity (Kumar et al. 2011). The coriander administered at the dosage of 40–100 mg/kg in Wistar rats could promote diuresis by increasing the excretion of chloride, potassium, and sodium. The potential anxiolytic activity of coriander's aqueous extract has been observed at dosages of 50, 100, and 500 mg/kg (Emamghoreishi et al. 2005). The coriander seeds can be used to treat peptic ulcers and gastric ulcers as it can kill *Helicobacter pylori* which cause ulcers (Önder 2018). The coriander extract can increase procollagen type I production thus slowing down the aging process and it is known to decrease the MMP-1 levels in the normal human dermal fibroblasts (NHDF) when exposed to UV rays (Hwang et al. 2014). Rich in antioxidants, the coriander has been effective in inhibiting malignancy and suppressing the oncogenes actively involved in cancer. It is actively involved in the prevention of colony formation, cell invasion, and migration (Elmas et al. 2019). *C. sativum* are effective metal detoxifiers by aiding the expulsion of metals naturally. The amphoteric electrolytes, phytic acid, and citric acid in the coriander chelate the metals and expel them out of the body (Mehrandish et al. 2019).

The essential oil in coriander has numerous applications in the food industry. It is used as a flavoring agent in the liquor industry and chocolate industries. The residues or the oil press cakes obtained after oil extraction can be used for livestock feeding by farmers. The oil press cakes from coriander also have applications in the production of antibiotics, biopesticides, industrial enzymes, vitamins, and other biochemicals (Sriti et al. 2012).

3.4 Economic Importance of Coriander

The economic importance of coriander varies in different countries. India holds the distinction of being the top producer and exporter of coriander accounting to 64.35% of total production and 24.97% of exportation worldwide. India produces 1.43 metric tons of coriander per year according to the report by Tridge (2020). However, in some countries with tropical climates, coriander cultivation is restricted to high-altitude areas, and therefore very less economic importance is being given to coriander cultivation in such regions. With respect to the market value of coriander oil, it has been observed that currently, the oil costs $12 US/kg to $50 US/kg. This fluctuation and volatility of international trading can be addressed by establishing transparency between the producers and the industry. Attributing its nutritional and pharmaceutical importance, there is a need for cultivation on a large scale in order to improve the economic status of the country. The promotion of coriander cultivation would open up new markets and help improve the economic status of the country (Diederichsen and International Plant Genetic Resources Institute 1996).

Despite its wide range of applications such as spices, flavoring agents in food, cosmetics, and most importantly medicine, the cultivation of coriander on a large scale has been given less importance and it still remains an underutilized crop (Kalidasu et al. 2015). There is lack of knowledge about the genetic diversity of coriander, genetic relationships within the species, and intra-specific variability. Furthermore, there is very limited research conducted on coriander breeding around the globe. Due to inadequate knowledge about genetic diversity, the coriander breeding strategies have decelerated. Therefore, it is imperative to investigate and comprehend the genetic diversity of coriander, as this serves as a fundamental requirement for crop enhancement initiatives. In this regard, the present review discusses the current cultivation practices of coriander, the traditional breeding methods, and advanced breeding methods such as molecular breeding and mutation breeding methods that have been developed for producing new coriander cultivars with desired genotypes. The examination also delves into the contribution of plant tissue culture and genetic engineering towards developing new varieties/germplasms.

3.5 Current Cultivation Practices and Challenges

Coriandrum sativum is a tropical crop that flourishes in frost-free and moderately dry tropical climates. It is also found in subtropical regions of the world. It is a commercial crop that grows in various habitats which include agricultural lands, plains, hills, gardens, open areas, roadsides, and railroads (Datiles 2020). It prospers in areas with temperatures typically ranging from 15 to 28 °C and necessitates an average rainfall of 75–100 mm. The plant can adjust to any soil type and preferably grows well in well-drained loamy soil and it does not grow in sandy and alkaline soil. The soil pH has to be maintained at 8–10 (Kheti 2022). The optimal seasons for coriander cultivation are between June–July and October–November.

Coriander is propagated through seeds, and for optimal yield, these seeds should be sown during the specific timeframe of October and November. Since not all coriander seeds will germinate, it's advisable to split them open into halves before sowing and soak them in water for 8–12 h. The seeds also have to be pretreated with Azospirillum for better growth. The field has to be prepared by ploughing 3–4 times post-monsoon and the seeds are sown immediately to avoid excess moisture. In the case of an irrigated field, ploughing twice would be sufficient. The seeds are sown at the rate of 8–10 kg per acre of land and are placed at a distance of 15 cm apart, depth of 3 cm and a row distance of 30 cm, by the Pora sowing method (Kheti 2022). The soil is supplemented with 40 kg phosphorus, 10 kg nitrogen, and 20 kg potassium. It is irrigated immediately after sowing followed by 5–6 irrigation in the later sessions depending on the climate (Sandesh 2017). The crop will be ready to harvest in 90–110 days after sowing. The harvesting for leaves is done once it reaches 6 inches tall and it is harvested for seeds when the fruits ripen from green to brown. It is then pulled off and arranged into smaller stacks. The stacks are then winnowed, cleaned, and dried. The coriander yield varies from 400 to 1200 kg/ha depending on the irrigation (Indiaagronet.com 2022). The different parts of the coriander plant is represented in Fig. 3.3.

One significant challenge encountered in coriander cultivation is its heightened susceptibility to insect infestation and diseases caused by pests. Thus it is essential for the seed treatment prior to sowing. The coriander is also prone to aphids and whitefly attack and wilting. The plant is susceptible to Powdery mildew caused by *Erysiphe polygoni* which would destroy the whole crop if unchecked in the beginning

Fig. 3.3 Illustration of different parts of *Coriandrum sativum* L. **a** whole plant, **b** leaf of coriander, **c** coriander dried seeds, **d** coriander grown indoor **e** flower of coriander

stages. Fungi such as *Curvularia* sp., *Alternaria* sp., and *Fusarium* sp. can cause grain mould (Indiaagronet.com 2022). The seed has to be treated with carbendazim and Mancozeb which would add up to the expenses for coriander cultivation. Coriander is also susceptible to nematodes like armyworm, cutworm, caterpillars that eat leaves and semi loopers and diseases like bacterial leaf spot and carrot motley dwarf. The other main challenge in coriander cultivation is bolting where the plant enters the flowering stage, resulting in a reduction in foliage (AllThatGrows 2022). Unfavorable climatic and soil conditions are other threats to coriander cultivation (Diederichsen and International Plant Genetic Resources Institute 1996).

3.6 Germplasm Biodiversity and Its Conservation

Since time immemorial, humans have been conserving plants by preserving the seeds and vegetative propagules from one season to the next and from one place to another. This approach has been pivotal in the advancement of agriculture and the introduction of genetic variability into the population. This method is currently being established as germplasm conservation. Germplasm encompasses all the allelic sources of diverse genes found in both wild and cultivated species, influencing a multitude of traits. It serves as a vital resource, offering insights into the genetic makeup of a species, and plays a critical role in conserving plant diversity (Priyanka et al. 2021). Germplasm includes breeding stocks, cultivated crops, landraces, modern cultivars, obsolete cultivars, and wild forms that could be indigenous or exotic. Germplasm is preserved and maintained by botanical gardens, centers of diversity, cryopreservation, farmer's fields, gene banks, gene sanctuaries, and pollen and seed companies (TNAU 2015). The protection of germplasm is an important strategy to rescue endangered plants, preserve all the important plants, and promote food security (Wilkes 1989; Priyanka et al. 2021).

3.6.1 Gene Banks

Despite its significance, there is a lack of available information regarding the conservation of coriander germplasm (Popova et al. 2010). However, there are various research institutes that have volunteered to preserve the germplasm. One of the major institutions that collect and preserve the coriander germplasm is Gatersleben Gene bank, Gatersleben, Germany. All the characterization and evaluation data for coriander are documented at Gatersleben Gene bank. The herbarium of each collected accession is available and is further used to identify the gene bank accession after field reproduction. Each accession is reproduced, characterized, and documented by Gatersleben Gene bank (Diederichsen and International Plant Genetic Resources Institute 1996).

3.6.2 *Cryopreservation*

The seed germplasm of coriander collected requires periodic reproduction and thus susceptible to diseases, pests, and insect attacks. The in vitro conservation results in a transition to a mature phenotype over years. Thus cryopreservation of somatic embryos serves as an alternative efficient long-term method for germplasm conservation (Popova et al. 2010). The somatic embryo has an ability to be cultured to form embryo clumps at a higher rate which in turn are capable of plant regeneration. Somatic embryos can be preserved through cryopreservation for an extended duration without requiring any genetic modifications. The air desiccated somatic embryo has been cryopreserved without any investment in cryoprotectant solutions (Popova et al. 2010). Embryo clumps with 3–4 heart-shaped globular somatic embryos were desiccated under sterile air in the laminar airflow. The desiccated were cryopreserved in liquid nitrogen for 1 h and the cryopreserved tubes were rewarmed by submerging them in a water bath kept at a constant temperature of 41 °C. The cultures were then recovered by incubation at 25 °C and the growth was determined in 4 weeks. The embryo was subjected to maturation and regeneration by being placed on an MS medium supplemented with 30 g/L sucrose and 2.2 g/L Gelrite and germination was observed in 1.5 months. The relative DNA content of the regenerated plant was evaluated using flow cytometry. The cryopreservation method could retain the main cultural characteristics (Popova et al. 2010).

Along with the maintenance and conservation of germplasm, it is important to generate genetic diversity for sustainable agriculture. This genetic diversity can be achieved by intra and inter-specific crossing by traditional breeding or by molecular breeding methods. The efficacy of these breeding techniques hinges on a comprehensive understanding of the plant genome's cytogenetics. Chromosomal manipulation is a pivotal phase in the creation of novel cultivars. However, compared to other members of the Apiaceae family, coriander has not been extensively studied in terms of cytogenetic manipulation and karyotype analysis. Karyomorphological analysis indicates that coriander possesses 11 chromosomes and is diploid (Yilmaz and Turan 2022).

3.7 Traditional Breeding

The production of high-quality coriander can be achieved by the cultivation of qualified cultivars under appropriate growth conditions and the application of suitable cultivation methods. High-yielding, disease-resistant crop varieties can be achieved by choosing suitable breeding methods. The selective breeding method for coriander is the most employed breeding method. The suitable breeding method for coriander is dependent on the pollination biology of the plant. It has been observed that the selfing of the plants has failed to show inbreeding depression and thus emasculation and fertilization by selfing have been unsuccessful (Kalidasu et al. 2016). Nonetheless, a male plant containing a marker gene can be employed for crossbreeding purposes. Outcrossing can be achieved by using the pollen of other plants and the stigma of two different plants (Yilmaz and Turan 2022). Although the breeding of coriander is an age-old practice, the studies on coriander breeding in Egypt have been reported by El-Ballal and Abou El-Nasr (1987). To examine the phenological earliness and medicinal attributes of oil within the fruits. Numerous coriander breeding initiatives have been undertaken, aiming to cultivate high-yielding varieties with elevated oil content. These programs entail careful selection of appropriate locations and climatic conditions. The Simferopol institute has been involved in selecting famous cultivars and conducting research and breeding to develop high-quality oil-yielding coriander varieties (Cernodubov and Berestovaja 1986). The institute has also developed varieties that can survive winter (Moskalenko 1972), showed resistance to *Ramularia coriandri* (Cislova 1988), and resistance to drought (Karpov et al. 1990). The major limitation in traditional coriander breeding is that there is no room for the addition of new desirable traits to the already existing traits in the species that are involved in cross breeding. The other disadvantage of traditional breeding is that undesirable traits are passed on along with the traits of interest when plants are crossed (ISAAA Inc. 2006). The large-scale production of coriander is arduous with traditional breeding methods as it is time-consuming and laborious. Thus, the intervention of biotechnological strategies is necessary for coriander production on a large scale.

3.8 Plant Tissue Culture

Plant tissue culture stands as a key biotechnological approach poised to augment secondary metabolite production and overall yield. Among the various tissue culture methods, somatic embryogenesis emerges as the most prevalent for generating and maturing somatic embryos into plants. The embryogenic callus of coriander was achieved until the cotyledonary stage through somatic embryogenesis established on the media supplemented with 2,4-D. The embryos on transfer to 1/2 strength MS media further germinated into seedlings (Murthy et al. 2008). The callus on regeneration shows chromosome abnormalities, aneuploidy, polyploidy, and more somaclonal variation, however the somatic embryos exhibit genetic integrity. Further, it has been observed that plants regenerated via indirect somatic embryogenesis have shown genetic differentiation which is determined by flow cytometry. Preserving somatic embryos for extended periods can safeguard coriander against genetic alterations. Additionally, somatic embryos have been noted for their utility in synthetic seed formulation, facilitating enhanced in vitro plant production (Yilmaz and Turan 2022).

The somatic embryo has an ability to be cultured to form embryo clumps at a higher rate which in turn are capable of plant regeneration. Somatic embryogenesis in coriander has been investigated using zygotic embryos, hypocotyls, and petioles as initial explants (Popova et al. 2010). Cryopreservation of somatic embryos enables long-term maintenance without genetic alterations. The air desiccated somatic embryo has been cryopreserved without any investment in cryoprotectant solutions (Popova et al. 2010). Embryo clumps with 3–4 heart-shaped globular somatic embryos were desiccated under sterile air in the laminar airflow. The desiccated samples were cryopreserved in liquid nitrogen for one hour, followed by re-warming through immersion of the cryopreserved tubes in a water bath held at 41 °C. The cultures were then recovered by incubation at 25 °C and the growth was determined in 4 weeks. The embryo was subjected to maturation and regeneration by being placed on an MS medium supplemented with 30 g/L sucrose and 2.2 g/L gelrite and germination was observed in 1.5 months. Flow cytometry was then employed to assess the relative DNA content of the regenerated plant. The cryopreservation method could retain the main cultural characteristics (Popova et al. 2010).

3.9 Molecular Breeding

The primary objective of plant breeding methods is to enhance crop yield and enhance nutritional quality and other traits with commercial importance. The genes coding for these phenotypes can be altered and selectively transferred to the plant of interest and the genes with undesirable traits can be knocked off by using DNA recombination technology (Moose and Mumm 2008). The plant tissue culture method has not only contributed to retrieving valuable genetic variants (somaclones) but also has developed new variants by altering specific genes. This genetic manipulation at the DNA level and molecular marker-assisted selection, genomic selection to develop plants with desirable traits is known as molecular breeding. This involves utilizing molecular markers alongside linkage maps and genomics to genetically manipulate and create new plant varieties. Advanced molecular breeding techniques such as Genome-wide selection (GWS), Marker-assisted backcrossing (MABC), Marker-assisted recurrent selection (MARS), and Marker-assisted selection (MAS) are employed for this purpose (Jiang 2013).

Currently, a variety of molecular marker technologies have been established over years and are tremendously employed in various crop breeding projects (Gupta and Rustgi 2004). The molecular marker technologies include AFLP (Amplified Fragment Length Polymorphism), ISSR (Inter-Simple Sequence Repeat), CAPS (Cleaved Amplified Polymorphic Sequences), RFLP (Restriction fragment length polymorphism), SNP (Single Nucleotide Polymorphism), RAPD (Randomly Amplified Polymorphic DNA), SSR (Simple Sequence Repeat), and SRAP (Sequence Related Amplified Polymorphism) (Yilmaz and Turan 2022). A lot of studies with respect to molecular markers has been carried out in other plants and not much has been worked in *coriander* (Galata et al. 2014) and it has been reported that the assessed molecular attributes cannot be linked to the associated phenotypic traits (Galata et al. 2014). In the characterization of coriander varieties, RAPD molecular markers have been highly efficacious (Singh et al. 2013). The RAPD markers are employed to understand the genetic diversity in coriander in various studies (Tomar-Rukam et al. 2014). RAPD is also used in combination with ITS markers to understand the genetic diversity in Indian coriander varieties. The other markers such as ISSR alone (Mustafina et al. 2017), ISSR with RAPD (Choudhary et al. 2019a), and ISSR in combination with RAPD have been engaged in studying the genetic diversity in coriander. AFLP markers are utilized for the molecular characterization of coriander germplasm (López et al. 2008). SSR, a popular marker, has not been explored in coriander. Furthermore, creating SSR markers in coriander is a laborious, time-consuming, and costly process, involving the isolation, cloning, sequencing, and characterization of microsatellite loci. As a result, SSR markers have been developed through primary screening of microsatellite flanking regions in closely related species (Schnell et al. 2005). An attempt to develop SSR for coriander has been made by Choudhary et al. (2017) by investigating the transferability of SSR loci of carrot to coriander. Of 100 SSR loci, thirty-nine primers amplified SSR for coriander. Four out of thirty-nine SSRs were polymorphic and have the ability to furnish an effective tool for coriander breeding and genetics (Choudhary et al. 2017).

The examination of gene organization, control mechanisms, and interplay across cellular and organismal levels can be obtained with functional genomics. The critical functions of genes and the link between genes can be obtained based on the genome projects and genome-wide approaches such as RNA sequencing, Microarray, Expressed Sequence Tags (EST), etc. With the advancements in sequencing technologies, a large number of genomes are sequenced and released. The complete genomic sequences of the coriander plant have been developed by Coriander Genomics Database (CGDB) by using Illumina, Pacific Biosciences, 10X Genomics, and HiC next-generation sequence technologies. The Coriander Genomics Database (CGDB) has 3 datasets; metabolic, expression, and genomic data sets. Researchers are afforded the opportunity to explore and retrieve omics datasets from the database. These datasets have been compiled through standardized experiments and bioinformatics analyses. The information on the protein sequence, gene annotation, genome sequence, gene sequence, and repeat sequence data are available in the genomic dataset. The gene expression data for coriander from three different growth stages are available in the expression dataset. The information on metabolism genes from coriander from three different growth stages is available in the metabolic data set. The database also provides the information on the collinear blocks, gene annotations, repeat sequences, and transcription factors. The sequence data available on the CGDB aids in the analysis of gene expression in coriander (Song et al. 2020). This makes it possible to characterize the genomic expressions under different stress conditions and to establish signal networks for stress perception and tolerance. Thus, the stress-tolerant plants can be developed by illuminating the genes responsible for stress factors (Ramegowda and Senthil-Kumar 2015). Using the available data on CGDB, a transcriptomic profiling in *C. sativum* L. has been done to identify the stem gall resistance genes. For this investigation, samples from stem gall-resistant (ACr-1) and susceptible (CS-6) leaf tissues of coriander cultivars were employed. RNA was pooled from these samples, and transcriptome sequencing was conducted using the Illumina NextSeq500 platform (Choudhary et al. 2019b). In the study by

Wu et al. (2021), five Vitamin E metabolic pathway-related genes (VEMPRGs), six porphyrins and chlorophyll metabolic pathway-related genes (PCMPRGs) and ten tyrosine metabolic pathway-related genes (TMPRGs) in coriander have been identified through transcriptomic and metabolomic profiling in three developmental stages of coriander. These pathways in which these genes are part of are found to be involved in production of phenols. Furthermore, the study has made an attempt to construct an interaction network involving genes associated with pathways and transcription factors (TFs) involved in phenol metabolism. Various stress resistant varieties have been tabulated in Table 3.2.

3.10 Micropropagation of *Coriandrum sativum*

Coriander is mostly used for its rich flavor and taste throughout the major parts of the world. In vitro propagation of good quality coriander plants will help in deriving a large number of individuals with similar genetic makeup. There are various factors or parameters, like source of explant, growth medium, plant growth regulators, and culture conditions, that affect the efficiency of in vitro propagation of coriander. Essential oil, oleoresin, dry leaf, and seed powders are just a few of the value-added products that may be generated from coriander, and their quality can be improved using tissue culture techniques (Gantait et al. 2022). Tissue culture studies carried out on coriander helped in the production of the optimum biomass and made the crop rich in value-added products. Somatic embryogenesis is one of the successful strategies for plant regeneration in coriander. Somatic embryos were developed from callus using hypocotyl explants and zygotic embryos when inoculated on Murashige and Skoog (MS) medium supplemented with 1.0 mg/L 2,4-D. When sub-cultured in a liquid medium containing 0.1 mg/L 2,4-D and 2% sucrose, induced callus developed into globular embryos after plating on basal medium (Kim et al. 1996). The MS medium supplemented with H (Hildebrandt) vitamins supplemented with 2,4-D or NAA helped in the induction of embryogenic callus and their germination was carried on when sub-cultured onto the half-strength MSH medium supplemented with 0.05 mg/L of BAP (Stephan and Jayabalan 2001). The repeated production of somatic embryos from cotyledon and hypocotyl explants was seen when inoculated on MS medium supplemented with 4.52 μM 2,4-D (Murthy et al. 2008). Root explants of coriander when inoculated onto the MS medium supplemented with 1.0 mg/L 2,4-D and 0.2 mg/L BA showed the highest callus induction percentage. The highest embryogenesis on callus was recorded on the MS medium supplemented with 1.0 mg/L 2,4-D and later when subculture into the media supplemented with 0.5 mg/L NAA along with 0.2 mg/L IBA, heart, torpedo, and cotyledonary embryos were developed (Ali et al. 2017) (Fig. 3.4).

Indirect organogenesis was observed in coriander when a callus was initiated from seeds upon inoculation on MS medium supplemented with 1.0 mg/L benzyladenine (BA) in combination with 0.5 mg/L 2,4-dichlorophenoxyacetic acid (2,4-D) and later the green branches were induced from the callus with maximum efficiency (Hassan

Table 3.2 Stress resistant varieties of *Coriandrum sativum*

Stress agent	Genotype/ varieties	References
Powdery mildew caused by *Erysiphe polygoni*	CIMAP 2096, CIMAP 2053, Sindhu, DH 220, Ajmer Coriander (ACR-3),	(Kalra 2003; Praveena 2019)
Stem gall diseases caused by *Protomyces macrosporus* L.	CIMAP 2096, CIMAP 2053, RCr -41, Hisar Sugandh, Hisar Bhoomit, Ajmer Coriander (ACR-2)	(Kalra 2003; Praveena 2019)
Wilt caused by *Fusarium oxysporum*	CS 287, Karan, Sindhu	(TNAU 2016; Praveena 2019)
Grain mould caused by *Helminthosporium, Fusarium, Curvularia, Alternaria*	CS 287, Co. 1, Co. 3	(TNAU 2016)
Aphid and mites	Sadhana	(Indiaagronet.com 2022)
Whitefly	Swathi	(Praveena 2019)
Drought resistance	Co. 2, Rcr-446 and Rcr-20, Sindhu	(TNAU 2016; Bajya et al. 2022) (Praveena 2019)
Salinity resistance	Verdão SF 177 and Português Pacífico'	(da Silva Sá et al. 2016)
Moisture resistance	Sadhana	(Praveena 2019)

2019). Greater response for direct organogenesis from shoot tips was established when cultured on SH (Schenk and Hildebrandt) medium supplemented with 0.2 mg/L NAA along with 0.5 mg/L GA_3 (Leaves: 9 ± 0.6, rooting: 8 ± 0.9). For the same in vitro generated plantlets, the best flowering response and seed formation was observed when cultured on SH medium enriched with 0.1 mg/L naphthaleneacetic

Fig. 3.4 In vitro regeneration of *Coriandrum sativum* L. **a, b** Micro callus and embryogenic callus from protoplast. **c, d** Heterogeneous mixture and isolated somatic embryos. **e, f** In vitro grown coriander. *Source* Mujib et al. (2014) https://doi.org/10.1007/s11240-014-0442-3, accessed on 23 July 2022; reproduced with permission from the publisher; License No: 5354681420571

acid (NAA) in combination with 0.5 mg/L gibberellic acid (GA_3) (20 days for flowering and 27 days for the opening of flowers) and SH supplemented with 0.15 mg/L NAA along with 0.5 mg/L GA_3 (18.3 flowers per plantlet and 16.2 seeds per plantlet) (Stephen and Jayabalan 1998).

Protoplasts derived from coriander callus were cultured on MS medium containing 1.0 mg/L 2,4-D and 1.0 mg/L kinetin (KN). This resulted in the formation of callus, which subsequently underwent differentiation into somatic embryos. The highest efficiency of germination of 31.3% was observed from these embryos when cultured on MS medium enriched with 1.0 mg/L BA in combination with 0.5 mg/L NAA (Mujib et al. 2014).

3.10.1 Synthetic Seed Production

Synthetic seeds are plant propagation materials that are artificially encapsulated and can be utilized as one of the strategies for crop improvement. It is seen in coriander; studies reported for the preparation of artificial seeds. Preparation of synthetic seeds with suitable encapsulating material and storage at an optimal temperature will help maintain the viability of the encapsulated explants (Fig. 3.5). Sodium alginate at 3%

Fig. 3.5 Synthetic seed development and plantlet formation in *Coriandrum sativum*. **a** Encapsulation of somatic embryos. **b** Cultivation of synthetic seeds on germination medium. **c** Initiation of germination in synthetic seeds. **d** Development of rooted plantlets; and **e** Growth of plants derived from synthetic seeds under outdoor conditions. *Source* Ali et al. (2018): https://doi.org/10.13128/ahs-22287 Accessed on 20 March 2024

along with 100 mM $CaCl_2$ showed great efficiency in the regeneration of the plantlets from explants of somatic embryo (Ali et al. 2018). Similarly, somatic embryos developed from 0.5 mg/L 2,4-D enriched MS medium were encapsulated using 100 mM $CaCl_2$ and 4% sodium alginate. Synthetic seeds showed varied germination efficiencies, but among all, Schenk and Hildebrandt medium (SH medium) showed improved germination (Stephen and Jayabalan 2000).

3.11 Mutation Breeding

In the global agricultural world, mutation breeding is the most important technique to develop new varieties with high yields. It is reported that about 3278 varieties and 210 plant species were developed by using mutation breeding techniques (Yilmaz and Turan 2022). The breaking of chromosomes and rearrangement of them play an important role in induced mutagenesis. Thus, chemical agents and ionizing radiation are employed for successful mutation. For plant breeding programs, gamma rays are extensively used (Sengupta and Datta 2004). Usually, mutagens help to alter the DNA of targeted organisms. Chemical mutagens like sodium azide (SA), methyl methanesulfonate (MMS), ethyl methanesulfonate (EMS) and colchicine are treated in different varieties of coriander named Surabhi, Cross-91, Green wonder, and Mrudul. Seed treated with colchicine and sodium azide showed better results and proved as better mutagenic agents (Kolhe et al. 2020). Various chemical, physical, and biological mutagens like ethyl methanesulfonate (EMS), UV-rays, and agrobacterium are utilized extensively in the mutation breeding of coriander. EMS and gamma irradiation showed correlation and viable mutation frequency in coriander (Datta and Sengupta 2002). Some recent studies investigated that a combination of EMS and gamma rays shows mutagenic sensitivity in coriander. When the dose of combined mutagens increased; shoot length, germination percentage, the number of leaves, and seeding height decreased (Yilmaz and Turan 2022). It is also noted that, when coriander was treated with gamma rays from lower to higher doses, the germination percentage gradually decreased. Gamma irradiation causes cytological and physiological disorders (Pramanik et al. 2018). When coriander is treated with heavy metals like $CdCl_2$ and $Pb(NO_3)_2$ showed chromosomal abnormalities. In comparison to both heavy metals, $Pb(NO_3)_2$ showed high genotoxicity in the coriander root meristem (Kumar and Pandey 2015). Another study showed that, because of chromosomal abnormalities, the fertility of pollen was declining. Salve and More (2019) reported the effect of EMS and gamma rays on various biochemical elements of coriander. They found that the passel of carbohydrate content was the merest in dwarf mutants whereas it was most in tall mutants (Salve and More 2019).

3.12 Genetic Engineering and Gene Editing

The whole genetic engineering technique is fundamentally the same for plants where the desired gene of interest is placed in the organism of interest or involves the gene modification. The time required to successfully transfer the gene and execute the desired traits depends upon the available resources, genes, and crop species. It will take around 6–15+ years to release in the market. With respect to coriander, no genetic manipulation studies have been reported till date. There are studies which discuss the biosynthetic pathway of essential oil production but no studies have made an attempt to elucidate the genes involved in the biosynthetic cycle. Thus, there is a need for development of reliable technology for developing coriander varieties with novel features such as enhanced essential oils and resistance to pests, diseases and drought tolerance using genetic engineering.

3.13 Conservation Techniques

In the Indian subcontinent coriander has become widespread production. Coriander is cultivated widely but most of the time it is cultivated in small gardens/spaces. Exceptionally for commercial purposes, it is cultivated in large fields. Germplasm conservation is a process in which the entire genetic material of a plant can be preserved for a long-term period for the extraction of genetic information which will be further useful for different breeding techniques (Gantait et al. 2022). Germplasm can be conserved in the form of pollen, seeds, and complete plant or plant tissues. Two different methods of coriander germplasm conservation include in-situ conservation under natural conditions and ex-situ conservation. The in-situ conservation of coriander includes home gardens and large fields (commercial production). The ex-situ conservation of coriander germplasm includes gene banks and the taxonomy departments (Diederichsen and International Plant Genetic Resources Institute 1996). In vitro conservation includes cryopreservation, low oxygen conditions, and cold storage (Engelmann 2012).

3.13.1 In-Situ Conservation in Natural Habitat

Preservation and improvement of populations of different species can be done in their natural environment whereas ex-situ conservation can be done in artificial habitat. National Bureau of Plant Genetic Resources (NBPGR, New Delhi), India works primarily on the collection and maintenance of genetic resources of plants. Upto date, in the national gene bank around 1128 accession of coriander was maintained and all are used as a base collection (Gantait et al. 2022). During conservation of coriander isolated gene bank accession is required from each other as coriander shows high chances of cross-pollination. It is a very important step in the conservation technique for genetically identical reproduction. Thus, this approach connects the protection and maintenance of species within their native environments, allowing them to undergo natural processes of reproduction, adaptation, and evolution.

3.13.2 Seed Storage

In coriander, it is important to dry seeds to a final moisture content between 7 and 9% by shade drying or by using artificial dryers completely before storage. At high temperatures, coriander seeds show exceptional resistance. For long-haul seed conservation, the temperature should not be more than 40 °C. At room temperature, coriander seeds can be stored for 2–3 years without losing genetic viability (Coşkuner and Karababa 2007).

3.13.3 In Vitro Conservation

Tissue culture techniques have gained importance due to their rapid multiplication and storage of germplasm. In aseptic conditions, propagation of plant material can be done, which shows a high multiplication rate. In short and medium-term conservation, reducing growth and increasing the time interval between subcultures is the primary aim (Engelmann 2012). In long-haul conservation, only cryopreservation is the current method. In cryopreservation storage at ultra-low temperature (– 196 °C in liquid nitrogen), all metabolic and cell division processes stop. Thus, plant material can be stored for an immersive time without any modification or development (Engelmann 2012).

3.14 Hybridization in Coriander

There are significantly fewer studies related to hybridization in *Coriandrum sativum*. Since wild plants are recognized for their disease resistance and stress-tolerant abilities, they are explored for breeding purposes to develop new varieties to enhance economic needs. Unfortunately, in the case of coriander, there is no significant report on its wild varieties, other than *C. tordylium,* and there is no breeding project reported with these plants. However, an intergeneric hybridization of coriander with its closely related genera was conducted and was not effective because of incompatibility (Mujib et al. 2014). The crossing programs in coriander were difficult because of the delicate inflorescence and protandrous nature of the flowers (Kalidasu et al. 2016). Since the conventional breeding methods were unsupportive, the demand for future research like somatic cell hybridization is required. Mujib et al. (2014) developed a method for protoplast isolation and culture from coriander hypocotyl-derived callus cells.

3.15 Conclusion and Prospects

Coriander has long been utilized as a culinary ingredient as well as for a variety of medical purposes by people from all over the world. The plant's essential oil, abundant in linalool, possesses a diverse array of medicinal properties including antioxidant, anticancer, anthelmintic, anxiolytic, antiulcer, and immunomodulatory activities. It is also used as a food supplement across Asia and the Middle East. Regardless of its medicinal and economic importance, the plant is not extensively cultivated, remains underutilized, and there are limited studies on its variety development and agronomy. The dearth of information underscores the need for the establishment of breeding programs geared towards creating higher-yielding, disease-resistant, and drought-tolerant varieties. This review delves into the exploration of molecular markers, quantitative trait locus QTLs, and functional genomics associated with these desirable traits. The study discusses the several genetic resources and tools employed in current breeding practices that aid in understanding the properties of coriander for the development of new varieties in the future. New breeding methods can be introduced using MAS (Marker Assisted Selection) applications, genome-wide association studies, and multidisciplinary studies. The new varieties with desired traits could also be developed by genetic manipulation in the future as there are no gene editing studies that have been conducted till date in coriander. Furthermore, the complete data on the transcriptome, metabolome, and proteome sequences and phenomics of coriander has to be established.

Appendix 1: Research Institutes Working on Coriander Germplasm Conservation and Management

Institution	Specialization research activities	Website
Banco Portugues de Germplasm Vegetal (BPGV), Portugal	Germplasm collection	https://www.iniav.pt/bpgv
Center for Plant Breeding and Reproduction Research (CPRO-DLO), Netherlands	Germplasm collection, breeding	https://bv.fapesp.br/en/instituicao_exterior/1052/centre-for-plant-breeding-and-reproduction-research-cpro-dlo/
Genetic Resources Center, Bangladesh Agricultural Research Institute, Bangladesh	Exploration, collection, characterization, utilization, preservation and regeneration of genetic resources, Documentation and training	http://pgrcbari.org/index.php?script=home
Horticultural Research Section, Agricultural Research Corporation, Sudan	Collection, conservation and characterization of the local genetic resources	http://arc.gov.sd/#
The Indian Council of Agricultural Research-National Research Center on Seed Spices ICAR-NRCSS, India	Collection, evaluation, conservation, and cataloging of germplasm	https://nrcss.icar.gov.in/
Institute of Crop Germplasm Resources (CAAS), China	Germplasm storage	https://www.cgris.net/icgr/icgrgb.html
Institute of Crop Science, Federal Research Center for Agriculture (FAL), Germany	Germplasm collection	http://www.fal.de/

(continued)

(continued)

Institution	Specialization research activities	Website
Institute for Plant Genetics and Crop Plant Research (IPK) Genebank, Gatersleben, Germany	Collection, conservation and characterization, documentation and distribution of PGR	https://www.gbif.org/dataset/8634d730-f762-11e1-a439-00145eb45e9a
National Bureau of Plant Genetic Resources, India	Conservation of germplasm	http://www.nbpgr.ernet.in/
N. I. Vavilov All-Russian Research Institute of Plant Industry (VIR), Russian Federation	Characterization, preservation and utilization of local cultivars	http://www.vir.nw.ru/
North Central Regional Plant Introduction Station, USDA-ARS (NCRPIS), USA	Collection and conservation of germplasm, storage of information, utilization of germplasm for research, crop improvement and product development	https://www.cals.iastate.edu/ncrpis
Plant Genetic Resources Center, Ethiopia	Collection of germplasm	https://ssl.fao.org/glis/wiews/detail?wiewsCode=ETH001
Plant Genetic Resources Program, Pakistan	Allocation and conservation of plant genetic resources are agronomical and bio-chemical analysis of traits, conservation distribution and data management	http://www.parc.gov.pk/index.php/en/pgrp-home
Plant Genetic Resources Dept., Aegean Agricultural Research Institute, Turkey	Conservation of plant genetic resources	https://arastirma.tarimorman.gov.tr/etae/Sayfalar/EN/AnaSayfa.aspx

Source Diederichsen and International Plant Genetic Resources Institute (1996)

Appendix 2: List of cultivars of *Coriandrum sativum*

Cultivars	Institute/place	Characteristic and seed yield	References
Ajmer Coriander 1	National Research Center on Seed Spices (NRCSS), Ajmer	Tall erect variety, 12.5 q/ha	(AgriStudyHub 2021)
APHU Dhania 1	APAU, Guntur (AP)	Medium sized variety, 10–12 g/ha	(AgriStudyHub 2021)

(continued)

(continued)

Cultivars	Institute/place	Characteristic and seed yield	References
Azad Dhania 1	Chandrashekhar Azad University of Agriculture and Technology	Erect variety	(AgriStudyHub 2021)
Co.1	Tamil Nadu Agricultural University (TNAU), Coimbatore	Dwarf variety, the yield of 4.0 q/ha	(Sandesh 2017) (TNAU 2016)
Co.2	TNAU, Coimbatore	Erect-growing variety, the yield of 5.2 q/ha	(Sandesh 2017) (TNAU 2016)
Co.3	TNAU, Coimbatore	Dwarf variety, the yield of 6.5 q/ha	(Sandesh 2017) (TNAU 2016)
CIMPOS 33	TNAU, Coimbatore	Tall, erect variety	(AgriStudyHub 2021)
CS 287	TNAU, Coimbatore	Early-maturing variety, 6.0 q/ha yield	(Sandesh 2017) (TNAU 2016)
DH 5 (Hisar Anand)	CCS Haryana Agricultural University (HAU), Hisar	Medium-tall variety, 18–20 q/ha	(Sandesh 2017)
DH 36 (Hisar Sugandh)	CCS HAU, Hisar	Dual purpose variety 18–20 q/ha	(AgriStudyHub 2021)
DH 228 (Hisar Bhoomit)	CCS HAU, Hisar	Late flowering variety, 16–18 q/ha	(AgriStudyHub 2021)
DH 246 (Hisar Surbhi)	CCS HAU, Hisar	Bushy erect variety, 18–20 q/ha	(AgriStudyHub 2021)
DWA3	–	Dual purpose variety, 4 q/ha	(AgriStudyHub 2021)
GC 1	Spices Research Station, Jagudan, (Gujarat)	Erect-growing variety, 11 q/ha	(Sandesh 2017)
GC 2	Spices Research Station, Jagudan, (Gujarat)	Medium-tall 14.5 q/ha	(Sandesh 2017)
Pant Haritima	G.B. Pant University of Agriculture and Technology (GBPANT)	Tall, erect, dual-purpose, 12 q/ha yield	(AgriStudyHub 2021)
Rajendra Sonia	Rajendra Agricultural University (RAU), Bihar	Tall, medium size, 12 q /ha yield	(AgriStudyHub 2021)
RD 44 (Rajendra Swathi)	RRC. (RAU) Dolia, Bihar	Medium variety, 13 q/ ha yield	(Sandesh 2017)
RCr 20	SKN College, Jobner (RAU), Rajasthan	Bushy variety, 10 q/ha yield	(AgriStudyHub 2021)

(continued)

(continued)

Cultivars	Institute/place	Characteristic and seed yield	References
RCr 41	SKN College, Jobner (RAU), Rajasthan	Tall, erect, 9.2 q/ha yield	(AgriStudyHub 2021)
RCr 435	SKN College, Jobner (RAU), Rajasthan	Bushy variety, 10.5 q/ha yield	(AgriStudyHub 2021)
RCr 446	SKN College, Jobner (RAU), Rajasthan	Leafy erect, 10.5 q/ha yield	(AgriStudyHub 2021)
RCr 480	SKN College, Jobner (RAU), Rajasthan	Bushy erect, 13.25 q/ha yield	(AgriStudyHub 2021)
RCr 728	SKN College, Jobner (RAU), Rajasthan	Bushy erect, 13.70 q/ha yield	(AgriStudyHub 2021)
Sindhu	Acharya N. G. Ranga Agricultural University (APAU), Guntur (AP)	Dwarf variety, 10.5 q/ha yield	(AgriStudyHub 2021)
Sadhna	APAU, Guntur (AP)	Semi erect variety, 10.3 q/ha yield	(AgriStudyHub 2021)
Swathi	APAU, Guntur (AP)	Semi erect variety, 8.89 q/ha yield	(AgriStudyHub 2021)

References

Abascal K, Yarnell E (2012) Cilantro—culinary herb or miracle medicinal plant? Focus Altern Complement Ther 18(5):259–264

AgriStudyHub (2021) Cultivation of coriander (*Coriandrum sativum*). Agri study hub. https://www.agristudyhub.com/coriander-coriandrum-sativum.html. Accessed 22 May 2022

Aissaoui A, Zizi S, Israili ZH, Lyoussi B (2011) Hypoglycemic and hypolipidemic effects of *Coriandrum sativum* L. in Meriones shawi rats. J Ethnopharmacol 137(1):652–661

Ali M, Mujib A, Tonk D, Zafar N (2017) Plant regeneration through somatic embryogenesis and genome size analysis of *Coriandrum sativum* L. Protoplasma 254(1):343–352

Ali M, Mujib A, Zafar N, Tonk D (2018) Somatic embryogenesis, biochemical alterations and synthetic seed development in two varieties of coriander (*Coriandrum sativum* L.). Adv Hortic Sci 32(2):239–248

AllThatGrows (2022) Grow coriander at home with this simple guide. AllThatGrows. https://www.allthatgrows.in/blogs/posts/grow-coriander. Accessed 5 June 2022

Bajya M, Kakralya BL, Bajaya T, Choudhary M (2022) Evaluation of yield contributing traits in coriander (*Coriandrum sativum* L.) varieties under drought conditions. Legume Res. https://doi.org/10.18805/lr-4941

Cernodubov AI, Berestovaja MM (1986) Breeding coriander. Sel Semenovod 23–24

Choudhary S, Jethra G, Sharma R, Verma AK (2017) Microsatellite in coriander: a cross species amplification within Apiaceae family. Int J Curr Microbiol Appl Sci 6(5):2714–2721

Choudhary S, Sharma R, Jethra G, Vishal MK (2019a) Molecular diversity in coriander (*Coriandrum sativum*) using RAPD and ISSR markers. Indian J Agric Chem 89:193–198

Choudhary S, Naika MBN, Sharma R et al (2019b) Transcriptome profiling of coriander: a dual purpose crop unravels stem gall resistance genes. J Genet 98. https://doi.org/10.1007/s12041-019-1064-7
Cislova LS (1988) Useful types of coriander. Sel Semenovod 33
Cooking HW (2022) Cooking with Coriander—including health benefits and recipes. Help with cooking. https://www.helpwithcooking.com/herb-guide/coriander.html. Accessed 18 May 2022
Coşkuner Y, Karababa E (2007) Physical properties of coriander seeds (*Coriandrum sativum* L.). J Food Eng 80(2):408–416
Darughe F, Barzegar M, Sahari MA (2012) Antioxidant and antifungal activity of Coriander (*Coriandrum sativum* L.) essential oil in cake. Food Chem Toxicol 19(3):1253–1260
da Silva Sá FV, Souto LS, de Paiva EP (2016) Tolerance of coriander cultivars under saline stress. Afr J Agric Res 11:3728–3732
Datiles MJ (2020) *Coriandrum sativum* (coriander). CABI. https://www.cabi.org/isc/datasheet/15300. Accessed 26 May 2022
Datta AK, Sengupta K (2002) Induced viable macromutants in coriander (*Coriandrum sativum* L.). Indian J Genet Pl Br 62(03):273–274
Delaquis PJ, Stanich K, Girard B, Mazza G (2002) Antimicrobial activity of individual and mixed fractions of dill, cilantro, coriander and eucalyptus essential oils. Int J Food Microbiol 74(1–2):101–109
Diederichsen A (1996) Coriander: *Coriandrum sativum* L. Bioversity International
El-Ballal ASI, Abou El-Nasr TH (1987) Selection for earliness and phytotherapeutic value in coriander. Acta Hort 208:179–196
Elmas L, Secme M, Mammadov R et al (2019) The determination of the potential anticancer effects of *Coriandrum sativum* in PC-3 and LNCaP prostate cancer cell lines. J Cell Biochem 120(3):3506–3513
Emamghoreishi M, Khasaki M, Aazam MF (2005) *Coriandrum sativum*: evaluation of its anxiolytic effect in the elevated plus-maze. J Ethnopharmacol 96(3):365–370
Engelmann F (2012) Germplasm collection, storage, and conservation. In: Altman A, Hasegawa PM (eds) Plant biotechnology and agriculture. Academic Press, San Diego, pp 255–267
Galata M, Sarker LS, Mahmoud SS (2014) Transcriptome profiling, and cloning and characterization of the main monoterpene synthases of *Coriandrum sativum* L. Phytochemistry 102:64–73
Gantait S, Sharangi AB, Mahanta M, Meena NK (2022) Agri-biotechnology of coriander (*Coriandrum sativum* L.): an inclusive appraisal. Appl Microbiol Biotechnol 106(3):951–969
Guerra NB, de Almeida Melo E, Filho JM (2005) Antioxidant compounds from coriander (*Coriandrum sativum* L.) etheric extract. J Food Compost Anal 18(2):193–199
Gupta PK, Rustgi S (2004) Molecular markers from the transcribed/expressed region of the genome in higher plants. Funct Integr Genom 4(3):139–162
Harten AM (1974) Coriander: the history of an old crop. Dutch Land Bowkd Tijds Chr 86:58–64
Hashim MS, Lincy S, Remya V et al (2005) Effect of polyphenolic compounds from *Coriandrum sativum* on H_2O_2-induced oxidative stress in human lymphocytes. Food Chem 92(4):653–660
Hassan S (2019) Shoots regeneration in callus tissue of coriander plant (*Coriandrum sativum* L.) in vitro. Plant Arch 19(1):1203–1205
Hwang E, Lee D-G, Park SH et al (2014) Coriander leaf extract exerts antioxidant activity and protects against UVB-induced photoaging of skin by regulation of procollagen type I and MMP-1 expression. J Med Food 17(9):985–995
Indiaagronet.com (2022) Coriander crop cultivation guide. In: Indiaagronet.com. http://www.indiaagronet.com/indiaagronet/horticulture/CONTENTS/Coriander.htm. Accessed 26 May 2022
ISAAA Inc (2006) Conventional plant breeding. ISAAA Inc. https://www.isaaa.org/resources/publications/pocketk/13/default.asp. Accessed 31 May 2022
Ivanova KV, Stoletova EA (1990) The history of culture and intraspecific taxonomy of *Coriandrum sativum* L. Russ Eng Bot Genisel 133:26–40
Jiang G-L (2013) Molecular markers and marker-assisted breeding in plants. In: plant breeding from laboratories to fields. InTech

Kaiser A, Kammerer DR, Carle R (2013) Impact of blanching on polyphenol stability and antioxidant capacity of innovative coriander (*Coriandrum sativum* L.) pastes. Food Chem 140(1):332–339

Kalidasu G, Suryakumari S, Sarada C et al (2015) Exploiting genetic divergence for crop improvement in Coriander (*Coriandrum sativum* L.): a neglected and underutilized crop. Indian J Plant Gen Resourc 28(1):222

Kalidasu G, Surepeddi S, Sarada C, Naram NL (2016) Crop improvement of coriander (*Coriandrum sativum* L. subsp. indicum var. indicum.) through crossing. J Spices Aromat Crops 25:1–6

Kalra (2003) Screening of *Coriandrum sativum* accessions for seed and essential oil yield and early maturity. Plant Gen Resour Newslett 113:19–21

Karpov GY, Popovtsev IL, Mel'nikova VA (1990) Physiological and biochemical features of the drought resistance of coriander in relation to breeding problems. Tr VNII efirnomasli n kul'tur 21:40–47

Keskin D, Toroglu S (2011) Studies on antimicrobial activities of solvent extracts of different spices. J Environ Biol 32(2):251–256

Khan IU, Dubey W, Gupta V (2014) Taxonomical aspect of coriander (*Coriandrum sativum* L.). Int J Curr Res Rev 6(11):9926–9930

Kheti A (2022) Coriander farming guide. ApniKheti. https://www.apnikheti.com/en/pn/agriculture/horticulture/spice-and-condiments/coriander. Accessed 26 May 2022

Kim SW, Park MK, Liu JR (1996) High frequency plant regeneration via somatic embryogenesis in cell suspension cultures of coriander (*Coriandrum sativum* L.). Plant Cell Rep 15(10):751–753

Kolhe PN, Varne MD, Deshmukh SN (2020) Assessment of mutagenicity induced by different mutagens in coriander (*Coriandrum sativum* L.). Biotechnol J Int 24(5):12–21

Kumar G, Pandey A (2015) Heavy metal induced genomic distortion in root meristems of coriander (*Coriandrum sativum* L.). Int J Res Method Educ 5(4):47–53

Kumar CH, Ramesh A, Kumar JNS (2011) A review on hepatoprotective activity of medicinal plants. J Pharm Sci Res 2(3):501–515

Lal AAS, Kumar T, Murthy PB, Pillai KS (2004) Hypolipidemic effect of *Coriandrum sativum* L. in triton-induced hyperlipidemic rats. Indian J Exp Biol 42(9):909–912

Lalitha V, Kiran B, Raveesha KA (2011) Antifungal and antibacterial potentiality of six essential oils extracted from plant source. Int J Eng Sci Technol 3(4):3029–3038

López PA, Widrlechner MP, Simon PW et al (2008) Assessing phenotypic, biochemical, and molecular diversity in coriander (*Coriandrum sativum* L.) germplasm. Genet Resour Crop Evol 55(2):247–275

Manniche L (1989) An ancient Egyptian herbal. University of Texas Press, Austin, Texas, p 176

Maroufi K, Farahani HA, Darvishi HH (2010) Importance of coriander (*Coriandrum sativum* L.) between the medicinal and aromatic plants. Adv Environ Biol 4(3):433–436

Mehrandish R, Rahimian A, Shahriary A (2019) Heavy metals detoxification: a review of herbal compounds for chelation therapy in heavy metals toxicity. J Herbmed Pharmacol 8(2):69–77

Moose SP, Mumm RH (2008) Molecular plant breeding as the foundation for 21st century crop improvement. Plant Physiol 147(3):969–977

Moskalenko VS (1972) Autumn sowing of coriander (*Coriandrum sativum* L.). Me dunar kongr po efir maslam, Tibilisi 2:120–122

Mujib A, Tonk D, Ali M (2014) Plant regeneration from protoplasts in Indian local *Coriandrum sativum* L.: scanning electron microscopy and histological evidence for somatic embryogenesis. Plant Cell Tissue Organ Cult 117(3):323–334

Murthy HN, Hahn EJ, Paek KY (2008) Recurrent somatic embryogenesis and plant regeneration in *Coriandrum sativum* L. Sci Hortic 118(2):168–171

Mustafina FU, Kim EH, Son S-W et al (2017) Assessment of genetic diversity of Prangos fedtschenkoi (Apiaceae) and its conservation status based on ISSR markers. Sigmul Bunryu Hag-Hoeji 47(1):11–22

Nadeem M, Muhammad Anjum F, Issa Khan M et al (2013) Nutritional and medicinal aspects of coriander (*Coriandrum sativum* L.): a review. Br Food J Hyg Rev 115(5):743–755

Nagella P, Thiruvengadam M, Ahmad A, Chung I-M (2014) UHPLC analysis of polyphenol composition and antioxidant activity from different solvent extracts of *Coriandrum sativum* seeds cultivated in Korea. Asian J Chem 26(19):6351–6356

Önder A (2018) Coriander and its phytoconstituents for the beneficial effects. In: El-Shemy HA (ed) Potential of essential oils. IntechOpen, Rijeka

Our Herb Garden (2013) Coriander. In: Our Herb Garden. http://www.ourherbgarden.com/herb-history/coriander.html. Accessed 16 May 2022

Patel DK, Desai SN, Gandhi HP et al (2012) Cardio protective effect of *Coriandrum sativum* L. on isoproterenol induced myocardial necrosis in rats. Food Chem Toxicol 50(9):3120–3125

Pimenov MG, Leonov MV (1993) The genera of the umbelliferae: a nomenclator. Royal Botanic Gardens, Kew and Botanical Garden of Moscow University, Russia, p 156

Popova E, Kim H-H, Paek K-Y (2010) Cryopreservation of coriander (*Coriandrum sativum* L.) somatic embryos using sucrose preculture and air desiccation. Sci Hortic 124(4):522–528

Prakash V (1990) Coriander In: Prakash V (eds) Spices: leafy spices. CRC Press Inc., Boca Raton p 126

Pramanik A, Datta AK, Gupta S et al (2018) Gamma irradiation sensitivity in *Coriandrum sativum* L. (Coriander). Cytologia 83(4):381–385

Praveena R (2019) Coriander—good agricultural practices. Indian Institute of Spices Research, Kozhikode, 1–24

Priyanka V, Kumar R, Dhaliwal I, Kaushik P (2021) Germplasm conservation: instrumental in agricultural biodiversity—a review. Sustain Sci Pract Policy 13(12):6743

Ramegowda V, Senthil-Kumar M (2015) The interactive effects of simultaneous biotic and abiotic stresses on plants: mechanistic understanding from drought and pathogen combination. J Plant Physiol 176:47–54

Ramezani S, Rasouli F, Solaimani B (2009) Changes in essential oil content of coriander (*Coriandrum sativum* L.) aerial parts during four phonological stages in Iran. J Essen Oil Bear Plants 12:683–689

Salve KM, More AD (2019) Biochemical investigation of tall and dwarf mutants in coriander. Int J Res Anal Rev 6:644–673

Sandesh K (2017) Cultivation of coriander. In: Sandesh K (ed) Agriculture in India. https://www.krishisandesh.com/cultivation-of-coriander/. Accessed 22 May 2022

Schnell RJ, Olano CT, Quintanilla WE, Meerow AW (2005) Isolation and characterization of 15 microsatellite loci from mango (*Mangifera indica* L.) and cross-species amplification in closely related taxa. Mol Ecol Notes 5(3):625–627

Sengupta S, Datta AK (2004) Desirable macromutants induced by chemical mutagens in sesame (*Sesamum indicum* L.). Cytologia 69:291–295

Shahwar MK, El-Ghorab AH, Anjum FM et al (2012) Characterization of coriander (*Coriandrum sativum* L.) seeds and leaves: volatile and non volatile extracts. Int J Food Prop 15(4):736–747

Singh RK, Verma SS, Meena RS, Kumar R (2013) Characterization of coriander (*Coriandrum sativum* L.) varieties using SDS-PAGE and RAPD markers. Afr J Biotechnol 12(11):1189–1195

Song X, Nie F, Chen W et al (2020) Coriander genomics database: a genomic, transcriptomic, and metabolic database for coriander. Hortic Res 7:55

Sriti J, Neffati M, Msaada K et al (2012) Biochemical characterization of coriander cakes obtained by extrusion. J Chem Chem Eng 2013:1–6. https://doi.org/10.1155/2013/871631

Stephan R, Jayabalan N (2001) Propagation of *Coriandrum sativum* L. through somatic embryogenesis. Indian J Exp Biol 39(4):387–389

Stephen R, Jayabalan N (1998) In vitro flowering and seed setting formation of coriander (*Coriandrum sativum* L.). Curr Sci 74:195–197

Stephen R, Jayabalan N (2000) Artificial seed production in coriander (*Coriandrum sativum* L.). Plant Tissue Cult Biotechnol 10(1):45–49

Tang ELH, Rajarajeswaran J, Fung SY, Kanthimathi MS (2013) Antioxidant activity of *Coriandrum sativum* and protection against DNA damage and cancer cell migration. BMC Complement Altern Med 13:347

TNAU (2015) Plant genetic resources. In: TNAU agritech portal: crop improvement. https://agritech.tnau.ac.in/crop_improvement/crop_imprv_plantgeni.html. Accessed 22 May 2022
TNAU (2016) Coriander (*Coriandrum sativum* L.). In: TNAU agritech portal. https://agritech.tnau.ac.in/horticulture/horti_spice%20crops_corinader.html. Accessed 22 May 2022
Tomar-Rukam S, Kulkarni GU, Parakhia MV (2014) Genetic diversity analysis in coriander (*Coriandrum sativum*) genotypes through morphological and molecular characterization. Res J Biotechnol 9:1–3
Tridge (2020) Coriander global production and top producing countries. In: Tridge. https://www.tridge.com/intelligences/coriander/production. Accessed 5 June 2022
Vavilov NI, Dorofeev VF (1935) The phyto-geographical basis for plant breeding—the origin, variation, immunity and breeding of cultivated plants. Chron Bot 13:1–366
Wei A, Shibamoto T (2007) Antioxidant activities and volatile constituents of various essential oils. J Agric Food Chem 55(5):1737–1742
Wilkes G (1989) Germplasm preservation: objectives and needs. In: Knutson LV, Stoner AK (eds) Biotic diversity and germplasm preservation, global imperatives. Springer, Netherlands, pp 13–41
Wu T, Feng S-Y, Yang Q-H et al (2021) Integration of the metabolome and transcriptome reveals the metabolites and genes related to nutritional and medicinal value in *Coriandrum sativum*. J Integr Agric 20(7):1807–1818
Yepez B, Espinosa M, López S, Bolaños G (2002) Producing antioxidant fractions from herbaceous matrices by supercritical fluid extraction. Fluid Phase Equilib 194–197:879–884
Yilmaz Y, Turan C (2022) Biotechnological advancements in coriander (*Coriandrum sativum* L.). Avrupa Bilim Ve 35:203–220
Zare-Shehneh M, Askarfarashah M, Ebrahimi L et al (2014) Biological activities of a new antimicrobial peptide from *Coriandrum sativum*. Int J Biosci 4(6):89–99

Chapter 4
Advances in Dill (*Anethum graveolens* L.) Molecular Breeding Strategies

Mai M. Wahba, Mohamed A. Abdein, Haifa A. S. Alhaithloul, Suliman M. S. Alghanem, and Maha M. Elshamy

Abstract Chibitt (dill) (***Anethum graveolens*** **L.**) is a popular aromatic herb and spice with a lengthy history of use dating back more than 5000 years. It is a cross-pollinated crop in the Apiaceae (Umbelliferae) family. It is the sole species of the genus Anethum grown for both culinary and decorative reasons in herb, vegetable, and flower gardens. In addition to being harvested for its aromatic leaves and seeds, it also has beautiful foliage and blooms for ornamental display. Globally, Dill's inclusion of vitamins, minerals, and antioxidants makes it both a food and a medicine. Dill is one of the medicinal plants that are severely threatened due to improper farming methods, the destruction of plant habitats, and the indiscriminate and illegal gathering of plants from various locations. Anethum seeds are shipped to European nations because they are commonly employed in the flavouring and pharmaceutical industries. The supply of herb oil and seeds is necessary for the pickle and fragrance businesses, as well as aromatherapies. For the majority of Apiaceae spices, a significant amount of traditional breeding work has been done, particularly in the area of phenotypic-based variability evaluation.

The original version of the chapter has been revised. A correction to this chapter can be found at https://doi.org/10.1007/978-3-031-81838-7_18

M. M. Wahba · M. M. Elshamy
Department of Botany, Faculty of Science, Mansoura University, Mansoura 35516, Egypt
e-mail: maiwahba@mans.edu.eg

M. M. Elshamy
e-mail: mahammy@mans.edu.eg

M. A. Abdein (✉)
Seeds Development Department, El-Nada Misr Scientific Research and Development Projects, Turrell, Mansoura 35511, Egypt
e-mail: abdeingene@yahoo.com

H. A. S. Alhaithloul
Department of Biology, College of Science, Jouf University, Sakaka 72388, Saudi Arabia
e-mail: haifasakit@ju.edu.sa

S. M. S. Alghanem
Department of Biology, College of Science, Qassim University, Buraydah, Saudi Arabia
e-mail: su.alghanem@qu.edu.sa

J. M. Al-Khayri et al. (eds.), *Biodiversity and Genetic Improvement of Herbs and Spices*, Advances in Plant Breeding Strategies 8, https://doi.org/10.1007/978-3-031-81838-7_4

Keywords *Anethum graveolens* · Cultivation · Phytochemistry · Biodiversity · Germplasm conservation · Breeding techniques

4.1 Introduction

According to Delaquis et al. (2002), dill is a common, fragrant edible plant of the Apiaceae (Umbelliferae) family. It is known by several different vernacular names, including shevid in Iran, Suva bhaji and Sadakuppai in India, shwit in Iraq, shi luo in China, kopar in Bulgaria, sbt in Turkey, and enduro in Southern Brazil (Oshaghi et al. 2015). It is the only species in the genus Anethum, typically growing to a height of 40–150 cm, with thin stems and finely split, delicately soft leaves (Jana and Shekhawat 2010). The umbels are tiny, 2–9 cm in diameter, and the blooms range in color from white to yellow (Mirhosseini et al. 2014).

An annual with a 150 cm height limit, *Anethum graveolens*. It has alternate, sheathing leaves and tapering, dark-to-blue-green stalks. It is well recognized in both traditional and modern medicine that A. *graveolens*, sometimes known as the dill plant, is a highly significant shrub that may relieve colic discomfort and flatulence (Benlembarek et al. 2022).

Ambrozja, Kronos, and Lukullus are commercial dill cultivars with a variety of features, including leaf density and color, lowering time, nutritional value, and aroma strength (Biesiada et al. 2019). On the other hand, regional organizations in Greece cultivate traditional dill landraces that have been chosen and tailored to the unique climatic and environmental circumstances of the region. Although many of these landraces have been preserved in the Greek Gene Bank, it is yet unknown how effective they will be. Previous studies on the phenological classification of a few landraces of Greek dill have shown that morphological characteristics may effectively divide genotypes into groups according to where they are native (Ninou et al. 2017).

4.2 Taxonomic Studies

Anethum graveolens L., Sp. Pl., ed. 1:263 (1753).

4.2.1 Suprafamilial Classification

4.2.1.1 Aaron Goldberg's System (1986)

Classis: Dicotyledoneae
Order: Apiales
Family: Apiaceae

4.2.1.2 Angiosperm Phylogeny Group (APG) Classification (APG I 1998, APG II 2003, APG III 2009 and APG IV 2016)

Clade: Angiosperms
Clade: Eudicots
Clade: Asterids
Clade: Euasterids II
Order: Apiales
Family: Apiaceae

4.2.2 Infrafamilial Classification

Erdtman (1952a) studied 85 species belonging to 62 genera of Apiaceae. He was showing little variations in pollen morphology. Pollen with typically radially symmetric tricolporate, prolate. Rarely bilateral symmetrical dicolporate pollen. Cerceau-Larrival (1971) studied over 1500 species of Apiaceae. She proposed a new classification of the family depend on the type of cotyledon, type of pollen grain and other morphological characters, she claimed that pollen characters correlate with seedling, vegetative system, inflorescence, and fruit. She divided the family into five groups, and she used Drude's tribes. Her system was supported by Guyot (1971) who studied the stomatal type in taxa of these tribes. Hutchinson (1967) used floral and inflorescence characters to divide the family into seven tribes. Thorne (1968) included the Apiaceae in the Araliaceae and recognized four subfamilies viz., (*Hydrocotyloideae*, *Saniculoideae*, *Apioideae* and *Aralioideae*), three of them are of Drude's subfamilies.

Thorne (1972) explained why he included *Apioideae* in Araliaceae due to three rules: [(1) consistency in the size of phyletic gaps between related taxa, (2) avoiding polyphyletic taxa in the subfamilies that seem to have originated in different parts of the world, (3) a taxon should be monophyletic]. Cronquist (1968) and Takhtajan (1969) agree that Araliaceae and Apiaceae are related. Eyde and Tseng (1971) proposed that none of the treatments of the family show a real evolutionary pathway.

From the Greek word for dill, the name "Genus" was created. The Norse term dilla, which meaning to soothe or lull, is said to have inspired the popular name of the herb dill. Because they were periodically given to children to chew on during drawn-out church meetings, dill seeds were known as meetinghouse seeds in colonial America.

4.3 Synonymy

Peucedanum graveolens (L.) Benth. & Hook. F. Gen. Pl. 1:919. (1867).

Pastinaca anethum Spreng., Syst. Veg., ed. 15 bis [Roemer & Schultes] 6:587 (1820).

Peucedanum anethum Jess., Deut. Excurs.-Fl. 180 (1879).

Selinum graveolens Vest, Man. Bot. [Vest] 501 (1806).

Selinum anethum Roth, Tent. Fl. Germ. 1:134 (1788).

Anethum arvense Salisb., Prodr. Stirp. Chap. Allerton 168 (1796).

Anethum sowa Roxb., Asiat. Res. 11:156 (1810).

Ferula graveolens Spreng., Neue Schr. Halle ii. (1813) 14.

Angelica graveolens Steud., Nomencl. Bot. [Steudel], ed. 2. 2:555, in syn. (1841).

Peucedanum sowa Kurz, J. Asiat. Soc. Bengal, Pt. 2, Nat. Hist. 46(2):116 (1877).

4.4 Origin and Distribution

A. graveolens has been used for more than 5000 years and is a well-known fragrant herb and spice (Gunalan et al. 2020). The term "dill plant" or "chibitt" in Arabic refers to a fragrant herb of the Apiaceae family that is frequently grown in the Mediterranean, Southeast Asia, and Europe (De Guzman and Siemonsma 1999). It belongs to the genus *Anethum*, which has two species: *A. graveolens* L., the type species for the European variation, and *A. sowa* Roxb, the type species for the Indian form. Dill is one of the minor spice seed crops that is mostly used for food and medicine. One of the first seed spices to be developed, dill is most likely a native of the Mediterranean region of West Asia. The origins of the dill crop are attributed to Europe, Africa, and Asia. Dill was first used as a flavor plant in dynastic Egypt, and then by the Greeks and Romans for both flavoring and therapeutic purposes. As early as 1000 A.D., when Roman legions transported it to Europe, it was widely recognized as dill. By the Middle Ages, it was a common herb and spice for pickling. The Indian states of Jammu and Kashmir were intentionally introduced to European dill as a potential crop at the beginning of the 1950s. Later, it expanded to other suitable locations. According to Setorki et al. (2013), dill is cultivated in numerous nations throughout the world, including South Europe, Bulgaria, India, Iran, China, Pakistan, Germany, Hungary, the Netherlands, the United States, and Estonia.

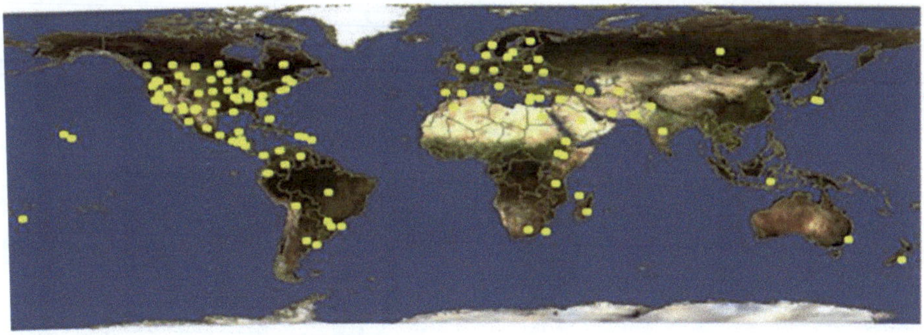

Fig. 4.1 Map showing the locations where dill (*Anethum graveolens* L.) is grown. Yellow dots indicate the places that have been planted. This number was made public online

In Egypt, the plant grows everywhere, cultivated, and escaped from cultivation (Boulos 2000) (Fig. 4.1).

4.5 Plant Description

The single species in the genus *Anethum* is *Anethum graveolens* L. (Fig. 4.2), while some botanists categorize it as *Peucedanum graveolens* (L.) in the closely related genus *Peucedanum.* East Indian dill, often referred to as Sowa, is a kind of dill that naturally grows in India and is cultivated for its leaves in Japan, the Malaysian archipelago, and the Indian subcontinent.

4.6 Morphological Aspects

The blue-green leaves of dill are aromatic, lacy, delicate, and pinnately split into tiny, thread-like segments. Dill normally grows to a height of 3–5 ft (De Guzman and Siemonsma 1999). Taller plants need shelter from severe winds since they can topple over. In mid-summer, fragrant, yellow blooms appear in huge, flattened, complex umbels (each up to 10″ in diameter). Flowers come after fragrant seed. Around the time the flowers first open, the flavor of the leaves is usually at its peak. Even though leaves can be preserved or stored in the refrigerator for later use, fresh leaves from the garden have the best flavor.

It is commonly known as Shabat. Glabrous biennial or perennial herb up to 50 cm long. **Stems**: tall, erect stems are hollow, cylindrical with a thin, spindle-like taproot with few terminal branches. **Basal leaves** 7–18 × 2–18 cm including petiole; sheath with membranous margin; blade ovate in outline, 3–4 pinnatisect filiform linear segments, 0.3–0.7 × 0.01–0.03 cm with entire margin and acute apex.

Fig. 4.2 The single species in the genus *Anethum* is *Anethum graveolens* L.

Upper leaves 1.3–2.6 × 0.2–2 cm, sessile; blade oblong in outline, 2-pinnatisect filiform segments with entire margin. The leaves have a characteristic slightly harsh, spicy, and long-lasting flavor. **Inflorescence** umbel, umbel peduncle 12–15 cm long; involucres absent; rays 7–15 umbellule, unequal; involucel absent. Nine **Flowers** per umbellule; pedicel 0.6–2.5 mm, sepals absent, petals yellow, ovate with inflexed apex. **Fruit** ellipsoid ovoid, glabrous, brown, with straight ribs, fruit pedicel 0.6–3.7 mm long (Figs. 4.3 and 4.4).

Fig. 4.3 *A. graveolens* L. has branching stems

Fig. 4.4 *A. graveolens* L. blooms

4.7 Anatomy Aspects

Stem Anatomy: **Stem** terete in cross section. **Epidermal cells** are tangentially thick cuticles. **Trichomes** absent. **Cortex** complex of four types of tissues, collenchyma (angular at corners, 6-layered cells and lamellar inbetween, 2-layered cells), parenchyma (4 layers) of usually oval cells, chlorenchyma (4–5 layers) and sclerenchymatous fibers. **The aspect of vascular bundles** is distinct. **Pith** wide and fill in with thin parenchyma cells.

Lamina Anatomy: **Lamina** cordate in cross section. **Trichomes** e-glandular (unicellular, unbranched). **Mesophyll** dorsiventral type. **Epidermal cells** arranged tangentially. **Cuticle** thick. **Palisade tissue** one row, extended at midrib region. Collenchyma absents. Parenchyma present at midrib. **Vascular system** in crescentiform and 1 of vascular bundles. **Crystals** absent.

4.7.1 Lamina Abaxial Epidermal Characteristics (LM & SEM)

Cell shape polygonal with glebulate blade surface sculpture. **Anticlinal walls** straight, narrow, raised with granular surface. **Periclinal walls** depressed with granular surface. **Stomata shape** oblong elliptic. **Stomata** diacytic, **stomatal aperture** length 6.401 μm (Fig. 4.5).

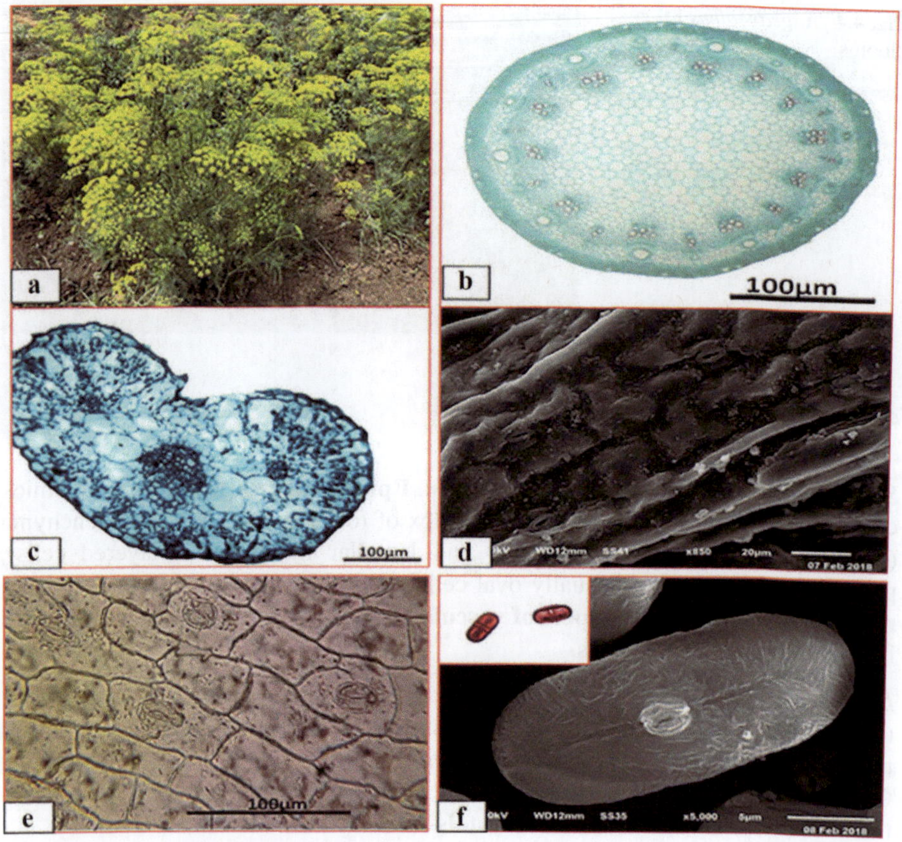

Fig. 4.5 Some morphological and anatomical aspects of *A. graveolens*

4.8 Pollen Morphology (LM & SEM)

Pollen grains are isopolar, radially symmetrical, prolate in form, with polar axes of 21.184–21.186 mm and equatorial axes of 12.030–12.032 mm, P/E = 1.760, and an outline with a regulated exine that is elliptic in the equatorial optical section and triangular in the polar optical section (Fig. 4.6).

4.9 Phytochemistry

There have been many pharmacological effects attributed to *A. graveolens*, including antimicrobial, antihyperlipidemic, and antihypercholesterolemic effects. Mice's stomach ulcers are prevented by dill seed extracts, which also have anti-ulcer, antisecretory, and anti-HCl and ethanol-encouraged effects. Seed extracts also significantly

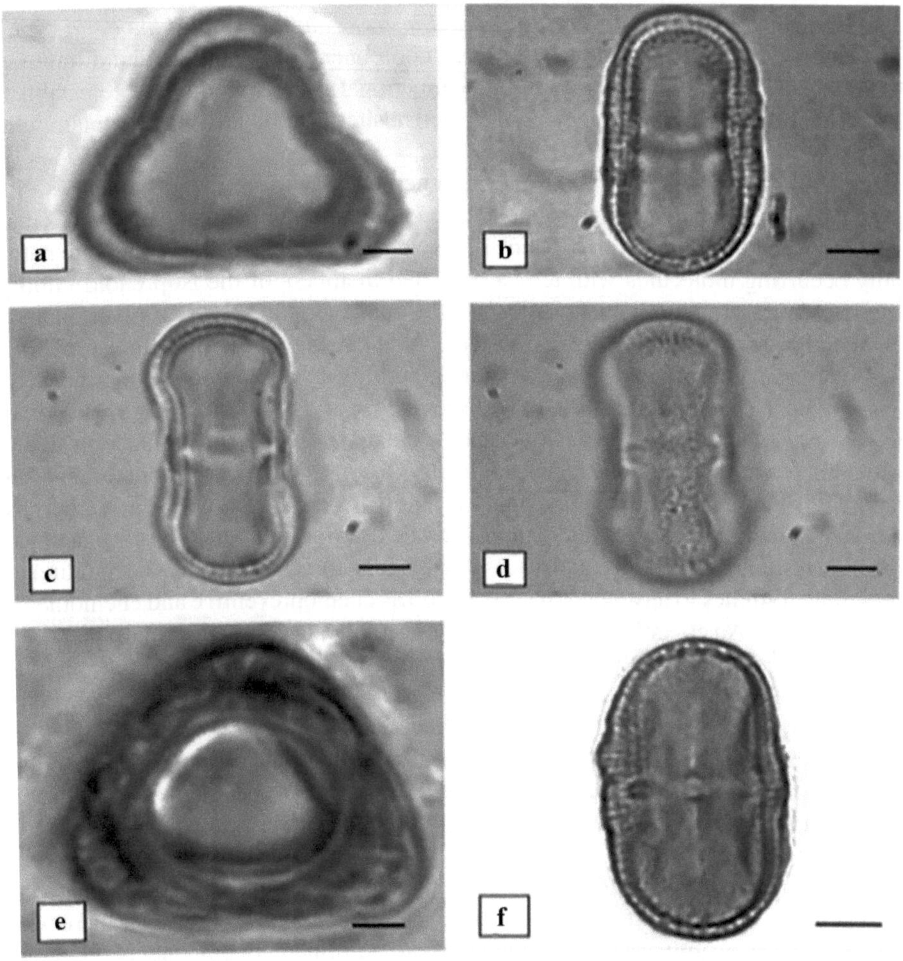

Fig. 4.6 Apiaceae pollen grains photographed under a microscope. Equatorial view (**c**) and surface (**d**) of coriander pollen grain (*Coriandrum sativum* L.); polar view (**e**) and equatorial view (**f**) of dill pollen grain (*Anethum graveolens* L.); and polar view (**a**) and equatorial view (**b**) of fennel pollen grain (*Foeniculum vulgare* Mill.)

preserve the mucosa. Seeds of *A. graveolens* have yielded two flavonoids, isoharmentin and quercetin, with antioxidant characteristics and may be utilised to battle free radicals. This finding could aid in the prevention of peptic ulcers. Since the dill fruit is a potent relaxant of spasms induced by a variety of spasmogens in the rat ileum, the HCl extract of the dill fruit supports its usage in traditional medicine for digestive problems. Crude extracts of *A. graveolens* can enhance the biological antioxidant status of rats fed a high-fat diet while also considerably lowering hepatic lipid peroxidation and modulating the activities of antioxidant enzymes. Aqueous extracts of *A. graveolens* have been shown to have broad-spectrum antibacterial activity against *Shigella flexneri*, *P. aeruginosa*, *S. aureus*, and *Salmonella typhi*. The chemical structure of the extract's key components, as anethole and dillapiole, which feature aromatic nuclei with polar functional groups recognised to establish hydrogen bonds with the target enzyme's active sites, can be utilised to explain why the extract is more active.

Dill oil from grapes contains the monoterpenes carvone and limonene as its primary ingredients. Dill ether, α-phellandrene, and myristicin are the molecules that give dill weed its unique aroma. Many plants include monoterpenes, which are naturally occurring molecules with ten carbons and members of the isoprenoid family. Monoterpenes are mostly responsible for the characteristic scents of plants. These compounds are thought to serve mostly ecological purposes, including repelling herbivore grazing, protecting against fungi, and luring pollinators. Indian dill leaf has seventeen chemicals that have been discovered. Carvone has many uses, including as a flavour and aroma, a preventative against potato sprouting, an antibacterial agent and component of biochemical environments. One of the most prevalent terpenes in nature is d-limonene. Due to its exceptional ability to dissolve cholesterol, Numerous citrus oils, including orange and lemon, include d-limonene, which has been shown in clinical studies to dissolve gallstones that contain cholesterol. It has been found in pre-clinical studies to have low toxicity as well as chemopreventive and chemotherapeutic actions. An essential oil's myristicin is a significant component and a naturally occurring pesticide. Anethole is a terpenoid that is found in essential oils such as anise and fennel as well as *Anethum*. It serves as a flavour agent. P-anisaldehyde is a crucial ingredient in medicines and fragrances and has a potent aroma.

GC-MS was used to ascertain the chemical makeup of the essential oil hydrodistilled from the aerial parts of *A. graveolens*. A total of twelve elements, or 96.8% of the oil, were found. The most prevalent compounds were p-cymene (5.3%), α-phellandrene (31.8%), apiole (15.3%), dill ether (13.2%), limonene (11.8%), and geraniol (10.6%). The chemical makeup of the oil was discovered to be comparable to that which has been informed from India and other countries, however there was a considerable quantitative difference between the principal components observed (Rana and Blazquez 2014).

In the extraction of fennel grass with tetrafluoroethane, an aromatic product is obtained with a high content of D-carvone (53.1%) and lemon (37.1%), with a characteristic taste and smell. The extract exhibits antimicrobial activity more pronounced compared to Gram-positives bacteria, as well as highly expressed antioxidant action. It can find application as traditional aromatic products in different food and cosmetic products.

To explore the chemical components and antibacterial activity of the medicinally important *A. graveolens* volatile oil. GC-MS analysis revealed thirty components. D-limonene (18.13%), D-carvone (37.80%), and apiol (16.16%) are the main ingredients. *Escherichia coli*, *Candida albicans*, and *Pseudomonas aeruginosa* are Gram-negative bacteria, whereas *Staphylococcus aureus* and *Bacillus subtilis* are Gram-positive, and these five frequent human infections were tested using the disc diffusion method. The oil was found to be effective against *B. subtilis*, *P. aeruginosa*, *E. coli*, and *S. aureus*, among other bacteria (Abdalkarim et al. 2021) (Fig. 4.7).

Compounds	Molecular formula	References
Carvone 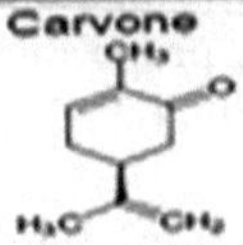 d-carvone	$C_{10}H_{14}$	[46,37]
Limonene 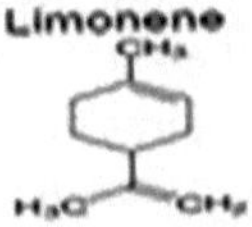 d-limonene	$C_{10}H_{16}$	[15,43]
α-phellandrene 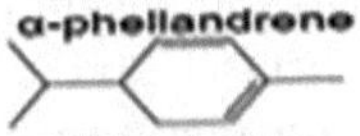 alpha phellandrene	$C_{10}H_{16}$	[15,34,43]
β-phellandrene 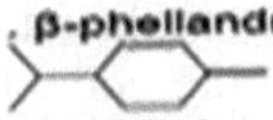 beta phellandrene	$C_{10}H_{16}$	[15,34,43]
Myristicin 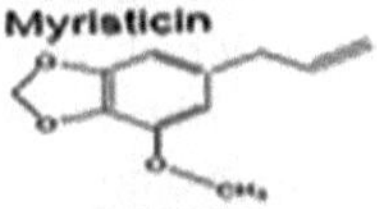 myristicin	$C_{11}H_{12}O_3$	[34,37,44]
Apiole 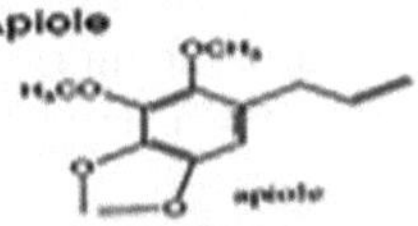	$C_{12}H_{14}O_4$	[3,31,45]
Umbelliferone 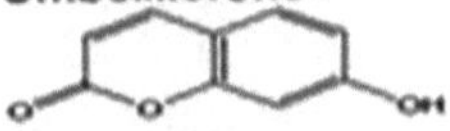 umbelliferone	$C_9H_6O_3$	[37]
Anethole 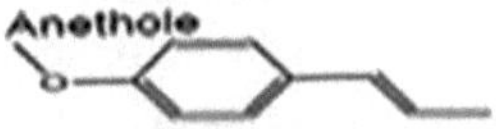 anethole	$C_{10}H_{12}O$	[38]
P- anisaldehyde 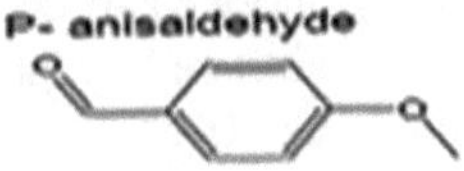 p-anisaldehyde	$C_8H_8O_2$	[46]

Fig. 4.7 Showing important compounds extracted from dill according to (Rana and Blazquez 2014)

4.10 Economic Important, Uses and Health Benefits:

A. graveolens has been used for more than 5000 years and is a well-known fragrant herb and spice (Gunalan et al. 2020). Dill draws a wide range of beneficial insects to the garden, including wasps, bees, lacewings, butterflies, hover flies, tachinid flies, and lady beetles. The black swallowtail butterfly caterpillar lives on dill plants. Previous research has shown that *A. graveolens* has antibacterial, analgesic, gastric mucosal protecting, antisecretory, and smooth muscle relaxant properties., hypolipidemic, and many other effects. In Traditional *Siddha* literatures, this medicinal plant is mentioned as *Sadhakuppai* and indicated for anti-inflammatory properties. Dill is grown, among other locations, in Iran, India, South Europe, Bulgaria, China, and Estonia (Setorki et al. 2013).

Plant volatile oils and the main ingredients in them may be antimicrobial, antifungal, and antioxidant. They can prevent food contamination as a result. Also, these oils are a superb source of natural and secure preservatives since they are commonly acknowledged as acceptable materials for use in the food sector (Najaran et al. 2016).

Due to its pleasant peppery fragrance, dill is frequently used in cuisine and the food industry to flavor meals and beverages. Salads, dips, pickled vegetables, sour cream, seafood dishes, meats, sauces, soups, and cream cheese all benefit from the flavoring and seasoning of dill seeds and leaves (Small and Deutsch 2001).

Dill seeds have long been used as a typical home cure for digestive problems like gas, indigestion, and stomachaches, in addition to the leaves being used in meats, eggs, soups and salads. Babies are given dill water to cure colic, hiccups, and gripes because it is said to have a calming effect. The galactagogues in dill are also known to boost milk production in nursing women, which benefits the infant by preventing colic. Bad breath is lessened by chewing the seeds. There have been claims that dill possesses antihyperlipidemic and antihypercholesterolemic properties (Yazdanparast and Bahramikia 2008).

The plant is used as an unguent for headaches and used as a mouth rinse (Hosni 1988). Fruits are used as flavoring foods and pickles (Sievers 1947), draw down milk and stop vomiting (Boulos 1983).

It is frequently employed in the conventional medicinal, food, cosmetic, and nutraceutical sectors due to its qualities. According to Garcez et al. (2017), seed extract exhibits antibacterial, antifungal, anti-inflammatory, and antioxidant characteristics that help preserve food and guard against food pathogens. Additionally used as a diuretic and cholesterol-lowering medication, dill also has lactagogue, anticonvulsant, antiemetic, antispasmodic, and antidiabetic properties. It is also used to treat many gastrointestinal problems. Herbal extracts and oils are more easily grown and kept for human consumption than the chemical preservatives utilized by large-scale companies. Their appeal has grown as a result (El-Zaeddi et al. 2016). According to Ayurveda, dill seeds have diuretic, stomachic, and carminative properties. Dill seeds primarily contain carvone, whereas dill plants primarily include α-phellandrene, limonene, dill ether, and myristicin. Dill seeds and herbs include a variety of volatile components. Other compounds found from seeds include flavonoids, coumarins, steroids, and phenolic acids (Jana and Shekhawat 2010). It is a popular plant that produces essential oil and is used widely as a spice.

Gram-negative bacteria, which are resistant to these oils, were significantly impacted by dill essential oils. Carvone and α-phellandrene, the two primary constituents of *A. graveolens* essential oils, also have a variety of pharmacological qualities, including a low potential for cytotoxicity and a high potential for accelerating wound contraction (Benlembarek et al. 2022).

Although it inhibits the growth of carrots, the herb grows nicely with lettuce, corn, onions, and cabbage. If grown to maturity near carrots, dill diminishes their yield. The young plant, on the other hand, will help to keep carrot root fly at bay. Coriander and dill have a great pest control benefit when planted together (Carrubba et al. 2008). In agroecosystems, intercropping with blooming herbaceous plants promotes parasite survivability, fecundity and retention, and pest reduction. Potential hosts for parasitoids include *Pediobius foveolatus Crawford*, *Cotesia glomerata*, and *Edovum puttleri Grissell* (Wanner et al. 2006).

According to studies by Goodarzi et al. 2016; Al-Sheddi et al. 2019; Tanveer et al. 2020; Yazdanparast and Alavi (2001), the dill essential oils of have pharmacological activities that include antidiabetic, anticancer, antisecretory, and antihyperlipidemic effects. In the culinary and cosmetic sectors, it is also used to season scents, soaps, and cosmetic products (Said-Al Ahl et al. 2015).

4.11 Nutritional Studies

Numerous studies have shown that dill includes special and beneficial ingredients including niacin, vitamin A, and essential oils as well as proteins, fiber, fatty acids, carbs, and macroelements like calcium, potassium, magnesium, and sodium (Ghassemi-Golezani et al. 2011).

4.12 Recent Cultivation Procedures *A. graveolens* L.

An annual, dill grows best on rich, sunny, well-drained soils that receive plenty of direct sunshine. Plants are more prone to fall over in partial shade. Climates with chilly summers promote the best growth. Plants value regular soil moisture. It's important to prevent soil from drying out. When circumstances are dry, plants often bolt. Just before the last spring frost date, directly plant seeds in the ground. Seeds are typically not started indoors since seedlings can be challenging to transplant. To extend the time during which new leaves may be harvested, extra seeds may be sown every two weeks up to the beginning of the summer. About two to three weeks after flowering, the seed heads of dill can be picked and dried; doing so will also encourage the plant not to self-seed in the garden. In St. Louis, dill frequently self-seeds and produces new plants every spring. This species may be grown in large pots as well because it needs them for its thick roots (Boeckmann 2023). Ideally, dill needs six to eight hours per day of direct sunlight.

Dill loves soil that is loose, fertile, and well-drained. It can resist pH levels between 5.3 and 7.8. It needs summers that are warm to hot with plenty of sunshine; even a little shade will significantly lower output. In dry conditions, the plant soon runs out of seeds. It frequently self-sows when grown in a favorable environment. Plants reproduce using seeds, which can be stored for three to ten years. When the seed

Fig. 4.8 Showing fruits and vegetables and compound umbrellas of Dill

begins to ripen, the flower inflorescences are removed from the stalks and the seeds start to mature then are harvested (Jana and Shekhawat 2010).

The use for which dill is produced determines when to harvest it. It can be harvested when it has 4–5 leaves when planted for vegetables. When grown for seed, the crop is harvested when it is fully mature. The crop can be harvested in April after maturing in 130–150 days.

When the seeds of the main umbel become brown, the plants are harvested by cutting them with sickles 40 cm above the ground. The field is invaded by grain that hasn't yet been harvested. The gathered dill plant bundles should be dried in the shade using a stick or an automated seed and spice thresher. To prevent losses during storage, threshed seeds should be cleaned, dried, and stored in ventilated containers.

Yield: Under irrigated conditions, the yield is typically 10–15 q/ha, while under rainfed circumstances, it is 6–7 q/ha. Dill seed output at this level can be achieved with better crop management (Fig. 4.8).

4.12.1 Diseases

Leaf eating caterpillar: The plant's leaves frequently suffer damage from it. Spraying endosulfan 0.05% once or twice can control it, depending on how many caterpillars are present.

4.13 Current Agricultural Challenges

It is a non-traditional seed spice that has been found as a promising crop for growing in saline circumstances. It is also a resilient drop that thrives in low moisture environments. It is typically grown in moderate, cool climates for 112–128 days. In dill cultivated in sodic soils, Gautam et al. (2011) discovered genotypes that were sodicity tolerant and investigated the features of genetic diversity as well as the nature of correlations between several factors that affected seed output. Eleven dill genotypes were tested during the winter in three different conditions: normal, moderate sodic, and high sodic. 'CSS 1' was found to produce the most seeds out of all collections over a three-year period in various alkalinity locals.

Plant height, seed production, primary and secondary branches, and plant stand followed this trait's great sensitivity in almost both stress regimes. There was a positive and substantial link between the number of secondary branches, primary branches, inflorescences, and seeds produced in the moderately sodic soil. These characteristics also had a larger immediate effect. There was a positive and substantial link between the number of secondary branches, primary branches, inflorescences, and seeds produced in the moderately sodic soil. Additionally, these traits had a stronger immediate impact.

To generate high-quality seeds, Ghassemi-Golezani et al. (2018) identified the ideal time to harvest three *A. graveolens* ecotypes (Isfahan, Malayer, and Varamin). The results unequivocally showed that seed development on mother plants enhanced seed quality for all ecotypes for all watering intervals. Depending on the ecotype and irrigation intervals, maximum seed quality was reached between 2 and 12 days after mass maturity. The electrical conductivity of seed leachates, germination rate and percentage, and seedling dry weight were used to make this determination. The quality of the seeds may be improved by watering dill plants, particularly during seed filling, it was found as a result. The most robust seeds were produced by the Varamin and Malayer ecotypes, according to germination rate and seedling dry weight. The Isfahan ecotype generated seeds with the least amount of vigor. The genetic variations between the ecotypes may be related to these variations in seed vigor. Protect plants from windy conditions. Stakes, cages, or close planting can all be employed as supports.

Only up to the 60 kg ha^{-1} N treatment did the good effect on biomass production result in an increase in seed yield because raising the nitrogen (N) rate had a negative impact on the harvest index. The highest levels of carvone yields were attained in the nitrogen rate treatments (30–60 kg ha^{-1}) because N rate had a negative impact on seed carvone content. The following facts help to support this conclusion: (1) Despite the dry conditions, the crop effectively utilized the available N; (2) The relationship between N rate and carvone content was consistent with other research findings; (3) Lodging happens (at a greater N rate in a year with more rainfall), which lowers the yield of seeds and carvone; (4) An increase in N per seed was associated with a decrease in carvone per seed. It was discovered that the difference in soil type between the two trials (19 vs. 43% clay) was not relevant since the effect of N rate on seed and carvone production was the same in both trials (Wander and Bouwmeester 1998).

Numerous fungi that cause plant pathology can harm dill, but root rot and foliar diseases like powdery mildew and blight cause the most damage. Climate and soil conditions affect the prevalence of disease and the harm it causes to dill crops. Conditions that are wet and muggy generally encourage the development and spread of illnesses. Important dill-related illnesses include:

Erysiphe polygoni is a fungus that is responsible for powdery mildew on dill. The disease manifests as a white powdery substance on the leaves, stems, and inflorescence of all green plant parts. Later, the problems spread to other plant parts, including the seeds, which have an impact on both yield and seed quality.

Sulphur dusting at a rate of 20–25 kg/ha can successfully manage the sickness when it is initially identified. Spraying dinocap (0.1%) or wettable sulphur (0.2%) as a disease treatment method is also beneficial. As soon as symptoms appear, the first spray should be used, and depending on how rapidly the illness is progressing, a second spray may be required after 10–15 days.

Fusarium spp. is a fungus that causes root rot. All growth phases can experience illness signs; however, the early vegetative growth stage is more susceptible. The plant prematurely dries out and dies once the chlorotic leaves first turn yellow. Plant roots that are harmed display browning and decaying. If the illness develops early in the season, it may result in significant yield losses.

Controlling the disease requires phytosanitary practices and seed treatment. By treating seeds with *Trichoderma* (4g/kg seed) or carbendazim (2g/kg seed), two biofungicides, the disease can be successfully controlled.

4.13.1 Other Diseases

Several additional diseases of limited regional significance also damage dill crops. Following is a list of ailments and the germs that cause them:

Pythium species, *Rhizoctonia species*, seedling blight, and *Alternaria radicina* all cause "damping off".

- *Ascochyta anethicola* (leaf spot)
- Head rot: *Alternaria spp.*, *Botrytis* spp.
- Scabbing—*Fusicladium* spp.

If these mild illnesses manifest, regional or local Agricultural Research Stations can provide control strategies.

Insect pests: Dill is frequently attacked by insects that target other Umbelliferae crops and locally significant vegetable crops, but the damage is relatively minor. Insects including aphids, cutworms, and leaf-eating caterpillars typically damage the crop.

Aphid: It feeds on the sensitive parts of the plant and the blossoms, causing the crop to yellow and the grains to shrivel. Endosulfan 500–600 L/ha solution, a comparatively less dangerous insecticide that should be applied in the evening when insect activity is at its lowest, can be used to manage it. The optimum planting area should have well-draining soil that is rich in organic matter and has a pH between slightly acidic and neutral because dill's tall leaves are easily blown away by high winds.

According to Naeem et al. (2013), a water shortage or drought is known to be a significant factor that restricts the growth and productivity of aromatic and medicinal plants. Farmers need their crops to be successfully established in such conditions. Especially in unfavorable environmental conditions, seed quality is crucial for the sustainable and lucrative production of crops (Finch-Savage and Bassel 2016). According to Ghassemi-Golezani (1992) and Finch-Savage (1995), seed quality can have a direct impact on the rate of seedling emergence as well as an indirect impact on the percentage of fields that emerge. In order to guarantee the production of high-quality seeds, commercial seed businesses tend to choose areas with favorable temperatures. According to reports, several conditions, including drought, insects, diseases, and extremely high and low temperatures throughout the maturity and production stages, can affect the quality of seeds (França-Neto et al. 2000).

A field experiment was carried out in the winters of 2001–2002 and 2002–2003 to ascertain how the genotypes of European dill (*A. graveolens*) and Indian dill (*Anethum sowa* Roxb. ex. Flem.) responded to various agrotechniques in semi-arid environments. On October 1 and October 15, respectively, the European 'NRCSS-AD-1' and Indian 'NRCSS-AD-2' dill kinds were sowed, and these dates led to noticeably different plant heights and numbers of main branches per plant. The Indian variety 'NRCSS-AD-2' that was seeded earlier on September 16 produced considerably the greatest essential oil concentration at 3.41%, followed by sowing on October 15 (3.26%). Dill genotypes from Europe and India responded favorably to a 4 kg/ha seed rate. The leading high plant height and total number of main branches of the Indian and European species of dill followed a pattern that was comparable when rows were separated 30 and 45 cm apart, but when rows were 60 cm apart, they clearly displayed a difference. The contributing metrics for umbels/plant, umbellets/umbel, seeds/umbel, and seed output were all very high with a 45 cm row spacing. The amount of essential oil that was gathered was substantial (3.28 and 3.6% in both varieties) at the dose of 100 kg N, 40 kg P_2O_5, and 25 kg K_2O/ha (Malhotra 2007).

The N.E. Borlaug Crop Research Centre at the G.B. Pant University of Agriculture and Technology in Pantnagar (Uttarakhand) examined the performance of European dill during a two-year period in the field and in the lab. The treatments in a factorial randomized block design with three replications were three fertility levels (0, 15, and 30 t FYM ha^{-1}) for the sub plots and four cumulative pan evaporation irrigation levels (50, 100, 150, and 200 mm) for the main plots. Following harvest, crop productivity was evaluated, and the oil content of the seeds was examined. *A. graveolens* Linn's growth, seed and oil output, and oil quality were all significantly boosted by the treatment combination, according to data gathered over a two-year period. However, yields were also increased by increasing the quantity of irrigation and farmyard manure application. Almost all growth parameters for A. graveolens, including plant height, branch count, and dry matter accumulation, as well as yield and yield characteristics, such as the number of umbels and umbellets on different branches, the number of fruits per umbel, and test weight, were significantly positive straightforward type of correlation (Rathore 2020).

4.14 Conventional Breeding of *A. graveolens* L.

Dill is a fragrant plant that is widely utilized in the culinary business, and various commercial cultivars with varying qualitative features are available. In 2023, Kadoglidou et al. investigated the morphological, genetic, and chemical biodiversity of 22 Greek landraces and nine modern/commercial cultivars. Greek landraces could be identified from modern cultivars at the molecular, phenological, and chemical levels, according to a multivariate analysis of polyphenol content, essential oil, molecular markers, and morphological descriptors. The height of landraces increased along with the size of their umbels, foliage, and leaves. Several dill landraces, such as GRC-1348/04 and T538/06, which were equal to or better than several commercial cultivars, were found to have acceptable plant height, leaf density, feathering density, and fragrance characteristics. For landraces and modern cultivars, respectively, the polymorphic loci for inter-simple sequence repeat (ISSR) and start codon targeted (SCoT) molecular markers were 76.47% and 72.41% and 68.24% and 43.10%. Despite considerable gene flow between landraces and cultivars, there was some genetic divergence but not total isolation. α-Phellandrene was the component that was present in all dill leaf essential oils most frequently (54.42–70.25%). α-Phellandrene and dill ether concentrations are greater in landraces than in cultivars. The predominant polyphenolic component found in two dill landraces was chlorogenic acid. Also, this study discovered Greek landraces with desired features in terms of quality, production, and harvest time, which might be used in breeding programmes to create new dill cultivars with superior attributes.

4.14.1 Genetic Improvement Objectives Dill

Greek dill landraces' secondary metabolite diversity and phenotypic analysis might yield useful data for further breeding initiatives. To discover and characterize genetic diversity within and among aromatic and medicinal plant species, molecular approaches have been widely used (Solouki et al. 2008). DNA-based markers can be used to evaluate the genetic diversity and population structure of crop germplasm (Suresh et al. 2013). No survey that looks at the genetic, phenotypic, and chemical diversity of dill landraces in the literature is known to them.

Because of improper growing methods, the degradation of plant ecosystems, and the indiscriminate and illegal gathering of plants from numerous locations, the extinction of many medicinal plants is a serious issue. For application in the food and pharmaceutical industries, anethum seeds are shipped to Europe. The availability of its herb oil and seeds is necessary for the pickle and fragrance businesses, as well as aromatherapy. In their material assessment of France, the United Kingdom, the Netherlands, and Germany, the International Trade Centre estimated that less than 300 tonnes of freeze-dried herbs would be required yearly to meet industrial demand. European and worldwide genetic resources are the main topics of the current study.

Accessions from the Nordic germplasm collection were subsequently assessed to monitor trait trends. Both gaps and overlaps in gene bank holdings were discovered by the survey. Wild material, especially crop wild cousins, was woefully underrepresented in international collections outside of Europe. The analysis discovered connections between significant agro-botanical traits including umbel size, plant height, and other size factors. Furthermore, the content of essential oils was impacted by severe heat (Solberg and Yndgaard 2018).

4.15 Germplasm Diversity and Conservation

The best way to preserve the genetic characteristics of threatened and economically important species is through germplasm conservation. For every gene found in a given plant, germplasm serves as a living information source that may be preserved for extended periods of time and renewed as needed in the future. The preservation of genetic diversity among plants also results in the creation of a gene pool that serves as a resource (gene bank or library) for novel or undiscovered species (Bhati 2015).

4.15.1 Conserving Plant Germplasm

The creation of effective conservation measures was made possible using RAPD markers to examine population structure and genetic diversity as well as the integration of data from molecular studies. High polymorphism, Nei's gene diversity, Shannon's information index, total gene diversity values, and UPGMA analysis all contributed to the situation's high levels of genetic diversity within and between populations, leading to a negligible correlation between geographic distance and genetic diversity. The findings demonstrate the need for sizable germplasm collections to accurately represent *A. graveolens* genetic diversity. Additionally, the genetic data gathered during this experiment may offer precise inputs for the creation of a practical and suitable instrument for controlling dill germplasm breeding projects and choosing representative genotypes (Suresh et al. 2013).

4.15.2 In-Situ Conservation

The wide range of plants seen in backyard gardens is significant from both a social and ecological standpoint. The prevention of environmental degradation typically associated with monocultural production systems, sites that generate income (Hoogerbuugge and Fresco 1993), and the in-situ conservation of agrobiodiversity (Watson and Eyzaguirre 2002) are all related to the production of food and other goods like fodders, spices, medicinal plants, firewood, and ornamentals. Home gardens serve as refuges for a variety of plant species, particularly those that are less frequently grown in larger agroecosystems.

4.15.3 *Ex-Situ Preservation*

Overharvesting of biological resources in the wild, altered, fragmented, and lost natural ecosystems, rapid growth of residential and commercial development, pollution, and climate change are the main human hazards to the genetic erosion of plant species. Agriculture and modern plant breeding are also major threats. The threat that modern agriculture and people posed to (plant) agrobiodiversity led to the development of plant introduction centers, which eventually evolved into gene banks, in several different nations beginning at the turn of the twentieth century. These programmes were developed in response to the increasing demand for widespread genetic variety among plant breeders, which is necessary for the creation of well-adapted, high-yielding cultivars with resilience to biotic challenges and tolerance to abiotic pressures (Engels and Ebert 2021).

Since preventing biodiversity loss is strongly linked to ensuring the security of the global food supply, protecting plant genetic resources from genetic erosion has become crucial (Alexandratos 1995). According to Terzopoulos and Bebeli (2010), phenotypic variety needs to be researched in order to make the best use of landraces, which exhibit significant variability and are among the most significant genetic resources (Zeven 2002). The study and use of regional landraces as a starting point for breeding programmes is crucial for plant breeders since it opens new possibilities for outlook agriculture seeking to increase crop yield and flexibility (Ulukan 2011). Recent research has explored the variety of medicinal and aromatic plants utilizing agromorphological features or molecular markers (Patel et al. 2015).

4.15.4 *Cryopreservation*

Young dill plants are in the stage of rapid development at harvest time, and they only have their edible qualities for a brief period. Dill may be preserved using a variety of techniques to increase its longevity and meet demand for the crop beyond the vegetative stage. The idea of employing completely young dill plants as well as their leaves in the dehydration industry was proposed by Kmiecik et al. (2001).

The aim of the work was to compare the contents of nitrates, nitrites, and oxalates in the leafy part of dill and in whole plants (leaves with petioles and stems) of dill harvested at the 25 cm stage of growth. Changes in the levels of these compounds in the technological processes of freezing and refrigerated storage were also determined (Kmiecik et al. 2004) The technological methods of freezing and cooled storage make it possible to measure variations in these compounds' concentrations. The study concentrated on two distinct raw material categories (leafy sections and whole plants), several raw material freezing techniques (blanching or non-blanching of the raw material), various raw material freezing temperatures, and a 12-month storage period. Three months later, the frozen items were inspected. Fresh dill leaves alone had a substantially lower level of nitrates (54%), a higher level of nitrites (albeit less than 1 mg/1000 g), and a higher level of oxalates (26%), compared to whole plants (stems, leaves, and petioles). NO_3^-, NO_2^-, and oxalate concentrations were all dramatically decreased by blanching. Except for oxalates, this had a greater impact on whole plants than just leaves. Apart from intact, unblanched plants, where the level of nitrites dramatically increased compared to the raw material, freezing had no impact on the concentrations of the investigated chemicals in frozen goods. No matter the refrigerated storage temperatures employed, the number of analyzed compounds was somewhat different from what was seen in the sample just after it had been frozen. (Kmiecik et al. 2004).

4.15.5 Cytogenetics

Dill is a common out-crossing species that is diploid (2n = 22), according to Snell and Aarssen (2005). All the examined metaphases included 2n = 22 chromosomes. Capraru et al. (2004) ordered 11 chromosomal pairs from the best metaphase in decreasing order of their overall length. The first chromosome pair total length media were discovered to be 2.96 m and last chromosome pair total length media to be 1.81 m (long arm + short arm + centromere). The homologous chromosomal pair's establishment was verified by the closeness of the variability limits. The difference (La–Sa) between the chromosomal arms was 0.72 (pair I) and 0.03 (pair X) respectively. For the creation of homologues, the arms ratio (La / Sa) was observed to range between 1.00 (pairs III and IX) to 1.66 (pair I). The centromeric index ranged from 61.14 for pair I to 49.62 for pair III. The 11 chromosomal pairs ranged in relative length from 10.80 (pair I) to 6.67 (pair XI). Two morphological kinds, M (pairs III and IX) and m (nine pairs), were identified among the 22 chromosomes of *A. graveolens*.

Eight genotypes of garden dill ('Moravan', 'Kozak', 'Szmaragd', 'Turkus', 'Herkules', 'Lukullus', 'Krezus', 'Ambrozja') were examined in the study by Biesiada et al. (2019). The research demonstrated that genotype influences yield, bioactive component concentration, and antioxidant activity. 'Krezus' and 'Szmarag' produced the highest yields, while 'Herkules' produced the lowest. 'Krezus' had the highest dry matter while 'Ambrozja' had the lowest. The had the lowest levels of both carotenoids and chlorophyll a + b. "Szmaragd" and "Lukullus," respectively. On these colors were concentrated to the greatest extent. "Moravan." Only a small amount of nitrates was accumulated. The 8 cultivars had no effect on the Mg and P contents. 'Moravan' had the highest K content, whereas 'Turkus' had the highest Ca content. The concentration of these macroelements was lowest in the 'Kozak'. 'Lukullus' and 'Ambrozja' have different total oil contents. The sugar concentration of 'Lukullus' was greatest, while 'Herkules' was lowest. While "Szmaragd" and

"Krezus" had the lowest levels of L-ascorbic acid and antioxidant activity, "Turkus" had the highest values. Due to the notable changes in composition, it is possible to select a dill variety for its intended use.

The analysis of various cytogenetic data revealed that caffeine administration of *A. graveolens* resulted in a mild mutagen impact on root apex cells. In addition, cells with aberrations in ana-telophases, specifically ana-telophases with lagging and expulsed chromosomes, as well as in metaphases (metaphases with expulsed chromosomes, C-metaphases, and picnotic chromosomes), showing disruptions of mitotic spindle function after caffeine treatment, accumulate (Maxim et al. 2009).

4.16 Traditional Breeding of *A. graveolens*

In comparison to a few commercial cultivars, Kadoglidou et al. (2023) looked at the physical characteristics, genetic diversity, and chemical composition of dill landraces from a Greek Gene Bank collection. The variety found showed that some landraces may be usefully exploited in breeding efforts to produce cultivars with greater biomass yields, the ability to be harvested sooner than present commercial cultivars, or a specific scent or nutritional quality. Also emphasized was the need of considering all traits, whether they be phytochemical, morphological, agronomic, or genetic, in breeding schemes.

4.16.1 Inbreeding Depression and Selection

Another plant grown in Moldova that is both therapeutic and contains essential oils is dill. Dill oil was shipped each year between 1996 and 2000, when this country had lost the markets for essential oils and the essential oil production plantation had been destroyed. To improve the productivity of this species' cultivation and processing, the Ambassador variety of early-ripening dill has been created. Average yearly yields of 10.5 t/ha in raw materials and 88.9 kg/ha in essential oils with a concentration of 39.8% carvone are used to describe it. In 2004, the variety was registered. According to Gonceariuc and Balmush (2006), the dill cultivars that have been grown up to this point can only generate 50–60 kg/ha of essential oil, with only 29.5% of that quantity coming from carvone. Genotypes with novel traits and features have been created using germ plasma resources from various genetic and geographic origins.

4.16.2 Heritability

Dill is a temperate summer crop that has evolved to grow in warmer regions. It is grown on the Northern Indian plains during the rabi season. During the rabi season, it is grown on the plains of Northern India. Several other characteristics that are also connected to yield have an impact on this intricate trait. Given that many yield-attributing characteristics are quantitatively transmitted and have a significant environmental influence, figuring out whether the observed variability is heritable can be difficult. The study of Raut et al. (2020) set out to measure the variety seen in active dill germplasm and examine the degree to which yield, and yield-contributing traits are correlated. In order to study the genetic variation on yield and yield-related thirteen attributes, including plant height (cm), number of branches, umbels per plant, umbellets per umbel, seed yield per plant (g), seeds per umbel, seeds per umbellets, test weight (g), oil content (%), days to first flowering, and days to 50%, sixteen dill genotypes with checks were evaluated.

Heritability, genotypic and phenotypic coefficients of variation, and genetic progress were all calculated for the entire range. The genotypes demonstrated substantial differences for all traits. In almost every respect, it was found that the genotypes AKDIL-13 and AKDIL-04 were superior. Among the characteristics that showed additive gene action were high genetic and heritability, genetic advancement, and GA as a percentage of mean. High plant height, days until maturity, and number of seeds produced per plant were all considered (Raut et al. 2020).

High heritability traits (% of mean) suggest additive gene effects in their inheritance and can be improved by direct selection, according to Panse and Sukhatme (1954). These characteristics include seed output in both normal and high sodic soil, as well as plant stand in high sodic soil. Although plant standing only showed moderate to higher heredity under high stress, plant height, main branches/plant, and seed output exhibited moderate to higher heritability in both high sodic and non-stress environments. Under normal and moderate sodicity, plant height and secondary branches demonstrated significant heritability but only modest genetic advancement (% mean), demonstrating non-additive gene action of these variables (Gautam et al. 2011).

4.16.3 Genotypic and Phenotypic Correlation

In addition to foliar leaf yields, which differ depending on genotype genetic background, dill is largely farmed for its seeds. For more adaptation and culture, it is therefore highly desirable to choose genotypes with high leaf yield, as well as magnitude of heritability and genetic progress. After the local genotypes were gathered and analyzed, it was discovered that there was a sizable level of genetic variation among these genotypes. The features appeared to be highly heritable in nature, and selected breeding is thus most successful, according to the high heritability and the strong genetic advancement expressed as a percentage of the mean for the economic characteristic. The genotypic relationship between the genotypes was stronger than the phenotypic correlation. To increase the leaf yield of the dill genotype, it would be good to either apply simple selection or mass selection of desirable traits and their associated qualities (Sinhasane et al. 2022).

4.17 Molecular Breeding

According to Ninoua et al. (2017), the diversity of landraces was connected to several characteristics particular to their region of origin, such as early flowering and high stem waxiness in landraces from the Aegean Islands. Greek dill landraces have demonstrated advantageous diversity in yield component traits related to fresh herb weight and seed production, such as many leaves per plant and a big primary umbel diameter, which can be exploited right away in breeding plans.

RAPD markers were chosen as the preferred method for determining genetic diversity over conventional morphological and biochemical markers because they are completely free of interference from environmental factors. Solouki et al. (2012) used the very reliable method of amplified fragment length polymorphism (AFLP) to effectively investigate dill genetic diversity.

4.17.1 Molecular Marker-Assisted Breeding

One of the most important and economically significant seed spices that are planted and grown all throughout India is dill, according to widespread consensus. The majority of earlier research, however, focused on physiological and biochemical studies rather than molecular breeding because of a lack of genetic and genomic resources. The objective of the current effort was to create highly informative microsatellite markers using computational and in silico methods. It was the first time that Dill EST and genomic sequences had been used to make microsatellite markers. 1556 SSRs were found in the 589.390 kb sequence under investigation at a density of 1 SSR/0.38 kb. Overall, 1265 (81.30%) of the SSRs were flawless, but only 291 (18.70%) of these were compound SSRs. Many nucleotides were mononucleotides (1089, 69.98%), which were subsequently accompanied by trinucleotides (254, 16.32%), dinucleotides (183, 11.76%), tetranucleotides (22, 1.41%), hexanucleotides (6, 0.38%), and pentanucleotides (2, 0.13%) (Jethra et al. 2017). In-silico mining of SSRs has aided the traditional laborious and time-consuming procedures by automating the development of SSR markers and categorization for Dill. It is the first time that Dill EST and genomic SSR sequences have been used to make SSRs and microsatellite markers. 1556 SSRs were found in 589.390 kb sequence searches at a density of 1 SSR/0.38 kb.

4.17.2 Genomic Resources

Numerous antioxidants can be found in dill seeds, leaves, and essential oil. Using random amplified polymorphic DNA (RAPD) markers, this study examined the genetic diversity structure of 135 *A. graveolens* accessions from several continents.

In all the accessions having Nei's genetic diversity (H) values, the genetic diversity was high. For primer OPB18, it was discovered that the values of H and I were the highest. The 135 RAPD accessions were divided into two main groups following a cluster analysis using the UPGMA approach based on Nei's genetic similarity matrix. Even though multiple clusters can be found, the relationships between the clusters and the location of the material are tenuous at best (Suresh et al. 2013).

4.17.3 Tissue Culture and Applications

The significance of *A. graveolens* secondary metabolites as a medicinal plant was emphasized by Jana and Shekhawat (2010). The tissue culture methods employed for *A. graveolens*' propagation and preservation may aid in its improvement and proliferation (Sharma et al. 2004). Due to the commercial significance of these substances as flavorings, fragrances, and medications, several attempts have been made to boost monoterpene synthesis in plants using in vitro technologies Additionally, suspension culture may be utilized to evaluate several physiological and biochemical characteristics of this economically significant plant species. This can be done by employing RDT, bioprocess technologies, plant cell and tissue culture techniques, and suspension culture. Humanity has had a huge potential to enhance tissue culture methods, RDT, bioprocess technologies, and other sectors by utilizing the totipotent biosynthetic and biotransformation capacities of plant cells in vitro. These tools have provided people with a significant opportunity when used in conjunction with the most sophisticated analytical tools, including as NMR, HPLC, GC-MS, and LC-MS (Stockigt et al. 1985). Techniques using plant cells and tissues provide an alternate source of research materials, particularly for investigations involving development and metabolism that would be challenging to conduct in entire plants.

Santos et al. (2002), reported that *Agrobacterium rhizogenes* carrying the plasmid pRi 1855 was inoculated into aseptically grown seedlings to induce transformed root cultures of *A. graveolens*. The parent plant's fruits and roots had the highest concentrations of carvone, whilst the aerial portions and hairy roots had higher concentrations of α-phellandrene and apiole, respectively. The parent plant's fruits, aerial parts, and roots contained 2%, 0.3%, and 0.06% (v/w) of the essential oils, whereas cultures of hairy roots only contained 0.02% (v/w). After 50 days, the hairy root cultures had grown to 600 mg dry weight/50 ml media. The essential oils' chemical makeup didn't change appreciably as they were being developed.

An effective in vitro regeneration method was developed for the aromatic plant *A. graveolens*, which is significant in medicine. To create nodal segments, a Murashige and Skoog (MS) basal was employed. The best shoot regeneration (85.7%) was attained when callus was subcultured on MS medium for two weeks with 12.86 shoots per explant. The length of the regenerated shoots, on average, ranged from 3.15 to 4.8 cm. On 14 MS enhanced, rooting of the regeneration shoots was greater than 100%. Plantlets that were successfully acclimated survived in 60% of cases. During organogenesis, the activities of catalase and ascorbate peroxidase rose whereas those of superoxide dismutase fell. Using ten distinct decamer primers and randomly amplified polymorphic DNA, the clonal fidelity of in vitro produced plants was examined. According to predictions, plant regeneration has been observed (Jana and Shekhawat 2012).

4.18 Gene Editing and Genetic Engineering

Abiotic stressors, such as heavy metals and drought, significantly lower agricultural productivity worldwide. Transgenic technologies and conventional breeding methods have been extensively employed to reduce the hazards associated with these environmental pressures. Thanks to its ease of use, versatility, and broad range of applications, the CRISPR/Cas genome-editing technology has completely changed the gene editing landscape. The development of crop varieties with improved resistance to abiotic stressors is a promising use of this technology (Kumar et al. 2023).

CRISPR-Cas9 (clustered regularly interspaced short palindromic repeats/CRISPR-associated protein 9) is currently being used in plant systems (Venegas-Rioseco et al. 2022). Increasing crop resilience to biotic and abiotic stress is a simple and affordable process (Pandita 2021). It is composed of spacer sequences, which are translated into CRISPR RNA (crRNA). CRISPR RNA then interacts with trans-activating crRNA (tracrRNA) to form a mature crRNA/tracrRNA complex, or guide RNA or gRNA. When the Cas nuclease is instructed by the gRNA to produce a DNA double-strand break (DSB) in the target DNA sequence, the target gene is eliminated by cell repair procedures (Venegas-Rioseco et al. 2022). As a result, gRNA-guided-Cas9 systems may employ gene deletion, gene expression modulation, and transcription factor regulation to enhance the tolerance of a variety of plants to heavy metal stress and phytoremediation (Bao et al. 2019). Transcriptomics can be used to identify significant genes involved in metal stress tolerance in phytoremediator plants like *Arabidopsis halleri*, *Hirschfeldia incana, Noccaea caerulescens*, and *Pteris vittata*, which are suitable for producing the best hyperaccumulator plants (Kumar and Trivedi 2018). The development of phytoextraction technology will also be aided by contemporary gene editing techniques like CRISPR-Cas9 (Thakur et al. 2020). It is feasible to alter the expression of a specific gene or group of related genes of interest by combining transcription factors with dCas9 (Miglani 2017). Without introducing new genes, CRISPR-Cas9 is an efficient way to improve a plant's innate ability to grow, accumulate, and endure heavy metal stress.

Insufficient N availability is one of the key factors limiting plant growth and development, and adequate agricultural yields require considerable dosages of N fertilizers, claim Lebedev et al. (2021). However, the majority of N is not utilized by plants, and as a result, it pollutes the environment. By improving plants' nitrogen utilization efficiency (NUE), this situation can be changed. Understanding the important plant N metabolism phases of uptake, absorption, and remobilization is essential for optimizing NUE interactions, which can be impacted by a variety of genetic and environmental factors. Although RNAi and genome editing technologies were also

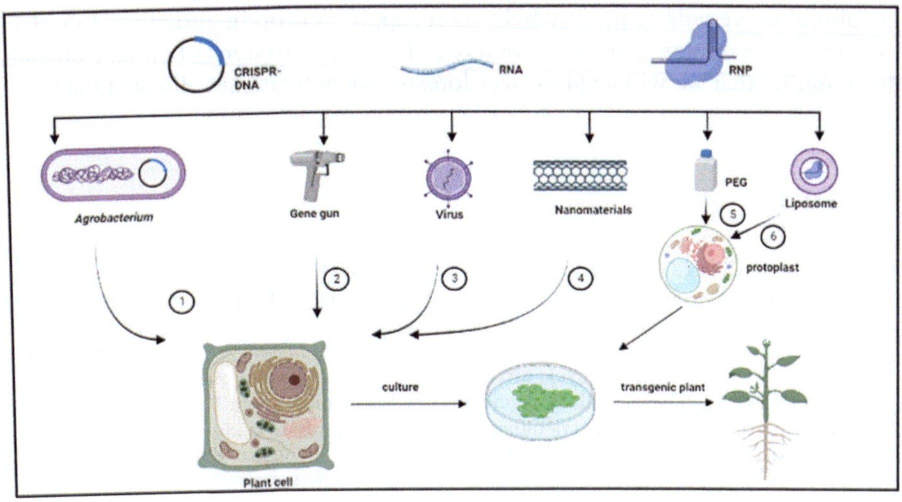

Fig. 4.9 CRISPR/Cas system delivery technique, using CRISPR/Cas9 to transmit foreign recombinant DNA. Agrobacterium tumefaciens, a liposome conveying the targeted gene into the plant cell, PEG, nanoparticles, viral vector, and gene gun are shown from left to right. The plant cell is then cultivated (Min et al. 2022)

used, overexpression methods were more frequently used to alter the expression of structural and regulatory genes. Due to their significance in commerce, critical role in ecosystems, and fundamental distinctions from herbaceous species, woody plants have garnered a lot of research. The article also covers unexpected effects of transgenic plants with altered N metabolism, like early flowering—a research area that hasn't gotten much attention up to this point. The prospects for improving NUE in crops, which is essential for the expansion of sustainable agriculture, are explored in the context of global climate change and utilizing a variety of approaches (Hirel et al. 2011) (Fig. 4.9).

4.19 Conclusion and Prospects

The creation of pathogen-resistant cultivars and genetic preservation of diverse landraces, especially dill (*A. graveolens*), one of the most significant herb crops worldwide, should receive a lot of attention. Each country should have a gene bank that works with the agriculture ministry to develop and test innovative genetic engineering methods for enhancing crop productivity. Karyotype, as well as genetic mapping to develop quantitative trait loci (QTL), should be used to advance qualitative and quantitative traits. According to the analysis, individuals already use roughly half of the plant arrive for nutrition generation, and as global population levels rise, agricultural arrive will be in extremely short supply. A drop in agricultural productivity across the tropics is often one or more of the repercussions of climate change, which implies that we will need to clear forests and cultivate uncultivated land.

References

Abdalkarim SYH, Chen LM, Yu HY, Li F, Chen X, Zhou Y, Tam KC (2021) Versatile nanocellulose-based nanohybrids: a promising-new class for active packaging applications. Int J Biol Macromol 182:1915–1930

Al-Sheddi ES, Al-Zaid NA, Al-Oqail MM, Al-Massarani SM, El-Gamal AA, Farshori NN (2019) Evaluation of cytotoxicity, cell cycle arrest and apoptosis induced by *Anethum graveolens* L. essential oil in human hepatocellular carcinoma cell line. Saudi Pharm J 27(7):1053–1060

Alexandratos N (ed) (1995) World agriculture: towards 2010: an FAO study

APG (1998) An ordinal classification for the families of flowering plants. Ann Missouri Bot Gard 85:531–553

APG (2003) An update of the angiosperm phylogeny group classification for the orders and families of flowering plants: APG II. Bot J Linn Soc 141:399–436

APG (2009) An update of the angiosperm phylogeny group classification for the orders and families of flowering plants: APG III. Bot J Linn Soc 161:105–121

APG (2016) An update of the angiosperm phylogeny group classification for the orders and families of flowering plants: APG IV" (PDF), Bot J Linn Soc 181(1):1–20. https://doi.org/10.1111/boj.12385. Retrieved 11 June 2016

Bao A, Burritt DJ, Chen H, Zhou X, Cao D, Tran L-S P (2019) The CRISPR/Cas9 system and its applications in crop genome editing. Crit Rev Biotechnol 39:321–336. https://doi.org/10.1080/07388551.2018.1554621

Benlembarek K, Lograda T, Ramdani M, Figueredo G, Chalard P (2022) Chemical composition and biological activities of *Anethum graveolens* L. essential oil from Algeria. J Essen Oil Bear Plants 25(4):728–740

Bentham G (1867) Umbelliferae. In: Bentham G, Hooker D (eds) Genera Plantaram, vol 1. Reeve & Co, London, pp 859–931

Bhatia S (2015) Application of plant biotechnology. In: Modern applications of plant biotechnology in pharmaceutical sciences, pp 157–207 (Chapter 5)

Biesiada A, Kedra K, Godlewska K, Szumny A, Nawirska-Olszanska A (2019) Nutritional value of garden dill (*Anethum graveolens* L.), depending on genotype. Not Bot Horti Agrobo Cluj-Napoca 47(3):784–791. https://doi.org/10.15835/nbha47311395

Boeckmann C (2023) Yankee Publishing, Inc., An Employee-Owned Company 1121 Main Street, P.O. Box 520, Dublin, NH 03444

Boulos L (1983) Medicinal plants of North Africa. Medicinal plants of North Africa

Boulos L (2000) Flora of Egypt (Geraniaceae-Boraginaceae), vol II. Al-Hadara Publishing, Cairo, Egypt, pp 107–109

Capraru G, Cimpeanu M, Cimpeanu C, Toma C (2004) Comparative karyotype analysis in members of apiaceae (Umbelliferae) family. In: Proceedings from the 3rd CMAPSEEC. ISBN 86-909717-0-X

Carrubba A, La Torre R, Saiano F, Aiello P (2008) Sustainable production of fennel and dill by intercropping. Agron Sustain Dev 28:247–256

Cerceau-Larrival MT (1971) Morphologie pollinique et corrèlations phylogènetiques chez les Ombellifères. In: Heywood VH (ed) The biology and chemistry of the umbelliferae. J Linn Soc Bot Suppl 109–156

Cronquist A (1968) The evolution and classification of flowering plants. Houghton-Mifflin Company, Boston

De Candolle AP (1813) Théorie élémentaire de la botanique. Paris
De Candolle AP (1830) Umbelliferae. De Candolle, prodromus systematis naturalis regni vegetabilis, vol 4, pp 55–250
De Guzman CC, Siemonsma JS (1999) Spices. PROSEA. Backhuis Publishers, Leiden, 400pp
Delaquis PJ, Stanich K, Girard B, Mazza G (2002) Antimicrobial activity of individual and mixed fractions of dill, cilantro, coriander and eucalyptus essential oils. Int J Food Microbiol 74(1–2):101–109
El-Zaeddi H, Martínez-Tomé J, Calín-Sánchez Á, Burló F, Carbonell-Barrachina ÁA (2016) Volatile composition of essential oils from different aromatic herbs grown in Mediterranean regions of Spain. Foods 5(2):41
Engels JM, Ebert AW (2021) A critical review of the current global ex situ conservation system for plant agrobiodiversity. I. History of the development of the global system in the context of the political/legal framework and its major conservation components. Plants 10(8):1557
Erdtman G (1952a) Pollen morphology and plant taxonomy: angiosperm. Almquist and Wiksell, Stockholm
Eyde RH, Tseng CC (1971) What is the primitive floral structure of Araliaceae? J Arnold Arboretum 52:205–239
Finch-Savage WE (1995) Influence of seed quality on crop establishment, growth and yield. In: Basra AS (ed) Seed quality: basic mechanisms and agricultural implications. Haworth Press, New York, USA, pp 361–384
Finch-Savage WE, Bassel GW (2016) Seed vigour and crop establishment: extending performance beyond adaptation. J Exp Bot 67(3):567–591
França-Neto JB, Krzyzanowski FC, Henning AA, Costa NP (2000) Tecnologia de produção de sementes. In: A cultura da soja no Brasil. Embrapa Soja. 1CD-ROM, Londrina
Garcez JJ, Barros F, Lucas AM, Xavier VB, Fianco AL, Cassel E, Vargas RMF (2017) Evaluation and mathematical modeling of processing variables for a supercritical fluid extraction of aromatic compounds from *Anethum graveolens*. Ind Crops Prod 95:733–741
Gautam RK, Chauhan MS, Panesar B, Nayak AK, Sharma DK, Qadar A, Singh G (2011) Stress tolerance, heritability in dill (*Anethum graveolens*) lines from central and western India under contrasting soil sodicity environments. Indian J Agric Sci 81(4):353
Ghassemi–Golezani K (1992) Effects of seed quality on cereal yields. PhD Thesis, University of Reading, UK, pp 205–222
Ghassemi-Golezani K, Zehtab-Salmasi S, Dastborhan S (2011) Changes in essential oil content of dill (*Anethum graveolens*) organs under salinity stress. J Med Plants Res 5(14):3142–3145
Ghassemi-Golezani K, Rezaeipoor L, Dalil B (2018) Changes in seed quality of dill (*Anethum graveolens*) ecotypes in response to water limitation and harvest time. J Plant Physiol Breed 8(2):99–107
Goldberg A (1986) Classification, evolution and phylogeny of the families of dicotyledons. Smithson Contrib Bot 58:1–314
Gonceariuc M, Balmush Z (2006) Studies of genetics and breeding of aromatic and medicinal plants carried out in the Republic of Moldova. In: Proceedings 4th conference on medicinal and aromatic plants of South-East European Countries, pp 112–116
Goodarzi MT, Khodadadi I, Tavilani H, Abbasi Oshaghi E (2016) The role of *Anethum graveolens* L. (Dill) in the management of diabetes. J Trop Med
Gunalan G, Rathinamala R, Kumar AR (2020) Gas chromatography-mass spectroscopy fingerprinting and in silico molecular docking analysis of secondary metabolites from *Anethum graveolens* L. seeds for anti-inflammatory activity. Drug Invent Today 13(5)
Guyot M (1971) Phylogenetic and systematic value of stomata of the umbelliferae. In: Heywood VH (ed) The biology and chemistry of the umbelliferae. Academic Press, London, UK, pp 199–214
Hirel B, Tétu T, Lea PJ, Dubois F (2011) Improving nitrogen use efficiency in crops for sustainable agriculture. Sustainability 3:1452–1485. https://doi.org/10.3390/su3091452
Hosni HA (1988) A conspectus of Tribulus in Arabia. Taceckholmia 11:1–18

Hutchinson J (1967) The Genera of flowering plants, vol 2. Oxford University Press, Oxford. Hydrocotyloideae. Pollen et spores 6:479–514
Jana S, Shekhawat GS (2010) Anethum graveolens: an Indian traditional medicinal herb and spice. Pharmacogn Rev 4(8):179
Jana S, Shekhawat GS (2012) In vitro regeneration of *Anethum graveolens*, antioxidative enzymes-during organogenesis and RAPD analysis for clonal fidelity. Biol Plant 56(1):9–14. https://doi.org/10.1007/s10535-012-0009-2
Jethra G, Choudhary S, Sharma V (2017) Development of new set of EST-SSR markers for evaluating genetic diversity and fingerprinting local celery (*Apium graveolens* L.) cultivars. Curr Biomark 7:52–63 (Print)
Kadoglidou K, Cook C, Boutsika A, Sarrou E, Mellidou I, Aidonidou C, Xanthopoulou A (2023) Evaluation of a dill (*Anethum graveolens* L.) gene bank germplasm collection using multivariate analysis of morphological traits, molecular genotyping and chemical composition to identify novel genotypes for plant breeding. PeerJ 11:e15043
Kmiecik W, Lisiewska Z, Jaworska G (2001) Effect of storage conditions on the technological value of dill (*Anethum graveolens* L.). Folia Hortic 13:33–43
Kmiecik W, Lisiewska Z, Słupski J (2004) Effects of freezing and storing of frozen products on the content of nitrates, nitrites, and oxalates in dill (*Anethum graveolens* L.). Food Chem 86(1):105–111
Kumar S, Trivedi PK (2018) Glutathione S-transferases: role in combating abiotic stresses including arsenic detoxification in plants. Front Plant Sci 9:751
Kumar M, Prusty MR, Pandey MK, Singh PK, Bohra A, Guo B, Varshney RK (2023) Application of CRISPR/Cas9-mediated gene editing for abiotic stress management in crop plants. Front Plant Sci 14:1157678. https://doi.org/10.3389/fpls.2023.1157678
Lebedev VG, Popova AA, Shestibratov KA (2021) Genetic engineering and genome editing for improving nitrogen use efficiency in plants. Cells 10(12):3303
Linneaeus C (1735) Systema naturae. Leiden
Linneaeus C (1753) Species plantarum. Stockholm
Malhotra SK (2007) Response of Indian dill (*Anethum sowa*) and European dill (*Anethum graveolens*) varieties to different agro-techniques. Indian J Agric Sci 77(8):519–522
Maxim E, Capraru G, Campeanu MM (2009) Some citogenetic effects of sodium azide treatments in caraway root meristems. J Exp Mol Biol 10(1):9
Miglani GS (2017) Genome editing in crop improvement: present scenario and future prospects. J Crop Improv 31(4):453–559
Min T, Hwarari D, Li D, Movahedi A, Yang L (2022) CRISPR-based genome editing and its applications in Woody plants. Int J Mol Sci 23:10175. https://doi.org/10.3390/ijms231710175
Mirhosseini M, Baradaran A, Rafieian-Kopaei M (2014) Anethum graveolens and hyperlipidemia: a randomized clinical trial. J Res Med Sci 19(8):758
Morison R (1672) Plantarum umbelliferarum distribution nova. Oxford, 99 pp
Naeem M, Khan MN, Khan MMA, Moinuddin (2013) Adverse effects of abiotic stresses on medicinal and aromatic plants and their alleviation by calcium. In: Plant acclimation to environmental stress, pp 101–146
Najaran ZT, Hassanzadeh MK, Nasery M, Emami SA (2016) Dill (*Anethum graveolens* L.) oils. In: Essential oils in food preservation, flavor and safety. Academic Press, pp 405–412
Ninou EG, Mylonas IG, Tsivelikas AL, Ralli PE (2017) Phenotypic diversity of Greek dill (*Anethum graveolens* L.) landraces. Acta Agric Scand Sect B Soil Plant Sci 67(4):318–325. https://doi.org/10.1080/09064710.2016.1276957
Nyman C (1878) Umbellatae. In: Conspectus Florae Europaea Plantarum phanerogamarum, Örebro, pp 276–319
Oshaghi EA, Tavilani H, Khodadadi I, Goodarzi MT (2015) Dill tablet: a potential antioxidant and anti-diabetic medicine. Asian Pac J Trop Biomed 5(9):720–727
Pandita D (2021) Cas9 technology: an innovative approach to enhance phytoremediation. In: Plant-microbe dynamics: recent advances for sustainable agriculture. CRC Press, pp 221–248

Panse VG, Sukhatme PV (1954) Statistical methods for agricultural workers. In: Statistical methods for agricultural workers

Patel RP, Kumar RR, Singh R, Singh RR, Rao BRR, Singh VR, Lal RK (2015) Study of genetic variability pattern and their possibility of exploitation in *Ocimum germplasm.* Ind Crops Prod 66:119–122

Rana VS, Blazquez MA (2014) Chemical composition of the essential oil of *Anethum graveolens* aerial parts. J Essent Oil Bear Plants 17(6):1219–1223

Rathore V (2020) Correlation study between growth parameters, yield and oil quality of *Anethum graveolens* Linn. Indian J Pharm Sci 82(5):773–780

Raut AU, Ghawade SM, Mali VV, Dahatonde KA, Sawant BS (2020) Genetic variability, heritability and genetic advance in Dill (*Anethum graveolens* L.) genotypes. Int J Curr Microbiol Appl Sci 11:2293–2298

Said-Al Ahl H, Sarhan A, Abou Dahab AD, Abou Zeid ES, Ali M, Naguib N (2015) Flavonoids, essential oil and its constituents of *Anethum graveolens* L: herb affected by nitrogen and bio-Fertilizers. Agric Biol Sci J 1(3):105–109

Santos PA, Figueiredo AC, Lourenço PM, Barroso JG, Pedro LG, Oliveira MM, Scheffer JJ (2002) Hairy root cultures of *Anethum graveolens* (dill): establishment, growth, time-course study of their essential oil and its comparison with parent plant oils. Biotech Lett 24:1031–1036

Setorki M, Shahinfard N, Ansari R, Forouzandeh Z, Asgharzadeh S, Rafieian KM (2013) Comparison between the effect of hydroalchohlic extract of dill and statins on lipid profile. J Kerman Univ Med Sci 20(2):158–168

Sharma RK, Wakhlu AK, Boleria M (2004) Micropropagation of *Anethum graveolens* L. through axillary shoot proliferation. J Plant Biochem Biotechnol 13:157–159

Sinhasane SR, Shinde US, Dhumal SS, Aher AR, Manjunathagowda DC, Benke AP (2022) Understanding of genetic variables for growth and yield traits of dill (*Anethum graveolens* L.). Genet Resour Crop Evol 69(7):2575–2584. https://doi.org/10.1007/s10722-022-01409-3

Small E, Deutsch G (2001) Culinary herbs for short-season gardeners. NRC Research Press

Snell R, Aarssen LW (2005) Life history traits in selfing versus outcrossing annuals: exploring the 'time-limitation' hypothesis for the fitness benefit of self-pollination. BMC Ecol 5(1):1–14

Solberg SØ, Yndgaard F (2018) Trait patterns and genetic resources of dill (*Anethum graveolens* L.). ASM Handb 15:468–481

Solouki M, Hoseini SB, Fakheri BA, Tavassoli A (2012) Genetic diversity in dill (*Anethum graveolens* L.) populations on the basis of morphological traits and molecular markers. Afr J Biotechnol 11(15):3649–3655. https://doi.org/10.5897/AJB11.2169

Solouki M, Mehdikhani H, Zeinali H, Emamjomeh AA (2008) Study of genetic diversity in Chamomile (*Matricaria chamomilla*) based on morphological traits and molecular markers. Sci Hortic 117(3):281–287

Sprengel K (1820) Umbelliferae. In: Römer JJ, Schultes JA (eds) Systemata vegetabilium, vol 6, pp 315–628

Stöckigt J, Obitz P, Falkenhagen H, Lutterbach R, Endress S (1995) Natural products and enzymes from plant cell cultures. Plant Cell Tissue Organ Cult 43:97–109

Suresh S, Chung JW, Sung JS, Cho GT, Park JH, Yoon MS, Baek HJ (2013) Analysis of genetic diversity and population structure of 135 dill (*Anethum graveolens* L.) accessions using RAPD markers. Genet Resour Crop Evol 60:893–903

Takhtajan A (1969) Flowering plants. Origin and dispersal (trans. C. Jeffrey). Oliver & Boyd, Edinburgh

Tanveer M, Richhariya B, Khan RU, Rashid AH, Khanna P, Prasad M, Lin CT (2020) Machine learning techniques for the diagnosis of Alzheimer's disease: a review. ACM Trans Multimed Comput Commun Appl (TOMM) 16(1s):1–35

Terzopoulos PJ, Bebeli PJ (2010) Phenotypic diversity in Greek tomato (*Solanum lycopersicum* L.) landraces. Sci Hortic 126(2):138–144

Thakur S, Choudhary S, Majeed A, Singh A, Bhardwaj P (2020) Insights into the molecular mechanism of arsenic phytoremediation. J Plant Growth Regul 39:532–543

Thorne RF (1968) Synopsis of a putatively phylogenetic classification of the flowering plants. Aliso 6(4):57–66

Thorne RF (1972) Inclusion of the Apiaceae (Umbelliferae) in the Araliaceae. Notes from the royal botanical cardens. Edinburgh 32:151–159

Tournefourt JP (1694) Element de Botamque, vol 3. Paris

Ulukan H (2011) The use of plant genetic resources and biodiversity in classical plant breeding. Acta Agric Scand Sect B Soil Plant Sci 61(2):97–104

Venegas-Rioseco J, Ginocchio R, Ortiz-Calderón C (2022) Increase in phytoextraction potential by genome editing and transformation: a review. Plants 11(1):86

Wander JGN, Bouwmeester HJ (1998) Effects of nitrogen fertilization on dill (*Anethum graveolens* L.) seed and carvone production. Ind Crops Prod 7(2–3):211–216

Wanner H, Gu H, Günther D, Hein S, Dorn S (2006) Tracing spatial distribution of parasitism in fields with flowering plant strips using stable isotope marking. Biol Control 39(2):240–247. https://doi.org/10.1016/j.biocontrol.2006.06.001

Watson JW, Eyzaguirre PB eds (2002) Homegardens and in situ conservation of plant genetic resources in farming systems. In: Proceedings of the second international home gardens workshop, 17–19 July 2001, Witzenhausen, Federal Republic of Germany. International Plant Genetic Resources Institute (IPGRI), Rome, Italy, pp105–124

Yazdanparast R, Bahramikia S (2008) Evaluation of the effect of *Anethum graveolens* L. crude extracts on serum lipids and lipoproteins profiles in hypercholesterolaemic rats. DARU J Pharm Sci 16(2):88–94

Yazdanparast R, Alavi M (2001) Antihyperlipidaemic and antihypercholesterolaemic effects of Anethum graveolens leaves after the removal of furocoumarins. Cytobios 105(410):185–191

Zeven AC (2002) Traditional maintenance breeding of landraces: 2. Practical and theoretical considerations on maintenance of variation of landraces by farmers and gardeners. Euphytica 123:147–158

Chapter 5
Advanced Breeding Strategies of Lemon Grass (*Cymbopogon citratus*)

Chole Pranjali Bajrang, Wudali Narasimha Sudheer, Akshatha Banadka, V. L. Vasantha, Praveen Nagella, and B. T. Manjunath

Abstract *Cymbopogon citratus* belongs to the Poaceae (Gramineae) family and is distributed worldwide in tropical and subtropical countries. When the green leaves are crushed, it gives a lemony aroma hence known as lemongrass. It is cultivated for its medicinally crucial essential oil. Worldwide, lemongrass is cultivated in over 16,000 ha, producing around 1000 tons of oil per year. The oil is rich in citrus and is a good source of secondary metabolites, which shows various pharmacological activities like antibacterial, antifungal, anti-diarrheal, and anti-inflammatory properties. Other than this, properties like anti-malarial, anti-oxidants, hypoglycaemic, anti-mutagenicity, anti-mycobacterial, and neurobehavioral have also been studied. The traditional breeding technique is used to develop new varieties, but as it is time-consuming and the rate of success is not high, advanced breeding techniques have gained importance. Advanced breeding techniques such as plant tissue culture and genetic engineering helped achieve a high yield of oil and genetically improved varieties in a short time. With this, the current study helps explore different studies related to modern breeding programs, genetic variability, and germplasm. This chapter delves into the cultivation of lemongrass, providing in-depth guidance on techniques aimed at optimizing both yield and quality. It covers essential aspects such as ideal climate conditions, soil preparation, and methods for propagation, whether through seeds or root division. Furthermore, the chapter introduces advanced breeding strategies like selection, hybridization, and tissue culture, presenting promising avenues for the advancement of superior lemongrass varieties.

Keywords *Cymbopogon citratus* · Pharmacological activities · Traditional breeding · Advanced breeding · Plant tissue culture

---

C. P. Bajrang · W. N. Sudheer · A. Banadka · V. L. Vasantha · P. Nagella · B. T. Manjunath (✉)
Department of Life Sciences, School of Sciences, CHRIST (Deemed to be University), Bengaluru, Karnataka 560029, India
e-mail: manjunath.bt@christuniversity.in

J. M. Al-Khayri et al. (eds.), *Biodiversity and Genetic Improvement of Herbs and Spices*, Advances in Plant Breeding Strategies 8, https://doi.org/10.1007/978-3-031-81838-7_5

5.1 Introduction

Cymbopogon citratus is commonly known as lemon grass. Word Cymbopogon is derived from two Greek words, 'kymbe' (boat) and 'pogon' (beard), which refer to flower arrangement (Shah et al. 2011). Lemon grass is known for its aromatic oil worldwide. It is a perennial plant belonging to the Poaceae (Gramineae) family. The Gramineae family includes about 500 genera and 8000 species of plants (Magotra et al. 2021). When the green leaves are crushed, it gives a citrus/lemony aroma, commonly known as lemongrass. In English, it is also called 'Citronella' or 'Squinant' (Haque et al. 2018). The aromatic oil present in grass shoots also provides a citrus aroma. Cochin port exports about 90% of its oil; therefore, commercially lemongrass oil is known as Cochin oil (Joy et al. 2006). Worldwide, lemongrass is cultivated in over 16,000 ha, producing around 1000 tonnes of oil per year. In India, the annual production of essential oil is 250 tonnes over 4000 ha of cultivated area (Joy et al. 2006). The well-grown root system of lemongrass helps in water and soil conservation.

5.1.1 *Origin and Distribution of Lemongrass*

Lemon grass is distributed worldwide in tropical and subtropical countries like Asia, Australia, South America, North America, and Europe as shown in Fig. 5.1 (Haque et al. 2018). Some reports linked its origin to America, Africa, and Asia (Malaysia, Indochina, Indonesia) (Olorunnisola et al. 2014). In India, it is found in all regions, from an altitude of 4200 m to extending sea level. It is commonly cultivated in Kerala, Maharashtra, Assam, Uttar Pradesh, Madhya Pradesh, and Sikkim. On a large scale, lemongrass is cultivated at Chinnar Wildlife Sanctuary in Western Ghats (Joy et al. 2006).

5.1.2 *Taxonomical Classification*

Kingdom	Plantae
Division	Magnoliophyta
Class	Liliopsida
Order	Poales
Family	Poaceae
Genus	*Cymbopogon* Spreng
Species	*Citratus*

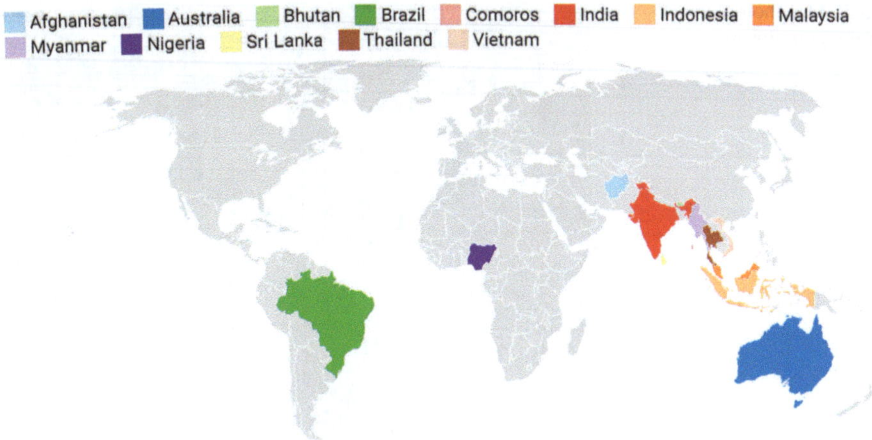

Fig. 5.1 Geographical distribution of *Cymbopogon citratus* across diverse continents and climatic zones

5.1.3 Plant Description

Lemon grass is a perennial and clumped grass. With a short rhizome, it grows up to 2–2.5 m in height. Leaves of lemongrass are non-articulate, glabrous, linear, and venated with an acute apex and narrow base (Fig. 5.2) (Kamaruddin et al. 2021). Lemongrass produces flowers at its mature stage of growth. Due to quick harvesting, flowering has never been observed in lemongrass cultivation. Flower panicles are found rarely in lemongrass (Manvitha and Bidya 2014). The inflorescence is around 30–60 cm long, and long spikes and nodding are also present (Sumeet Magotra et al., 2021). Different sources like leaves, fiber, and essential oil from the plant are used as medicine. It is an aromatic plant known for its citrus aroma.

5.1.4 Species and Varieties of Lemongrass

Genus *Cymbopogon* belongs to the Poaceae (Gramineae) family. Genus *Cymbopogon* includes around 55 species in various countries across the world. Table 5.1 lists a few species included in the genus *Cymbopogon*. Different species show different chemical composition and properties.

Most common three species are found in India (Joy et al. 2006);

(1) ***Cymbopogon flexuosus***: This species is known as Malabar grass, Cochin, or East Indian grass. The leaves are lanceolate and linear.

(2) ***Cymbopogon citratus***: This species is known as American grass or West Indian grass. It contains about 75–80% citral in essential oil and shows poor solubility in alcohol.

Fig. 5.2 Morphology of *Cymbopogon citratus*

Table 5.1 List of a few species included in the genus *Cymbopogon*

Cymbopogon species	Origin country
Cymbopogon annamensis	Thailand, Vietnam, Yunnan
Cymbopogon bombbycinus	Australia
Cymbopogon bhutanicus	Bhutan
Cymbopogon calciphilus	Thailand
Cymbopogon caesius	Africa, Afghanistan, Comoros, Madagascar, Sub-Saharan Indian subcontinent, Yemen
Cymbopogon citratus	Southeast Asia, Northeast, and Southern India, Sri Lanka, Nigeria
Cymbopogon coloratus	Vietnam, Myanmar
Cymbopogon dependens	Australia
Cymbopogon densiflorus	Central and south-central Africa

Source Kew world checklist of selected plant families

(3) ***Cymbopogon pendulus***: It is a dwarf, white stemmed commonly known as Jammu grass. It shows medium solubility in alcohol.

5.1.5 Traditional Uses of Lemongrass

Leaves of lemongrass are highly known for their essential oil. The essential oil shows significant activities against antibacterial, antifungal, antidepressant, carminative, fungicidal, astringent, and antiseptic properties (Joy et al. 2006). Due to its antiseptic and antibiotic properties, it is used to cure athletes' foot disease and ringworm. Most of the time, it is used to treat gastrointestinal disorders. Lemongrass tea is used as a

Table 5.2 Traditional uses of *Cymbopogon citratus*

Country	Traditional uses	References
Asia	Lemongrass leaf decoction is consumed for sore throat and empacho	(Filipov 1994)
Brazil	Leaf tea is commonly used as an anti-inflammatory, diuretic, antipyretic, and seductive	(Calil et al. 1986)
India	The entire fresh plant is used as a snake repellent. To overcome gastric problems, 2–3 drops of essential oil in hot water are added and consumed. Lemongrass tea is used as a sedative for the central nervous system	(Shah et al. 2011)
USA	Hot water extract of the entire plant helps bone fracture and wound healing	(Spring 1989)
Thailand	Dried root hot water extract is taken for diabetes. Fresh leaves are used as condiments	(Shah et al. 2011)
Egypt	Dried leaves and stem hot water extract is taken as a renal antispasmodic and diuretic	(Shah et al. 2011)

home remedy for fever, illness, and pneumonia (Magotra et al. 2021). The traditional uses of *Cymbopogon citratus* with country is mentioned in Table 5.2.

5.1.6 *Phytochemistry and Pharmacological Properties of Lemongrass*

A primary and secondary bioactive component derived from the roots, leaves, and stems of *Cymbopogon citratus* shows various pharmacological activities. These chemical compositions vary according to the extraction method, geographical origin, harvesting season, plant part used, genetic difference, and plant age. Factors like water content and soil salinity (Agro-climatic elements) could also influence oil quality, growth, and yield from lemongrass (Idrees et al. 2012). Alkaloids (octacosanol, 1.8–cineole and menthol, dotriacontanol, triacontanol, Methaheptanol, isolated hexacosanol, nerolic acid, ester laurate, geranic acids), flavonoids (apigenin, chlorogenic acid, cynaroside, caffeic acid, hydroquinone, kaempferol, luteolin, orientin, O-rhamnosyl isorientin, quercetin, isoscoparin, swetiajaponin), steroids (β-sitosterol, fucosterol, tracontanol, proanthocyanidin), sugar and ketone has identified from the root and leaves extracts of *C. citratus* (Negrelle and Gomes 2007; Aftab et al. 2011; Bharti et al. 2013). At the same time, essential oils of lemongrass contain citronellal, geranyl acetate, borneol, geraniol, citral α (geranial), citral β (neral), α-Pinene, β-Pinene, β-caryophyllene, linalyl acetate, myrcene, methyl heptenone, nerol, limonene, terpinolene, terpineol (Majewska et al. 2019). *C. citratus* shows a high percentage of citral (about 80%), a mixture of geranial, terpenoids, and neral. The high percentage of citral shows more cultivation of *C. citratus* for commercial purposes.

Nutritionally, lemongrass shows the presence of different electrolytes and minerals (Mn, Cu^{2+}, Na^{+}, Fe^{2+}, K^{+}, Ca^{2+}, Mg^{2+}, P, Se, Zn^{2+}), macronutrients (carbohydrate, proteins, and small amount of fat), and vitamins (vitamins A, C, E, niacin, folate, riboflavin, pyridoxine) (Aftab et al. 2011).

Cymbopogon citratus shows antibacterial, antifungal, anti-diarrheal, and anti-inflammatory properties. Other than this, properties like anti-malarial, anti-oxidants, hypoglycaemic, anti-mutagenicity, anti-mycobacterial, and neurobehavioral have also been studied. With that, lemongrass works as a good cleanser that helps detoxify the kidney, pancreas, liver, bladder, and digestive tract (Manvitha and Bidya 2014). It helps to reduce skin problems like acne and pimples and works as muscle toner. A study conducted by the Food and Nutrition Research Institute of the Department of Science and Technology (DOES) has concluded that lemongrass can help prevent cancer (Manvitha and Bidya 2014).

5.1.7 *Economic Importance of Lemongrass*

Essential oil components like geraniol, myrcene, and citral separated from lemongrass play an important role as raw materials in food, beverage, detergent, soap, perfume, and cosmetic industries (Solomon et al. 2019).

5.2 Current Cultivation Practices in Lemongrass

Lemongrass, an aromatic perennial sedge, is predominantly grown in subtropical and tropical regions of Australia, South Asia, and Southeast Asia. In India, it is raised in Arunachal Pradesh, Karnataka, Sikkim, Tamil Nadu, Uttar Pradesh, and Western ghats of Kerala and Maharashtra. This tall grass preferably grows in warm, sunny, and humid habitats at an elevation of 750–900 m above sea level (NHB 2022). It requires a temperature ranging from 23 to 30 °C and can tolerate upto 18–34 °C with an annual rainfall of 1500–3000 mm and can tolerate up to 700–4200 mm (Rojas-Sandoval 2016; NParks 2019). It grows in clusters during the summer and rainy seasons but fails to grow in cold conditions. The ability of this grass to grow in a wide variety of soil types (loamy to poor laterite soil) makes it unique from the rest of the plants. Lemongrass prefers sterile, well-drained loamy soil with a pH ranging from 6.5 to 7.0 and can tolerate upto pH of 5–8.4 (Bonnie Plants 2022). *C. citratus* can be cultivated from seeds or by vegetative propagation.

5.2.1 Seed Propagation

The grass can be grown indoors from seeds by placing the seeds in seedling trays with a starting mix containing compost, cocopeat, and abrasive in equal ratios. The seeds must be sown into the starting mix about 1/4–1/2 of an inch deep. The seeds are watered, and the seed tray is then covered with plastic wrap and placed on the window sill to receive a good amount of light for its growth. The seed requires moist conditions to germinate and is kept moist by watering them once a week. The germination starts within 10–30 days. When the plant height reaches 2.54 cm, it can be transferred to pots for further growth. Outdoors, the seeds are sown at the onset of the rainy season on raised beds of 1–1.5 m width and covered with a thin layer of soil with regular watering. The seeds germinate within 5–6 days and are then allowed to grow up to 60 days to be transplanted (NHB 2022). However, when seeds germinate slowly or fail to germinate, vegetative propagation is preferred.

The vegetative propagation can be achieved by the division of existing clumps into slips or from stalks of established plants. The bulb of the stalks of lemongrass can be simply placed in a jar with 1–2 in. of water for the development of roots. Once the roots of 1/2–1 in. or leaves develop, they can be transferred to rich soil for further growth (Winger 2014). Propagation by dividing stalks from the rhizome is the primary method. The lemongrass plants should be divided in such a way that it has at least 1 in. of root and 2 in. of a blade by splicing through the rhizome using a sharp spade or trowel. The plant has to be divided into at least 6 in., which is then planted (NHB 2022). The slips have to be planted on ridges at a distance of either 40 × 30 cm, 40 × 40 cm, or 40 × 60 cm. apart and 15 cm deep. It has to be watered regularly in the summer, and in the rainy season, it can be watered at an interval of 3 days. The soil has to be fertilized with the fertilizer containing NPK (Nitrogen, Phosphorus, and Potassium) in a 1:1:1 ratio once in two weeks (NHB 2022).

The lemongrass gets into a dormant stage during the winter season. Hence the ornamental grass is pruned to about 6 in. at the end of winter or beginning of spring season to develop new shoots. The lemongrass can be harvested once it reaches 12 in. tall with a 1/2-in. thick stem base. A hand trowel can be used to remove individual stalks and roots from the clump (Bonnie Plants 2022). During its cultivation, lemongrass is prone to pests such as *Tylenchorhynchus vulgaris* and stem boring caterpillar (*Chilotrea* sp.) that feeds on the stem resulting in stunted growth. The plant also gets infected with red leaf spot disease caused by *Colletotrichum graminicol*a where the leaves exhibit brown spots and dry out. It can be controlled by treating with 0.1% of Bavistin. *Curvularia andropogonis* causes leaf blight disease, which causes reddish-brown spots and premature drying of leaves; *Balansia sclerotic*a causes grassy shoot where there is stunted growth and little leaf formation. Leaf blight disease and grassy shoot can be controlled by treatment with 2% Dithane M-45 (NHB 2022).

Table 5.3 List of *Cymbopogon* species with collection site

Species	Collection site (India)	Major chemical component(s) present
C. martinii (Roxb.)	Madhya Pradesh, India	Geraniol
C. citratus (DC.) Stapf	Madhya Pradesh, India	Citral (72–75%)
C. pendulus (Nees ex Steud.) Wats	Himalayan regions of India	Citral (75–80%)
C. winterianus Jowitt	Assam, India	Citronellol, Geraniol + Citronellal
(*C. khasianus* × *C. pendulus*) Hybrids	Jammu, India	Citral (75–80%)
C. flexuosus (Steud.) Wats	Punjab, India	Citral (80–82%)
Mutant Jamrosa	Jammu, India	Geraniol (75–78%)

5.3 Germplasm Diversity

The term germplasm describes different parts of a plant, seeds that are useful for conservation or crop breeding, or research purposes. It includes both wild species as well cultivated varieties. The genetic pool helps to study genetic diversity as well as genetic variability in crop plants. Germplasm conservation helps in the maintenance of biological diversity and food security. Other than this, it is essential to conserve existing species of medicinal plants. With this, many tissue and cells of various plants can be conserved and maintained for several years. Aromatic and Medicinal Plants Research Station, Odakkali, of the Kerala Agricultural University, is known for the largest collection of lemongrass germplasm (Thomas et al. 1998). About 423 *Cymbopogon* varieties have been collected from different places and maintained here. All accessions are appropriately catalogued and described. Few species and their major chemical components are listed in Table 5.3 (Thomas et al. 1998).

Lemongrass is widely cultivated for its aromatic and essential oil across the world. Varieties that are commonly cultivated in India, their specific characters, and variety of developed institution information are listed in Table 5.4 (Joy et al. 2006).

5.4 Traditional Breeding

Traditional breeding or conventional breeding is a technique in which a new variety of plants is developed by using natural processes. The crop improvement with the help of the introgression of traits will help increase the accessibility of genetic resources. In recent years, traditional breeding has been established with various crosses, from mutagenesis, hybrid cross, and double haploid technology (Ahmar et al. 2020). Traits that are developed using traditional plant breeding will have their specific agronomic and natural phenotypic properties. A trait can be enhanced by

Table 5.4 List of different varieties with their specific characters

Varieties	Institute	Specific characters
Sugandhi (OD-19)	Aromatic and Medicinal Plants Research Station (AMPRS), Odakkali, Kerala, India	Red stemmed variety, height of 1–1.75 m, most popular in India
Pragati (LS-48)	Central Institute of Medicinal and Aromatic Plants (CIMAP), Lucknow, India	Tall, with dark purple leaf sheath, with 86% citral content
RRL-16	Regional Research Laboratory (RRL), Jammu, India	Evolved from *C. pendulus*, citral content is 80%
Praman (Clone 29)	Central Institute of Medicinal and Aromatic Plants (CIMAP), Lucknow, India	Tetraploid type, erect, medium-size leaves
Kavery and Krishna	Central Institute of Medicinal and Aromatic Plants (CIMAP), Bangalore, India	Long and narrow leaves are present
RLJ-TC-7, RLJ-TC-8, RLJ-TC-11	Regional Research Laboratory (RRL), Jorhat	Higher percentage of oil and geraniol
Chirharit	Systematic breeding for genetic improvement at Pant Nagar, Chirharit, India	High-yielding variety, frost resistant

using the cross-breeding technique. In cross-breeding, one should select the best hybrid progeny. Desired traits should be introduced into the selected recipient line by using the backcross technique. Long-term traditional breeding helps to reduce genetic variability. In the case of *Cymbopogon citratus,* there is no report on specific cross-breeding or old traditional breeding techniques. As it is a time-consuming method, the introduction of a new gene is necessary to enhance the desired trait by mutation, rapid generation advance (RGA), and speed breeding (Gaur et al. 2020). The standard plant breeding innovative cycle is illustrated in Fig. 5.3.

5.5 Mutation Breeding

Mutation breeding is a technique in which a plant or plant part will be exposed to radiation, enzymes, or chemicals, which results in the formation of mutant varieties in cultivation practices. It plays an essential role in plant biotechnology. Mutational breeding is mainly based on mutation detection or mutation induction. Mutation breeding is an alternative and important technique which helps to increase crop productivity in plant breeding. For successful mutation breeding, different types of mutagens can be used. In Table 5.5, a list of different types of mutagens is mentioned. Physical mutagens like X-ray, neutron, gamma rays, alpha particles, beta particles, UV rays, protons, etc., can be used. Chemicals like Methyl methanesulfonate,

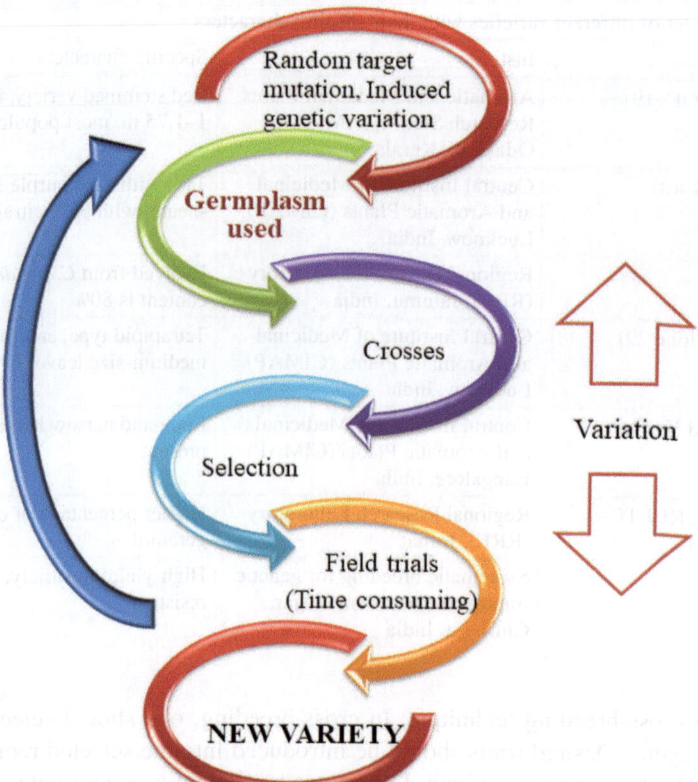

Fig. 5.3 The plant breeding innovation cycle

Ethyl methanesulfonate (EMS), Bromo uracil, Tetramethyl Urea (TMU), aminopurine (2AP) will be used in chemical mutagenesis. In a few mutagenesis techniques, biological agents are used as a tool (Agrawal and Kumar 2021). It includes T-DNA, transposons, and retrotransposons. Clustered regularly interspaced short palindromic repeat (CRISPR) and transcription activator-like effector nucleases (TALENS) cause target mutation (Agrawal and Kumar 2021). It is also used as an alternative mutagen. In *C. citratus* mutation, studies were mainly carried out to increase the oil content. Most of the mutant varieties of *C. citratus* were achieved with the help of physical mutagen. When *C. citratus* is exposed to Ultraviolet-B, it stimulates the oil-producing cell quality and percentage of essential oil contents. The yield of essential oil has increased to 117.56% compared to the control (Sharifah et al. 2013). By using different doses of gamma irradiation, Jor Lab L-14 variety has been developed (Lal et al. 2018). This mutated variety shows 75–80% of citral content and 0.8–1.35% of essential oil. Sharifah et al. (2012) treated West Indian lemongrass with different doses of gamma radiation 40, 60, and 80 Gy. In all three, the mutation rate was 25.8%,

Table 5.5 List of different types of mutagens

Mutagen	Types of the mutagen	References
Methyl methanesulfonate, Ethyl methanesulfonate, Bromo uracil, Tetramethyl Urea, aminopurine	Chemical mutagens	(Sharifah et al. 2013; Agrawal and Kumar 2021)
X-ray, neutron, gamma rays, alpha particles, beta particles, UV rays, protons	Physical mutagens	(Agrawal and Kumar 2021; Lal et al. 2018)
T-DNA, transposons, and retrotransposons	Biological mutagens	(Agrawal and Kumar 2021)

36.4%, and 69.2%, respectively. Dose with 80 Gy, i.e., LD_{50}, showed new characteristics. Mutant characters included long ground stem (not stalk) with the presence of nodes and internodes together with flowers. It is also noticed that flowering was frequent in mutant clones (Sharifah et al. 2012). Nearly 1411 varieties have been established by using radiation as a mutagen (Sharifah et al. 2013).

5.6 Molecular Breeding

Plant breeding reports the selection, creation, and fixation of superior phenotypes in plant development programs. This helps for the development of improved varieties for plant cultivators. Conventional breeding techniques are not adequate for developing new varieties. Various modern breeding techniques include marker-assisted recurrent selection (MARS), marker-assisted backcrossing (MABC), marker-assisted selection (MAS), and genomic selection (GS). Molecular breeding is an important, developed technique. The primary aim of molecular breeding is to improve the nutritional value, the yield of a crop, and the commercial value of the trait. With that molecular breeding;

- Expands the genetic diversity in plant breeding.
- Improves favourable gene action
- Increases the efficiency of selection.

Molecular-assisted breeding helps in the linkage of genome and maps. Different markers like Random Amplified Polymorphic DNA (RAPD), Restriction Fragment Length Polymorphism (RFLP), Amplified Fragment Length Polymorphism (AFLP), etc. can be used for successful breeding.

C. citratus is wild grass and is mainly cultivated for its essential oil worldwide. Most of the breeding studies have been done to increase the essential oil content. There is no specific report on a genetically engineered developed trait. In *C. citratus* most of the molecular studies have been done for the analysis of genetic diversity. For the diversity study, they have used different markers like RAPD, Inter-simple sequence repeat (ISSR), and Simple sequence repeats (SSR) (Bishoyi et al. 2016;

Adhikari et al. 2015). Kumar et al. (2009) used DNA markers to analyze the genetic diversity of 25 accessions of *Cymbopogon* species. It includes mutant, hybrid as well as different species of *Cymbopogon.* The study concluded that it might be helpful in developing a perfect marker with the association of gene mapping involved in developing improved traits.

5.7 Tissue Culture Applications

Cymbopogon citratus is conventionally propagated through suckers and the efficiency for the propagation is comparatively low. Tissue culture studies can help on rapid regeneration and mass production of the lemon grass. Linsmaier and Skoog medium (LS medium) supplemented with 0.1 μM NAA along with 10 μM BAP induced multiple shoots from shoot tip explants. Rooting was best recorded on LS medium supplemented with 1 μM NAA (Mizukami et al. 1989). Temporary immersion system can be an alternative approach for in vitro production and multiplication of lemon grass as shown in Fig. 5.4. The temporary immersion system revolutionizes plant micropropagation by employing periodic cycles of immersion in a liquid medium followed by exposure to a gaseous environment. With shorter immersion periods and prolonged air exposure, this system optimizes humidity and nutrient supply while reducing hyperhydricity in cultured tissues. The optimised models even offer CO_2 enrichment during air exposure, further promoting tissue multiplication and also for secondary metabolite accumulation (Georgiev et al. 2014). It is seen that MS medium supplemented with 0.89 μM BAP helped in establishment of shoots from shoot tip explants and later the shoots were immersed 6 times per day showed better multiplication efficiency of the shoots and showed highest fresh and dry weights (i.e., biomass) (Quiala et al. 2006). Similarly, scaling up of the biomass production was monitored, and it is seen that highest biomass was achieved from in vitro shoots when temporary immersion was done once in 4 h in 10 L of culture medium (Quiala et al. 2014). In addition to temporary immersion systems, traditional methods for in vitro production of secondary metabolites remain viable options. These methods include callus, cell culture, and various organ cultures such as shoot, adventitious roots, and hairy root cultures. These techniques offer flexibility and can be further explored to optimize secondary metabolite production in lemon grass. By leveraging these classical modes, researchers can enhance the efficiency and yield of secondary metabolites from lemon grass, catering to diverse industrial and pharmaceutical needs.

Fig. 5.4 In vitro propagation of *Cymbopogon citratus* in temporary immersion system, **a** the apical region of the stem from field plants (named spindle); **b** apical shoot culture in MS-I medium (MS + 0.89 μM BA); **c** 2.5–3.0 cm long individual shoots; **d**, **e** shoots growing in RITA under 1 min of immersion every 4 h, 3 weeks after culture on MS-I medium (MS + 0.89 μM BA); **f**, **g** shoots growing in 1 L TIS under 2 min of immersion. *Source* Reproduced from Quiala et al. (2016). https://doi.org/10.1007/978-1-4939-3332-7_30 with permission License No: 5355470641386

5.8 Conservation Techniques

Conservation of germplasm is an important phenomenon that helps to maintain genetic variability as well as food security. Conservation of genetic resources can be done by using ex-situ or in-situ conservation.

5.8.1 *In-Situ Conservation*

In-situ conservation is commonly known as on-site conservation. In this type of conservation, genetic resources can be conserved in forests, sanctuaries, and reserved areas in their natural state. It is a more realistic type of conservation. The different species of *Cymbopogon* are conserved in National Botanical Garden, Chinnar Wildlife Sanctuary in the Western Ghats in their natural state (Joy et al. 2006).

5.8.2 *Ex-Situ Conservation*

Ex-situ conservation is another type of conservation in which germplasm can be conserved outside of its natural state. In this, preservation of seeds, pollen, embryo, etc. included. The seed bank, in-vitro technique, and cryopreservation are all included in the ex-situ conservation technique. For the in-vitro conservation of *Cymbopog*on species, Murashige and Skoog medium has been used with its 50% concentration and mannitol (Corona et al. 2020). Preserved plants can be transferred to a culture medium for the multiplication and obtain immense plant materials.

5.8.3 *Seed Bank*

A seed bank is the simplest and most economical method for the conservation of genetic resources. It helps maintain 5% of seed moisture for its viability (Pérez-García et al. 2006). Lemongrass harvested for seed purposes is rare. A healthy plant gives 12–20 g of seeds (Joy et al. 2006). Matured inflorescence should be cut and dried completely and thrashed on the floor. Seeds of *Cymbopogon* species lose viability after six months; hence fresh seeds are always recommended (Joy et al. 2006).

5.9 Hybridization in Lemongrass

There are remarkably fewer studies related to hybridization in *Cymbopogon citratus*. Wild plant species are always explored for breeding purposes to evolve new varieties because they are known for their stress tolerance and disease-resistant abilities (Watson et al. 2018). But unfortunately, in the case of *C. citratus,* there are no significant studies on its wild varieties. *C. citratus is* mainly cultivated for its aromatic, essential oil. Mutagenic and in vitro studies showed promising results in improving oil quantity, which is easier, cost-effective, and time-saving than the hybridization technique (Joy et al. 2006). This could be one reason for fewer studies related to hybridization in *C. citratus.*

5.10 Conclusion and Prospects

Lemongrass has been used as a food condiment in most countries. Essential oils like geraniol, myrcene, and citral, separated from lemongrass, have potential antibacterial, antifungal, antidepressant, carminative, fungicidal, astringent, and antiseptic activity. It is widely used in various industries like food, beverage, and cosmetics as raw materials. Besides its medicinal and economic properties, lemongrass is not cultivated extensively. Hence, there are very few reports available on the development of its traits and genetic engineering. This lack of information provides an opportunity to study and develop new varieties using advanced breeding technology. With the advent of advanced breeding techniques, such as mutation breeding, there is an opportunity to enhance desirable traits in lemongrass, including improved yield, enhanced quality of essential oil, and resistance to biotic and abiotic stresses. Furthermore, the emergence of gene editing technologies, such as CRISPR-Cas9, presents a promising avenue for precise manipulation of the lemongrass genome. Although there is currently a lack of reported research on gene editing in lemongrass, the application of this technology holds great potential for targeted trait improvement, including enhanced disease resistance, increased essential oil production, and modification of biochemical pathways to optimize medicinal properties. In addition to genetic enhancement, advances in agronomic practices, such as optimized cultivation techniques, irrigation management, and post-harvest processing, can further contribute to maximizing the productivity and quality of lemongrass crops.

References

Adhikari S, Saha S, Bandyopadhyay TK, Ghosh P (2015) Efficiency of ISSR marker for characterization of *Cymbopogon* germplasms and their suitability in molecular barcoding. Plant Syst Evol 301(1):439–450. https://doi.org/10.1007/s00606-014-1084-y

Aftab K, Ali MD, Aijaz P, Beena N, Gulzar HJ, Sheikh K (2011) Determination of different trace and essential elements in lemon grass samples by X-ray fluorescence spectroscopy technique. Int Food Res J 18:265–270

Agrawal L, Kumar M (2021) Improvement in ornamental, medicinal, and aromatic plants through induced mutation. J Appl Biol Biotechnol 9(4):162–169. https://doi.org/10.7324/JABB.2021.9422

Ahmar S, Gill RA, Jung KH, Faheem A, Qasim MU, Mubeen M, Zhou W (2020) Conventional and molecular techniques from simple breeding to speed breeding in crop plants: recent advances and future outlook. Int J Mol Sci 21(7):1–24. https://doi.org/10.3390/ijms21072590

Bharti SK, Kumar A, Prakash O, Krishnan S, Gupta AK (2013) Essential oil of *Cymbopogon citratus* against diabetes: validation by in vivo experiments and computational studies. Bioanal Biomed 5:194–203

Bishoyi AK, Sharma A, Kavane A, Geetha KA (2016) Varietal discrimination and genetic variability analysis of *Cymbopogon* using RAPD and ISSR markers analysis. Appl Biochem Biotechnol 179(4):659–670. https://doi.org/10.1007/s12010-016-2022-y

Bonnie Plants (2022) Growing & planting lemongrass. In: Bonnie plants. https://bonnieplants.com/blogs/how-to-grow/growing-lemongrass. Accessed 12 Mar 2022

Corona HR, Rosaria MM, Trinidad JD (2020) Propagation and conservation germplasm of lemongrass (*Cymbopogon citratus* stapf) through biotechnology techniques. Int J Agric Environ Biores 5(1):108–123. https://doi.org/10.35410/IJAEB.2020.5464

Filipov A (1994) Medicinal plants of the Pilagá of Central Chaco. J Ethnopharmacol 44(3):181–193. https://doi.org/10.1016/0378-8741(94)01185-0

Georgiev V, Schumann A, Pavlov A, Bley T (2014) Temporary immersion systems in plant biotechnology. Eng Life Sci 14(6):607–621

Haque ANMA, Remadevi R, Naebe M (2018) Lemongrass (*Cymbopogon*): a review on its structure, properties, applications and recent developments. Cellulose 25(10):5455–5477. https://doi.org/10.1007/s10570-018-1965-2

Idrees M, Naeem M, Khan MN, Aftab T, Khan MM, Moinuddin (2012) Alleviation of salt stress in lemongrass by salicylic acid. Protoplasma 249(3):709–720. https://doi.org/10.1007/s00709-011-0314-1. Epub 1 Sept 2011. PMID: 21882051

Joy PP, Skaria BP, Mathew S (2006) Lemongrass. Ind J Arecanut Spices Med Plants 2:55–64

Kamaruddin ZH, Jumaidin R, Selamat MZ, Ilyas RA (2021) Characteristics and properties of lemongrass (*Cymbopogon citratus*): a comprehensive review. J Nat Fibers 00:1–18. https://doi.org/10.1080/15440478.2021.1958439

Kumar J, Verma V, Goyal A, Shahi AK, Sparoo R, Sangwan R, Qazi GN (2009) Genetic diversity analysis in *Cymbopogon species* using DNA markers. Plant Omics 2(1):20–29

Lal M, Dutta S, Munda S, Pandey SK (2018) Identification and registration of a high essential oil yielding variety (Jor Lab L-14) of lemongrass (*Cymbopogon flexuosus* L.) through mutation breeding technique. J Essent Oil Bear Plants 21(6):1604–1611. https://doi.org/10.1080/0972060X.2018.1559105

Leite J, Maria De Lourdes VS, Maluf E, Assolant K, Suchecki D, Tufik S, Klepacz S, Calil HM, Carlini EA (1986) Pharmacology of lemongrass (*Cymbopogon citratus* Stapf). III. Assessment of eventual toxic, hypnotic and anxiolytic effects on humans. J Ethnopharmacol 17(1):75–83

Magotra A, Singh AP, Singh AP (2021) A review on pharmacological activities of *cymbopogon citratus*. Int J Pharm Drug Anal 9(2):151–157

Majewska E, Kozlowska M, Gruszczynska-Sekowska E, Kowalska D, Tarnowska K (2019) Lemongrass (*Cymbopogon citratus*) essential oil: extraction, composition, bioactivity and uses for food preservation—a review. Polish J Food Nutr Sci 69(4):327–341

Manvitha K, Bidya B (2014) Review on pharmacological activity of *Cymbopogon citratus*. Int J Herb Med 1(6):5–7

Mizukami H, Okada Y, Ohashi H (1989) Clonal propagation of Lemongrass (*Cymbopogon citratus* Stapf) through shoot tip culture. Plant Tissue Cult Lett 6(1):22–24. https://doi.org/10.5511/plantbiotechnology1984.6.22

Negrelle RRB, Gomes EC (2007) *Cymbopogon citratus* (DC.) Stapf: chemical composition and biological activities. Rev Bras Plantas Med 9(1):80–92

NHB (2022) Lemongrass. http://nhb.gov.in/model-project-reports/Horticulture%20Crops/Lemongrass/Lemongrass1.htm. Accessed 12 Mar 2022

NParks (2019) *Cymbopogon citratus* (DC.) Stapf. In: A Singapore government agency website. https://www.nparks.gov.sg/florafaunaweb/flora/1/9/1918. Accessed 12 Mar 2022

Olorunnisola SK, Hammed AM, Simsek S (2014) Biological properties of lemongrass: an overview. Int Food Res J 21(2):455–462

Pérez-García F, Huertas M, Mora E, Peña B, Varela F, González-Benito ME (2006) *Hypericum perforatum* L. seed germination: inter-population variation and effect of light, temperature, presowing treatments and seed desiccation. Gen Resour Crop Evol 53(6):1187–1198

Quiala E, Barbón R, Jiménez E, De Feria M, Chávez M, Capote A, Pérez N (2006) Biomass production of *Cymbopogon citratus* (D.C.) Stapf., a medicinal plant, in temporary immersion systems. In Vitro Cell Dev Biol Plant 42:298–300. https://doi.org/10.1079/IVP2006765

Quiala E, Barbón R, Capote A, Pérez-Alonso N, Chávez M, de Feria M, Jiménez E (2014) Scaling-up the biomass production of *Cymbopogon citratus* L. in temporary immersion system. Biotecnol Vegetal 14:67–71

Quiala E, Barbón R, Capote A, Pérez N, Jiménez E (2016) In vitro mass propagation of *Cymbopogon citratus* Stapf., a medicinal gramineae. Methods Mol Biol 1391:445–457. https://doi.org/10.1007/978-1-4939-3332-7_30

Rojas-Sandoval J (2016) *Cymbopogon citratus* (lemongrass). In: CABI. https://www.cabi.org/isc/datasheet/17377. Accessed 12 Mar 2022

Samineni S, Sen M, Sajja SB, Gaur PM (2020) Rapid generation advance (RGA) in chickpea to produce up to seven generations per year and enable speed breeding. Crop J 8(1):164–169

Shah G, Shri R, Panchal V, Sharma N (2011) Scientific basis for the therapeutic use of *Cymbopogon citratus*. Stapf (Lemon Grass) 2(1):3–8. https://doi.org/10.4103/2231-4040.79796

Sharifah NRSA, Mahir AM, Jusoff K (2012) Unique flowers produced from West Indian Lemongrass, *Cymbopogon citratus* (DC.) Stapf. through induced mutation. World Appl Sci J 17:71–74

Sharifah NRS, Mahir AM, Jusoff K, Hanina MN, Salhah OS, Noor IM (2013) Prospective effects of induced mutation by gamma radiation in essential oil production of lemongrass (*Cymbopogon citratus*). Int J Agric Syst 1(1):1–21

Solomon O, Enitan F, Temitope D, Abraham K (2019) Phytochemistry and pharmacological activities of *Cymbopogon citratus*: a review. Sci Afr 6:e00137. https://doi.org/10.1016/j.sciaf.2019.e00137

Spring MA (1989) Ethnopharmacologic analysis of medicinal plants used by laotian hmong refugees in minnesota. J Ethnopharmacol 26(1):65–91. https://doi.org/10.1016/0378-8741(89)90114-1

Thomas J, Mathew S, Joy P et al (1998) Genetic resources and diversity in Lemongrass in Kerala. In: Golden jubilee national symposium on spices, medicinal and aromatic plants: biodiversity conservation and utilisation, Calicut, Kerala. At: Calicut, Kerala Volume: Spices, Medicinal and Aromatic Plants: Biodiversity Conservation and Utilisation, 10–12 Aug 1998, pp 941–944

Watson A, Ghosh S, Williams MJ, Cuddy WS, Simmonds J, Rey MD, Asyraf Md Hatta M, Hinchliffe A, Steed A, Reynolds D, Adamski NM (2018) Speed breeding is a powerful tool to accelerate crop research and breeding. Nature Plants 4(1):23–29. https://doi.org/10.1038/s41477-017-0083-8

Winger J (2014) Lemongrass—how to grow it and use it. In: The Prairie Homestead. https://www.theprairiehomestead.com/2014/11/lemongrass-grow-use.html. Accessed 13 Mar 2022

Chapter 6
Long Pepper (*Piper longum*) Breeding Strategies for Improved Production

Juhi Puthukulangara Jaison, Jaya Gangwar, and Kadanthottu Sebastian Joseph

Abstract *Piper longum* belongs to the family Piperaceae and is an important spice in India. The dried female spikes and roots of this plant have high commercial value because it is utilized in several formulations for treating different diseases like fever, common cold, indigestion, bronchitis, asthma, cancer, leprosy etc. *P. longum* contains more than fifty alkaloids and several essential oils. The cultivation of *P. longum* is not very common and it is extensively collected from the forest due to high demand from pharmaceutical industries, threatening the existence of the plant. The seeds of *P. longum* have a low germination rate so, vegetative propagation is the most commonly used to cultivate it. Viswam was the first reported improved variety of *P. longum*. To fulfil the demand, aromatic and medicinal species must be duplicated more quickly while causing the least amount of damage to their natural environments. Micro-propagation technique allows for the rapid multiplication of planting material may be extremely beneficial. Advancements in molecular biology, gene technology, and cell biology have significantly contributed to our understanding of diseases at the molecular and genetic levels. High throughput screening (HTS) techniques have evolved from novel target-directed screening assays and automation significantly enhancing the industrial drug development process. Furthermore, in vitro, gene banks can play a critical role in providing feedstock for this revolution. Using current biotechnological methods, new pepper varieties with higher production and adaptability must be produced. In this chapter, we have discussed the origin, distribution, economic importance, traditional cultivation, role of biotechnology, germplasm diversity, conservation approaches, and molecular breeding.

Keywords *Piper longum* · Breeding · Piperine · Cultivation · Hybrid · Traditional methods

J. P. Jaison · J. Gangwar · K. S. Joseph (✉)
Department of Life Sciences, Christ University, Bengaluru, Karnataka 560 029, India
e-mail: joseph.ks@christuniversity.in

J. P. Jaison
e-mail: juhi.pj@res.christuniversity.in

J. Gangwar
e-mail: jaya.gangwar@res.christuniversity.in

J. M. Al-Khayri et al. (eds.), *Biodiversity and Genetic Improvement of Herbs and Spices*, Advances in Plant Breeding Strategies 8, https://doi.org/10.1007/978-3-031-81838-7_6

6.1 Introduction

Piper longum L. is commonly known as pippali or long pepper and belongs to the Piperaceae family utilized as a medicinal plant in ayurvedic treatment particularly for respiratory tract diseases (Khandhar et al. 2010). There are four types of pippali reported in ayurvedic literature Rajanighantu i.e. pippali, saimhali, gajapippali and vanapippali. The National Medicinal Plant Board gives it a priority for the cultivation and development of *P. longum* since it is one among 14 medicinal plants which have high demand in the drug industries. It is native to countries in South Asia, including India. Ancient India utilised it as a spice (Yadav et al. 2020). In addition to being used as a spice to flavouring food, *P. longum* is widely utilized around the world in several conventional medical practices. Ancient books like the Charaka Samhita, Susruta Samhita, Vagbhata's Astangya Hrdayam, etc. have discussed the significance of *P. longum* in the Indian traditional medical system (Gajurel et al. 2021). There are claims that the entire plant, including the stem, roots, leaves, and fruits, can be utilised to treat a variety of diseases. *P. longum* is a key component of the Ayurvedic, Siddha, and Unani medical systems and is used to treat multiple illnesses (Bhatt and Deshpande 2021). In the Indian Ayurvedic System, the fruits and roots of *P. longum* were utilised in more than 100 crude medicine formulations. *P. longum* is a perennial low-creeping root climber. This species has distinct male and female individuals since it is dioecious. The nature of the branches is dimorphic, with fertile branches standing upright with swelling nodes and no adventitious roots, and vegetative branches creeping and rooting at nodes. The blooms are borne on upright spikes and are unisexual. Fruiting spikes can grow as long as 5 cm. Drupes have a highly aggregated, globular shape that turns black as they mature with a strong flavour (Manoj et al. 2004).

6.2 Origin and Distribution

P. longum is a native of Indo-Malaya and southeast Asia. It is found all over South Asia, including Burma, Malaysia, Nepal, and Singapore, but is most prevalent in India, where it is a native of peninsular India as it is shown in Fig. 6.1. In India, it mostly occurs in the hotter parts and from the central Himalayas to Assam, Khasi and the mikir hills, the lower hills of West Bengal, and the evergreen forests of western Ghats from Konkan to Travancore (Ayisha 1997). Globally the species is also distributed throughout the Sri Lanka and Indo-Malaysian regions (Fig. 6.1). It is found throughout northeast India i.e. from Assam, and Bengal to the southern edge of Nepal. The majority of *P. longum* comes from the wild variety, which is primarily grown in Kerala, Uttar Pradesh, Assam, Nepal, West Bengal, the North Eastern area, and Andhra Pradesh. It is also cultivated in the Car Nicobar Islands. (Yadav et al. 2020). In some areas of Bengal, Assam, Maharashtra, Tamil Nadu, Orissa, Andhra

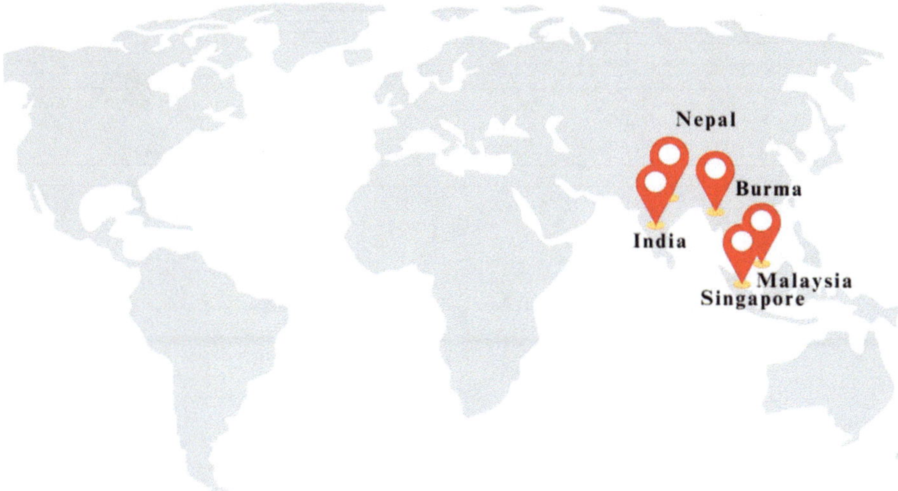

Fig. 6.1 Geographical distribution of *P. longum*

Pradesh, Arunachal Pradesh, Meghalaya, and Manipur (Fig. 6.2), it is also sparingly farmed (Babu et al. 2006).

6.3 Botanical Description

P. longum is an aromatic, trailing, slender, dioecious plant in the hotter region of India. It is a ground spreading perennial creeping underbrush. Creeping stems are jointed with erect fruiting branches and young shoots are downy (Fig. 6.3). In Creeping shoots, leaves are alternate, simple, sessile or petiolate, dimorphic, ovate, cordate with rounded wide lobes at the base, entire, subacute, glabrous and in fruiting branches leaves are oblong, lanceolate, and have an unevenly cordate base. Female spikes are shorter and slenderer than males, measuring between 1.3 and 2.5 cm long and 4–5 mm in diameter. Fruits are ovoid and submerged in the fleshy spike; when ripe, they transform from green to black Flowers are dioecious and bloom all year round. Inflorescence is a spike, that bears unisexual, tiny, densely clustered flowers and grey berries. The female spikes have a short, thick stalk that ranges in length from 1.5 to 2.5 cm and thickness from 0.5 to 0.7 cm. The spikes of males are slender and have 2.5 to 7.5 cm long stalks, partially elongated. Fruits are small, ovoid drupes fully sunk in the fleshy spikes, laterally fused, aromatic, pungent, and spicy. When the fruits are

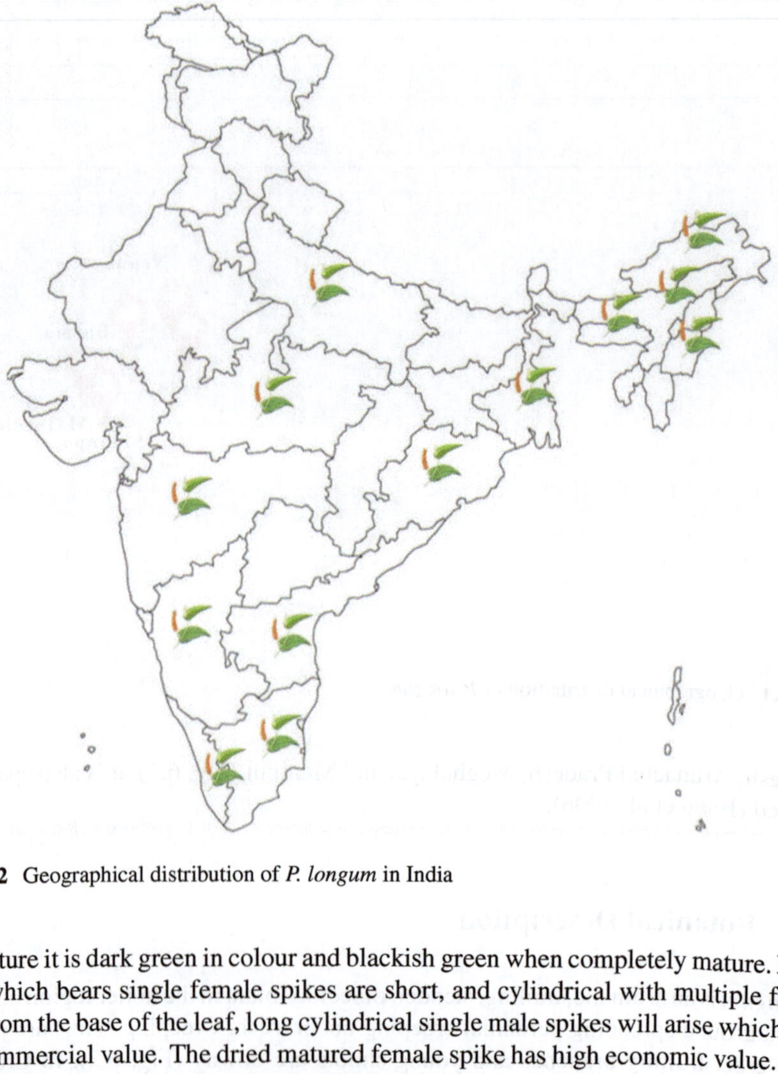

Fig. 6.2 Geographical distribution of *P. longum* in India

immature it is dark green in colour and blackish green when completely mature. Leaf axil which bears single female spikes are short, and cylindrical with multiple fruits and from the base of the leaf, long cylindrical single male spikes will arise which has no commercial value. The dried matured female spike has high economic value. The long Indonesian pepper is a slender climber. The leaves alternate and have swollen branches at the nodes (Yadav et al. 2020).

6.4 Economic Importance

The root and stem of the *P. longum* have been utilised in different Ayurvedic and Unani medical practices (Atal and Ojha 1965). For parturition, an infusion of the root helps in the removal of the placenta. It is also used as an alternate tonic in

Fig. 6.3 Habit of *P. longum*

cases of paraplegia, chronic cough, enlargement of the spleen, and other abdominal viscera (Khare 2008). Heart problems are treated using *P. longum* roots. It appears to partially absorb the stimulating components of the fruit. *P. longum* has been boiled with mustard oil, ginger, buttermilk, and curds and used to make various kinds of medicinal formulations. The *P. longum* fruit is used as a laxative, abortifacient, pungent, aphrodisiac, anti-diarrheic, anti-asthmatic, anti-bronchitis, leprosy, pain, inflammation, hiccoughs, jaundice, sleeplessness, tumours, spleen illnesses, and urine discharges. The fruit is known to have a heating effect, be useful against parasites like helminths, cleanse mucus from the lungs and other respiratory organs, and function as a carminative to relieve flatulence and boost metabolism in general. Additionally, *P. Longum* increases stomach appetite and taste and is very effective in treating fever, inflammatory issues, haemorrhoid infections, piles, and discomfort or pain in the belly (Salehi et al. 2019). The stem was utilised to make decoctions that were used to treat diarrhoea and rheumatoid arthritis aches as well as postpartum pain in mothers. The fruit was externally applied to cure toothaches. Also, the fruit was a significant treatment for gastrointestinal trouble, therefore tonic produced from the ripe fruits was utilised as medication (Chahal et al. 2011). The root and fruits of *P. longum* were commonly used in India as an antitoxin or nullifier against the venoms injected into the body through stings in the case of snake bites and scorpion stings (Shenoy et al. 2013). *P. longum* fruits were used in the Chinese medical tradition to treat stomach chills, headaches, acid reflux, nausea, and allergic rhinitis. The paste of leaves was applied to the chest who has breast cancer (Taufiq-Ur-Rahman et al. 2005; Choudhary and Singh 2018). The dried fruits of *P. longum* were given prominence in Thai traditional medicine for their ability to relieve flatulence, have antidiarrheal

Table 6.1 Medicinal uses of *Piper longum*

Parts used	Uses	References
Fruits	Colds, arthritis, rheumatism, asthma, sciatica, epilepsy, indigestion, lumbago, flatulence, antibiotic, improves decongestion analgesic effects, gynaecological diseases, toothache, eases chronic bronchitis, and intestinal distress	(Salehi et al. 2019)
Roots	Stimulant, Diuretic, sudorific, scorpion stings and antidote against snakebite	(Salehi et al. 2019)
Leaves	Breast cancer, body pain and joint ache	(Taufiq-Ur-Rahman et al. 2005; Choudhary and Singh 2018)
Stem	Body pain and joint ache	(Chahal et al. 2011)

activity, cleanse mucus, and reduce postpartum discomfort because of their oxytocic qualities. The medicinal uses of different part of *P. longum* were given in Table 6.1.

6.5 Phytochemicals Composition of *P. longum*

The *P. longum* was regarded as a valuable medicinal herb. The phytocompounds present in the *P. longum* are health-promoting chemicals. The *P. longum* is used as an adulterant of *P. nigrum* due to similar phytochemical components and it is less expensive compared with *P. nigrum.* The approximate phytochemical composition of the *P. longum* is given in Table 6.1. The fruit tested positive for proteins, alkaloids, carbohydrates, starch, volatile oils, saponins and amygdalin but not for tannins (Tharakan and Madhavan 2017). The approximate phytochemical composition of the *P. longum* is given in Table 6.2 (Babu et al. 2006). It contains many minerals, including Betaine, calcium, choline, folate, iron, magnesium, manganese, niacin, pantothenic acid, riboflavin, potassium, selenium, salt, thiamine, vitamins A, B, C, D, K, vitamin B12, vitamin B6, and zinc. Piperine, piperlonguminine, piperlongumine, and methyl-3,4,5-trimehoxycinnamate are the major alkaloids present in *P. longum.* The spike of *P. longum* contains alkaloids like piperine, piperolactum A, piperolactum B, piporadione and piplatin. The alkaloids present in the roots are piper longuminine (0.2–0.25%), piperlongumine (0.02%) and piperine. The *P. longum* contains essential oils which are made up of long chain hydrocarbons, caryophyllene, monoterpenes and sesquiterpenes. It also contains Piplartine, pipernonaline, pipercide, piperundecalidine, sesamine, beta-sitosterol, dieudesmin, and dihydrostigmasterol (Pandey 2018). The seeds of the *P. longum* contain dieudesmins, palmitic, stearic, hexadecenoic, syvatine, linoleic, linolenic, oleic, arachidic acid, behenic acid and high saturated acids (Choudhary and Singh 2017). The phytochemical present in the P. longum were mentioned in Table 6.3.

Table 6.2 Approximate composition of *P. longum*

Parameters	Percentage of components (%)
Moisture	9.5
Protein	12.2
Starch	39.5
Fiber	5.8
Total ash	5.9
Insoluble acid	4.2
Volatile oil	1.5
Fixed oil	6.6
Piperine	4.5

Table 6.3 Phytochemicals present in *Piper logum*

Chemical compounds	Plant part	References
Proteins, Alkaloids, Carbohydrate, Starch, Volatile oils, Saponins, Amygdalin, Dieudesmins, palmitic, stearic, hexadecenoic, syvatine, linoleic, linolenic, oleic, arachidic acid, behenic acid and high saturated acids	Fruit	(Tharakan and Madhavan 2017)
Piperine, Piperolactum A, Piperolactum B, Piporadione, Piplatin	Root	(Babu et al. 2006)
Caryophyllene, monoterpenes and sesquiterpenes. Piplartine, pipernonaline, pipercide, piperundecalidine, sesamine, beta-sitosterol, dieudesmin, and dihydrostigmasterol	Essential oil	(Pandey 2018)

6.6 Current Cultivation Practices and Challenges

6.6.1 *Current Cultivation Practices*

6.6.1.1 Soil and Climate

P. longum is naturally grown in tropical humid climates and prefers shady moist areas, well-drained sandy soil with a pH range of 5.5–8.5 with rich humus. Laterite soils with organic matter content and good water-holding capacity are suitable. Areas with good rainfall and high relative humidity are suitable. Areas with good rainfall and high relative humidity are suitable for successful growth. It is a shade-loving plant but for better fruiting 50% shade is best. It grows well in the shade of trees in areas of abundant rainfall and can also be grown as an intercrop in coconut plantations in the plains, though an altitude of 900–1500 m above sea level is recommended (Manoj et al. 2004).

6.6.1.2 Propagation Material

P. longum seeds have a low germination rate so, vegetative propagation is the most commonly used to cultivate it. Normally, root tiller or stem cuttings are chosen as propagation material. Propagation via stem cuttings commonly by suckers or by layering older branches. Planting material includes three to five nodded cuttings that are 15–20 cm long and have three nodes that are 5 cm in length. Terminal shoots are also used for planting. Cuttings can simply be rooted in a potting mixture, planted in nursery beds or polythene bags, and watered on alternate days. After planting, rooting takes roughly 15–20 days. In two months, the rooted cuttings will be suitable for transplanting; cutting establishment of 100% has been seen. The ideal time for raising in the nursery is from March to April. At the start of the rainy season, cuttings can either be planted there immediately or they can be rooted in a nursery first before being transplanted to the field. In the nursery, mealy bugs attack the roots. Mealy bug assault is lessened by spraying or soaking with 10% Aldrin (Babu et al. 2006).

In Arunachal Pradesh, an experiment was conducted on plant propagation using stem cutting and the results were promising. Cutting with three-node, two-node and one-node stems are the three different types of cuttings made from the stem. In the nursery, the cuttings were grown in plastic bags filled with sand, soil and cow dung. The planted *P. longum* cuttings needed frequent attention and irrigation. After one month, all three cuttings started to successfully develop new buds. After two months, the observation revealed that the stem cuttings had a good survival rate of at least 70% in all three types of cuttings, demonstrating the stem's ability for regeneration like in other Piper species. The three-node cutting should be used for commercial cultivation even if all three cutting types had positive results. This is because it will promote the growth of more shoots that can produce fruiting spikes in greater numbers. During the selection process, one female plant must be chosen. Piper had earlier recorded the production of fruits without their male counterparts, and *P. longum* had also seen this. However, at least 5% of male plants should be grown to determine whether the fruits are healthy. The well-developed seedlings were transferred in the field at a 2 × 2 m distance under mixed cropping or 1 × 1 m in mono-cropping. During the first year of the plantation, before the start of the monsoon, the plantation area can be fertilised with Farm Yard Manure (FYM) or any other organic manure. Depending on the quality of the soil and the weather, irrigation is typically performed two to three times per week. It's important to prevent weed growth on the plantation. Cuttings should be raised between March and April so they can be planted in the ground between May and July (Gajurel et al. 2021).

P. longum is planted for its roots in certain of Vishakhapatnam district's hilly regions in Andhra Pradesh. It is planted as a perennial in small plots of 25–50%, with the first harvest starting 18 months after planting. The roots are harvested for 10–30 years. The stem near the ground is removed and the roots are dug up, washed, and placed in the shade for a day before they are cut into pieces that measure 2.5–5 cm long. Each hectare yields 500 kg of roots on average (Babu et al. 2006).

6.6.1.3 Land Preparation and Planting Method

The field requires two to three ploughings, one or two harrowing and levelling based on the slope of the land, to allow for the drainage of excess water. In flooded conditions, the crop cannot survive. In Kerala and West Bengal, the best time to plant *P. longum* is in May and June. In between plant to plant and row to row 60 × 60 cm space should be maintained. If the plant is grown in a nursery then it should be planted before the actual time. In Maharashtra, the planting is usually done in February and March. When planting 3–4 nodes are retained inside the soil and 2 nodes continue to grow above the soil line. Typically, sprouts begin within 15–20 days after plantation. To provide shade and support to the sprouted cuttings of *P. longum* hard wood cutting of *Erythrina varigata* and *Sesbania grandiflora* were planted near the cuttings. In Kerala, It was cultivated as an intercrop in areca nut and coconut groves with irrigation (Babu et al. 2006).

6.6.1.4 Irrigation, Manuring and Inter Culture Operations

Irrigation is most important during the summer seasons. Depending on the soil's ability to retain water, irrigation should be done once or twice a week. Irrigated crops can still produce fruit during the summer. During land preparation, FYM or another organic manure is applied at a rate of 20 tonnes per hectare. Chemical fertilizers are not advised for this crop. Weeding should be done as needed during the first year. Usually, 2 to 3 weeding is enough. Once the plant has grown and covered the field, no significant weed issue has been seen (Babu et al. 2006).

6.6.1.5 Diseases and Pests

P. longum was mainly harmed by stem rot, phytophthora leaf, and anthracnose. The loss brought on by these diseases is substantially reduced by spraying 0.5% of Bordeaux mixture every 15 days and soaking the soil with 1.0% of Bordeaux mixture every month. When *Helopeltis theivora* (mealy bugs) are damaging delicate foliage and spikes, 0.25% neem seed kernel extract or any other neem-based insecticides can be applied as a spray.

During the monsoon season, *Colletotrichum* can cause the rotting of vines and leaves. In the summer season, necrotic spot and blight of leaves were also observed in this crop which can result in whole or partial crop loss. The causative organism for these diseases are *Colletotrichum* and *Cercospora* and spraying a Bordeaux mixture (1%) throughout May and the next rainy season can control the loss caused due to these organisms (Babu et al. 2006).

6.6.1.6 Harvest and Yield

The vines begin to flower after 6 months of planting and they continue to produce flowers for almost the entire year. Two months after flowering, the spikes are collected when they are fully grown but not yet ripe, and they are sun-dried. If unpicked, they ripen and lose a significant amount of their pungency. Harvesting ripe or over-matured fruit decreases the quality of the crop and makes it difficult to break after complete drying. Indian long pepper is often grown as a four- to five-year crop because, after the fifth year, when output starts to decline and gradually becomes unprofitable, it needs to be replaced. In these circumstances, fruits, roots, and thicker basal stem parts are additionally collected before crop abandonment.

In Kerala, depending on the fruit's maturity, three to four pickings are made. When dry spike was irrigated, the output was 400 kg/ha the first year, grows to 1.0–1.25 t/ha in the following years, and then reduced. A crop that is rain-fed has a shorter flush of fruiting, which lowers yield. A hectare typically yielded 500 kg of dried roots. Economic conditions can change from year to year on the market for medicinal plants (Babu et al. 2006).

6.6.1.7 Varieties and Cultivars

P. longum is a multipurpose medicinal plant that is thriving in several nations. The major purpose of growing *P. longum* is to utilise it as a spice (Weragoda 1980). Viswam was the first improved variety in *P. longum* was done through clonal selection by Kerala agriculture university in 1996. Viswam's potential yield was reported as 472 kg ha^{-1}. The variety produces an acceptable yield for 240–270 days out of every year. Fruits have 2.83% alkaloid and 20% dry matter, respectively. The other important varieties of *P. longum* for production are Vellanikara-1 (Visvum Tippili) and Yercaud (PI)-9. Goli thippili, pipal nan sori, Asali suvali, and cheemathippalli are the other local varieties. Table 6.4 shows the list of varieties and cultivars of *P. longum*.

Table 6.4 List of *P. longum* varieties and cultivars

Varieties	State
Goli thippili	West Bengal
Pipal nan sori	Maharashtra
Asli and Suvali	Assam
Vishwam	Kerala Agriculture University and CIMAP
Vellanikara-1 (visvum tippili)	–
Yercaud (PI)-9	–
Cheemathippalli	–

6.7 Current Agriculture Problems and Challenges

P. longum cultivation is adversely affected by the effects of the recent economic boom, industrialisation, and urbanisation as well as the resultant inevitable decline of viable agricultural lands, as it must compete with other crops for the remaining arable ground. *P. longum* grows in both humid and warm climatic conditions (Basak and Mohapatra 2015). Extreme fluctuations in the weather can have a bad impact on growth and productivity. It prefers to grow in well-drained and fertile soil highly rich in organic matter. pH of soil is very important and it prefers to thrive in neutral pH. Additional stressors on new kinds include drought brought on by global warming, the emergence of new pests and insects, and the shifting host preferences of pests and insects. In light of this, attempts have been made to create novel types suited for increased cultivation. Weed management is necessary to maintain the growth condition for *P. longum*, weeds compete with *P. longum* for sunlight, water, and nutrients which decrease the quality and yield. The unpredictable changes in prices and market demands are another challenge faced by growers and distributors. Agricultural land has decreased due to land degradation and urbanisation, which is a major challenge that limits suitable land availability (Kubiak et al. 2022).

6.7.1 Improvement Objectives

To improve the problems faced in traditional methods, tissue culture techniques could be used to reduce this difficulty. In vitro, micropropagation has many benefits over traditional methods of producing medicinal plants, including the ability to produce large quantities of desired plants, identification and production of clones with desired traits, conservation of threatened plant species, and the development of new and improved genetically modified plants, cryopreservation of genetic material, and production of secondary metabolites (Manoj et al. 2004).

Compositing, organic matter and suitable nutrients incorporation will help to improve the fertility of soil. Ensure the nutrient balance and optimum pH level by conducting regular soil test. Minimize the chemical fertilizers by developing integrated pest and diseases management process which includes intercropping with pest repellent plants, crop rotation, biological control methods and by using resistant cultivars. Climatic adaptation can be controlled by installing irrigation system to ensures the continues supply water during summer seasons. Agroforestry practices are adapted to regulate the shade and moisture content and by selecting weather resistant cultivar. Implement weed controlling techniques like manual weeding, cover cropping and mulching to maintain the controlled growing condition and to reduce the competition for sunlight, water, and nutrients (Rana and Rana 2019).

6.7.2 *Role of Biotechnology*

Biotechnology plays a crucial role in breeding *P. longum.* Techniques like tissue culture enable the mass propagation of plant varieties (Hasnain et al. 2022). Molecular markers and DNA sequencing help in the identification and selection of desirable features such as high piperine content and disease resistance (Kumari et al. 2019). Genetic engineering allows for the introduction of novel features into *P. longum*, hence improving its therapeutic benefits. Biotechnology also aids in the identification and preservation of plant genetic resources. Overall, biotechnology speeds up the breeding process enhances trait selection, and leads to the generation of superior *P. longum* types with higher therapeutic value (Choudhary and Singh 2018). Genetic resources are crucial global assets that can help improve food security. The decline in crop variety due to agricultural technology is known as genetic erosion. This effect is visible in spice crops, where the loss of cultivated variety represents a bottleneck in diversity levels that occurred throughout the crop's history. This bottleneck covers the initial replacement of traditional landraces with contemporary cultivars, as well as long-term declines in diversity caused by modern breeding procedures. (Salgotra and Chauhan 2023).

6.7.3 *Germplasm Diversity and Conservation*

Genetic resources are crucial global assets that can help improve food security. The decline in crop variety due to agricultural technology is known as genetic erosion. This effect is visible in spice crops, where the loss of cultivated variety represents a bottleneck in diversity levels that occurred throughout the crop's history. This bottleneck covers the initial replacement of traditional landraces with contemporary cultivars, as well as long-term declines in diversity caused by modern breeding procedures (Puneeth et al. 2023). Plant breeders use germplasm, which is made up of a variety of collections, as the starting point for creating new varieties. These collections include rudimentary cultivars, naturally occurring hybrids, wild and weedy relatives, wild species, outdated varieties, elite lines, breeding lines, mutants and polyploids, and systematically developed interspecific and intergeneric hybrids (Parthasarathy et al. 2006).

6.7.3.1 Germplasm Diversity

Piper longum, or long pepper, is a significant plant species recognised for its medicinal and culinary properties. Its genetic variety is of great interest at both the national and international levels, owing to its numerous uses and possible economic advantages. Nationally, nations with a strong herbal medicine legacy, such as India and Indonesia, recognise the need of conserving Piper longum's genetic variety for medical and cultural purposes (Pandey 2018). *Piper longum's* genetic variety is recognised as a worldwide asset, and efforts are being made to protect and utilise its unique genetic resources for sustainable agricultural and medicinal reasons (Gajurel et al. 2021). Using Random Amplified Polymorphic DNA (RAPD) markers, the genetic diversity in *P. longum* germplasm gathered from India's North East states was examined. For the RAPD analysis, a total of 16 *P. longum* germplasms were taken into account, and they produced a sizable amount of high-quality DNA (A260/A280 = 1.8). The results showed that differential polymorphism in 16 distinct germplasm samples varied in percentage of polymorphism in twenty decamer random primers (from 25.01 to 60.00) (Sen et al. 2010).

The RAPD-PCR using primers which showed following results OPB 1 (57.14%), OPA08 (50.5%), OPB02 (50.0%), OPB20 (50.0%), and OPA 11 (50.0%), OPB19 (60.00%) showed the highest amount of polymorphism. One hundred and five (105) bands were amplified in all, with forty-one (41) of them being polymorphic (Sharmah et al. 2022a).

6.7.3.2 Genetic Resources Conservation Approaches

To conserve forest genetic resources, ex-situ (off-site) and in-situ (on-site) conservation techniques are primarily utilised. The in situ conservation strategy protects and maintains species populations in their natural habitat (Maxted 2001).

6.7.4 *In-Situ Conservation*

Since the species being protected must co-evolve with climatic changes, in situ biodiversity conservation is preferable to ex situ. It is critical to build sanctuaries for wild relatives and other commercially significant species in the Western Ghats and North Eastern India–the two distinct locations of the genesis of Piper species in India. Because the species variety is greater in Silent Valley, Velayudhan et al. (1999) proposed establishing a gene reserve for Piper species there. Parthasarathy et al. (2006) named the Western Ghats region from Wayanad to Palghat District, as well as Kollam, Tirunelveli, and Thiruvananthapuram, as Piper hotspots in South India. It will be acceptable to construct Piper gene sanctuaries in these locations (Yadav et al. 2020).

6.7.5 *Ex Situ Conservation*

The Indian Institute of Spices Research (IISR) germplasm accessions are preserved in both the field gene bank and the nursery. Two backup locations, one at Central Plantation Crops Research Institute (CPCRI), Seed Farm, Kidu and the other in

Cardamom Research Centre (CRC) Appangala, Karnataka, were chosen to safeguard the germplasm accession for cultivars and wild germplasm, respectively (Sharmah et al. 2022a).

6.7.6 In Vitro Conservation

P. longum, often known as long pepper, in-vitro conservation is the growing and preservation of plant tissues, cells, or organs in a controlled laboratory setting. The sterilisation of explants, such as shoot tips or axillary buds, is followed by their placement on a nutrient-rich agar media. Explants develop into callus under sterile circumstances, which can be stimulated to differentiate into shoots or roots. The resultant plantlets are acclimatised to greenhouse conditions before being moved to the field to grow. *P. longum* may be stored for a long-time using cryopreservation, a process that includes freezing plant material at ultra-low temperatures and utilising cryoprotective chemicals. Other techniques of conservation, such as slow growth storage and meristem culture, are also used (Soniya and Das 2002). Cryopreservation and slow growth approaches have different benefits and drawbacks for maintaining Piper longum. Cryopreservation allows for the long-term storage of genetic material at extremely low temperatures while preserving its genetic integrity and therapeutic qualities (Acheampong Amankwaah et al. 2023). This technology allows for easy transportation and can be cost-effective over time. However, it necessitates specialised equipment and knowledge, poses a danger of contamination, and can result in cell damage and viability loss over time. Slow growth procedures, on the other hand, retain Piper longum‘s genetic integrity while eliminating the hazards associated with freezing and thawing. These approaches provide greater flexibility in growing circumstances, a lesser danger of contamination, and simplicity of maintenance once established. However, they are sluggish and resource-intensive, which may result in genetic drift and restricted application for specific kinds (Goulet et al. 2017).

A total of 35 accessions are now held in the in vitro gene bank thanks to standardised slow growth procedures for in vitro conservation. Aside from the IISR at Calicut, the National Bureau of Plant Genetic Resources in New Delhi has an in vitro conservatory for Piper species (Krishnamurthy et al. 2008).

P. longum in vitro cultures were established using three explants: nodal, internodal, and petiolar segments. On semi-solid Murashige and Skoog (MS) medium with 1 mg/l 6-benzyladenine (BA) supplement, several shoots were induced. The medium and tissue browned significantly less with the addition of ascorbic acid (40 mg/l). Petiolar explants showed the best shoot regrowth and were thus employed in all subsequent trials. An indexing approach for assessing bacterial contamination in well-established shoot multiplication cultures was developed (Rani and Dantu 2012).

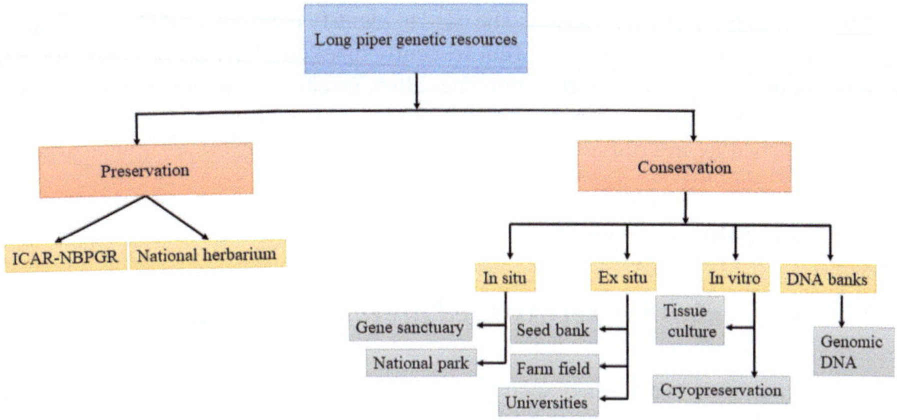

Fig. 6.4 Strategies for conservation of *P. longum* genetic resources

6.7.7 DNA Banks

Another method for maintaining plant genetic resources throughout time is the genetic data contained in a plant's DNA. Since DNA can be isolated and preserved, it is possible to effectively preserve the genetic material of extinct plant species as well as fossilised plant specimens (Rogers and Bendich 1988). Depending on the materials and species, plant genomic DNA may be extracted utilising a variety of regular approaches (Vijayan et al. 2004). Pure DNA may be preserved in a variety of ways. The most often used procedure is to dissolve the DNA in TE buffer (10 mM Tris–HCl and 1 M EDTA) and store it at 80 °C. Because DNA is exceedingly durable, pure DNA may be preserved in alcohol forever; it is believed that completely hydrated DNA can be held at ambient temperature for 10,000 years before depolymerizing into minute pieces (Schlumbaum et al. 2008). Figure 6.4 shows the Strategies for conservation of Piper genetic resources describes the overall Strategies for conservation of Piper genetic resources.

6.7.8 Cytogenetics

The flowering plant *P. longum* has chromosomes in each of its cells, just like most other plants do. Chromosomes are thread-like structures found in the nucleus of cells that transport DNA, which contains genetic information. Scientists may learn about *P. longum*'s chromosomal makeup and organisation by examining its cytogenetics, which is helpful in knowing its genetics and evolutionary links. *P. longum* has a basic chromosome number of $2n = 24$ to $2n = 96$. Karyotyping, the process of ordering and analysing chromosomes according to their size and shape, can reveal more about the structure and variety of *P. longum*'s chromosomes. Fluorescence in situ hybridization (FISH) may also be used to identify particular DNA sequences or genes on *P. longum* chromosomes. Fluorescently labelled DNA probes that attach to complementary sequences on chromosomes allow researchers to observe the position and distribution of certain genetic markers (Babu et al. 2006).

6.8 Molecular Breeding

6.8.1 Molecular Marker-Assisted Breeding (MMAB)

Molecular markers are distinct sections of DNA that differ across people and can be connected to certain features of interest. Breeders may select individuals with desired qualities more efficiently and precisely than traditional breeding methods by analysing the genetic makeup of distinct *P. longum* plants using molecular markers. The procedure usually consists of two major parts. To begin, a mapping population is established by crossing several *P. longum* plants with known features of interest. The offspring are then genotyped using molecular markers to determine which markers are related to the target characteristics (Kumari et al. 2019). MMAB in Piper longum to select for high piperine content, a major bioactive component responsible for the plant's therapeutic qualities. The researchers found molecular markers associated with the piperine production pathway and utilised them to screen a population of *Piper longum* plants. This strategy allowed them to discover plants with high piperine content early in the breeding process, saving time and resources over traditional breeding approaches (Tripathi et al. 2022). MMAB for *Piper longum* entails applying this method to additional key qualities, such as disease resistance, abiotic stress tolerance, and agronomic characteristics. Furthermore, combining MMAB with other breeding procedures, such as marker-assisted recurrent selection (MARS) and genomic selection (GS), may improve the efficiency and precision of *Piper longum* breeding programmes.

6.8.2 Functional Genomics

Primers based on the flanking sequences of the aforementioned ESTs were employed to screen the genomic DNA of male and female *P. longum.* Primers were created in the following ways. FADPA1-Forward and FADPA1-Reverse primers were developed from Gene Bank sequence AY168429, FADPD6-Forward and FADPD6-Reverse primers were developed from Gene Bank sequence AY168431, FADPC6-Forward and FADPC6-Reverse primers were developed from Gene Bank sequence AY168430, FADPE19-Forward and FADPE19-Reverse primers were developed from Gene Bank sequence AY168432, FADPC forward and reverse primers from Gene Bank sequence AY168435, FADPB forward and reverse primers from Gene

Table 6.5 Molecular markers used for genetic characterization in *P. longum*

Marker type	Purpose	Country	References
RAPD	Genetic diversity of germplasm	India	(Sharmah et al. 2022b)
RAPD, SCAR marker	Sex determination in plant	India	(Manoj et al. 2005)
RAPD, RFLP, ISSR, AFLP	Selection of microprogramed plants	India	(Joshi et al. 2023)
SSR	Understanding piperine pathway	India	(Dantu et al. 2021)
RAPD	selection of micro propagated plants	India	(Parani et al. 1997)

Bank sequence AY168434, and MADP5 forward and reverse primers from GenBank AY168438, sequences. With the advancement of genome sequencing technology, more SAP (Stress associated proteins) genes have been identified in a variety of plant species since the discovery of the first one, the OsiSAP1 (OsSAP1) gene in rice. However, no SAP genes have been identified from the pepper plant. SAPs with somewhat good characterization have been discovered to feature an A20 domain at the N terminus and/or an AN1 domain at the C terminus, both of which are substantially conserved across species (Sharmah et al. 2022a). The Molecular markers used for genetic characterization in Piper longum as depicted in Table 6.5.

6.8.3 *Bioinformatics*

Long pepper bioinformatics have not grown much as other agricultural crops. ICAR-NBPGR Gene bank which collects, preserves the indigenous varieties of long pepper. NBPGR works to determine the genetic and agricultural value of germplasm accessions through field and laboratory research as the caretaker of these precious resources. A dataset of compounds found in *P. longum* was created by conducting an exhaustive literature review and mining public database resources. Relevant research publications were chosen and manually reviewed from PubMed-NCBI (https://www.ncbi.nlm.nih.gov/pubmed/). Three UNPD (Universal Natural Products Database) (pkuxxj.pku.edu.cn/UNPD), TCMSP (Traditional Chinese Medicine Systems Pharmacology), and PCIDB (PhytoChemical Interactions DB) (http://www.genome.jp/db/pcidb) databases were searched for potentially active phytochemicals found in *P. longum* (George et al. 2022).

6.8.4 Genetic Engineering

P. longum is a vital medicinal herb. It possesses antibacterial, anti-inflammatory, and antioxidant properties. *P. longum* genetic alteration has the potential to increase its present characteristics for the benefit of mankind. When *P. longum* callus tissue was in vitro cultured and transformed with *Agrobacterium tumefaciens*, Gus gene expression was found. Agrobacterium strain with plasmid pBECK2000 harbouring GUS and kanamycin genes is employed (Banerjee and Mallick 2014). The callus of *P. longum* and the *Agrobacterium* strain were co-cultivated at 28 °C for 2–3 days of incubation, and infection times varied from 5 to 10 min depending on the acetosyringone concentration (50–200 M). The infected callus genomic DNA was extracted, and polymerase chain reaction was used to do a molecular study of the plant to detect the expression of the Gus gene, which revealed a stable transformation (Sen et al. 2010).

Differences in genotype and morphology between three female *P. longum* variations, one each from Assam and Calicut, and one released variety, were discovered. Viswam was studied in order to provide a standard and effective technique for plant regeneration. These types are genetically distinct, according to RAPD research utilising random oligonucleotide primers. In comparison to the Assam variety, the Viswam and Calicut types are genetically similar (95% similarity). The potential of node, internode, and leaf explants from all three kinds was compared. Leaf explants from several types had the greatest potential for regeneration. Among the kinds evaluated, the Viswam variety showed the best morphogenetic response, followed by varieties from Calicut and Assam. An effective approach for regeneration from leaf calli of all three genotypically distinct types has been developed. Plants regenerated from leaf explants were rooted, hardened, and grown on soil under normal growth conditions.

6.8.5 Mutagenesis

Through tissue culture techniques, long pepper can be exposed to mutant breeding. Plant tissues, such as shoot tips or callus cultures, are treated with mutagenic chemicals in this method. The mutagenized tissues are subsequently regenerated into complete plants, which are then evaluated for acceptable characteristics (EFSA Panel on Genetically Modified Organisms (GMO) et al. 2021).

P. longum clonal multiplication protocols were standardized using shoot tip explants. *P. longum* has been shown to convert root meristems into shoot meristems, which then grow into plantlets. Plants were grown from *P. longum* leaf and stem explants using direct and indirect organogenesis. Root explants from *P. longum* were immediately regenerated into plantlets (Sarasan et al. 1993). *P. longum* microprogramed effectively on McCown's Woody Plant Medium (WPM) enriched with BAP and kinetin. In both species, WPM with 3 mg l^{-1} BAP and 1 mg l^{-1} kinetin was shown to be excellent for shoot regeneration and subsequent development from both leaf and stem explants, with or without an intervening callus phase. Several (10–100) shoot primordials in the form of organogenesis could be acquired in another 20–30 days, and more than 40% of these primordials had excellent elongation and maintained normal development. When growth regulators were removed from the culture media, these shoots formed strong root systems. When the growth regulators were taken out of the culture media, the roots developed healthy root systems. When these grounded plantlets were cultured in culture media with 3 mg l^{-1} BAP and 1 mg l^{-1} kinetin, root meristems were converted to shoot meristems, which then grew into shoots and eventually plantlets. More than 90% of the regenerated plantlets were easily grown in soil (Babu et al. 1997, 2006). Shoot tips could be preserved in WPM with annual subculture under low growth conditions without the use of growth regulators. After one year of storage, the plantlets proliferated normally, and the rooted plantlets were successfully planted out. This contributes to the preservation of long term genetic information in in-vitro gene banks (Nirmal et al. 2012).

6.8.6 *Hybrid Cultivars*

Pippali Shakti: The resulting variety is noted for its excellent yield potential, immunity to common illnesses, and enhanced plant height and fruit size consistency. It is grown for its therapeutic virtues and is popular among farmers for its agronomic benefits.

Pippali Suvarna: The hybrid variety known for its high essential oil content and exceptional quality. It has long spikes, consistent fruit size, and an excellent market appeal. This hybrid is regarded for its therapeutic benefits and is grown for both residential and commercial use.

Pippali Madhur: The hybrid variety is cultivated especially for its exceptional sweetness and flavour. It is well-liked by customers who want a long pepper's gentler and sweeter flavour because of its lengthy spikes with sweet-tasting fruits.

6.9 Conclusion and Prospects

P. longum geographic range, botanical description, economic value, and phytochemical composition were all addressed in this study. This book chapter focuses on current cultivation practises and problems. It also provides information on the types and cultivars. It is a plant with significant economic and commercial value, and it can be used in a variety of formulations to increase bioavailability. Aromatic and medicinal species must be reproduced more quickly while inflicting the least amount of damage to their natural surroundings to meet demand. The micro-propagation technology allows for the fast multiplication of planting material, which might be immensely useful. The advancement of the fields of molecular biology, gene technology, and cell biology has benefited in the molecular/gene level knowledge of disorders. HTS approaches have grown from innovative target-directed screening assays and automation, considerably improving the industrial drug development process. Furthermore, in vitro gene banks have the potential to play a crucial role in supplying feedstock for this technological advancement. New pepper cultivars with increased productivity and adaptability must be developed using current biotechnological technologies. Biotechnological approaches, notably molecular marker-assisted breeding (MMAB), have greatly improved crop productivity in pepper and other spices. In pepper (*Capsicum* spp.), MMAB has helped to improve essential features including productivity, disease resistance, and fruit quality. For example, researchers have employed genetic markers related to genes influencing fruit size and shape to select for desirable features in pepper, resulting in the generation of superior cultivars with increased commercial appeal. Furthermore, biotechnology methods have contributed to the development of pepper types that are more resistant to diseases and pests. Genetic engineering techniques have been used to add genes that give resistance to bacterial wilt and pepper mild mottle virus (PMMoV), lowering production losses due to these infections. Other spices, such as turmeric (*Curcuma longa*) and cardamom (*Elettaria cardamomum*), have benefited from biotechnology treatments to improve crop yield. Tissue culture techniques, for example, have been employed in turmeric to mass multiply disease-free planting material, resulting in a steady supply of high-quality turmeric. Molecular markers have been used in cardamom to select for features such as high essential oil content and disease resistance, resulting in better cultivars with higher market value.

References

Acheampong Amankwaah V, Naa Ashiokai Prempeh R, Dorcas Quain M (2023) Plant cryopreservation importance, approaches and future trends. In: Quain M (ed) Biomedical engineering. IntechOpen

Atal CK, Ojha JN (1965) Studies on the genus piper. Part IV. Long peppers of Indian commerce. Econ Bot 19(2):157–164. https://doi.org/10.1007/BF02862827

Ayisha TP (1997) Yield & quality of *Piper longum* L. under differential spacing & manural regimes in coconut gardens. Thesis. Department of Agronomy, College of Horticulture, Vellanikkara

Babu N, Ravindran PN, Peter KV (1997) Protocols for micropropagation of spices and aromatic crops

Babu KN, Divakaran M, Ravindran PN, Peter KV (2006) Long pepper. In: Peter KV (ed) Handbook of herbs and spices. Woodhead Publishing, pp 420–437 (Chapter 25)

Banerjee S, Mallick MA (2014) Genetic transformation Gus gene expression in Piper longum. Int J Sci Res (IJSR) 3(7):661–663

Basak U, Mohapatra M (2015) Comparative appraisal of piperine content in various female spikes of *Piper longum* Linn., a RET medicinal plant of Odisha. World J Pharm Res 4:1048–1058

Bhatt N, Deshpande M (2021) A comparative review on medicinal plants used for the treatment of liver disorders as in Ayurved, Siddha and Unani [ASU] systems of medicine-part I-contextual and clinical aspects. https://doi.org/10.47191/ijahm/v11i5.04

Chahal J, Ohlyan R, Kandale A, Walia A, Puri S (2011) Introduction, phytochemistry, traditional uses and biological activity of genus piper: a review. Int J Curr Pharm Rev Res 2(2):130–144

Choudhary N, Singh V (2017) *Piper longum* Linn: a review of its phytochemicals and their network pharmacological evaluation, 169763

Choudhary N, Singh V (2018) A census of *P. longum*'s phytochemicals and their network pharmacological evaluation for identifying novel drug-like molecules against various diseases, with a special focus on neurological disorders. PLoS ONE 13(1):e0191006. https://doi.org/10.1371/journal.pone.0191006

Dantu PK, Prasad M, Ranjan R (2021) Elucidating biosynthetic pathway of piperine using comparative transcriptome analysis of leaves, root and spike in *Piper longum* L. Plant Biol

EFSA Panel on Genetically Modified Organisms (GMO), Mullins E, Bresson J, Dalmay T, Dewhurst IC, Epstein MM, Firbank LG, Guerche P, Hejatko J, Moreno FJ, Naegeli H, Nogué F, Sánchez Serrano JJ, Savoini G, Veromann E, Veronesi F, Casacuberta J, Lenzi P, Munoz Guajardo I, Raffaello T, Rostoks N (2021) In vivo and in vitro random mutagenesis techniques in plants. EFS2 19(11). https://doi.org/10.2903/j.efsa.2021.6611

Gajurel P, Kashung S, Nopi S, Panmei R, Singh B (2021) Can the *Ayurvedic pippali* plant (*Piper longum* L.) be a good option for livelihood and socio-economic development for Indian farmers? Curr Sci 120. https://doi.org/10.18520/cs/v120/i10/1567-1572

George JK, Fayad MA, Shelvy S, Sheeja TE, Eapen SJ, Charles S, Rai A, Kumar D (2022) Dataset of De Novo hybrid berry transcriptome profiling and characterization of Piper species (*Piper nigrum* and *Piper longum*) using Illumina and Nanopore sequencing. Data Brief 42:108261. https://doi.org/10.1016/j.dib.2022.108261

Goulet BE, Roda F, Hopkins R (2017) Hybridization in plants: old Ideas, new techniques. Plant Physiol 173(1):65–78. https://doi.org/10.1104/pp.16.01340

Hasnain A, Naqvi SAH, Ayesha SI, Khalid F, Ellahi M, Iqbal S, Hassan MZ, Abbas A, Adamski R, Markowska D, Baazeem A, Mustafa G, Moustafa M, Hasan ME, Abdelhamid MMA (2022) Plants in vitro propagation with its applications in food, pharmaceuticals and cosmetic industries; current scenario and future approaches. Front Plant Sci 13:1009395. https://doi.org/10.3389/fpls.2022.1009395

Joshi K, Chavan P, Warude D, Patwardhan B (2023) Molecular markers in herbal drug technology

Khandhar A, Patel S, Patel A, Zaveri M, Lecturer S (2010) Chemistry and pharmacology of *Piper Longum* L. Int J Pharm Sci Rev Res 5:67–76

Khare CP (2008) Indian medicinal plants: an illustrated dictionary. Springer Science & Business Media

Krishnamurthy KS, Prasath D, Kandiannan K, Suseela Bhai R, Saji K, Parthasarthy V (2008) Piperaceae crops-technologies and future perspectives

Kubiak A, Wolna-Maruwka A, Niewiadomska A, Pilarska AA (2022) The problem of weed infestation of agricultural plantations vs. the assumptions of the European biodiversity strategy. Agronomy 12(8):1808. https://doi.org/10.3390/agronomy12081808

Kumari R, Wankhede DP, Bajpai A, Maurya A, Prasad K, Gautam D, Rangan P, Latha M, John KJ, Suma A, Bhat KV, Gaikwad AB (2019) Genome wide identification and characterization of microsatellite markers in black pepper (*Piper nigrum*): a valuable resource for boosting genomics applications. PLoS ONE 14(12):e0226002. https://doi.org/10.1371/journal.pone.0226002

Manoj P, Soniya EV, Banerjee NS, Ravichandran P (2004) Recent studies on well-known spice, *Piper longum* Linn. NPR 3(4)

Manoj P, Banerjee NS, Ravichandran P (2005) Development of sex specific molecular markers in dioecious *Piper longum* L. plants by differential display. Plant Gen Resour Newslett 141:44–50

Maxted N (2001) Ex situ, in situ conservation. In: Encyclopedia of biodiversity. Elsevier, pp 683–695

Nirmal K, Yamuna G, Praveen K, Minoo D, Ravindran PN, Peter KV (2012) Cryopreservation of spices genetic resources. In: Katkov I (ed) Current frontiers in cryobiology. InTech

Pandey DK (2018) *Piper longum*: a concise review on Botany, Phytochem Pharmacol 5(12)

Parani M, Anand A, Parida A (1997) Application of RAPD fingerprinting in selection of micropropagated plants of *Piper longum* for conservation

Parthasarathy U, Saji KV, Jayarajan K, Parthasarathy VA (2006) Biodiversity of piper in South India—application of GIS and cluster analysis. Curr Sci Assoc 91(5):652–658

Puneeth GM, Gowthami R, Katral A, Laxmisha KM, Vasudeva R, Archak S (2023) On-farm conservation of crop diversity: a case study of landraces from Western Ghats of Karnataka, India (in Review)

Rana S, Rana MC (2019) Principles and practices of weed management, 3rd edn

Rani D, Dantu PK (2012) Direct shoot regeneration from nodal, internodal and petiolar segments of *Piper longum* L. and in vitro conservation of indexed plantlets. Plant Cell Tiss Organ Cult 109(1):9–17. https://doi.org/10.1007/s11240-011-0068-7

Rogers OS, Bendich AJ (1988) Extraction of DNA plant tissue. In: Gelvin SB, Schilpe, RA, Verna DS (eds) Plant molecular biology manual, Kluwer Academic Publishers, Dordrecht

Salehi B, Zakaria ZA, Gyawali R, Ibrahim SA, Rajkovic J, Shinwari ZK, Khan T, Sharifi-Rad J, Ozleyen A, Turkdonmez E, Valussi M, Tumer TB, Monzote Fidalgo L, Martorell M, Setzer WN (2019) Piper species: a comprehensive review on their phytochemistry, biological activities and applications. Molecules 24(7):1364. https://doi.org/10.3390/molecules24071364

Salgotra RK, Chauhan BS (2023) Genetic diversity, conservation, and utilization of plant genetic resources. Genes 14(1):174. https://doi.org/10.3390/genes14010174

Sarasan V, Thomas E, Lawrence B, Nair GM (1993) Plant regeneration in *Piper longum* L. (Piperaceae) through direct and indirect adventitious shoot development. J Spices Aromat Crops 2(1&2):34–40

Schlumbaum A, Tensen M, Jaenicke-Després V (2008) Ancient plant DNA in archaeobotany. Veget Hist Archaeobot 17(2):233–244. https://doi.org/10.1007/s00334-007-0125-7

Sen S, Skaria R, Muneer PMA (2010) Genetic diversity analysis in piper species (Piperaceae) using RAPD markers. Mol Biotechnol 46(1):72–79. https://doi.org/10.1007/s12033-010-9281-6

Sharmah AK, Sharmah D, Sen P (2022a) Determination of molecular diversity of ex situ conserved germplasm of *Piper longum* L. using RAPD markers

Sharmah AK, Sharmah D, Sen P (2022b) Determination of molecular diversity of ex situ conserved germplasm of *Piper longum* L. using RAPD markers

Shenoy PA, Nipate SS, Sonpetkar JM, Salvi NC, Waghmare AB, Chaudhari PD (2013) Anti-snake venom activities of ethanolic extract of fruits of *Piper longum* L. (Piperaceae) against Russell's viper venom: characterization of piperine as active principle. J Ethnopharmacol 147(2):373–382. https://doi.org/10.1016/j.jep.2013.03.022

Soniya EV, Das MR (2002) In vitro micropropagation of Piper longum—an important medicinal plant. Plant Cell Tissue Organ Cult 70(3):325–327. https://doi.org/10.1023/A:1016561521050

Taufiq-Ur-Rahman M, Ahmad Shilpi J, Ahmed M, Faiz Hossain C (2005) Preliminary pharmacological studies on Piper chaba stem bark. J Ethnopharmacol 99(2):203–209. https://doi.org/10.1016/j.jep.2005.01.055

Tharakan ST, Madhavan M (2017) Therapeutic potential of piper longum. Int J Pharm Biol Sci

Tripathi AK, Ray AK, Mishra SK (2022) Molecular and pharmacological aspects of piperine as a potential molecule for disease prevention and management: evidence from clinical trials. Beni-Suef Univ J Basic Appl Sci 11(1):16. https://doi.org/10.1186/s43088-022-00196-1

Velayudhan KC, Amalraj VA, Abraham Z, John KJ, Nizar MA, Asha KI (1999) Wild crop genetic resources of Silent Valley with special reference to in situ conservation of Piper species. In: Manoharan TM et al (ed), Silent Valley: whispers of reason, Kerala For. Dept, Thiruvananthapuram, India

Vijayan K, Awasthi AK, Srivastava PP, Saratchandra B (2004) Genetic analysis of Indian mulberry varieties through molecular markers. Hereditas 141(1):8–14
Weragoda PB (1980) The traditional system of medicine in Sri Lanka. J Ethnopharmacol 2(1):71–73. https://doi.org/10.1016/0378-8741(80)90033-1
Yadav V, Krishnan A, Vohora D (2020) A systematic review on *Piper longum* L.: bridging traditional knowledge and pharmacological evidence for future translational research. J Ethnopharmacol 247:112255. https://doi.org/10.1016/j.jep.2019.112255

Chapter 7
Breeding of Nutmeg (*Myristica fragrans* Houtt.)

Hasirumane Chinnappagowda Vikram and Nalumakkal Mini Raj

Abstract Nutmeg (*Myristica fragrans* Houtt.) is a vital perennial spice that stands dioecious in nature; rarely noticed as monoecious in cultivation. Nutmeg yields two types of spice, viz. nut and mace. The wild and related species of the genus *Myristica* are noticed in India, and the magnitude of germplasm distribution and its conservation by different research agencies are commendable. Nutmeg is a cross-pollinated crop; wide variability exists for morphological and biochemical characteristics. Many studies have been conducted to characterise the dioecy in nutmeg, resulting in no defined outcome to detect male and female plants in the juvenile phase. Biochemical constituents like myristicin, elemicin, sabinene and safrole are present in appreciable amounts in the volatile oil of kernel and mace. There is huge scope for exploring the variability to identify high-yielding genotypes and resistant sources to fungal pathogens. The multiple problems in nutmeg breeding are the prolonged juvenile phase, the difficulty in multiplying vegetatively in large numbers, the segregation in the gender of seedling populations into male and female plants, and the single ovule in the female flower. These issues would also impact the breeders in developing new nutmeg varieties. Few released varieties and farmers' selections are ruling in commercial cultivation of nutmeg. This chapter reviews the botany, flower characters, genetic sources, variability studies, high-yielding varieties developed, and crop improvement strategies of nutmeg.

Keywords Nutmeg · *Myristica fragrans* · Variability · Dioecious · Crop improvement · Clonal selection

H. C. Vikram (✉)
Department of Plantation Crops and Spices, College of Agriculture, Vellanikkara, Kerala Agricultural University, Thrissur 680656, India
e-mail: vikram.hc@kau.in

N. M. Raj
College of Agriculture, Vellanikkara, Kerala Agricultural University, Thrissur 680656, India
e-mail: n.miniraj@kau.in

J. M. Al-Khayri et al. (eds.), *Biodiversity and Genetic Improvement of Herbs and Spices*, Advances in Plant Breeding Strategies 8, https://doi.org/10.1007/978-3-031-81838-7_7

7.1 Introduction

Nutmeg is a well known perennial species that yields twin spices, nutmeg, and mace. The nut is a hard, brown kernel encased in a thin, brittle shell and is surrounded by a scarlet coloured aril (mace). The genus *Myristica* is derived from the Greek word *Myron*, meaning a sweet liquid extracted from the plant (Everett 1981). Nutmeg is an important spice, commonly used in kitchens and the traditional medicine. The spice has got significant potential in the industry due to its flavour in food and bakeries worldwide and as a component in many value-added products. Due to the therapeutic properties, it remains a staple in traditional pharmacopoeia. The nutmeg tree is native to Indonesia, particularly the Moluccas Islands, where it has a rich history spanning over 3500 years and it is now being cultivated in nearly about 12 countries, with some secondary domestication centers overlapping. Centres of domestication overlap the centre of origin in nutmeg. The genus *Myristica* is comprised of evergreen trees growing in Africa, Asia, the Pacific Islands and the Americas.

The nutmeg found its way from its original home to South-East Asia. Later the spread of the tree was closely associated with the colonialist history of the seventeenth and eighteenth centuries (Fig. 7.1). The history of the translational distribution of nutmeg is troubled with its share of competitors, clandestine voyages, sea as well as bio-piracy, and a lack of cogency on the chronology and false identity of specimen *Monodora myristica* Dun also known as Tobago nutmeg or Jamaica nutmeg or African nutmeg and *Pichurim* nut botanically called as *Nectandra pichurim* instead of true nutmeg. In 1843, nutmeg was introduced into Granada in the West Indies, where large-scale plantations were subsequently established. Eventually, Granada became one of the largest producers and exporters of nutmeg till the pre hurricane. This spice tree is now naturalised in Granada, Sri Lanka, India, Malaysia, Mauritius, Zanzibar, Seychelles, Reunion, Guatemala, and the Solomon Islands.

Nutmeg was introduced to India by the British in the eighteenth century. In India, it is cultivated in Kerala, certain pockets of Karnataka, Tamil Nadu, Goa, Maharashtra, North East India and the Andaman Islands (Krishnamoorthy 2000). India's most aged nutmeg habitats in the Pala and Kalady regions of Kerala; they are believed to be over 150 years old. Nutmeg is customarily grown along riversides, where it flourishes in the nutrient rich silt heaped by the rivers.

Owing to its therapeutic properties, nutmeg is used immensely in the traditional pharmacopoeia. The quality of the nutmeg varies with habitat and genotypes (Lewis et al. 1976). The significant features of nutmeg are due to presence of volatile oil, oleoresin and fixed oil which are collected from dried nutmeg and mace. The flavour and therapeutic activities are due to the presence of volatile oil, whose content range from 6 to 16% in nut and 4–15% in mace. Myristicin, elemicin, sabinene and safrole are the principal compounds present in nutmeg volatile oil; which constitutes about 80% of the oils. Greater variation is found in the composition as well as proportion of these constituents of oils isolated from nut and mace (Rema et al. 2015). Nutmeg fixed oil is obtained by pressing the floor and cooked or steamed kernels. The outcome is 24–30%. It is a concrete, colour of yellowish-red fat possessing nutmeg odour. The

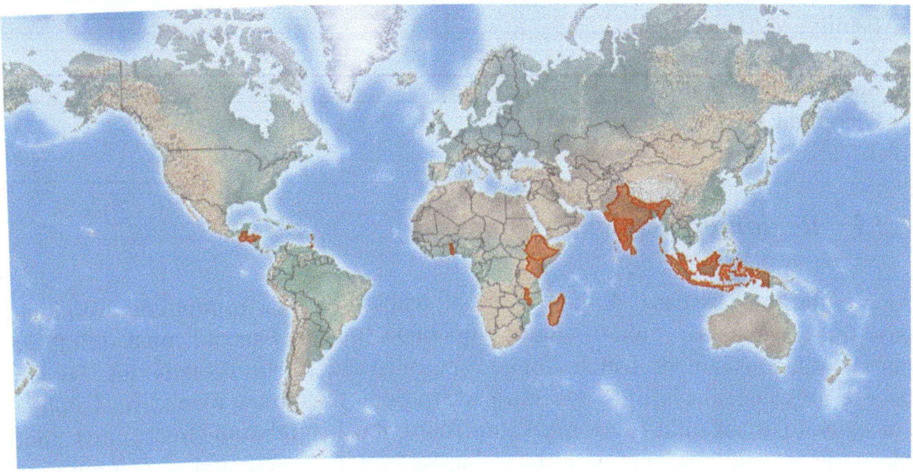

Fig. 7.1 Distribution of nutmeg cultivation in the world (highlighted in the colour)

crude fixed oil consists of essential oil up to 10–12%. The significant components present in the fixed oil are trimyristin. It also contains eight fatty acids, with myristic acid comprising the most prominent element. Fixed oil consists primarily of saturated fats (90%) and 10% unsaturated fats (Leela 2008).The oleoresin of nutmeg and mace is predominantly used in food flavouring within processing industries. Oleoresin is typically sold on a neutral base such as salt, dextrose, flour or husk. The oleoresin content in nutmeg and mace ranges from 20 to 30%. To improve the yield and quality of nutmeg, clonal selection is an effective method, particularly an added advantage of vegetative propagation. Although nutmeg is a perennial crop, this approach can still be successfully applied. Traditionally, new nutmeg varieties have been developed by identifying mother plants with elite characteristics and evaluating them over several years to understand their yield patterns. Those that consistently exhibit superior traits are eventually released as new varieties.

7.2 Botanical Classification

Kingdom: Plantae
Subkingdom: Viridiplantae
Infrakingdom: Streptophyta
Superdivision: Embryophyta
Division: Tracheophyta
Subdivision: Spermatophytina
Class: Magnoliopsida
Superorder: Magnolianae
Order: Magnoliales

Family: Myristicaceae
Genus: *Myristica* Gronov
Species: *fragrans*
Binomial: *Myristica fragrans* Houtt.

7.2.1 Regional/Vernacular Names

Arabic: basbas, basbasa, fuljan, josat al teeb; Chinese: roudoukou, roudoukouyi (mace), rouguo, roukou, yuguo, yuguohua (mace); Danish: muskatblomme, muskatnød, muskatnød; Dutch: foelie, nootmuskaat, nootmuskaatboom; French: fleur de muscade, macis, muscade, muscade (product); German: macis, muskatblüte, muskatblüte, muskatnut (nut), muskatnutbaum, muskatnutbaum (tree), muskatnut; Japanese: me-su, na-tu-me-gu, ni-kuzu-ku; Korean: me-i-seu, meisu, neo-teu-meg, Persian: basbaz, djushendi, ğwzhndý, jouzhendi, jouzboyah; Portuguese: bicuiba, flor de nozmoscada (Brazil), macis, moscadeira; Spanish: nuez de banda, macia, macis, moscada; Swedish: muskotblomma, muskotnöt, muskott (Katzer 2004; Ravindran 2017).

7.3 Botanical Description

Nutmeg is a member of the tiny and elementary family Myristicaceae with genus *Myristica* being the most extensive genus which encompassing approximately 120 species. The Myristicaceae family harbours about 18 genera and 300 species. The members of the family are pantropical, predominately found in the rainforests of Asia, Africa, Madagascar, South America and Polynesia (Sinclair 1958). The nutmeg is an evergreen tree growing up to a height of 20 m but is often much smaller. The leaves are alternately arranged, simple, exstipulate and chartaceous. The petiole is approximately 1.0 cm in length, and the leaf blade is elliptical to lanceolate, measuring between 5 and 15 cm length and 3–7 cm width. The base of the leaf blade is acute, the margin is entire and apex is acuminate. The leaves emit a scent when discoloured, and they feature 10–14 pairs of lateral nerves. Nazeem (1979) and Vikram et al. (2016b) observed cyclical shoot development in nutmeg, characterised by periods of growth followed by quiescence. Six growth flushes are produced annually, although not all shoots sprout at a time, resulting in continuous development.

Inflorescences are tiny, produced axillary in clusters (cymes). Male inflorescences are generally more in number produced in bunches, and females are solitary or rarely it may be produced in cluster with1–3 flowers. Joshy (1946) described the inflorescence of *M. fragrans* as an axillary raceme, but Joseph (1980) observed itas a branched raceme in male plants and a simple cyme in a female plant. However, the taxonomic description confirms that the inflorescence in both male and female trees is simple axillary cyme. Flowers are drooping, creamy yellow. Flowers are small

(< 1 cm) and visually inconspicuous but are fragrant when first opened. The prior attractant signal is furnished by floral odour, a strong, agreeable, definite aroma in *M. fragrans* and *M. insipida* (Armstrong and Drummond 1986). Nutmeg flowers are glabrescent with scant, very slight tomentum, faded yellow, with a three-lobed telpel (perianth). The blossom is bracteate and bracteolate (Nair and Bahl 1956). Sastri (1954) and Nair and Bahl (1956) described the perianth as obtaining ten vascular traces and hypothesised a pentamerous origin. Male or staminate flowers are 3–5 or it may produce even more in the axillary cymes, urceolate; the pedicels are dangling, bracts deciduous, persistent and the perianth three-lobed, tawny villous; number of stamens ranges from 9 to 12, synandrous, connate into a central stipitate column. The androecium possesses a solid column attached to 14–22 bilocular anthers. Pistillate flowers are urceolate and more prominent than the male with drooping pedicels; bracts are deciduous, bracteoles persistent, and the perianth is three-lobed, tawny villous; the ovary is superior, ovoid, one-celled and one ovule; the stigmas are connate and two-lobed. The pisital has a generally flask like shape, featuring either a short or absent style, and a bilobed stigma. It contains a single ovule (Joshy 1946; Sastri 1954; Nair and Bahl 1956).

The fruits are round, peach-shaped, or pear shaped, typically measuring 5–8 cm in length, with a yellow, drupe like appearance. The outer rind is succulent and yellowish in color. As the fruit matures, it splits open into two halves, revealing a single seed. The seed is oblong, measuring about 2–3 cm, with a shiny dark brown, hard, and stony shell that is furrowed and longitudinally wrinkled. This seed is encased in a red, lacelike aril attached to its base. The seed has a tiny embryo and ruminates endosperm with numerous veins consisting of essential oil. The mace of commerce is the dried aril, and the nutmeg is the dried kernel of the seed, often called the nut or kernel (Sasidharan 2006; Li and Wilson 2008). Sastri (1955) worked out the seed anatomy. There is a massive vascular supply to the testa, tegmen and aril. The testa and tegmen are thick and multiplicative. Nucellus enlarges by cell growth without division, which is absorbed and crushed by the enlarging endosperm tissue. The endosperm is oily and starchy. Ruminations formed by outgrowths of the inner brown vascular layer of the tegmen push the nucellar tissue into the endosperm, except at the pointed micropylar end. The aril arises from the outer integument, but Nair and Pillai (1959) considered it funicular.

Nutmeg is a dioecious or rarely monoecious crop (Shanmughavelu and Rao 1977; Verghese 2000). However, male trees of species *M. insipida* naturally produce a fewnumbers of female blossoms as well as fruits (Armstrong and Irvin 1989). The flowering style of male and female trees vary in *M. fragrans* (Nazeem 1979). In female nutmeg trees, flowering lasts for about seven months, while male trees exhibit flowering throughout the entire year. The peak blooming periods for both males and females occur in July, followed by a secondary peak in October. Female flowers take approximately 154 days to fully develop, whereas male flowers develop in about half that time. Anther dehiscence occurs roughly 24 h before anthesis, and the stigma remains receptive for about six days post-anthesis, with peak receptivity during the first three days. Wind is the primary pollination agent for nutmeg. The fruit set percentage varies among trees and directions, with the highest fruit set observed in trees oriented towards the East and West (Nazeem 1979). Fruits reach maturity 206–237 days after formation, following a sigmoid growth pattern (Nazeem and Nair 1981). Krishnamoorthy (2000) specified five types of trees: pure male and female, bisexual male and female, and hermaphrodite. The hermaphrodite nature is rare, whereas the bisexual types are observed in almost 3% of the trees in blossom.

The pollen germination studies by Nazeem (1979) reported the top most germination percentage of 96.90 in a nutrient medium possessing 4% sucrose, 25 ppm calcium nitrate and 75 ppm boric acid. Later, Krishnamoorthy (1986) studied the pollen viability and reported the maximum pollen germination of 57.80% with 5% sucrose, 100 ppm calcium nitrate and 200 ppm magnesium sulphate.

Studies at Kallar and Burliar estate in Tamil Nadu, where the climate is cool due to high altitude have shown that the pistillate and staminate buds open between 6 and 7 pm. Anther dehiscence happens 12 h earlier than the opening of flowers. The stigma was receptive and maintained the receptivity until 2.30 pm the day after the flower opening. The chief agent of pollination appeared to be wind, though this aspect was not substantiated. The percentage of fruit set differed among the genotypes or varieties and in different directions. The mean percentage of fruit drop after the fruit set was 74.40. The number of fruits harvested amounted to only 8.5% of the total flowers produced, and the period of the highest fruit drop after the set coincided with the period of maximum fruit development. Sharma and Armstrong (2013) noted that nutmeg trees, like many other dioecious tropical trees, have small, inconspicuous flowers and primarily rely on a pollen reward system. These trees interact with a variety of small, generalist insects, particularly beetles, thrips, and flies. Since pollen is the only significant reward offered by the flowers, pollination occurs through a combination of deceptive practices and foraging mistakes by the insects.

Climatic elements exert an essential role in the sex manifestation of nutmeg. Flach and Cruickshank (1969) observed a year-round display of flowers and fruits on nutmeg trees in West New Guinea; currently known as Papua or Indonesian Papua, with large production of flower in the month of April, May, November, and December.

Shinde et al. (2006) observed that the time span from floral bud emergence to anthesis ranged between 38.62 and 146.00 days. In their study, the formation of flower buds was solitary in female nutmeg trees, whereas hermaphrodite trees exhibited clusters of flower buds. The fruit set was reported to vary between 18 and 40 percent, with the highest fruit retention rate observed at 90%. The study, conducted in the Konkan region of Maharashtra, documented that the moderate-sized nutmeg fruits had an average length of 5.54 cm, a breadth of 4.85 cm, and a weight of 49.82 g. The nuts were found to be 2.84 cm long and 2.08 cm wide, with an average fresh weight of 7.28 g, a dry weight of 4.59 g, and a fresh mace weight of 1.87 g, which reduced to 0.78 g after drying. Additionally, the average pericarp weight of the genotypes under study was recorded at 40.08 g. The nutmeg genotypes measured had a kernel weight of 3.06 g and a mace weight of 1.53 g.

The phenological studies by Aarthy et al. (2018b) reported a distinction in the sex expression of nutmeg flowers across the months in a year at two different locations in India. Usually, at both the places, the male flowers were recorded throughout the year with variation in the course of their lean production. One location at Kozhikode, Kerala, having a typical hot tropical climate, the major female flowering was in the month of September. In contrast, the other location at Appangala, Kodagu, Karnataka flowering was observed in the month of December and March where weather condition is cool tropical with high altitude.

Variation in flowering pattern of monoecious nutmeg shows the three types of flowers: pistillate, staminate and hermaphrodite (Aarthy et al. 2018a). In the case of hermaphrodite flowers variation between flowers was evident. In hermaphrodite flowers, the filaments displayed either a fused or free shape, or both, and the number of androecia per anther varied from 1 to 4. Some anthers were fused with the gynoecium. Additionally, remnants of the stamens were found in the mature fruits of hermaphrodite flowers, and staminodes were also present in the flowers. The analysis of variance for different tree flower types and floral characteristics, such as flower type, number of anthers, anther length, filament length, and ovary size, showed significant results. The coefficient of variation was notably high for both hermaphrodite (183.84%) and pistillate (171.71%) flowers.

7.4 Cytogenetics

As per the studies, three chromosome numbers have been reported so far, viz. $2n = 38$ (Dhamayanthi and Krishnamoorthy 1999), $2n = 42$ (Simmonds 1954; Purseglove et al. 1981) and $2n = 44$ (Flach 1966) in nutmeg. Nair (2019) counted the total mitotic metaphase plates in nutmeg in different sex forms, and these vary with different sex forms of nutmeg. In the case of female plants, the count was recorded as 58.06%, and it was 72% in males and 83.33% in monoecious plants; among the different sex types of nutmeg plants, observed a somatic chromosome number of 44 (2n). Variations in cells of somatic chromosome numbers, i.e. 41, 42 and 45, were observed in different forms, and also, aneusomatic chromosome number variation has been documented in other nutmeg species. Further, Nair (2019) confirmed the findings of Flach (1966) that chromosome digit in diverse sex kinds and seedlings of nutmeg is $2n = 44$. Heikrujam et al. (2015) proposed that nutmeg exhibits a female heterogametic system, with males being ZZ and females ZW.

7.5 Germplasm Diversity and Conservation

Myristica is a wide genus of the family Myristicaceae, of which five species of economic importance have been described from India. *M. fragrans* Houtt. is introduced to India from Indonesia. Other species endemic to India are *M. malabarica* Lamk, *M. magnifica* Bedd., *M. beddomei* King and *M. trobogarii* Govind and M. Dan (Sasikumar 2021). Apart from species and cultivar diversity, ecosystem diversity is also a segment of the biodiversity of the genus *Myristica*. The species of the genus

Myristica are reported to be found in evergreen freshwater swamps comprising rich flora and fauna distinct to India (Sreedharan and Indulkar 2018). The wild and related species of *Myristica* originating in India are detailed in Table 7.1.

The Indian Institute of Spices Research is conserving genetic resources of nutmeg, with approximately 482 accessions is currently maintained. At Pechiparai (under Tamil Nadu Agricultural University), 22 collections are being maintained. Konkan Krishi Vidyapeeth (Dapoli) maintains 87 accessions, and Kerala Agricultural University holds 55 accessions, apart from high yielding nutmeg genotypes several unique genotypes or variants have been conserved (Ravindran 1999; Ravindran et al. 2005; Priyanka and Miniraj 2016). The elite nutmeg plants conserved as ex situ at Kerala Agricultural University is depicted in Fig. 7.2. Apart from the nutmeg of commerce (*M. fragrans* Houtt.), *M. malabarica, M. andamanica* Hook. f., *M. beddomei* King are used in the medicine and dye industry (Arunachalam and Subhasini 2011; Manjunatha et al. 2011). Fakfak nutmeg or Papua nutmeg (*Myristica argentea* Warb) is another economically essential species used in soap and perfume industries besides as a substitute for the cultivated nutmeg.

Three new subspecies *M. beddomei* King. described from India are *M. beddomei* subsp. *Beddomeia* King., *M. beddomei* subsp. *Sphaerocarpa* W. J. de Wilde and *M. beddomei* subsp. *Utsulata* W. J. de E. Wilde (Banik et al. 2017), in addition to the subspecies *M. fatua* var. *magnifica* also reported. Krishnamoorthy et al. (2007) reported a few other species in India as *M. elliptica* Wall. ex. Hook. f. et Thoms, *M. amygdalina* Wall., *M. glaucescens* Hook. F. Thoms, *M. prainii* King, *M. dactyloides* Gaertan. A new species has been recently reported from Kerala, i.e. *M. trobogarii* Govind & M. Dan (Govind et al. 2020), with larger leaves and glabrous egg-shaped fruit having yellow mace.

7.6 Economic Importance

Nutmeg is a perennial spice with tremendous economic importance. The global nutmeg volatile oil market is one of the leading segments within the essential oils trade. In 2021, nutmeg exports reached an estimated value of approximately USD 1441.39 million. The market is projected to grow from USD 2.60 billion in 2023 to USD 3.40 billion by 2028, reflecting a compound annual growth rate (CAGR) of 5.50%. Indonesia holds the position as the largest producer and exporter of nutmeg and mace globally, with Grenada following closely behind (Anandaraj et al. 2005). But, due to the impact of the hurricane, the Grenada nutmeg industry has lost the glory of being the second leading producer of nutmeg. Indonesia dominates the world nutmeg production with about 70% of world production, followed by India, Sri Lanka and Grenada. Indonesia has been in nutmeg production since the seventeenth century and has been dominating in the global trade of nutmeg till the day. The following two categories are recognized in Indonesian nutmeg based on its form of use and geographical origin.

Table 7.1 Wild and related species of nutmeg (*Myristica*) occurring in India

Name	Places of distribution	Brief description
**Myristica amygdalina* Wall	Nagercoil (Western Ghats)	Glabrous leaves, pale brown on both surface, nerves 8–12 pairs entire, yellow
M. andamanica Hook f.	Andaman Islands	Glabrous, leaves membraneous, nerves 12–15 pairs, bracteole large; apparently a unique species; but similar to the *M. malabarica*
**M. attenuata* Wall	Western Ghats	Branchlets and inflorescences furfuraceously tomentose; leaves pale brown above, nerves 14–20 pairs, fruit beaked, densely furfuraceously rustly tomentose, thin pericarp, aril entire, except towards the lobed apex
M. beddomei King *M. gibbosa* Hook f. and Thoms	Western Ghats, Khasia Mountains	Glabrous except the inflorscence and fruit, 15–30 pairs of nerves, aril entire, except at the top. This species finds application in dye industry and medicinal field
M. glabra Blume	Silhet (Assam), Tirunelvely (Tamil Nadu)	Glabrous, nerves 7–14 pairs, male much branched panicle, fruit 2.5 cm, deeply grooved on one side towards the base, valve thick
M. glaucescens Hook f. and Thoms	Tirunelvely, Andaman Islands	Branchlets and inflorescence rust-hairy, nerves 12–20 pairs, very small fruit, aril nearly complete
M. irya gartn. Frct.	Tirunelvely, South Andaman Islands	Leaves membranous at length, coriaceous dark brown above, lighter beneath, nerves 12–20 pairs, male panicles, many flowered, females shorter, perianth larger, fruit 2.5 cm in diameter, glabrous aril thin, entire yellow or reddish
M. kingii Hook. F.	Sikkim, Himalaya (3300 MSL)	Branches very stout with rough bark and large scars of fallen leaves, nerves 15–20 pairs spreading panicles sparsely rusty-tomentose, rather stout, thick perianth, staminal column small
M. longifolia Wall	Sikkim, Himalaya, Assam Silhet and the Khasia mountains	A lofty tree, leaves extremely variable usually linear, nerves 20–40 pairs, male flowers densely tomentose, scarlet within, style long, stigmas small toothed, ovary densely tomentose, fruit tomentose, aril pale, thin, lobed above the middle

(continued)

Table 7.1 (continued)

Name	Places of distribution	Brief description
* *M. magnifica Beddome*	Western Ghats, parts of Tirunelvely and South Karnataka	An immense gregarious tree 30 m in height, produced with multiple aerial roots, forming from the trunk at 3–6 m above ground and scattered along the ground rising in circles above it; bark purplish black, leaves thick and extensive with prominent nerves, thick reddish tomentose, soft above, nerves 20–26 pairs, blooms densely rusty tomentose, fruits densely tomentoseoblong arillus orange-red and deeply much laciniate
* *M. malabarica Lamk*	Western Ghats up to 300 m above MSL	It is called as Bombay mace used in medicine and dye industry. A tree reaching 15 m height, bark greenish black, leaves smooth, nearly glabrous, nerves about 10 pairs, fruit cylindrical, 5–7.5 cm long, aril yellow, cymes branched up to 20 males and 3 or 4 females; bigger than males

Fig. 7.2 An elite nutmeg germplasm maintained in Kerala Agricultural University (**a** fruiting nutmeg tree, **b** bold fruit **c** Dried bold nut, kernel and thick mace)

1. Tender nutmeg from Aceh province and West Java is the best suited for volatile oil extraction and for the industries' specific requirements.
2. Mature nutmeg is obtained from islands or the Sulawesi region, which is good for marketing as spice.

Table 7.2 Area and production of nutmeg in different countries (IFEAT 2021)

Country	Area (ha)	Production (Metric tonnes)
Indonesia	169,103	34,602
India	23,840	15,470
Sri Lanka	–	5000
Granada	–	2000

In Indonesia, nutmeg is categorized as a cash crop that improves the living standards of native farmers. It is estimated that the livelihood of more than 200,000 people depend on nutmeg crop. The production of nutmeg in India has been consistently increasing over recent years. The average price of Indonesian nutmeg oil on trade is close to US dollar 55 per kilo, varying from US dollar 40–75 per kilo. The sustainable price range for nutmeg oil is US dollars 50–60 per kilo; if it is lower, these prices would not be more economical. India stands in second position among the nutmeg producers globally. The majority of the farms in India are smallholdings, about 85–90%, and on an average of less than half a hectare. Also, nutmeg is a homestead crop component that would contribute a sustainable income to households. The world area and production of nutmeg are presented in Table 7.2.

In the international market, nutmeg and mace are grouped based on their origin (East Indian and West Indian nutmeg) and grade. The East Indian nutmeg comes from the production centres in the East, mainly from Indonesia and Sri Lanka. The second type is the West Indian nutmeg, primarily arriving from Granada. The whole nutmeg is grouped into three broad quality classifications in the trade:

i. Sound: Nutmeg is used majorly for grinding and, to a minimum extent, for oleoresin extraction.
ii. Substandard: Nutmeg is used for grinding, oleoresin, and volatile oil extraction.
iii. Distilling: Inferior quality nutmeg is used for the distillation of volatile oil.

The East Indian varieties of nutmeg are considered to be of higher quality compared to those from the West Indies. Major importing markets for nutmeg include the European Union, the United States, Japan, and India. Singapore and the Netherlands are key players in processing and re-exporting nutmeg (http://en.wikipedia.org/wiki/Nutmeg; May 2007). A small amount of wild "Bombay" nutmeg is imported into the United States, which represents the largest market for whole nutmeg. In the U.S., there is a preference for the East Indian varieties, which are characterized by deep brown, aromatic nuts and bright orange-red mace that is intact in form.

The US imports an average of 65% of its volume from Indonesia. Also, due to distinct volatile qualities, the European market prefers nutmeg from Indonesia as well as from India.

7.7 Health Benefits

Ravindran (2017) briefly reviewed the uses of nutmeg by various sectors. The seed and mace are used as traditional medicinal drugs. Nutmeg is a stimulant, astringent, carminative, hallucinogenic and aphrodisiac. In Indian traditional medicine, seed and mace are used in various formulations for internal and external use. According to Bhavaprakasha (a classical Ayurveda text), nutmeg is bitter, penetrating, hot in potency, increases food palatability, and is light in action. It is an appetizer, absorbent of water, suitable for voice, and reduces the body humour, *Kapha and Vata.* It cures distaste, clears tartar, purifies the oral cavity and normalizes the blackening of gums and tongue. It cures coughs, vomiting, dyspnoea, emaciation, chronic colds and cardiac problems. Nutmeg powder is given in curing respiratory diseases, treating premature ejaculation and treating gum diseases and is also used as an aphrodisiac (Sairam 1998; Khare 2007). In Arabic medicine, nutmeg and mace are used to 'strengthen the spirit" as a carminative, digestive, narcotic at higher doses and tonic for digestion and stomach cramps. In Chinese medicine, nutmeg and mace are used for their warming effects, to treat gastrointestinal infections and as palliatives. The German Commission E-listed nutmeg and mace for gastrointestinal disorders in European medicine. In modern Western herbal medicine, mace is included in the *Pharmacopeia of the United States.* In Unani medicine, nutmeg is used in several formulations for treating diarrhoea, premature ejaculation, as a stimulant and as an aphrodisiac (Anon 2013, 2014). Ross (2001) has compiled the traditional medicinal uses of nutmeg prevalent in various countries. Nutmeg and mace are essential items in the "grandma's" medicine chest. Mace is used for scenting cigarettes and chewing to suppress foul breath. Oil of nutmeg or mace is utilized for flavouring food by-products and spirits, soaps, tobacco, dental ointments and perfumery products. The volatile oil from the leaf has weed-suppressive properties. Nutmeg butter finds application in stimulative lotions and hair creams and in patients with rheumatism, paralysis and sprains. The succulent pericarp of the fruit is utilized for preparing pickles and jelly. Pruthy and Krishnankutty (1984) recovered graded pectin from nutmeg waste (rind).

Nutmeg and its many constituents have multiple biological and pharmacological properties. Leela (2008) and Jaiswal et al. (2009) offer short reviews on such effects. *M. fragrans* contains a variety of phytochemicals, including lignans, neolignans, diphenyl alkanes, phenylpropanoids, and terpenoids, which exhibit a range of pharmacological effects. Notably, compounds such as malignant (1), meso-dihydroguaiaretic acid (2), myristicin (111), and malabaricone C (Mal C, 104) are identified as key active constituents (Ha et al. 2020). Duke (2002, 2016) has compiled extensive data on the pharmacological properties of nutmeg, highlighting the biological and pharmacological effects of its individual components.

Kapoor et al. (2013) analysed the in vitro antioxidant nature of the volatile oil and oleoresins collected from nutmeg, introducing a unique natural source with potent antioxidant substances. The results revealed that these are as powerful as those of known antioxidants (BHA and BHT). According to Gupta et al. (2013), acetone extracts have shown significant antioxidant activity in the DPPH radical scavenging assay, attributed to the presence of terpenoids such as sabinene, myristicin, and eugenol. Similarly, Assa et al. (2014) demonstrated that methanolic extracts from nutmeg seeds exhibit notable antioxidant effects, as measured by the DPPH and FRAP assays. This activity is largely due to their high content of tannins, flavonoids, and terpenoids.

Singh et al. (2016) studied the antioxidant and antibacterial activity of nutmeg seed and mace. The antioxidant properties of nutmeg and mace were found to be the highest in ethanolic extract. Antibacterial activity of nutmeg and mace using 50% ethanolic concentration reduced the maximum growth of *Shigella*, followed by *S. aureus*, *Salmonella* and *E. coli*, whereas 70% reduced the maximum growth of *S. aureus* and *Salmonella*. Ethanolic concentration at 95% reduced the maximum growth of *E. coli*. Antibacterial activity of mace using 50% ethanolic concentration inhibited the maximum growth of *E. coli* followed by *S. aureus*, *Shigella* and *Salmonella*, whereas 70% reduced the maximum growth of *Shigella* followed by *Salmonella*, *S. aureus* and *E. coli*. Ninety-five per cent concentration exhibited inhibition against *S. aureus*.

According to Lestari et al. (2012), nutmeg extract exhibits dual agonist activity on peroxisome proliferator-activated receptors (PPAR) α/γ. Although its activity is less potent compared to full agonists of PPAR α and PPAR γ, nutmeg extract shows potential as a natural agent for managing type 2 diabetes due to its PPAR α/γ agonist properties.

Helen (2012) studied the effect of nutmeg oil on induced cytotoxicity to cancer cell lines such as MCF-7 (a breast cancer cell line) and A-357 (a skin cancer cell line). A methanolic isolate of the nutmeg seed induced cell death in a Jurkatleukaemia T cell line; the process involves down regulating SIRT1mRNA (Chirathaworn et al. 2007). SIRT-1 is the sirtuin (silent mating type information regulation two homologue)-1 of the yeast Saccharomyces cerevisiae, also learned as NAD-dependent deacetylase sirtuin-1; it is a protein coded by the SIRT-1 gene in humans.) Nutmeg seed shell isolates have remarkably lowered the incidence of papillomavirus-induced skin carcinoma in Swiss albino mice (Jannu et al. 1991).

Miniraj et al. (2017) conducted an analysis on the ripe rind of seventeen different nutmeg accessions, focusing on their biochemical properties. The study found that the ascorbic acid content varied between 87.1 and 89.1 mg per 100 g, pectin content ranged from 0.2 to 1.08% (as calcium pectate), and protein levels were between 0.21 and 1.85 g per 100 g. Additionally, starch content was observed between 0.3 and 1.23 g per 100 g, total phenol content ranged from 27.8 to 57.6 mg per 100 g, and tannin levels varied from 143.3 to 750.0 mg per 100 g. The total mineral content was between 2.01 and 2.57%, while crude fiber content ranged from 2.06 to 3.65%. These variations in biochemical composition are likely influenced by the genetic characteristics of the trees, as well as by geographic location and management practices. The findings suggest that, with appropriate value addition techniques, the nutmeg pericarp, which is often discarded as farm waste, could be repurposed for use in food, nutraceutical, and pharmaceutical industries.

7.8 Cultivation and Challenges

7.8.1 Current Cultivation Practices

Nutmeg is amenable for both seed as well as vegetative propagation methods. The mother plant selection is of utmost importance. Seeds should be sourced from healthy, high-yielding trees that produce fruits weighing around 30 g (wet weight), with approximately 1 g of mace and 10 g of nut per fruit. Ideal seeds are well-developed, of uniform size, shape, and light brown in color, and should be collected during the tree's optimal bearing stage. Typically, 125 medium to large seeds are selected per kilogram. Seeds should be sown immediately after extraction as they quickly lose viability. Germination usually begins within four to six weeks. Once the seeds have sprouted and reached the 2-leaf stage, with a height of 15–20 cm, they should be transplanted into polythene bags. These seedlings, after reaching about six months of age, can remain in polythene bags for 12–18 months before being planted in the field.

The sex of nutmeg trees can be confirmed only after 7–8 years of planting when they commence flowering. Generally, male and female plants are produced in a 1:1 ratio. A single male tree is enough for twenty female trees for pollination. By top working technique, the unproductive male trees can be converted to productive female trees.

Vegetative propagation is the only pragmatic approach to tackle the problem of dioecy in nutmeg. Patch-budding and epicotyl grafting, are practiced in nutmeg with the varying success of 38–80%. Epicotyl grafting has been standardized at Central Plantation Crops Research Institute (CPCRI), Kasaragod, by Mathew and Joseph (1982), employing *M. beddomei* as rootstock. About 48% survival of grafts was achieved in this method. Krishnamoorthy (1987) documented 80% of success in epicotyl grafting during August on *M. fragrans* rootstock.

Approach grafting trial at Dapoli, Maharashtra, showed that grafts could be produced throughout the year. A high percentage of the graft (30–100%) was accomplished on *M. malabarica*, the wild nutmeg rootstock. Use of *M. fragrans* as rootstock recorded about 40–90% success. However, approach grafting is a time-consuming and labour-intensive method. Kerala Agricultural University has standardized the techniques of patch budding and in situ budding, which are widely used in commercial nutmeg nurseries and plantations in Kerala (Lissamma et al. 2012; Beena 1994).

In situ budding is followed by farmers of Kerala state. This method performs budding on the field-planted nutmeg seedling after 2–5 years of planting. Orthotropic scion branches taken from superior mother trees (females) are patch budded on 2–5-year-old seedlings. Budding can be performed just above the first whorl of branches from the bottom. Investigations at Kerala Agricultural University confirmed that in situ budding on brown stem was successful. The best season for in situ budding in Kerala is the month of August.

Nutmeg plants are budded on wild (*M. beddomei*) and cultivated (*M. fragrans*) rootstock. But variation with respect to growth and yield between nutmeg on wild and cultivated root stocks were observed. Priyanka et al. (2023) have studied interaction of rootstock-scion in nutmeg budded on *Myristica* fragrans and *Myristicabeddomei* with respect to performance of growth and yield. Five each budded trees on wild and self-nutmeg rootstock were compared in the five different locations of Kerala state. Among the thirteen morphological and yield characteristics studied, eleven characteristics have showed significantly higher values in the nutmeg trees budded on self-rootstock. Significantly higher yield characteristics recorded in the trees budded on self-rootstock where high fresh and dry nut yield per tree as well as fresh and dry mace yield per tree. The overall performance of nutmeg trees budded on its own rootstock (*Myristica fragrans*) was superior than budded on wild rootstock (*Myristica beddomei*). Among all the five location and altitude self-rootstock had performed better than the wild rootstock.

7.8.2 Current Challenges of Nutmeg Farming

Present-day nutmeg cultivation confronts various issues. The scientific community works hard in addressing the existing challenges by adopting strategies to improve the production of nutmeg. Nutmeg improvement approaches should focus on the following aspects.

(a) Dioecy is a significant challenge in nutmeg cultivation. The seedling populations are segregated into male and female plants at near equal proportions. The identification of sex at the seedling stage is a difficult task. Studies conducted at various research institutes on identifying male and female plants in the seedling stage is yet to generate positive results. A first study by Krishnamoorthy et al. (1996) found an easy method to determine male and female seedlings by looking at morphological characteristics. This identification method failed in further confirmation studies. Babu et al. (2000) standardized the DNA isolation procedure as well as screened RAPD markers for sex determination in nutmeg. They have observed no polymorphism among the male and female nutmeg plants. Sudhamayee et al. (2014) focused on identifying sex-specific molecular and biochemical markers for distinguishing female nutmeg plants during their early growth stages. Their research highlighted that the RAPD marker OPK 01 produced a unique amplicon specific to female plants. In the biochemical marker assays, acid phosphatase did not exhibit polymorphism between male and female trees. Conversely, glutamate oxaloacetate transaminase showed a low-intensity polymorphic amplicon in female plants. However, the authors emphasized that further validation is necessary to confirm the repeatability of these findings.

(b) A few high-yielding varieties of nutmeg are now available for commercial cultivation. Apart from these elite varieties, few farmers varieties are also available for commercial cultivation. Breeding methods to develop varieties possessing more fruits per tree, bold nuts, and thicker mace are required which would significantly accelerate farmers' income and income from the export market.

(c) Tropism is a factor that has got a vital role in the vegetative propagation of nutmeg. While doing budding or grafting, if only an orthotropic shoot is used, the resultant graft/budling grows straight with the normal tree morphology. When the plagiotropic node is used for grafting/budding, the consequent graft/budling expands horizontally, giving a bushy appearance. Production of orthotropic shoots is generally scanty in nutmeg, which poses one of the challenges in multiplying high-yielding clones. At Kerala Agricultural University, an investigation was carried out on the effect of various physical treatments on the induction of orthotrops in mature nutmeg branches. Pruning the orthotropic branch close to the trunk leaving a stump of 30 cm, resulted in the production of few orthotropic shoots, but that is not a practically feasible approach (Rani 1994).

(d) Even though different methods of vegetative propagation, such as patch budding, in situ budding, epicotyl grafting and top working, have been standardised in nutmeg, low percentage of success only has been observed at field level. Success entirely depends on the season and skill of the budder/grafter.

(e) Nutmeg trees are prone to various stress conditions, mainly fungal diseases, viz. leaf spot, dieback, thread blight, fruit rot and leaf fall. In 2013, a severe outbreak of *Phytophthora* leaf fall disease was reported in major nutmeg-growing areas of Thrissur, Ernakulam and Kottayam districts of Kerala (Mathew and Beena 2012; Mathew and Miniraj 2013), where the experts estimated about a 35% reduced yield due to *Phytophthora* leaf fall of nutmeg (Sumbula and Mathew, 2015). The occurrence of dieback disease of nutmeg caused by *Lasiodiplodia theobromae* has been reported in the world over nutmeg growing areas (Biju et al. 2021). This disease exhibits dieback and declining symptoms in nutmeg. Developing integrated disease management approaches and tolerant varieties for major fungal diseases is essential in nutmeg.

(f) Climatic factors influence nutmeg production at a higher magnitude. Though released varieties and several elite genotypes perform well for yield attributes, their production, productivity and quality seem to be influenced by environment interactions; leading to inconsistent performance over the locations. The need of the hour is to develop climate-resilient varieties and agro techniques suitable to various agroecosystems.

(g) Market price depends on the distinct quality of the nutmeg. The natural variability existing in the crop have to be explored to meet the international standards. Breeding for agronomic traits and qualities is a vital aspect apart from post-harvest handling to be given priority in developing varieties and technologies. The global market demands high-quality nutmeg that meets specific criteria: it should contain an essential oil content exceeding 7%, be intact and unbroken, free from stalks, and should not include any woody or brittle fragments. The nutmeg's colour should range from reddish-brown to brown, with the base of the nut being light brown and closed. Nutmeg is graded according to the national standards of the producing country. Additionally, the ISO standard

6577:2002 outlines general guidelines for the grading, handling, and packaging of both whole and ground nutmeg.

(h) The delicate aroma describes the unique quality of nutmeg. "Ground nutmeg is sometimes subject to adulteration, where lower-quality materials like false nutmegs (*M. malabarica*), 'Bombay nutmegs', or spent, extracted material, along with ground shell or dried fruit pericarp, are blended in. These adulterants are often odourless and tasteless, diminishing the quality of the product. Additionally, nutmeg volatile oil may be adulterated with substances like turpentine oil or pinene. Similarly, mace can be adulterated with wild mace derived from *M. malabarica*, which is considered to be of inferior quality." A successful attempt by using analytical fingerprints (Ruth et al. 2018) and molecular-level (Sneha 2011) for detection of adulterants in traded nutmeg in powder form has been reported.

(i) Frequent reports of aflatoxin contamination due to *Aspergillus* spp. has been reported in processed and dried nutmeg. This would cause problems in the export of nutmeg. B1, B2, G1 and G2 are commonly found among the different aflatoxins. Aflatoxin B1 is reported to cause hepato-carcinogens in humans. The current Aflatoxin limits for nutmeg are 5 ppb for B1 and 10 ppb for the sum of B2 + G1 and G2 (Jayashree and Chempakam 2011).

7.9 Variability Studies

Only minimal work has been done on the crop improvement aspect of nutmeg. However, there has been significant variation observed in the shape, size, and color of the fruit, as well as in vegetative characteristics such as plant height, stem girth, number of branches, leaf shape, tree structure, canopy size, and yield attributes, including the dimensions of the fruit and the embryo (Shanmughavelu and Rao 1977; Sriram 1977).

Krishnamoorthy et al. (1991) observed that the number of fruits per tree varied, and they found a significant negative correlation between this number and the weight of mace. A significant and positive association was recorded between seed and mace weight. Haldankar et al. (1999) reported that the fruit characteristics like the weight of single fruit, weight of pericarp weight of fresh nutmeg and mace, and weight of dried nutmeg and mace exhibited positive and significant genotypic as well as phenotypic correlations on the yield. The high assessment of heritability for fruit characteristics with relatively high genetic advance indicated that these characteristics are controlled by additive gene action. Haldankar et al. (2004) noticed that canopy volume and surface area were positively correlated with yield. Plant height, spread, crown surface and crown volume are the noteworthy features of a nutmeg ideotype. Haldankar et al. (2005) reported a greater degree of variability in terms of the phenotypic and genotypic coefficient of variation, heritability, and genetic advance as a percentage of the mean for fresh seed weight.

Senthilkumar et al. (2010) reported significant variability in the fresh nut yield per tree, while leaf and nut characteristics exhibited low variability. Additionally, they observed moderate to high variability in mace traits in the high-altitude regions of Kodagu, Karnataka. Amman and Paramaguru (2016) assessed the performance of fourteen nutmeg ecotypes under the coconut ecosystem and found that accession MF1 performed satisfactorily under the coconut ecosystem under Tamil Nadu conditions.

Vikram et al. (2016b) reported that traits related to yield, including fruit weight, mace weight, mace volume, dry nut weight, kernel weight, fruit volume, and number of fruits per tree, exhibited high levels of both PCV (phenotypic coefficient of variation) and GCV (genotypic coefficient of variation). The heritability was found to be highest for these traits, except for shell thickness. The combination of high GCV with high heritability suggests that selection based on these traits can effectively improve yield. The greatest genetic gain was observed in the number of fruits per tree (144.3%), with significant gains also noted in mace characteristics such as fresh and dry weight and nut volume. Fruit weight showed a significant positive correlation with fruit breadth, volume, and pericarp thickness. An analysis of nutmeg data revealed that fruit weight, mace volume, and pericarp thickness have a direct impact on yield. Conversely, kernel volume, mace dry weight, and fruit set affect yield indirectly. For improving yield, selection should prioritize the principal factors: the number of fruits per tree, plant girth, canopy spread, flowers per 10 cm^2, fruit set, and fruit weight (Vikram et al. 2018).

Priyanka and Miniraj (2016) identified several unique nutmeg genotypes that differ from the standard varieties. These included yellow-maced, double-seeded, cluster-fruited, monoecious, wild, and grape-shaped nutmegs. Yellow-maced nutmegs are particularly valuable commercially due to their premium market price. The double-seeded variant was noted for its superior dry nut and kernel weights, while the seedless nutmeg variety had the highest fresh mace weight. The wild nutmeg (*M. malabarica*) is distinguished by its large, oval-shaped fruits with velvety brown rinds and yellowish-orange mace. Among all the genotypes, the cluster-fruited accession showed the most promising qualities, embodying the ideal characteristics of a superior nutmeg variety. Govind et al. (2022) also reported about the nutmeg variant with yellow mace which recorded high essential oil content. Mace obtained from the *Myristica fragrans* usually is red; the yellow mace variant was evaluated in comparison to red mace types for the morphological and biochemical attributes. The yield of mace was greater for yellow arils (2.5 kg per plant) compared to red arils (2.0 kg per plant). Additionally, the volatile oil content from yellow mace was higher (19.3% v/w) than that from red mace (9.2% v/w). GC–MS analysis revealed that while the distribution of monoterpenoids, sesquiterpenoids, and phenylpropanoids was similar between the two types, their relative proportions differed. Key components in both red and yellow mace included α-pinene and sabinene, with elemicin and safrole also being significant. These findings suggest potential for these distinct accessions in crop improvement programs.

Tremendous variation exists in the structure and proportion of the significant elements in nutmeg and mace oils. The scented steam volatile oil components define the spices' characteristic organoleptic properties. Significant differences were noted in both the chemical composition and aroma quality among the various nutmeg samples (Gopalam and Sayed 1988).

Krishnamoorthy et al. (2014) found a nutmeg genotype with high levels of sabinene and myrcene, with concentrations of 35.4% in the nut oil and 29.4% in the mace oil. Vikram et al. (2022) documented significant variation in the levels of volatile oil, oleoresin, fixed oil, and their constituents during their analysis of core nutmeg germplasm. They identified genotypes with notably higher or lower concentrations of specific volatile compounds, such as myristicin, elemicin, and safrole, while also noting higher levels of sabinene. For instance, the nutmeg germplasm A9/71, sourced from a farmer's field in Kallar, Ooty, Tamil Nadu, exhibited high sabinene levels in both nut and mace volatile oils, measuring 45.0 and 41.9%, respectively. Other components of the oil were found to be in the low to moderate range (Rema et al. 2013). Such qualitative variability in nutmeg could be utilized by the food and cosmetics sectors and traditional systems of medicine and pharmaceuticals.

The fat, fatty acid and amino acid quantum were assessed in *M. fragrans*, *M. malabarica*, *M. andamanica*, *M. beddomeii*, *M. magnifica*, *M. amygdalina*, *M. prainii* and *Knema andamanica* by Maya et al. (2006). Fat content in nutmeg varied from 28 to 38%, and in mace, it varied from 23 to 26%. While myristic acid is highest in the nut of *M. beddomeii* and *M. prainii*, the mace of *Myristica fragrans* possessed elevated palmitic acid. The total amino acid amount was highest in *M. prainii*. Among the amino acids, phenylalanine monopolises all *Myristica* species, and *Knema andamanica* has maximum alanine and threonine.

Fruits of *M. fragrans* collected from two different localities were subjected to hydro-distillation, and the oils obtained were analyzed by GC and GC–MS (Soni 2012). Both oils were comparable in terpenes, terpene alcohols and phenolic ethers. The principal components identified in the oil of nutmeg collected from the North-East were terpineol (23.39%), safrole (13.23%), ά-limonene (1.51%), m-cymene (4.03%), isohomogenol (5.14%) and myristicin (1.33%). In comparison, the oil from Kerala contained sabinene (38.79%), ά-pinene (15.21%), safrole (6.34%), limonene (4.51%), terpineol (4.15%), and myristicin (6.15%) as major constituents.

7.10 Traditional Breeding Methodologies and Limitations

In developing breeding strategies, all the research agencies have identified and conserved the elite nutmeg genetic stock as ex-situ through comprehensive surveys in the nutmeg-growing regions of India. Institutes involved in nutmeg research in India are Kerala Agricultural University, Thrissur, Kerala; Indian Institute of Spices Research, Kozhikode, Kerala; Tamil Nadu Agricultural University, Coimbatore, Tamil Nadu, ICAR-Central Coastal Agricultural Research Institute, GoA and Dr. Balasaheb Sawant Konkan Krishi Vidyapeeth, Dapoli, Maharashtra.

Early nutmeg breeding programs focused on evaluating progeny to assess the breeding characteristics of specific elite trees, particularly their sex ratio and prepotency. This assessment helped in selecting elite trees for seed and scion collection.

Significant variations were observed within the population regarding plant height, the number of main branches, the time required for flowering, fruit weight, and the mace-to-nut weight ratio (Krishnamoorthy and Rema 1994).

Few nutmeg varieties have been released following single tree selection method. The selected mother trees of seedling origin showing ideal characters of yield attributes were multiplied vegetatively and evaluated further. The best-performed accessions were further released as varieties. Dr. Balasaheb Sawant Konkan Krishi Vidyapeeth, Dapoli, Maharashtra, has released four varieties of nutmeg and ICAR—Indian Institute of Spices Research, Kozhikode, has released one variety of nutmeg following a single tree selection method.

Kerala Agricultural University has released five nutmeg varieties following a clonal selection method. These clones which were identified at the farmers' field were subjected to a step-by-step evaluation process with the help of the custodian farmers. Kerala Agricultural University has set a classic example in nutmeg, Participatory Varietal Selection (PVS) can be highly effective in identifying crop varieties when farmers are actively involved in the selection process. This approach leverages farmers' firsthand knowledge and preferences to identify promising varieties. Once farmers identify a variety they prefer, their claims and observations can be verified through rigorous scientific methods to ensure the variety's suitability and performance. This combined approach enhances the relevance and reliability of the selected varieties while aligning scientific validation with practical, farmer-driven insights.

In nutmeg, the Phenotypic Coefficient of Variation (PCV) is higher than the Genotypic Coefficient of Variation (GCV), indicating a significant environmental impact on trait expression (Krishnamoorthy and Ream 1994; Vikram et al. 2016b). Studies on trait associations in nutmeg show that fruit weight has a notable positive correlation with both seed weight and mace weight (Krishnamoorthy et al. 1991; Vikram et al. 2016b). A higher ratio of mace to seed is desirable. Therefore, selecting nutmeg trees that produce a good number of fruits with moderate seed weight can be financially beneficial. Additionally, the quality of nuts and mace should be a key consideration when choosing superior trees. Non-additive genetic factors are attributed to these traits' relatively low heritability estimates and genetic advances. Mace to nut weight ratio has an outstanding genetic advance, and this parameter could be very effectively considered for the selection criteria. Non-significant differences were noticed for canopy dimension, number of straight shoots, number of fruits per tree, plant girth etc.

Nutmeg breeding is complicated to plant breeders because of the long adolescent phase, difficulty in vegetative multiplication, male and female flowers produced on separate trees, and one ovule per female flower.

Therefore, there is need to identify and popularise elite types present in India. The future nutmeg improvement programme has to focus on identification of high-yielding mother trees, evaluation of mother trees with resistance or tolerance to biotic and abiotic stresses and their clonal multiplication. Further evaluation of their progenies in different locations can be a practical approach to evolve good varieties.

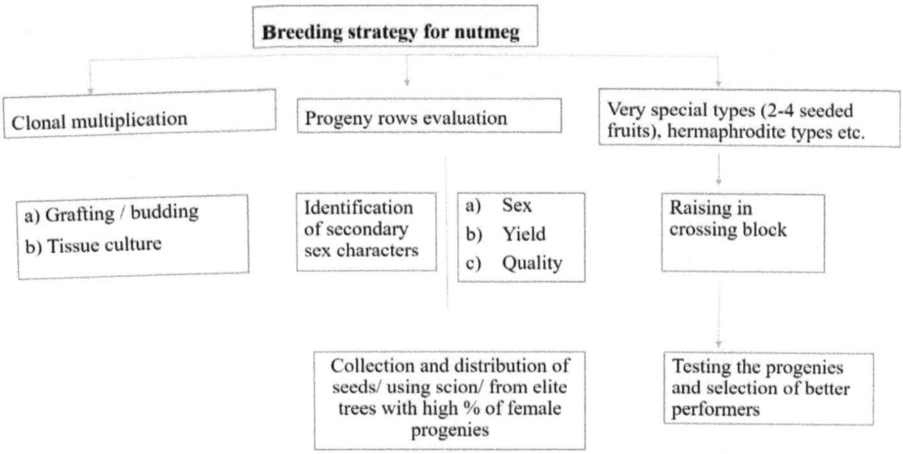

Fig. 7.3 Steps in nutmeg improvement (Krishnamoorthy and Rema 1994)

The general crop improvement steps followed in nutmeg as suggested by Rema and Krishnamoorthy (1994) is presented in Fig. 7.3.

Apart from evaluating high-yielding nutmeg accessions alone, another approach is suggested by Priyanka and Miniraj (2016) at Kerala Agricultural University. They have reported several nutmeg variants which differ from the normal nutmeg. These nutmeg variants have been characterized by morphological and quality characteristics with respect to canopy growth, nut and mace parameters and essential oil yield. Some of these variants could be very well utilized in the nutmeg breeding programme. A concept model was developed for utilizing these nutmeg variants according to their features (Fig. 7.4). In the concept model nutmeg variants were distributed into five categories based on their yield/quality/uniqueness.

1. Seed less (SL 1) having extremely thick mace with rudimentary and sterile seed.
2. Five accessions have been grouped under very good.
   (a) The wild nutmeg (*Myristica malabarica*), specifically accession WT 2, features a brown velvety rind and yellowish-orange mace. This variety is noted for its drought resistance and can serve as an effective rootstock for nutmeg cultivation in water-scarce regions. Utilizing wild nutmeg as a rootstock allows for successful planting even in areas with limited water availability (Krishnamurthy et al. 2008).
   (b) Double-seeded accession (DS 1) exhibited the transition of sex, which produces only minimum number of fruits that are slightly deformed. This type needs enhancement to increase both mace and nut yield per tree. A previous study by Miniraj et al. (2015) reported the double-seeded nutmeg fruits with multiple splitting, usually associated with monoecious types.
   (c) The enhanced accessions (IA 1 and IA 2) can be directly propagated for cultivation because of their high mace and nut weight.

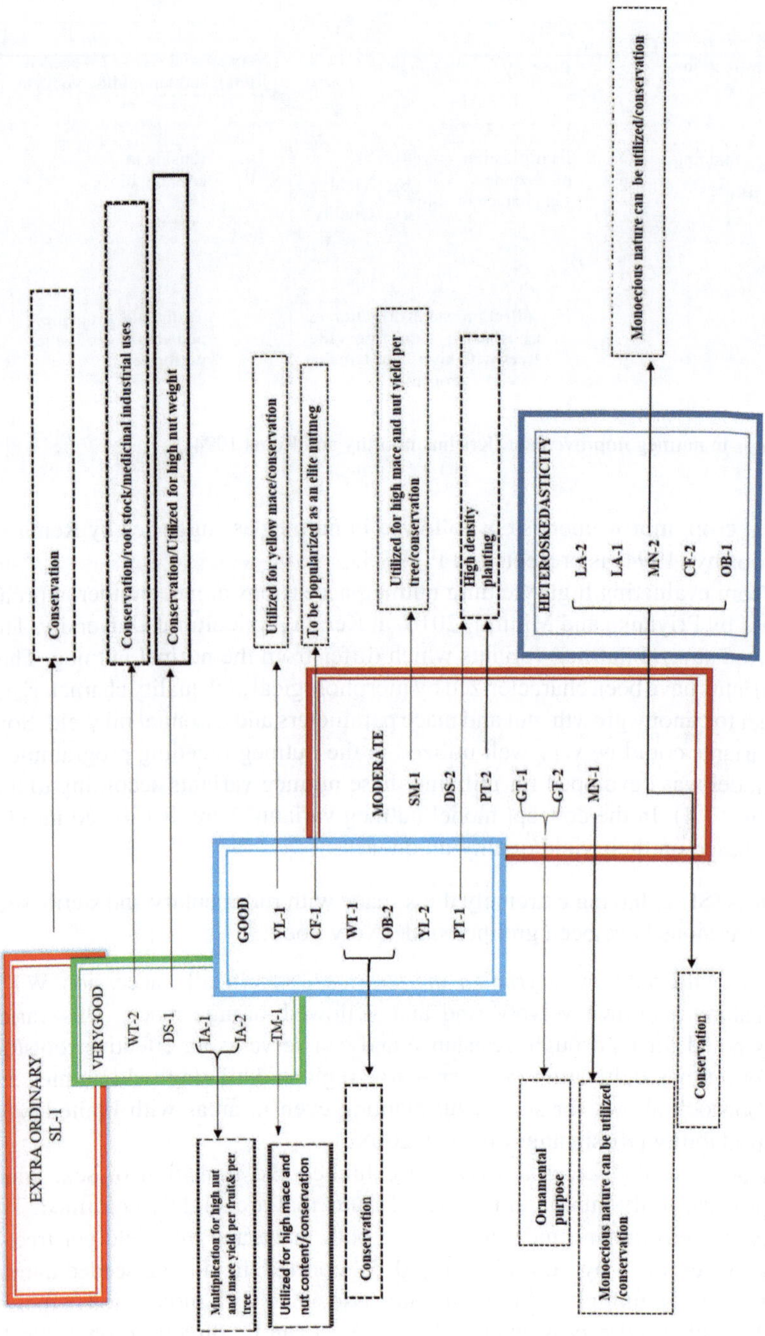

Fig. 7.4 Concept diagram for utilization of unique nutmeg accessions

   (d) Triangular nutmeg (TM-1) exhibited maximum mace and nut weight per fruit.

3. The next category included six good variants.
   (a) Yellow mace nutmeg varieties YL 1 and 2 are suitable for commercial farming due to their desirable yellow mace color, which is highly sought after and commands premium prices in the market.
   (b) The Cluster Fruited Accession (CF 1), which bears fruits in clusters and exhibits all the desirable traits of a top-quality nutmeg genotype, could be commercially developed and promoted as a new variety.
   (c) One wild type (WT 1) has been grouped under the good category based on its performance.
   (d) Oblong nutmeg accession (OB-2) produces oblong-shaped nut and mace, but it comprises of thin mace. Separation flower-grade mace from nut could be tedious work in this type, but the number of fruits per tree was significantly higher.
   (e) Pyramidal nutmeg accession (PT 1) characterised by pyramidal canopy architecture is highly suited for high-density planting.

4. Moderate category consisted of six accessions.
   (a) Small leaved accession (SM 1) possessed small elliptic leaves, dark green, which had high nut and mace yield per tree. This tree produced more orthotropic shoots.
   (b) One each of double seeded type (DS 2) and pyramidal type (PT 2) belonged to the moderate performing group.
   (c) Grape nutmegs (GT 1 and 2) which possessed very small fruits are apt for ornamental purpose.
   (d) Monoecious nutmeg (MN 1) is a desirable feature, and this type, if improved with yield characteristics, can solve the issues of dioecy in nutmeg. Many nutmeg researchers have documented the presence of monoecious nutmeg trees in Kerala (Miniraj et al. 2012) and Karnataka (Rema et al. 2015), highlighting the variability in nutmeg tree characteristics in these regions.

5. Five accessions belonged to the heteroskedastic category.
   (a) Low astringency accessions (LA 1 and 2) possessing lower acidity and tannin, as well as high pectin content in the rind, is ideal for value addition. One each monoecious (MN 2), cluster fruit (CF 2) and oblong nut (OB 1) have also been included in this group.

The study concluded that apart from superior nut and mace yield which are the economic parameters considered for an elite nutmeg clone, other unique features could also be utilised for special purposes or sometimes they could just be conserved for future breeding programmes.

7.11 Enhanced Traits and Improved Varieties

Both nutmeg and mace are equally important in nutmeg; breeding should emphasise on improving both the produce, and also the number of fruits produced per tree. A tree bearing both bold nut and mace is a valuable germplasm, and these characteristics could be considered in breeding programmes. The process of selecting germplasm or true seedling progeny has resulted in the creation of new plant varieties. Apart from improved varieties, many farmers' varieties, viz. Kadukanmakkal, Nova, Kallingal, Kinattukara, Kallivayal, Ceylon, Ceylon-1, Souriyamakkal, Nirappel etc. have been identified in Kerala. There are also several unique nutmeg types, like yellow maced, seedless, cluster fruited, double seeded and pyramidal canopy nutmegs (Priyanka and Miniraj 2016) reported from Kerala.

Miniraj et al. (2015) established criteria for classifying a nutmeg tree as a "plus tree," requiring it to produce approximately 2000–3000 fruits per year, with a minimum of 2 g of dry mace and 10 g of dry nut. Building on this, Vikram et al. (2016a) developed a statistical tool based on 13 key quantitative traits that influence yield, which can help identify elite nutmeg trees. According to their tool, an elite nutmeg tree is characterized by several exceptional attributes: a tree height of around 8 m, a canopy spread of 7 m east–west and 8 m north–south, about 6 flowers per 10 cm^2, a fruit set rate of 37%, 19 fruits per square meter, a fruit weight of 81 g, a pericarp thickness of 14 mm, dry mace weight of 2 g, dry nut weight of 10 g, kernel weight of 7 g, a nut-to-mace ratio of 6.6, and approximately 3342 fruits per tree. This set of characteristics provides a straightforward method for assessing and identifying superior nutmeg trees. Correlation analysis showed that seed weight positively associates with mace weight. Seed and mace weight are most important parameters to be considered in nutmeg crop improvement programme (Krishnamoorthy et al. 1991; Vikram et al. 2016b). The selection method will be adequate nutmeg breeding to choose high yielding trees with optimum fruit numbers and higher seed and mace weights (Krishnamoorthy et al. 1991; Vikram et al. 2016b).

The high-yielding varieties of nutmeg released through different methods of selection (Table 7.3) by the Kerala Agricultural University, Thrissur (Kerala); Indian Institute of Spices Research, Kozhikode (Kerala); and Dr. Balasaheb Sawant Konkan Krishi Vidyapeeth, Dapoli, Maharashtra are given below.

7.12 Molecular Breeding

Nutmeg has got abundant polyphenols and other secondary metabolites in its tissues; and hence isolating quality DNA from the genus *Myristica* is challenging. The challenge of obtaining high-quality DNA has been highlighted by low levels of DNA recovery and suboptimal absorbance ratios, often due to the presence of RNA and protein contaminants, as observed through spectrophotometric analysis. These high

Table 7.3 Varieties of nutmeg developed in India

S. No.	Variety	Breeding method and year of release	Yield and other characters
IISR, Kozhikode			
1.	Viswashree	Clonal selection from elite trees 2001	4800 kg mace and 31,220 kg dry nuts per ha Dark red shining mace, black bold seed and high yield
2.	Keralasree	Single tree clonal selection 2014	Weight of fruit: 75–100 g, Weight of seed: 13–16 g, Weight of aril: 4.5–6.0 g High yield, quality, extra bold fruit mace and nut. Suitable for Kerala and other nutmeg growing areas of India
KAU, Thrissur			
3.	KAU-Mundathanam	Clonal selection 2018	Mace yield/tree: 3.89 kg, Nut yield/tree: 19.7 kg Mace weight (dry): 2.49 g, Nut weight (dry): 12.6 g, Kernel oil: 6.9%, Mace oil: 11.0%
4.	KAU-Poothara	Clonal selection 2018	Mace yield/tree: 4.53 kg, Nut yield/tree: 22.0 kg Mace weight (dry): 2.06 g, Nut weight (dry): 10.0 g, Kernel oil: 4.16%, Mace oil: 7.0%
5.	KAU-Pullan	Clonal selection 2018	Mace yield/tree: 2.86 kg, Nut yield/tree: 22.79 kg Mace weight (dry): 1.36 g, Nut weight (dry): 10.85 g, Kernel oil: 4.6%, Mace oil: 12.06%
6.	KAU-Kochukudy	Clonal selection 2018	Mace yield/tree: 4.48 kg, Nut yield/tree: 20.88 kg Mace weight (dry): 2.49 g, Nut weight (dry): 11.60 g, Kernel oil: 6.9%, Mace oil: 11.0%
7.	KAU-Punnathanam	Clonal selection 2018	Mace yield/tree: 4.23 kg, Nut yield/tree: 19.39 kg Mace weight (dry): 3.02 g, Nut weight (dry): 13.85 g, Kernel oil: 4.5%, Mace oil: 9.5%
DBSKKV, Dapoli			
8.	Konkan Sugandha	Single plant selection 1998	600 fruits per tree. Bisexual flowers Higher fruit set, nut weight 5 g mace 1.2 g
9.	Konkan Swad	Single plant selection 2003	500–1040 fruits per tree. Female flowers, high fruit set, nut weight of 5.8 g
10.	Konkan Shrimanti	Single plant selection 2005	Dried nut of 9.370 kg and dried mace of 1.916 kg Bold nut, thick mace and high yield
11.	Konkan Sanyukta	Single plant selection 2019	500 fruits/tree. Bold nuts (9.20 g), higher nut (27%) and mace oil (17.75%), Bisexual flowers and high yield

levels of metabolites in tissue can negatively impact processes such as DNA restriction, amplification, and cloning (Bryant 1997). Sheeja et al. (2008) were the first to establish effective procedures for extracting high-quality DNA from *Myristica* species. They developed a method for isolating DNA from both wild and cultivated Myristica species, which are known to contain high levels of polysaccharides and polyphenols. The protocol utilized CTAB (3%), 1.5% PVP, and 0.3% β-mercaptoethanol for DNA extraction, along with RNase and phenol–chloroform for contaminant purification. Sudhamayee (2010) recommended using the third-position leaf from the shoot apex or newly emerging pale green leaves to obtain high-quality DNA from nutmeg. Lele and Deshpande (2011) developed a systematic protocol for extracting high molecular weight DNA from powdered nutmeg seeds, which can be useful in determining the geographical origin of the samples. Divyasree et al. (2014) evaluated different methods for isolating high-quality DNA from nutmeg, finding that the CTAB protocol by Doyle and Doyle (1987) yielded the best results in terms of DNA quantity and quality. Their method produced an average DNA recovery of 83.55 μg/g with a purity ratio of 1.79, especially from young leaf tissues. They also concluded that newly emerging pale green leaves are ideal for obtaining high-quality DNA with minimal protein and RNA contamination. Swapna et al. (2014) developed a standardized method for extracting DNA and RNA from Myristica fragrans with reduced contamination, which streamlines the process for initiating molecular studies on nutmeg.

Sheeja et al. (2006) examined the differences between vegetatively propagated and seedling-origin plants of a high-yielding nutmeg sample by using RAPD (Random Amplified Polymorphic DNA) profiles to identify variations and unique characteristics among the populations. The study generated a total of 405 amplicons across twenty polymorphic loci. The similarity index for the vegetative propagules ranged from 96 to 100%, while for seedling-origin plants, it ranged from 76 to 100%. Jaccard's Similarity Index values varied between 0.72 and 1.00, indicating a close genetic relationship between the offspring and the mother plants. In another study, Sheeja et al. (2013) used both RAPD and ISSR (Inter Simple Sequence Repeat) markers to analyze phylogenetic relationships among ten species and related genera of *Myristica.* This research utilized 20 RAPD and 16 ISSR primers, which produced 497 and 262 amplified DNA fragments, respectively. The average Polymorphic Information Content (PIC) was found to be 0.34 for RAPD markers and 0.35 for ISSR markers. The combined cluster analysis of RAPD and ISSR data divided the species into two major clusters: the first cluster contained *M. prainii*, and the second cluster included other species such as *M. fragrans.* The genetic similarities observed at the molecular level were largely consistent with morphological or geographical proximities.

Krishnan et al. (2019) conducted a study on 19 high-yielding nutmeg accessions, isolating their genomic DNA. Out of 34 ISSR primers tested, 16 primers generated reproducible banding patterns. These selected primers produced a total of 145 amplified bands, with 97 being polymorphic. The amplified band sizes ranged from 300 to 1500 bp. The percentage of polymorphism among the primers varied from 33% (UBC 860) to 100% (UBC 852), with an average of 65.73%. Among all primers tested, UBC 815 showed the highest Polymorphic Information Content (PIC) value of 0.48, while UBC 860 had the lowest PIC value of 0.10. The study reported a mean Marker Index (MI) value of 1.62. Jaccard's similarity coefficient for the accessions ranged between 0.565 and 0.939. The UPGMA cluster analysis dendrogram grouped the 19 nutmeg accessions into two main clusters. The results from this study can be valuable for future nutmeg breeding programs.

Aniedi et al. (2014) conducted a genetic diversity analysis of twenty-one African nutmeg accessions from South Eastern Nigeria using RAPD markers. Their study produced seventy-seven amplicons, ranging from 3 (OPB 17) to 13 (OPT 07), all of which were polymorphic. The average polymorphic data range and genetic diversity were found to be 0.763 and 0.697, respectively. Cluster analysis categorized the accessions into four groups, indicating significant genetic divergence among them.

In another study, Kusuma et al. (2020) utilized 48 nuclear microsatellite markers to study the genetic diversity of *Myristica fragrans*. Among these, 14 markers were found to be variable, with the number of alleles per locus ranging from 2 to 6. The heterozygosity level across loci varied between 0.038 and 0.929. Additionally, the transferability of these microsatellite markers was tested in other species of *Myristica* (such as *M. fatua*, *M. argentea*, and *M. crassipes*) and in the Myristicaceae family (e.g., *Horsfeldia palauensis*). The newly identified microsatellites are expected to be valuable for future studies on genetic variation, population structure of *M. fragrans*, and related species.

DNA barcoding was utilized to examine the evolutionary relationships among *Myristica* species and related genera within the Myristicaceae family. Various barcoding loci, such as *rbcL*, *matK*, *psbA-trnH*, ITS, and multilocus combinations, were evaluated to determine their effectiveness in assessing phylogenetic relationships. The *psbA-trnH* locus provided valuable insights into the associations of species within the *Myristica* genus. For instance, *M. fragrans* was closely related to *M. beddomei*, *M. amygdalina*, and *M. andamanica*, whereas *M. fatua* showed distinctiveness from *M. malabarica*. Additionally, species from the genera *Gymnacranthera* and *Knema* were identified as having a sister relationship with other Myristica species (Swetha et al. 2019). However, a phylogenetic analysis based on the *matK* gene showed a close relationship between *M. fragrans* and *M. fatua* (Nenlohy and Samuel 2020). An 18S rDNA study by Sheeja et al. (2013) also indicated a strong affinity between *M. beddomei*, *M. andamanica*, and *M. amygdalina*, as well as between *M. malabarica* and *M. fragrans*. The recently reported species, *M. trobogarii*, shares morphological similarities with *M. beddomei* subsp. *beddomei*, *M. malabarica*, and *M. magnifica*, although it differs in several other traits (Govind et al. 2020).

Vikram et al. (2018) noted that the primer OPE15 generated an extra band at 500 bp in two monoecious accessions, a feature that was absent in female accessions. Similarly, OPE16 had a characteristic band in monoecious accessions at 200 bp. The existence of additional bands in monoecious accessions throws light on their utilisation as candidate markers for seedling selection on a massive scale.

In a molecular study conducted by Krishnan et al. (2019) on 17 different genotypes of nutmeg, 145 bands were identified, out of which 97 were found to be polymorphic. The sizes of these bands varied between 300 and 1500 base pairs (bp). The percentage of polymorphism for the primers used in the study ranged from 33% for UBC 860 to 100% for UBC 852a, with an average value of 65.73%. The study reported a mean marker index (MI) value of 1.62. The Jaccard similarity coefficient for the genotypes analyzed was found to range between 0.565 and 0.939.

7.12.1 *In Vitro Shoot Development and Micro Grafting in Nutmeg*

Kerala Agricultural University attempted in vitro shoot development and micro grafting in nutmeg using nodal segments and shoot tips as explants (Antony 2008). The SH medium (Schenk and Hildebrandt) including Thidiazuron (0.03 mg/l) and activated charcoal (0.5%) gave the best result. In this medium bud expansion in 50% cultures occurred before nine days. The nodal pieces of length 1.5–2.5 cm without leaf blades documented the highest survival percentage of 46.70%. The condition of the culture growth was 26 ± 2 °C at a light supplementation of 1000 lx. The other inputs were carbon source; 2% Sucrose, 1% glucose or provided with 5% sucrose, this supported the bud elongation and leaf expansion. But experiment at Kerala Agricultural University on in vitro grafting in nutmeg ended with unsuccessful results. Callus formation from zygotic embryos has been previously documented by Iyer et al. (1993), while in vitro germination of adult embryos and the development of somatic embryo-like structures in in vitro plantlets were reported by Iyer et al. (1994, 1995). The process of direct somatic embryogenesis from isolated and ruptured zygotic embryos, as well as from in vitro-grown plantlets of nutmeg, was successfully achieved. The presence of activated charcoal and light were identified as crucial factors for inducing somatic embryogenesis in nutmeg. Histological studies revealed the development of globular and cotyledonary stages. After a lag period of six months, the somatic embryos showed partial germination, with a wide range of abnormal embryos being observed. These somatic embryos were capable of synthesizing chlorophyll and exhibited phenylalanine ammonia-lyase activity, along with the production of phenolic compounds. They could serve as a continuous source of commercially valuable secondary metabolites derived from nutmeg.

7.13 Current Research Initiatives to Combat Climate Change

The change in rainfall behaviour and high-intensity overflows could adversely influence nutmeg crops, when they are cultivated in ecologically susceptible hilly areas and river shores with high precipitation and humidity. This emphasizes the necessity to explore the associations between nutmeg genotypes and weather variables on growth and development. High moisture and distributed rainfall with well-drained productive soil are ideal for nutmeg growth. Adverse climatic conditions like higher wind speed and frequent torrential precipitation develop unhealthy environments like uprooting and the development of fungal diseases like root rotting in nutmeg, as they are shallow-rooted. Being perennial, species like nutmeg have to encounter the result of climate change even during a single year or in a standing plantation itself. In perennial crops, genotype-environment relations emerge as a significant issue for plant breeders in choosing and suggesting the superior genotype with the best yield performance and wider adaptability over diverse regions.

Nutmeg is a climate-sensitive perennial spice tree. Due to substantial environmental impacts, many of the existing superior accessions in nutmeg show varied performance across locations. Studies on genotype-environment interactions in nutmeg are limited due to the long pre-bearing period of the crop. The evaluation of nutmeg accessions across various locations, along with their responses to weather conditions, is crucial to understanding genotype-specific adaptation and broader adaptability in this crop. A recent attempt by Priyanka et al. (2022) looked into the genotype-environment relations in select six nutmeg genotypes. This study compared the growth and yield of three Kerala Agricultural University released varieties, two farmer varieties with a national check, IISR Vishwasree. The performance of select nutmeg genotypes was studied in three altitudinally varying locations. The study examined several weather factors, including maximum and minimum temperatures, relative humidity, precipitation, wind speed, and hours of sunshine. The study concluded that phenophases such as flushing phase, flowering phase, fruit development and maturity phase as well as fruit harvest phase in nutmeg showed variation, bothamong the genotypes and also across the locations. Among the genotypes, the regression coefficient R^2 value was the highest for KAU Mundathanam variety combined with the highest nut and mace yield per tree. In KAU Mundathanam, wind speed and rainfall positively influenced the yield, accounting for 85.50% of the impact. In KAU Pullan, the maximum temperature was the primary climate variable affecting yield, with a positive contribution of 40.10%. Meanwhile, in KAU Kochukudy, sunshine hours were found to have a beneficial effect on the dry nut yield per tree, contributing 62.6%. Relative humidity influenced on Kadukammakkal variety yield (26.3%). In contrast, lower minimum temperatures had a detrimental effect on the dry mace yield per tree, reducing it by 69.0%. Additionally, the amount of sunlight exposure, measured in sunshine hours, adversely affected all yield characteristics in both the IISR Vishwasree (with an impact of 81.6%) and Kinattukara varieties (with an impact of 47.2%). The variety, IISR Vishwasree even though recorded lowest yield at all the locations, was stable yielding and remained less affected by majority of weather variables. A favorable interaction between nutmeg genotypes and climatic conditions was observed across all locations. The impact of weather parameters on yield attributes varied significantly among different genotypes. This analysis may provide insights into how each genotype performs in the various studied locations. Thus, the knowledge on the gene—environment interactions of elite nutmeg genotypes become very useful. The expansion of area under nutmeg cultivation is fast happening due to the high demand for nutmeg worldwide for multiple applications.

To meet this requirement, nutmeg production is now being taken up in unconventional regions experiencing different biotic and abiotic stresses. Hence, the nutmeg breeder's priority should be to select elite stable yielding genotypes suitable for non-traditional areas.

7.14 Conclusion and Prospects

Nutmeg, a tree spice with dioecious nature, exhibits considerable variability in India. Formulation of breeding strategies is very vital for nutmeg which is a twin spice-yielding crop of commercial importance. Breeding priority should be on choosing ideotypes possessing desirable features viz. boldness, high dry nut and mace weights and greater number of fruits per tree. Breeding for quality assumes importance as the spice is finding myriad applications in food, cosmetic and pharmaceutical industries. Clonal selection and open pollination selection can yield tremendous outcome. However, to shorten the breeding cycle, farmer's participatory breeding/in situ evaluation can help to a greater extent. Breeding for biotic and abiotic stress management also warrants immediate attention. There is need to develop climate resilient nutmeg varieties that can better adapt to the fast-changing climate.

References

Aarthy S, Aravind S, Shivakumar MS, Rema J, Kandiannan K, Sasikumar B (2018a) Influence of weather parameters on sex expression and flowering pattern of nutmeg (*Myristica fragrans* Houtt.). J Agromet 20(2):155–156

Aarthy S, Rema J, Sasikumar B, Saji KV, Anandaraj M (2018b) Diversity in floral characters of monoecious nutmeg (*Myristica fragrans* Houtt.). J Spices Aromat Crops 27(1):66–70

Amman M, Paramaguru P (2016) Evaluation of growth and yield characters of nutmeg (*Myristica fragrans* Houtt.) ecotypes under coconut ecosystem. Environ Ecol 35(2A):808–810

Anandaraj M, Devasahayam S, Zachariah JT, Krishnamoorthy B, Mathew PA, Rema J (2005) Nutmeg (extension pamphlet). Indian Institute of Spice Research, Kozhikode, 2p

Aniedi EU, Umego C, Aikpokpodion PO (2014) Genetic diversity in African nutmeg (*Monodoramyristica*) accessions from South Eastern Nigeria. Afr J Biotechnol 13(42):4105–4111

Anonymous (2007) Available at: http://en.wikipedia.org/wiki/Nutmeg, May 2007

Anonymous (2013) Traditional uses for mace-Natural Standard: the authority on integrative medicine, www.naturalstandard.com. Accessed 15 June 2017

Anonymous (2014) Jatiphala (*Myristica fragrans*) http://www.herbalcureindia.com/herbs/myristica-fragrans.htm. Accessed 15 June 2017

Antony LZ (2008) In vitro shoot regeneration and micro grafting in nutmeg (*Myristica fragrans* Houtt.) M.Sc. (plant biotechnology) thesis. Kerala Agricultural University, Thrissur, p 106

Armstrong JE, Drummond BA (1986) Floral biology of *Myristica fragrans* (Myristicaceae), the nutmeg of commerce. Biotropica 18:32–38

Armstrong JE, Irvine AK (1989) Flowering, sex ratios, pollen-ovule ratios, fruit set, and reproductive effort of a dioecious tree, *Myristicainsipida* (Myristicaceae), in two different rain forest communities. Am J Bot 76:74–85

Arunachalam KD, Subhasini S (2011) Preliminary phytochemical investigation and wound healing activity of *Myristica* and *Amanica* leaves in *Swiss albino* mice. J Med Plants Res 5:1095–1106
Assa JR, Widjanarko SB, Kusnadi J, Berhimpon S (2014) Antioxidant potential of flesh seed and mace of Nutmeg (*Myristica fragrans* Houtt.). Int J Chem Tech Res 6:2460–2468
Babu TD, Achuthan CR, Nazeem PA, Nair RV (2000) Genetic DNA isolation and RAPD analysis in the tree spice-nutmeg (*Myristica fragrans*). In: Proceedings of the 87th Indian science congress, part IV. Indian Science Congress Association, Calcutta, p 38
Banik D, Bora PP, Sampathkumar V, Bezbaruah RL (2017) Conspectus on Indian *Gymnacrathera* and *Myristica*. Rheedea 27:1–12
Beena S (1994) Standardisation of top working in nutmeg. M.Sc. (Hort.) thesis. Kerala Agricultural University, Thrissur
Biju CN, Jeevalatha A, Peeran MF, Suseelabhai R, Basima F, Nissar MVA, Srinivasan V, Thomas L (2021) Association of *Lasiodiplodia theobromae* with die-back and decline of nutmeg as revealed through phenotypic, pathogenicity and phylogenetic analyses. 3 Biotech 11(9):422
Bryant JA (1997) DNA extraction. In: Dey PM, Harborne JB (eds) Methods in plant biochemistry, vol 10b. Academic Press, San Diego, pp 1–12
Chirathaworn C, Kongcharoensuntorn W, Dechdoungchan T, Lowanitchapat A, Sanguanmoo P, Poovorawan Y (2007) *Myristica fragrans* Houtt. methanolic extract induces apoptosis in a human leukaemia cell line through SIRT1 mRNA down regulation. J Med Assoc Thai 90(11):2422–2428
Dhamayanthi KPM, Krishnamoorthy B (1999) Somatic chromosome number in nutmeg (*Myristica fragans* Boutt.). J. Spices Aromat Crops 8(2):205–206
Divyasree P, Lokesh KA, John L, Joy T, Teena VR, Thomas A, Muttakulath TM, Manila VTK, Sandhya R, Deepu M (2014) A simplified protocol for the recovery of high quality DNA from nutmeg. J Trop Agric 52(1):79–85
Doyle JJ, Doyle JL (1987) A rapid DNA isolation procedure for small quantities of fresh leaf tissue. Phytochem Bull 19:11–15
Duke JA (2002) Handbook of medicinal herbs, 2nd edn. CRC Press, Boca Raton
Duke JA (2016) *Myristica fragrans* (Myristicaceae), Dr. Duke's phytochemical and ethnobotanical databases. https://phytochem.nal.usda.gov/phytochem/plants/show/1355?et=. Accessed 15 June 2017
Everett TH (1981) Myristica. Illustrated encyclopedia of horticulture, vol 7. Garland Publishing Inc., New York, pp 2264–2265
Flach M (1966) Nutmeg cultivation and its sex problem: an agronomical and cytogenetical study of the dioecy in *Myristica fragrans* Houtt. and *Myristica argentena* Warb. Ph.D. thesis. [Wageningen]: Mededligen Landbouwhode school Wageningen (66-1), Dissertation No Flach 397. Veenman, Wageningen
Flach M, Cruickshank AM (1969) Nutmeg (*Myristica fragrans* Houtt. and *Myristica argentea* Warb.). In: Ferwerda FP, Wit F (eds) Outlines of perennial crop breeding in the tropics. Landbho Gesch Wageningen, pp 330–336
Gopalam A, Sayed AAM (1988) Evaluating chemical and aromatic quality of nutmeg accessions *Myristica fragrans* L. Indian Spices 24(4)&25(1):9–11
Govind MG, Dan M, Rameshkumar KB (2020) *Myrisica trobograrii* (Myristicaceae), a new species from southern Western Ghats, India. Phytotaxa 437:206–212
Govind MG, Ananthakrishnan R, Dan M, Madhukrishnan M, Radhakrishnan KV, Rameshkumar KB (2022) Nutmeg with yellow arils-potential tree spice with high essential oil content. J Spices Aromat Crops 31(1):100–106
Gupta AD, Bansal VK, Babu V, Maithil N (2013) Chemistry, the antioxidant and antimicrobial potential of nutmeg (*Myristica fragrans* Houtt.). J Genet Eng Biotechnol 11:25–31
Ha MT, Vu NK, Tran TH, Kim JA, Woo MH, Min BS (2020) Phytochemical and pharmacological properties of *Myristica fragrans* Houtt.: an updated review. Arch Pharm Res 43(11):1067–1092. https://doi.org/10.1007/s12272-020-01285-4. Epub 18 Nov 2020. PMID: 33206347
Haldankar PM, Nagvekar DD, Desai AG, Patil JL, Gunjate RT (1999) The effect of season and the root stocks on approach grafting in nutmeg. Indian J Arecanut Spices Med Plants 1(2):52–55

Haldankar PM, Joshi GD, Jamadagni BM, Sawant VS, Kelaskar AJ (2005) Studies on germination and seedling vigour characters for genotype selection in nutmeg (*Myristica fragrans* Houtt.). J Spices Aromat Crops 14(2):137–144

Haldankar PM, Nagwekar DD, Khandekar PG (2004) Variability in nutmeg. Spice India 17:9–12

Heikrujam M, Sharma K, Prasad M, Agrawal V (2015) Review on different mechanisms of sex determination and sex-linked molecular markers in dioecious crops: a current update. Euphytica 201:161–194

Helen PM (2012) Phytochemical analysis and anticancer activity of essential oil from *Myristica fragrans*. Int J Curr Pharm 2:188–198

IFEAT (2021) Available at: https://ifeat.org/about-ifeat/

Iyer IR, Jayaraman G, Lakshmi GS, Gopinath PM (1993) Culture of various explants of nutmeg, *Myristica fragrans* Houtt. and factors affecting callogenesis. Indian J Plant Physiol 7(2):114–118

Iyer RI, Jayaraman G, Gopinath PM (1994) Morphogenetic responses of excised embryos of *Myristica fragrans* in culture. In: Second Asia-Pacific conference in agricultural biotechnology, 610 March, 1994, Madras, India, organized by National Academy of Agricultural Sciences, New Delhi, India & MS Swaminathan Research Foundation, Chennai. Abstract No. 117, p 200

Iyer RI, Gopinath PM, Lakshmi SG (1995) Tissue culture of nutmeg, *Myristica fragrans*. In: National symposium of tissue culture of medicinal plants. India organized by Department of Botany, University of Calcutta, Calcutta, 23–24 Feb 1995

Jaiswal P, Kumar P, Singh VK, Singh DK (2009) Biological effects of *Myristica fragrans*. Ann Rev Biomed Sci 11:21–29

Jannu LN, Hussain SP, Rao AR (1991) Chemopreventive action of mace (*Myristica fragrans* Houtt.) on DMBA-induced papillomagenesis in the skin of mice. Cancer Lett 56:59–63

Jayashree E, Chempakam B (2011) Afflation in spice control strategies. Spice India 24(8):21–24

Joseph J (1980) The nutmeg-its botany, agronomy, production, composition and uses. J Plant Crops 8(2):61–72

Joshy AC (1946) A note on the development of the pollen in *Myristica fragrans* and the affinities of the family Myristicaceae. J Indian Bot Soc 25:139–143

Kapoor IPS, Singh B, Singh G, De Heluan CS, De Lampasona MP, Catalan CAN (2013) Chemical composition and antioxidant activity of essential oil and oleoresins of nutmeg (*Myristica fragrans* Houtt.) fruits. Int J Food Prop 16:1059–1070

Katzer G (2004) Nutmeg and mace (*Myristica fragrans* Houtt.). http://gernot-katzers-spice-pages.com/engl/Myri_fra.html. Accessed 15 June 2017

Khare CP (2007) Indian medicinal plant. Springer India, New Delhi

Krishnamoorthy B (1986) Pollen germination and tube growth in nutmeg. Indian Cocoa Arecanut Spices J 10(1):10–11

Krishnamoorthy B (1987) Nutmeg. Plant Chron 82(3):83–84

Krishnamoorthy B (2000) Sex conversion in nutmeg. Spice India 13(5):11–12

Krishnamoorthy B, Rema J (1994) Tree spices crop improvement. Spice India, pp 24–30

Krishnamoorthy B, Sasikumar B, Rema J, Sayed AAM, Abraham J (1991) Variability and association in nutmeg. Indian Cocoa Arecanut Spices J 14(3):121–122

Krishnamoorthy B, Sasikumar B, Rema J (1996) Genetic variability and segregation of sex in nutmeg (*Myristica fragrans* Houtt.). J Plant Crops 24:468–472

Krishnamoorthy B, Rema J, Mathew PA (2014) A-9/69 of IISR (IC0537218; INGR 14039), a Nutmeg (*Myristica fragrans*) Germplasm with Bold Nut a, High Sabinene and Myrecene content. Indian J Plant Genet Resour 27(3):303–317

Krishnan AG, John R, George SV (2019) Marker based assessment of genetic diversity in nutmeg (*Myristica fragrans* Houtt.) using ISSR primers. J Exp Biol Agric Sci 7(2):108–116

Krishnamurthy KS, Rema J, Mathew PA, Krishnamoorthy B (2008) Identification of suitable *Myristica* species or related taxa as rootstock to combat drought in nutmeg. Indian J Hortic 65(2):204–208

Kusuma J, Scarcelli N, Couderc M, Mariac C, Zekraoui L, Duminil J (2020) Microsatellite markers development for Indonesian nutmeg (*Myristica fragrans* Houtt.) and transferability to other Myristicaceae spp. Mol Biol Rep 47:4835–4840

Leela NK (2008) Nutmeg and mace. In: Parthasarathy VA, Chempakom B, Zachariah TJ (eds) Chemistry of spices. CABI, Wallingford, UK, pp 165–189

Lele S, Deshpande S (2011) Isolation and PCR amplification of genomic DNA from traded seeds of nutmeg. J Biol Agric Healthc 1:1–6

Lestari K, Hwang JK, Kariadi SH, Wijaya A, Ahmad T, Subarnas A et al (2012) Screening for PPAR γ agonist from *Myristica fragrans* Houtt. seeds for the treatment of type 2 diabetes by in vitro and in vivo. Med Heal Sci J 12:7–15

Lewis VS, Namboodiri ES, Krishnamurthy N, Amma BSK (1976) Essential oils in spices. Arecanut Spices Bull 8(1):5–8

Li B, Wilson TK (2008) *Myristica fragrans* Houtt. In: *Flora of China*, vol 7. Myristicaceae. www.eFloras.org. Accessed 15 June 2017

Lissamma J, Anitha S, Aravindrakshan K (2012) Elite nutmeg clones: multiplication through budding [abstract]. In: Abstracts, mechanisation for sustainable productivity. UPASI, Coimbatore. Tamil Nadu, 12–15 Dec 2012. Abstract No. 19, p 16

Manjunatha BK, Mankani KL, Mukunda SK, Divakaran R, Sridar BK, Paul BK (2011) Antioxidant and hepato protective effect of *Myristica malabarica* seed aril extracts on carbon tetrachloride induced hepatic damage. Global J Biotech Biochem 6:25–30

Mathew SK, Beena S (2012) A new recording of *Phytophthora* sp. causing leaf fall and shoot rot of nutmeg (*Myristica fragrans* Houtt.). J Mycol Plant Pathol 42:529–530

Mathew PA, Joseph J (1982) Epicotyl grafting. J Plant Crops 10:61–63

Mathew SK, Miniraj N (2013) Leaf fall—a destructive disease of nutmeg. Spice India 26:11–12

Maya KM, Zachariah JT, Krishnamurthy KS, Rema J, Krishnamoorthy B (2006) Fatty acids and leaf amino acids in *Myristica fragrans* Houtt. and related taxa. Indian J Hort 63(3):316–318

Miniraj N, Mathew SK, Nybe EV (2012) Jathi krishiyum samskaranavum. Kerala agricultural university press, 62p

Miniraj N, Vikram HC, Manu P (2015) Variability of nutmeg in Kerala. Indian J Arecanut Spices Med Plants 17(2):6–14

Miniraj N, Vikram HC, Philip M, Deepa ET (2017) Fruit rind constituents in nutmeg (*Myristica fragrans*) morphotypes. J Plant Crops 44(3):153

Nair RR (2019) Chromosome number analysis in different sex types and open pollinated seedlings of nutmeg (*Myristica fragrans* Houtt. J Plant Crops 47(3):197–201

Nair NC, Bahl PN (1956) Vascular anatomy of the flower of *Myristica malabarica* Lank. Phytomorphology 6:127–134

Nair NC, Pillai NV (1959) Morphology of the flower and aril in *Myristica fragrans* Van Houtten. Sci Cult 25:211–213

Nazeem PA (1979) Studies on growth, flowering, fruit set and fruit development in nutmeg. M.Sc. (Hort.) thesis, Kerala Agricultural University, Thrissur, p 96

Nazeem PA, Nair PCS (1981) Growth and flowering in nutmeg. Indian Cocoa Arecanut Spices J 4:81–83

Nenlohy FN, Semuel MY (2020) Nutmeg (*Myristica fragrans*) of North Minahasa molecular entication based on chloroplast DNA of matK gene. Biodiv J 1:60–68

Priyanka SC, Miniraj N (2016) Morphological characterization of unique genotypes of nutmeg (*Myristica fragrans* Houtt.). J Trop Agri 54(2):120–128

Priyanka SC, Miniraj N, Lincydavis P, Vikram HC (2022) Genotype-environment Interactions in Nutmeg (*Myristica fragrans* Houtt.). Biol For Int J 14(4):394–403

Priyanka SC, Miniraj N, Vikram HC (2023) Growth and yield of nutmeg budded on wild (*Myristicabeddomei* King) and cultivated (*Myristica fragrans* Houtt.) rootstocks. Biol For Int J 15(2):1269–1275

Pruthy JS, Krishnamurthy S (1984) A study on factors governing the recovery and quality of pectin from nutmeg waste (rind). Indian Cocoa Arecanut Spices J 8(3):75–78

Purseglove JW, Brown EG, Green CL, Robbins SRJ (1981) Spices, vol I. Longman, London, pp 128–174

Rani TG (1994) Induction of orthotrops in vegetatively propagated nutmeg (*Myristica fragrans*). M.Sc. (Hort.) thesis, Kerala Agricultural University, Thrissur, p 99

Ravindran PN (1999) Genetic diversity of major spices and their conservation in India. In: Sasikumar B, Krishnamoorthy B, Rema J, Ravindran PN, Peter KV (eds) Biotechnology, conservation and utilization of spices. Medicinal and aromatic plants. IISR, Calicut, pp 16–44

Ravindran PN (2017) Nutmeg. In: The encyclopedia of herbs and spices, vol 2. CABI

Ravindran PN, Babu NK, Peter KV, Abraham Z, Tyagi AK (2005) Spices. In: Dhillon BS, Tyagi AK, Saxeena S, Randhawa GJ (eds) Plant genetic resources of horticultural crops. Narasa Pub. House, New Delhi, pp 190–227

Rema J, Krishnamoorthy B, Zachariah TJ, Mathew PA (2013) A9/71 (IC537220); INGR 10142, a nutmeg (*Myristica fragrans*) germplasm with high sabinene. Indian J Arecanut Spices Med Plants 15(4):12–13

Rema J, Saji KV, Sasikumar B, Anandaraj M (2015) Monoecious nutmeg in Uttarkannada and Shimoga districts of Karnataka. Indian J Arecanut Spices Med Plants 16(3):3–5

Ross IA (2001) Medicinal plants of the world, vol 2. Humana Press, New Jersey, pp 333–351

Ruth SMV, Silvis ICJ, Alewijn M, Liu N, Jansen M, Luning PA (2018) No more nutmegging with nutmeg: analytical fingerprints for distinction of quality from low-grade nutmeg products. Food Control 98:439–448

Sairam TV (1998) Home remedies, vol 1. Penguin Books, New York

Sasidharan N (2006) *Illustrated manual on tree flora of Kerala.* KFRI research report no. 282. Kerala Forest Research Institute, Peechi, Kerala, India

Sasikumar B (2021) Nutmeg—origin, diversity, distribution and history. J Spices Arom Crops 30(2):131–141

Sastri RLN (1954) On the vascular anatomy of the female flowers of *Myristica fragrans*. In: Proceedings of the Indian science congress. Part III, Indian Science Congress Association, Calcutta, pp 132–133

Sastri RLN (1955) Structure and development of the nutmeg seed. Curr Sci 24:172–173

Senthilkumar R, Krishnamoorthy B, Prasath D, Venugopal MN, Ankegowda SJ, Biju CN (2010) Variability in nutmeg (*Myristica fragrans* Houtt.) under high rainfall and high altitude Kodagu region of Karnataka. Indian J Plant Genet Resour 23:191–194

Shanmughavelu KG, Rao MVN (1977) Spices and plantation crops. Popular Book Depot, Madras-34, p 298

Sharma MV, Armstrong JE (2013) Pollination of *Myristica* and other nutmegs in natural populations. Trop Conser Sci 6(5):595–607

Sheeja TE, Rajesh Y, Krishnamoorthy B, Parthasarathy VA (2006) DNA polymorphism in clonal and seedling progenies of an elite nutmeg (*Myristica fragrans* Houtt.) by RAPD. J Plant Crops 34(3):558–561

Sheeja TE, George KJ, Jerome J, Varma RS, Syamkumar S, Krishnamoorthy B, Parthasarathy VA (2008) Optimization of DNA isolation and PCR parameters in *Myristica* spp. and related genera for RAPD and ISSR analysis. J Spices Aromat Crops 17(2):91–97

Sheeja TE, Anju PR, Shalini RS, Siju S, Dhanya K, Krishnamoorthy B (2013) RAPD, SCAR and conserved 18s rDNA markers for a red-listed and endemic medicinal plant species, *Knema andamanica* (Myristicaceae). Physiol Mol Biol Plants 19(2):245–250

Shinde SS, Haldankar PM, Khandekar RG, Haldavnekar PC (2006) Variability and correlation for fruit characters in nutmeg (*Myristica fragrans* Houtt.). J Plant Crops 34(3):152–154

Simmonds NW (1954) Chromosome behaviour in some tropical plants. Heredity 8(1):139–146

Sinclair J (1958) A review of the Malayan myristicaceae. Singap Cargens Bull 16:205–472

Singh S, Morya S, Saini P, Singh M, Rout S (2016) Studies on antioxidant and antibacterial activity of nutmeg seed and its mace. Adv Life Sci 5(8):3207–3210

Sneha DSL (2011) Isolation and PCR amplification of genomic DNA from traded seeds of Nutmeg (*M. Fragrans*). J Biol Agricul Healthcare 1(3):1–6

Soni DK (2012) Essential oils from *Myristica fragrans* Houtt (Nutmeg). Bioinfolet 9(2):196–197

Sreedharan G, Indulkar M (2018) New distributional record of the Northen most *Myristica* swamp from the Western Ghats of Maharashtra. Current Sci 115:1434–1435

Sriram TA (1977) Tree spices, retrospect and research needs. Arecanut Spices Bull 11(3):83–84

Sudhamayee M (2010) Sex determination in nutmeg (*Myristica fragrans* Houtt.) through molecular and biochemical markers. M.Sc. (Agri.) thesis. Kerala Agricultural University, Thrissur, p 102

Sudhamayee M, Valsala PA, Nazeem PA, Augustin A, Girija D (2014) Identification of sex-specific molecular and biochemical markers in nutmeg (*Myristica fragrans* Houtt.). [Abstract]. In: Abstracts, 2nd international conference on agricultural and horticultural sciences, Hyderabad, 3–5 Feb 2014. Abstract No. 4, p 202

Sumbula V, Mathew SK (2015) Management of *Phytophthora* leaf fall disease of nutmeg (*Myristica fragrans* Houtt.). J Trop Agric 53(2):180–186

Swapna S, Venkateshwararao Y, Mathew MK, Saideswararao Y (2014) Concurrent isolation of DNA and RNA in *Myristica fragrans* Houtt. Int J Trop Agric 32(4):419–421

Swetha VP, Sheeja TE, Sasikumar B (2019) DNA barcoding to resolve phylogenetic relationship in *Myristica* spp. J Spices Aromat Crops 28:131–140

Verghese J (2000) Nutmeg and mace-I, general back ground. Spice India 13(12):2–4

Vikram HC, Miniraj N, Krishnan S (2016a) Investigations on developing a key for identification of elite nutmeg tree. J Plant Crops 44(3):166–173

Vikram HC, Miniraj N, Presannakumari KT, Nybe EV, Deepu M (2016b) Variability in fruit characteristics of nutmeg (*Myristica fragrans* Houtt.) under Kerala conditions. J Spices Aromat Crops 25(2):187–194

Vikram HC, Miniraj N, Deepu M (2018) Candidate molecular markers for monoecy in dioecious tree spice nutmeg (*Myristica fragrans* Houtt.) and analysis of genetic variability in a core collection. Current Sci 114(1):23–25

Vikram HC, Mini Raj N, Deepu M, Priyanka SC, Sujatha VS (2022) Chemical profiling and constituent diversity in nutmeg genotypes (*Myristica fragrans* Houtt.). Indian J Agric Res A 5848:1–8

Chapter 8
Bulb Onion (*Allium cepa* L.) Breeding

Mohamed F. Mohamed and Azza A. Tawfik

Abstract The bulb or common onion is the most cultivated species of the genus *Allium* worldwide. The basal portion of the leaves has swollen forming the bulbs when a certain day-length is reached. The bulb composed of reduced, compacted, underground stems surrounded by modified fleshy scale (leaves). The scales encircle a central bud located at the tip of a reduced stem. Due to causing serious crop damage, the chief onion breeding objective is the pests and diseases resistance/tolerance. Onion improvement to cope with climatic changes is one of the core current breeding goals. Onions are cultivated and utilized around the year. Therefore, bulb adaptability and storability are important. In spite of existing useful inter-and intra-specific useful traits, their introgression is very limited due to genomic incompatibility resulting in partial or complete sterility. Generally, there is reluctance concerning onion breeding as it is open pollinated biennial crop and suffers inbreeding depression only after 2–3 cycles of selfing, the maximum. Male-sterility is available in onion, however, breeding highly inbred parental lines for F1 hybrid (heterozygous/homogeneous breeding) cultivar development is difficult if not possible at all. Consequently, population breeding (heterozygous/heterogeneous breeding) is still dominating in onion improvement. Double haploids can play an effective role to overcome the limitation of breeding pure lines. F1 hybrids industry is advantageous in terms of protecting the breeder rights but it can cause genetic erosion. Therefore, newly emerged omics technology may be beneficial for identification, characterization and conservation of the germplasm. Development of saturated genetic map and marker-assisted selection holds a promise to use in onion genetic improvement. Onion has a relatively, huge genome size and this is a disadvantage, especially for its functional genomic analyses and for its genetic transformation.

Keywords Allium · Dry bulb onion · Gene introgression · Heterogeneous · Intraspecific · Marker-assisted selection

---

M. F. Mohamed (✉)
Department of Vegetable Crops, Faculty of Agriculture, Assiut University, Assiut 71526, Egypt
e-mail: mofouad@aun.edu.eg

A. A. Tawfik
The Faculty of Agriculture, Assiut University, Assiut 71526, Egypt
e-mail: azza_at@aun.edu.eg

J. M. Al-Khayri et al. (eds.), *Biodiversity and Genetic Improvement of Herbs and Spices*, Advances in Plant Breeding Strategies 8, https://doi.org/10.1007/978-3-031-81838-7_8

8.1 Introduction

Onion (*Allium cepa* L.) is considered as one of the topmost important crops where it comes just after tomatoes. It is cultivated everywhere in the world for its dry bulbs. Beside its nutritional and culinary uses, it has medicinal properties. It has been called "the Queen of the kitchen". This is because of its loved aroma, flavor, and unique taste. Furthermore, onions have medicinal possessions of its compounds (Griffiths et al. 2002). Onions are possibly used in many forms including fresh, frozen, caramelized, pickled, canned and chopped. The dehydrated onion is obtainable as minced, chopped, granulated, kibbled, sliced, ring, and powder forms. Onions are used as spices and condiment and cooked with various other vegetables. It is also used as pickles, paste, and flakes. The spherical bulb was regarded, by the ancient Egyptians, as symbol of the universe. The name onion is perhaps taken from the Latin 'unus' which mean "one."

8.1.1 Distribution, Botanical Classification and Domestication History

Scientific classification	
Kingdom:	Plantae
Clade:	Angiosperms
Clade:	Monocots
Order:	Asparagales
Family:	Amaryllidaceae
Subfamily:	Allioideae
Genus:	*Allium*
Species:	***A. cepa***
Binomial name	
Allium cepa L.	

Source https://en.wikipedia.org/wiki/Onion#Varieties

Common or dry onion (*Allium cepa* L.) is the utmost prominent edible *Allium* crop. The genus *Allium* comprises a number of other species cultivated for food, such as the Japanese bunching onion (*A. fistulosum*), Egyptian onion (*A.* ×*proliferum*), and Canada onion (*A. canadense*). Variously these species are referred to as onions (Fritsch and Friesen (2002). As mentioned by Fritsch and Friesen (2002), *Allium* and related genera of the angiosperms were placed in the *Liliaceae.* Later, they were more repeatedly included in the Family *Amaryllidaceae* based on inflorescence structure. However, the most recent taxonomic treatment recognized *Allium* and its close relatives as a distinct family, the *Alliaceae* close to the Family *Amaryllidaceae.* (Takhtajan 1997). The genus *Allium* is widely distributed (Fig. 8.1). Especial high species diversity region stretches from the Mediterranean basin to Central Asia and Pakistan. Additional, less pronounced center of species diversity occurs in western North America. The genus is of great economic significance since it includes several important vegetable crops (including dry bulb onion) and ornamental species. Domestication did not change the ploidy status of onion and many other diploid species. The introgression of other species rarely had a role during the selection processes. Like many other ancient cultivated crops, limited amount of circumstantial evidences and no hard facts are available on the evolutionary history of cultivated Alliums. Sculptural and painted representations from ancient Egypt support the supposition that onion was already cultivated at that time.

The vast majority of *A. cepa* cultivars belong to the "common onion group" (*A. cepa* var. *cepa*). They are simply referred to as "onions". The species of the section *cepa* are diploid ($2n = 16$). In spite of the morphological and cytological resemblances among the species of section *cepa*, there are strong crossing barriers among them. This precludes interspecific gene flow. The bulb characteristics are widely different. This may reflect a practicing of an intensive selection. In contrast, organs which were not subjected to selection by humans (e.g. the flower and the capsule) have not been greatly affected by domestication and exhibit little or no prominent variations. The current natural distribution for this Alliance suggests that domestication of *A. cepa* most likely started in the Middle East (Hanelt 1990). *Allium cepa* is amongst the oldest cultivated vegetable crops; recorded for more than 4000 years ago. The earliest records come from Egypt. There, it was cultivated at the time of the 'Old Kingdom'. Onions look as carvings on pyramid walls and in tombs from the third and fourth families (2700 BC). This indicates the importance of these crops in the daily diet of those peoples. Furthermore, the biblical records of the Exodus (1500 BC) are also well-known. The onion was among the first taken cultivated plants from Europe to the Americas.

8.1.2 Economic and Nutritional Values and Health Benefits

Onion is a biennial bulb crop, lengthily grown with world production of 74,250,809 tonnes. That is from an area of 4,364,000 ha. China is the largest producer of dry bulb onion (Fig. 8.2) (FAOSTAT 2018; Hai 2015; Pareek et al. 2018). It followed by USA, Egypt, Iran, Turkey, Pakistan, Brazil, the Russian Federation, and the Republic of Korea (Pareek et al. 2018). The highest onion crop productivity is in the Republic of Korea (66.16 t/ha). It followed by the USA (56.26 t/ha), Spain (53.31 t/ha), and the Netherlands (51.64 t/ha). The average productivity across the world is 19.79 t/ha. The international trade in onion exports is estimated by 6.77 million tonnes. The Netherlands occupies the top as an onion exporter (1.33 million tonnes). It followed by India,

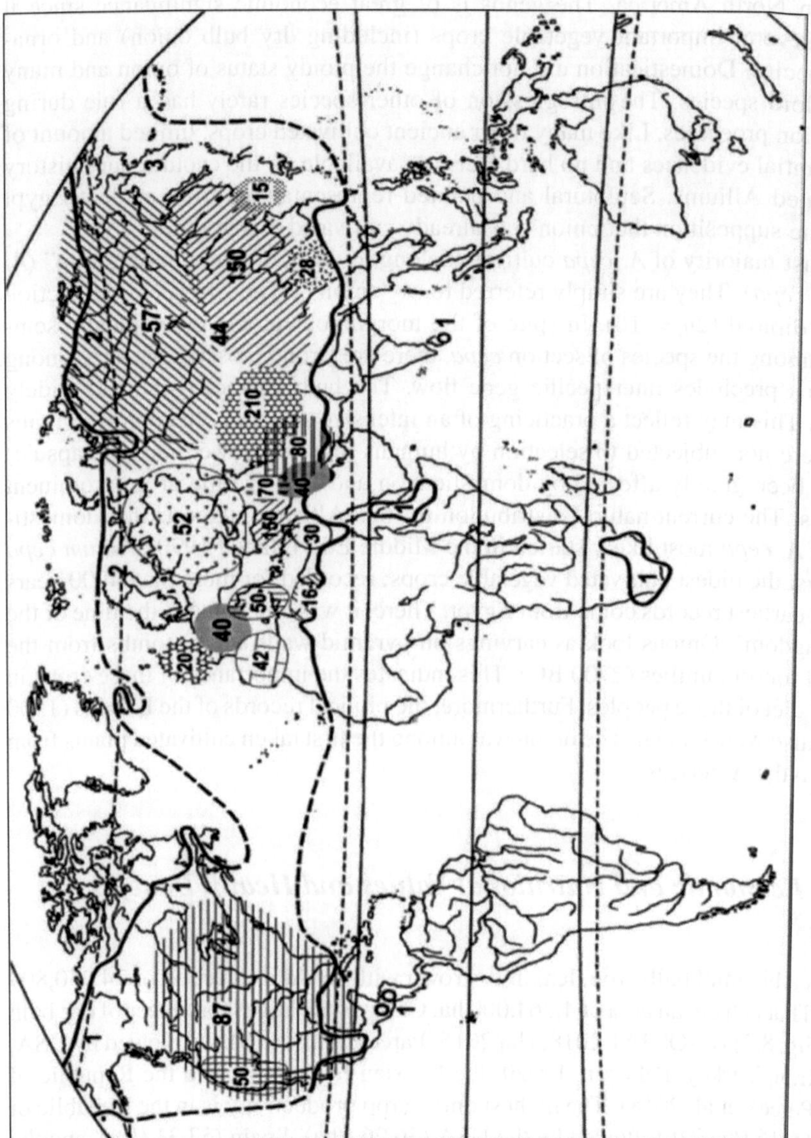

Fig. 8.1 World distribution of wild species of genus Allium. The numbers on the map indicate the number of species. https://www.google.com/search?q=map+of+onion+distribution+in+the+world&client=firefox, accessed 23 July 2022

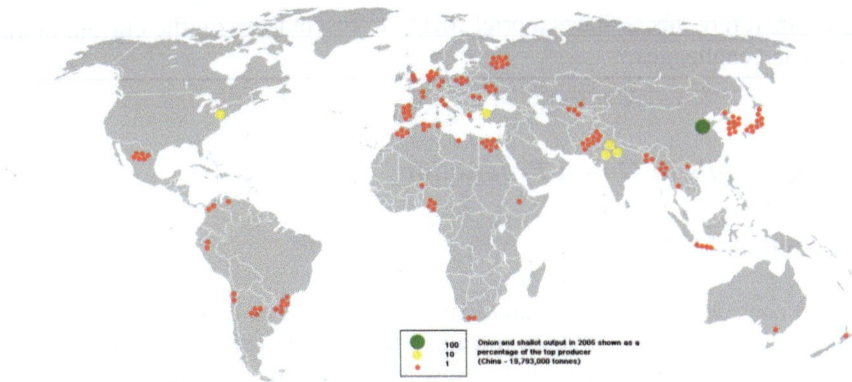

Fig. 8.2 Onion production as percentages of the top producer China

China, Egypt, Mexico, USA, Spain, and Argentina. The major onion importing countries in the world are Bangladesh, Malaysia, the Russian Federation, the UK, Japan, and Saudi Arabia.

The contents of bulb tissues in most onion cultivars are, in general, about 89% water, 9% carbohydrates (including 4% sugar and 2% dietary fiber), 1% protein, and negligible fat amount. Onions contain relatively low amounts of essential nutrients. It has an energy value of 166 kJ (40 cal) per 100 g (3.5 oz) bulb tissue amount. Onions contribute savory flavor to food dishes with no significant caloric content contributed ["History of onions" (https://www.onions-usa.org/all-about-onions/history-of-onions) US National Onion Association, Greeley, CO, 2011, accessed 23 July 2022].

This presentation consists of three sequential section concerning advances in onion breeding. In the first section, genetic improvement objectives were derived in light of current cultivation practices and challenges of dry bulb onion. The current status of conventional breeding of bulb onion was reviewed in terms of germplasm diversity and resources, and characterization and phylogeny, gene bank management and cross limitations and genome incompatibility constraints. Mutation approach for creating new variability for onion enhancement was considered. The role of newly emerged 'omics' and biotechnologies to overcome traditional breeding restrictions and accelerate its process was emphasized in the third section.

8.2 Current Cultivation Practices and Challenges

Many researchers have made a comprehensive presentation on cultivation practices in onion. This chapter is to highlighten the advances and current trend in onion breeding. The first principal in breeding is 'Know you plant'. Determination of current breeding objectives, therefore, necessitates a brief focused presentation on the current onion cultivation and challenges.

8.2.1 Principals in Onion Production

8.2.1.1 Culture Methods

Onions are considered as a labor-intensive crop. It requires careful skills in crop managing. Dry bulb onions are grown using either direct sowing or transplants or dry sets. This is decided based upon the growing conditions of the regions. Sets are small dry bulblets usually less than 2.5 cm in diameter. Transplanting is advantageous in terms of economic use of seeds and the chance to select superior (healthy and vigorous) seedlings. It saves weeding and watering effort during the initial weeks of onion growth. Further, it enables the farmer to attend the seedlings in a compact area. On the contrary, sowing directly in the production field requires about three times larger amount of seeds (12 kg) than transplants (3–4 kg). Besides, it needs land leveling and careful planting practices. Unlike transplanting seedlings and direct seeding, dry sets adapt to short growing seasons. Further, its plants can tolerate tough environmental conditions. They are earlier and more productive than those plants derived using direct seed sowing or seedling transplants. Onion producers have to decide on the most proper planting methods for the specific conditions in their regions.

Plants do not fully bulb unless exposed to certain day length. Severe freezes can damage onion crop although it is cold hardy. Low temperatures initiate flowering response. Prolonged periods of 40–50 °F may cause bolting (seed-stalk formation) especially if the cold weather is followed by a period of warm temperature. In the tropics, bulbs are generally harvested 90–150 days after transplanting; in the spring. In some countries, e.g. UK and New Zealand, both autumn and spring sown crops are produced. Bulb high yield can be attained when an early leaf canopy is produced and then maintained for a long period prior to setting the bulbs. Growing at higher plant densities can increase light interception. However, intensive culture produces smaller bulb size. Thus, it is vital to control plant density to obtain high bulb yield of a wanted size. Planting depth affects bulb shape. Shallow planting results in flatter bulbs, while deeper placement produces taller and sometimes top-shaped bulbs.

8.2.1.2 Main Field Practices

Onions rooting is generally poor as they are short sparse and few hairs root systems. Thus, it is important to keep nutrients and moisture accessibility within the rooting area. Studies have suggested that water uptake by the plants is limited to the top ten inches of the soil profile. Onions can, relatively, withstand long periods of drought stress. However, availability of the water is crucial for growth and yield of high quality bulbs. Onions are more likely to split or to form double and multiple bulbs under drought conditions. Onion plants have, relatively, a high water requirement; usually around three inches per week. Nonetheless, irrigation late in the season can delay maturity and causes bulb skin cracking. Soil pH is preferably to be in the range of 6 -7, but in organic soils, onions can tolerate pH down to 4. When pH drops below 5.5, magnesium and molybdenum availability reduces and above 6.5, zinc, manganese and iron become deficient. Onions are very sensitive to salinity. So, salty irrigation water and saline soils should be avoided.

Onions seedlings are slow to emerge in the nursery and to grow up following transplanting in the production field. Therefore, weed control is vital to reduce competition of the weed concerning moisture and nutrients. The onions do not compete effectively with weeds as they have erect growth habit and general slow growth nature. Pre- and post-emergence herbicides are usually employed in production of bulb onion. Interrow shallow cultivating and hand weeding are useful and are practiced in certain regions.

8.2.1.3 Bulb Formation

Formation of onion bulb is influenced by the photoperiod. Generally, onion is regarded as a long day plant since its bulb formation is encouraged by long photoperiod conditions. The day length requirements for bulbing are a quantitative trait and it differs according to the cultivar. Each cultivar needs a minimum photoperiod for bulb initiation and development. This photoperiod is known as the critical photoperiod value. The critical photoperiod in combination with the prevailing temperature control the establishment of bulb initiation. Some cultivars like "Cojumatlan" (Mexican cultivar) may start bulbing under short day condition. Long day cultivars, on the other hands, will not initiate bulbing unless its longer critical value of photoperiod is attained.

Under long day and high temperature, leaf growth was inhibited (Aoba 1967). It was completely ceased within 25–30 days of the start of long photoperiod treatment. Short day (8 h light) treatment, when given to the plants for 30 days before the long day treatment, growth was accelerated and large sized bulbs were formed. Kato (1964) found that the photoperiod impact was neutralized when long day treatment was interrupted by a period of short days. Indeed, the bulb-forming process was reverted to vegetative phase. It was experimentally possible to reverse the bulb formation phase, even after the tops were fallen over.

Interestingly, Heath and Hollies (1965) revealed that onion plants when they were kept under short photoperiod condition they continued growing for whole three years without bulbing. Working in Cuba, Kononkov et al. (1969) tried some cultivars from different regions in the world. They found that cultivars from the countries approximately of the same day length as Cuba, produced bulbs. Those cultivars from far northern region did not produce bulb. Furthermore, Austin (1972) showed that the response of different cultivars to day length remained in the same order, relative to each other over wide ranges of photoperiodic treatments (Sidhu et al. 2005).

Fig. 8.3 White rot (*Sclerotium cepivorum*) damage. *Source* google search, Sept. 7 2021. https://www.google.com/search?sxsrf=ACYBGNQcSft628ki89ba_v5aKTeT7hxaKA:1567805148510&q=onion+white+rot&tbm=isch&source

8.2.1.4 Pests and Diseases

Initially, seeds should be free of nematode. The control measurements recommended to reduce the buildup of nematode population include crop rotation and soil sterilization. Thrips is a serious pest worldwide, particularly, in warmer regions. Thrips attacks the young leaves and the plant neck. It decelerating plant growth and extensively injures the leaves. Onion fly or maggot is considered to be a foremost pest. It can be controlled using insecticides. Bacterial diseases (e.g. *Pseudomonas alliicola*, and *P. cepacia*) considerably reduce bulb quality. Key onion diseases include neck rot (Fig. 8.3) caused by (*Botrytis allii*), mildew caused by (*Peronospora destructor*) rust caused by (*Puccinia porri*) and leaf rot caused by (*Botrytis squamosa*). Onion crop may be sprayed up to 15 times with fungicide during the production season. Generally, soil sterilization, deep cultivation and wide rotations are important to minimalize disease problems.

8.2.1.5 Storage

Bulb onions are habitually starts when about 60–80% of the tops fall over. Bulbs need curing by drying. This will help to close the neck and consequently prevent invasion of diseases and rots causal agents. Curing in temperate regions is usually achieved in store under cool and shade conditions. Elsewhere, onions are lifted and left to dry in shade area or under straw in the field. Chlorophyll formation (green scales) may occur in the outer bulb skin. This happens when bulbs are left to cure for too long time or when bulb shoulders are exposed to sunlight in the field. Bulbs should be prudently handled to minimize wounding cuts to prevent disease agents to enter. Controlled atmosphere storage is increasingly practices. This is to keep bulb quality over longer periods of time in store and to extend the marketing time. Maintaining carbon dioxide at 5% level inhibits fungal sporulation. Oxygen at 3% lessens bulb sprouting. Temperatures should be around 34°F to help reduce respiration. Relative humidity is to be kept at 65–70% to inhibit fungal diseases spread.

8.2.2 *Current Genetic Improvement Goals*

As it can be concluded from the information presented here—about the status of onion production—bulb onion suffers some sever growing constraints. These include diseases, pests, storability, weeds, bolting and doubling. However, the most serious challenge seemed to be the diseases and pests control. This is in line with the assessment conducted by Haile et al. (2016) concerning constraints of bulb onion production especially in developing countries. The findings of the study of Haile et al. (2016), in their questioner for onion growers, revealed that difficulties the growers face are mainly the disease and insect pests attack. This was agreed by the majority of the sampled farmers (about 77%). Further, weeds, unavailability of quality seeds, and lack of pesticide, post-harvest technologies, community awareness on crop protection and inadequate demonstration of new technology are factors that seriously responsible for the yields and quality reduction in onion. An example, for the severity of the diseases attacks is the epidemic spread of the white rot (*Sclerotium cepivorum*) in the most productive onion (Fig. 8.3) area in Upper Egypt that was producing dry bulbs for exportation and dehydrated onion as well. The production area had to move totally up to the north in the Delta region of the country.

The storability is also a serious challenge where magnificent bulb yield loss has been reported. Cultivars breeding for soft neck and high dry matter may store well providing that bulb has been well-cured (Valenzuela et al., 1999) Sustainable production of onion that can provide an economic return to the grower is a major challenge. More research would be needed concerning organic production, especially from the viewpoint of breeding (Mousa and Mohamed, 2009). Hybrid cultivar production should be emphasized in targeting the motioned challenges. All the above mentioned constraints of onion production are considered potential current breeding goals. Goals remain the same but tools and strategies to realize them are changing.

8.3 Germplasm Biodiversity and Conservation

8.3.1 *Biodiversity Status, Types and Sources*

Genetic diversity of crop plants is vital resource for human life. The identification/classification, preservation and sustainable utilization for such resources are critical to plant crop production. This is because availability of genetic diversity is essential for establish of successful plant breeding program. Biodiversity, in general, comprises genetic diversity, species diversity and ecosystem diversity. The species diversity includes flora in wild as well as cultivated plants. The wide range of variation in bulb characteristics indicates an intensive selection. Under continuation of cultivation and selection by human, the onion bulbs have developed into large organs of various shapes. The bulbs are variable in color (yellow, red, white and the intermediate shades of these colors). On the other hand, traits like flower and the capsule traits have been slight affected by domestication because they have not been under artificial selection (by humans) (Brahim 2020; Hanci and Gokce 2016; Rivera et al. 2016). In addition to usual main sources of biodiversity, heirloom vegetables occasionally have been preserved in home gardens for generations (Galluzzi et al. 2010). These vegetables normally have different characteristics linking with commercial ones such easy propagation, storability and taste which are more valued than merely the elevated yield.

The genetic diversity denotes the variability in the genome ranging from rearrangement of single base pair to entire chromosome (as defined by Hamrick and Godt 1996). Noticeable alterations in diversity (Huang and Han, 2014) are due to variation in the genome interactions with environment. Due to their value as germplasm, landraces have received a special attention of researchers concerning molecular identification and diversty (Abdou et al. 2016; Brahim 2020; Hanci and Gokce 2016; Rivera et al. 2016). In onion the information on genetic resources and diversity is, relatively, still inadequate. This mainly is due to insufficiency of selective markers, germplasm resources, out breeding nature and biennial habit (Sudha et al. 2019).

8.3.2 Biodiversity Assessment Techniques

The information on genetic diversity within crops is prominent for germplasm collection. Various techniques (morphological, Biochemical and molecular markers) can be employed to assess the genetic diversity. Morphological markers are the visually measurable traits. These markers are affected by the environmental factors. This limits variable articulateness. Biochemical markers depend on gene product while molecular markers rely on a DNA assay. Molecular markers have proven as vital tools in the assessment and characterization of genetic diversity both within and between species (Ghodke et al. 2018). Simple, quick, economical and easily automated marker systems viz. SSR (Simple Sequence Repeats); and RAPD (Randomly Amplified Polymorphic DNA) are useful (Fischer and Bachmann 2000). The co-dominant markers SSR (simple sequence repeats) have been confirmed to be effectively distinguish tool for biodiversity and relationships analyses in onion germplasm.

The classifcation of germplasm and the genetic relationships among them is established using multivariate statistical algorithms (Ghodke et al. 2018; Hanci and Gokce 2016). There are many molecular markers currently routinely used to evaluate genetic diversity in plants (Rashid et al. 2013). Presently, various techniques have been developed to identify enormous varieties through molecular markers in onion, garlic and their related species (Fang et al. 2020; Fischer and Bachmann 2000).

As example for molecular diversity analysis utilizing RAPD and ISSR molecular markers is shown for nine local Indian cultivars (Sudha et al. 2019). Ten primers of 18 RAPD's exhibited specificity producing a total of 52 amplified polymorphic fragments. Six primers in ISSR analysis among 10 ISSR's, generated 28 polymorphic bands. The sixteen RAPD and ISSR primers distinguished 9 cultivars into two groups by the UPGMA method. Noticeably, single cultivar 'Agrifound rose' was in group-I while all remaining cvs were separated in the second group. The high significant variability suggested the existence of an opportunity to develop cultivars through breeding. Further the information gained may be beneficial for future genebank management programs.

8.3.3 *Intraspecific Classification*

Variability within the species is great and, therefore, different proposals for intraspecific groupings have been advised. The gene pool of onion is more than 20,000 all over the world and the grouping in onion genotypes is carried out by many scientists. The genotypes were sub divided into Common group, Aggregatum group and Ever-ready onion groups (Fritsch and Friesen 2002). Common onion group is considered the most economically significant *Allium* crops. It comprises open pollinated old and modern cultivars, in addition to hybrids and local races cultivated in the world. Plants of the common onion group produce bulbs that are large and normally single. These plants propagate by both seeds and seed-grown sets. The mainstream of cvs grown for dry bulbs belongs to this group. Maximum diversity exists in North India, Pakistan, former Soviet Union, Europe, Middle East, and in Mediterranean area (Astley et al. 1982).

8.3.4 *Status of Genetic Resources/Germplasm*

Establishing successful breeding programs for desirable trait(s) in crop plants requires existence of germplasm genetic diversity (Ghodke et al. 2018; Glaszmann et al. 2010). Onion has been grown and used since ancient times. Selection of plants possessing required traits has been practiced by growers in different regions according to their needs. As a result, landraces germplasm possessing a range of great genetic variation for various traits is available (Brewster 2008). Astley et al. (1982) made first overview of Allium genetic resources. They identified major Allium collections worldwide. The number of accessions per species and collection was also mentioned and they listed Allium descriptor together with collection significances.

Table 8.1 Number of accessions of the onion gene pools based on the International Plant Genetic Resources Institute (IPGRI) (www.ipgri.cgiar.org)

Genepools	Accession (no)
Primary	
A. cepa	12,740
A. vavilovill	20
A. galanthum	34
A. roylei	4
Secondary	
A. Fishtulosum	951
A.altaicum	121
Tertiary	
A. paskemense	21
A. oschaninii	41

Source www.ipgri.cgiar.org

Totally about 9000 accessions were reported to be present worldwide of which onion accessions was about 7000. The threat of possible genetic erosion as a result of the introduction and spread of F1 hybrids led to the collection of onion landraces and modern cultivars from different countries (Fritsch and Friesen 2002) in the late 1980s. The collected germplasm represents a wide range of onion gene pools and it is available in gene banks (Table 8.1). Development of core collections representing this diversity and evaluation of common set of lines under diverse environmental conditions would enhance the utilization of bulb onion germplasm.

A large numbers of landraces including some wild species are available in India, particularly, in North eastern states. According to Singh and Rana (1994), an extensive survey in different Allium-growing regions in India was directed by ICAR-National Bureau of Plant Genetic Resources (ICAR-NBPGR). Kale et al. (1994) started a detailed survey of traditional and non-traditional onion growing areas of the state of Maharashtra, and India in general, and collected148 red-skin and 33 white-skin types of onion, evaluated and identified some lines on the basis of maximum average bulb weight, high TSS and centeredness.

As per Singh and Rana (1994), a source of resistant/tolerant to purple blotch (*Alternaria* species) and *Stemphylium* blight was detected in cultivated Indian accessions. Many farmers in various parts of the country are growing old land races of onion. For example, Pune Fursungi, a red colored land race is being cultivated in Nashik and Pune areas of Maharashtra in late *kharif* and *rabi* seasons. The Junagadh, Saurashtra and Mehsana areas of Gujarat are dominated by Pili Patti, which is commonly grown in *rabi* season. Bellary Red, another red onion land race is prevalent in Karnataka and land race Sukhsagar is being cultivated in West Bengal. K.P. onion dominates in Andhra Pradesh, whereas Nirmal Local occupies large area in Madhya Pradesh. Further, the multiplier type of onion has been a unique feature in Tamil Nadu. This variability is being maintained in national germplasm collection of onion at ICAR-DOGR, which is the National Active Germplasm site for onion.

The present status of germplasm collection at ICAR-DOGR is given in Table 8.1. At present, nearly eighty varieties have been developed by different public as well as private institutes (Gopal 2015).

8.4 Traditional Breeding

8.4.1 Cytological Status

The domesticated Alliums have all a basic chromosome number of 8 and most of them are diploids (A. cepa, $2n = 2x = 16$) (McCollum 1976). The genus Allium is one of the largest plant genera as it has about 600–920 species (Peterson et al. 1988). The most important agricultural crops of this genus are bulb onion (*Allium cepa*), shallot (*A. cepa* syn. *A. ascalonicum*), Japanese bunching or Welsh onion (*A. fstulosum*), chive (A. schoenoprasum), garlic (*A. sativum*) and leek (*A. ampeloprasum* syn. *A. porrum*). In spite of the likenesses on both morphological and cytological levels among the species of section cepa, there are still serious crossing barriers (Romanov et al, 2021). This prevents gene flow between sympatric species. Besides, gene introgression from wild to cultivated species is critically difficult and marginal success has been reported in several hybridizations (Sidhu et al. 2005).

8.4.2 Flowering and Pollination

Flower induction is chiefly induced by cool temperature around 4.4 °C for about a week. Flower induction can be initiated in bulbs or in growing plants. Larger plants are more responsive to cool temperature for flower. The inflorescence (umbel) may comprise as much as 2000 flowers. The flowering stalk is an apical elongation of the stem but with no nodes and internodes. The inflorescence (umbel) may comprise as much as 2000 flowers. The flowering stalk is an apical elongation of the stem but with no nodes and internodes. Onion is an entomophily cross pollinated crop. Anthesis occurs at 06–07 am and anther dehiscence continues from 07 am to 05 pm. Floral nectarines located at lower part of ovary. All the four honeybee species acts as major pollinators in onion ecosystem. Maximum pollinator activity was observed at temperature 28–32 °C (Kavitha and Rami Reddy 2018). However, honeybees and drone flies are the most imperative. Noticeably, only the honeybees that can be employed in large scale onion seed production (Long and Morand 2011).

The number of seed stems produced per plant varies depending on lateral buds number on the hard plate exists on the bottom of the bulb. Plants grown during winter to the spring may produce bulbs and seed stems alongside. This is because one or more buds remain in vegetative status while other lateral buds are initiated to form a seed stem. Umbel terminates the stalk. The flower stalk of onion may be 1–2 m tall. The length of scape (flower stalk or seed stem) is under genetic control. The long is dominant over dwarf. Flowers open in a sequential order and it may last for over a month. The flower has usually six stamens, three carpels united with one pistil, and six perianth segments. The pistil comprises three locules each contains two ovules. Cross-pollination is dominating since the stigma becomes receptive late after shedding of pollen grains (delayed female maturity, protandry).

8.4.3 *Breeding F1 Hybrid Cultivars*

8.4.3.1 Male-Sterility and Its Role in Breeding F1 Hybrids

Flower of onion is hermaphrodite having both the staminate (pollen-producing) and carpellate (ovule-producing). Jones and Emsweller (1933) discovered genetic and cytoplasmic male sterility in a single-plant segregate of the onion cultivar Italian Red. Male-sterile plants produce normal flowers but the pollen does not develop into a viable stage. The inheritance was found to be controlled by a single recessive nuclear gene (ms/ms) and a cytoplasmic factor (Jones and Clarke 1943). Cytoplasm is considered either normal (N) or sterile (S). The genotype of the sterile plant should be Sms/ms. Onion is among the earliest crop plants in which heterosis was exploited. Availability of male sterility in onion is the core components for exploiting heterosis in this crop. Three lines are used to produce F1 onion variety. These are male-sterile line (S msms), inbred maintainer (N msms) and a pollinator male (N MsMs). The development and maintain S msms and N msms lines can be accomplished by backcrossing (S msms × N msms). N msms line is maintained by selfing. Pollinator line, i.e., (N MsMs) is development in parallel with male-sterile lines.

Selection from open pollinated population and promising crosses are utilized as base population for this purpose. Commonly, 3 generations of selfing with field selection are conducted. When male-sterile individuals are found within local adopted populations, proper pairs of male-sterile/maintainer lines can be established. If not, the breeder may make use of known male-sterile lines from other breeder's materials. These can be transformed into male-sterile lines by crossing and backcrossing with selected maintainer lines. Thus isogenic pairs of male-sterile and maintainer lines of agronomics value can be obtained. The lines S msms and N msms will be nearly genetically identical after 4–5 back crosses, except for male sterility in line S msms, while the other lines remain fertile.

8.4.3.2 Hybrid Cultivars Commercial Seed Production

The hybrid onion seed is produced in an open isolated field. The bulbs of S msms and N msms lines are grown in alternative rows; 4:1 respectively. (Swarup 1991).

The flowering of S msms and N msms lines must be synchronized. Three to four beehives/0.40 ha are sited. Rouging off-types and removing pollen shedding plants in line Smsms are operated via regularly field inspection. This is essentially to be conducted in early morning before dehiscence of anthers. Hybrid seeds are produced on the plants in the rows of line S msms but before harvesting these seed, seed produced in the pollinator rows (line N MsMs) should be excluded and the plants are destroyed. This is to retain purity of hybrid seeds through avoiding possible mechanical contamination.

Pollen collecting bees is less efficient as pollinator than strictly nectar collecting type of bee in seed production. The pollen collecting bees will not intensively visit male sterile lines (Lederhouse et al. 1972). Accordingly, the hybrid seed yield will be far less than the potential seed yield (Long and Morand 2011). This could be compensated using a heavier population of bees in the seed production field. In the lack of intense onions attractiveness, bees may neglect the crop. This, particularly occur when another highly attractive crop is in flower state. Usually, honeybees are effective pollinators for open pollinated onions as both pollen and nectar are available on all umbels.

8.4.4 Open-Pollinated (OP) Versus F_1 Hybrid Onion Varieties

The discovery of male-sterility in the onion variety Italian Red by Jones and Clarke (1943) has facilitated the commercial production of F1 hybrid varieties. Open-pollinated (OP) cvs dominated before this discovery. Hybrid varieties offer producing companies protection for breeder rights. These F1 cannot be maintained by the onion grower or sold by other seed companies as the identity of the inbred lines used to develop the hybrid is protected. F1 Hybrid varieties exhibit heterosis reflected in a hybrid vigor for horticultural traits comparing with their inbred lines used to develop them. According to Janik et al. (1999), crop uniformity is considered a beneficial character in modern agriculture because product uniformity permits efficient mechanical harvest. The uniformity of hybrids has two dimensions. The first is the genetic homogeneity as a result of the presence of identical genotypes. The second is the genetic stability leading to phenotypic uniformity (homeostasis) in different environments. Onion hybrid varieties, thus, offer benefits to both seed companies and growers. However, onion suffers severe inbreeding depression after few cycles of self-pollination. This depression is manifested as severe decreases in growth, bulb size, and seed production (Abdelmotaal et al. 2022). Due to the severe inbreeding depression, highly inbred lines and consequently the highly homogenous hybrids are hard to satisfactorily attain.

Hybrid varieties, on the other hands, have several drawbacks (Cramer 2001). These include more time and costs are needed to development the inbred parents used for hybrid production. Indeed, the high price of hybrid seeds has restricted their spread around the world. Additionally, onion F1 may not be beneficially superior as compared to OP varieties. In this context, Cramer (2001) compared open-pollinated vs. F1 hybrid cultivars using 23 open-pollinated and 18 hybrid varieties. He mentioned that hybrid varieties outperformed OP varieties for leaf number and plant height but not for percentage of seed talks, pink root incidence, bulb size, percentage marketable yield and marketable bulb yield. In his assessment, most OP cultivars performed as well or sometimes better than the majority of the studied hybrid cultivars. It is expected that new open pollinated varieties will continue to be required, especially, in those countries where the growers are unable to pay for high cost F1 hybrid seeds (Brewster 2008).

8.4.5 Interspecific Hybridization

Beneficial traits for improving bulb onion exist in Wild *Alliums* (Chuda and Adamus 2009; Romanov et al. 2021). These traits can be transferred through interspecifc hybridization from Alliums into bulb onion Different approaches to overcome species incompatibility including embryo culture in vitro, polyploidization, backcrossing and inbreeding of hybrid forms, cytological analysis and selection of recombinant forms to create initial forms of onion interspecific hybrids were developed (Romanov et al. 2021). Introgressions from *Allium roylei, Allium galanthum* and *Allium fstulosum* have been employed to improve onion (Scholten et al. 2007; Vu et al. 2012). Traits such as resistance to different diseases along with desirable production traits including early maturity and stronger root system can be found in these relative species. Further, *Allium cepa, A. fstulosum* and *A. roylei* species have been used to develop Alien Monosomic Addition lines (AMAL), which are useful for increasing genetic variability in cultivated Alliums and for comparative genetic studies (Vu et al. 2012). However, utilization of wild Alliums in breeding is a very enduring approach as shown, for example, in introgression of downy mildew resistance into bulb onion from *A. roylei* (Scholten et al. 2007). Also, biennial nature, high cross-pollination and sharp inbreeding depression in onion are still challenges for breeders with conventional approaches (Abdelmotaal et al. 2022). There may be an opportunity to use biotechnology mainly molecular tools and functional genomics to deal with onion breeding difficulties.

8.5 Molecular Breeding

8.5.1 Molecular Progress

Conventional breeding of onion is facing serious problems. This crop species has biennial life cycle, cross-pollination nature and great inhibition due to inbreeding depression. New technologies can offer genomic approaches to deal with onion breeding challenges. It could be possible to accelerate onion improvement by integrating genomic and molecular tools with conventional breeding methods. Molecular markers have presently been employed for germplasm analysis and mapping in onion. The large genomic size of onion, contrariwise, has hindered application of molecular tools.

8.5.1.1 Transcriptome and Genomic Sequencing as Genomic Resources

Applications of molecular techniques in onion breeding can be possibly valuable. The molecular methods can offer fast and cost-effective tools. Transcriptome and genome sequencing analyses can provide beneficial molecular-markers and exploring candidate genes (Khosa et al. 2016). Thus marker-assisted selection can be routinely employed (Kim et al. 2015) and with aid of these molecular tools, new onion cvs could then be advanced. RNAsequencing enables identifying genes and examining gene expression profile. Additionally, this is useful for variant detection and for identifying candidate genes. Also, probable markers of SSR and SNP to employ for marker-assisted onion breeding (Martin et al. 2013) are derived. The potential transcriptome sequencing has been recognized (Kuhl et al. 2004) as to generate expressed sequence tags (ESTs). By means of deep transcriptome sequencing, the available data sequencing has substantially been enlarged. Candidate sequences impacting a number of valuable traits can be brought (Kim et al. 2015). It has been anticipated that transcriptomic data of numerous different tissues can allow identifying stage-specific gene expression (Kamenetsky et al. 2015). In this context, DNA of short-and-long day onion lines was employed (Kim et al. 2015) in sequencing cDNA libraries and produced high-quality afresh assembled transcriptome.

Onion has relatively too large genome size comparatively (16 gigabase pairs of DNA) (McCallum 2007). It is about 15.9 Gb and that is 4.9x as much DNA as does a human genome (3.2 Gb). Mostly (75–80%), these genomes are repetitive sequences. Unique sequences can be detected only at specific genomic regions (Do et al. 2004). BAClibrary comprising 48 000 clones presenting 0.3x of onion genome has been already established (Suzuki et al. 2002). Advances are rapidly occurring in sequencing methodologies and bioinformatics.

The used method of genomic sequencing depends upon the size and complexity of the genome. Shotgun gDNA is the most used genomic sequencing comparing with BACs sequencing and reduced representation sequencing. Predictably, gene density of onion is very low (only one gene per 168 Kb) as by both BAC and whole genome shotgun sequencing. Large genome size coupled with a huge quantity of repetitive sequences makes simple shotgun DNA sequencing approaches are not appropriate (Jakse et al. 2003). Reduction noticeably was found for anonymous and transposon sequences using methyl filtration method while an increase of ~ 10-folds was recognized in sequences that encoding genes. Recently, TrueSeq genomic DNAlibraries were sequenced using Illumina HiSeq 2500 (Finkers et al. 2015). An improved sequence reads assembly would occur upon the current progress of long read sequencing technology including Oxford Nanopore or PacBio. Decreased representation genomic sequence should be monitored for onion using double haploids jointly with transcriptome sequencing (Khosa et al. 2016).

8.5.1.2 Genomic Database for Allium Species

Allium species genomic database has been established online and called 'Allium Map' https://allium.scinet.org.nz/. It involves genomic data and genetic maps (McCallum et al. 2012). Further, transcriptome database (http://onion.riceblast.snu.ac.kr/) was founded. It comprises transcriptome assembly and also gene remarks. Further, it gives basic bioinformatics tools. Indeed, web-based resources are really appreciated to store and regain existing data of genomic sequences.

8.5.2 Molecular Markers in Onion

Molecular markers in onion are remarkable tool for its breeding (McCallum 2007). Onion is biennial plant which is extremely affected by inbreeding depression after few selfing cycles. Application of molecular markers can compress breeding cycle and increase genetic gains. Isozyme markers were firstly employed but presently they were replaced by DNA-based markers such as, for instance, random amplified polymorphic DNA (RAPDs), restriction fragment length polymorphism (RFLP) and amplified fragment length polymorphism (AFLP). However, due to their minute polymorphism and low reproducibility plus being time-consuming, simple sequence repeat (SSR) markers have given preference. This is because they show hyper-variable and multi-allelic nature, high reproducibility, co-dominance inheritance and fitness for genotyping automation (Kalia et al. 2011).

A number of genomic SSR markers were established for onion via next-generation skim sequencing of a doubled-haploid population. As reported by Baldwin et al. (2012), a total of 921 SSR motifs were documented from 6.6 Mbp genomic sequences. EST-SSRs recognized from ESTs show higher transferability comparing with genomic SSRs (Khosa et al. 2014). Due to the double haploid onion lines availability, SNPs are readily identifiable (Finkers et al. 2015) using high-throughput technologies for genotyping. Occurrence of 1.7 SNPs/Kb has been gotten. This frequency seems similar to other plant species. The created onion SNPs have been certified through a wide range of germplasm (Duangjit et al. 2013) and recorded on present linkage maps.

8.5.2.1 Mapping

Various plant materials comprising F_2 segregates, near-isogenic lines (NILs) and diversity sets of recombinant inbred lines (RILs) have been used for mapping and trait analysis (Collard et al. 2005). Developing inbred populations is hard to achieve in onion due to occurrence of the severe inbreeding depression in this crop species. In onion, therefore, populations employed are frequently F2 families. It is suggested (Baldwin et al. 2014) that a beneficial method to get large trial stocks while avoiding deleterious inbreeding effects is by construction big multiple F_2 families by mass pollination of clonal hybrid bulbs from wide crosses to DH lines. Alternatively, a panel or core collections of diverse germplasm can be considered for association mapping. Multi-parent populations may be used for mapping of onion. For better understanding of polygenic traits, association mapping and genomic selection strategies should be employed.

A total of 116 RAPD and RFLP markers were utilized to construct the first genetic map for onion using BYG15-23 × AC43 F2 population. This map included 12 linkage groups at an average distance of nine cM. SSR and SNP markers have been sited on this linkage map giving ten linkage group at > LOD 4 (Duangjit et al. 2013. Genetic maps assist to identify marker–trait linkages and enabling marker-assisted selection. However Male-sterility, bulb color and flavor were located to different chromosomes. Linkage map was established also for F2 population (OH1 × 5225) and the used 597 SNPs determined ten linkage groups at LOD > 8. Their combined maps showed that markers remained similar except on chromosome two and six (Duangjit et al. 2013). It would be beneficial to expand maps with SNP or other sequence based markers towards developing a saturated interspecific linkage map.

8.5.2.2 Germplasm Identification, Characterization and Classification

Characterization of genetic variation within and among germplasm populations enabling their management and use in breeding programs (see also item 3 of the current text). Molecular marker have proven to be a potential tool in this regard. Much information is needed in onion within and among its different germplasm. RAPD, RFLP and SSR have been used in onion diversity analyses. The greatest understanding of onion diversity, however, was provided by SSR (Khosa et al. 2014). Several reports have revealed a genetic distinction of short- and long-day onion germplasm. These two onion groups are further subdivided based on geographical origins (Mallor et al. 2014). Indian short-day germplasm showed great diversity compared to other exotic germplasm. Therefore, the Indian short-day germplasm is considered a valued source of novel alleles for onion breeding.

8.5.3 *Further Significant Concrete Applications in Onion Molecular Breeding*

Chloroplast genome of N and S plants was sequenced and their polymorphisms were identified (Kohn et al. 2013). Further, research has been directed toward detection of markers linked with the fertility restoration locus (Ms). This is to assist finding new elite maintainer lines (N; msms). Marker-aided breeding in Allium was utilized also for breeding to downy mildew resistance. Marker-aided selection has speed up the development of resistant cultivars. This trait is regulated by two dominant genes and it was achieved by gene introgression from *A. roylei* into *A. cepa.* AFLP markers tightly linked with downy mildew resistance have been identified and they should be helpful for marker assisted selection (Scholten et al. 2007). SNPmarkers may be used for marker-assisted breeding toward developing onion cultivars resistant to thrips (Damon and Havey 2014). The quantity of wax affects the resistance that is influenced by two loci on chromosomes 2 and 5.

Several major QTLs control quality characteristics. Pungency and fructan content exhibit dominance and use marker-assisted selection for their breeding would be ideal (McCallum et al. 2007). Candidate genes for onion pungency have been mapped. A major locus on chromosome 8 (Frc) controls the variation in fructan contents. ACM033 and ACABE58 are two markers that have been detected to link with this locus. Functional markers have been established and may potentially be used in marker-assisted selection for bulb color (Duangjit et al. 2014).

8.6 Tissue Culture Applications

8.6.1 *Onion Gynogenic Haploids (Maternal Haploids) Challenges and Opportunities*

Protocols differ according to the objective of in vitro use. Somewhat details of in vitro goals and tissue culture methods available in onion have been shown under transformation subject elsewhere in this chapter. Herein, full focus has been given to the flower bud/ovary cultures to produce haploids and double haploids (Fig. 8.4).

Onion seriously suffers high inbreeding depression due to possessing deleterious recessive alleles. It can be selfed for only few (2–3) generations. The obtained inbreds are still highly heterozygous and lack homogeneity. This presents a challenge for a biennial crop such as onion. Obtaining highly homozygous inbred lines for F1 hybrid

Fig. 8.4 In vitro gynogenesis induction in onion. **a** An immature flower bud suitable for gynogenesis work. **b** An anthesis-stage flower in induction culture. **c** A flower bud with an enlarged ovary about 2 weeks after culture initiation. **d** A responsive flower bud. **e** A gynogenic onion plantlet. **f** A gynogenic onion in vitro plant. *Source* https://link.springer.com/protocol/10.1007/978-1-0716-1315-3_6/figures/1

production is difficult if not impossible at all using conventional breeding (Dhatt and Thakur, 2014; Khan et al. 2020). Gynogenesis has been developed as a potential tool to overcome this problem. Identify protocol and optimizing factors affecting gynogenic frequency in onion have been under study since 1980s (Khan et al. 2020). Doubled haploid (DH) production is a fast and efficient alternative to obtain onion inbred lines. Generally, in vitro gynogenesis induction has been reported in a wide range of crop species such as cucumber, squash, gerbera sunflower, wheat, barley, onion and sugar beet (Murovec and Bohanec 2012).

Commonly, obtaining haploid plants may be attained via androgenesis (from male gametes) and/or gynogenesis (female gametes). Response of either one is a species-dependent (Bohanec 2002). Androgenesis is considered unfeasible in onion (Keller and Korzum 1996). Onion response through maternal haploids (gynogenesis) at high incidence was detected (Bohanec et al. 1995), Usually, the key factors of haploid regeneration include genotype, physiological status of donor plants, developmental stage of the gametes (microspores, megaspore), pretreatment, composition of culture medium, and physical factors during tissue culture (Dhatt and Thakur 2014; Khar et al. 2019).

8.6.1.1 The Genotype of the Donor Plants

Gynogenesis is greatly influenced by both the genotype of donor onion plant and the surrounding growth conditions. In testing of onion diverse genetic material from different regions across the world, Geoffriau et al. (1997) reported that only two out of eighteen cvs showed gynogenesis. As noticed by Bohanec and Jakse (1999), accessions from diverse areas in the world expressed different gynogenesis response. Hybrids of crosses of responsive and non-responsive lines had lower gynogenic ability. In a different study, inbreeds regenerated significantly better than synthetics (Geoffriau et al. 1997) and regenerants from inbred lines were vigorous.

8.6.1.2 Explant Preparation

Onion has been shown to express recalcitrance reaction to regenerate haploids in anther culture (androgensis) (Khar et al. 2019; Keller and Korzum 1996). On the other hand, onion is responsive to gynogenic haploid induction thru in vitro culture of un-pollinated ovules, ovaries and whole flower buds. Ovule explants is laborious, in addition, to producing low number of haploid regenerants (Bohanec 2002). On contrary, flower bud culture is simpler and gave better response to gynogenic haploids induction (Keller, 1990). Ovary often shows response similar to flower-culture. Depending on the genotype, whole flower culture may be disadvantage. This is because of the possible somatic regeneration from basal callus. Ovary cultures (Bohanec 2002) is not so. Culture of whole onion flowers, without sub-culture, is considered the most efficient method for haploids induced from the female gametophyte (Bohanec et al. 2003; Ibrahim et al. 2016).

Use of flower bud explants that collected 3–5 days prior to anthesis are favorable as they are superiority over older and younger ones (Muren, 1989). In terms of size, it has been shown that flower buds 3.5–4.5 mm in length produced larger embryos than smaller ones (2.8–3 mm). Likewise, Alan et al. (2004) reported that unopened flower buds sized 2.25–4.5 mm gave the best results compared with smaller (< 2–2.5 mm) and larger (> 4.5 mm) ones. Small onion flowers comprising egaspore mother cells were less responsive than medium-sized ones with 2–4 nucleate embryo sacs (Musial et al. 2001). Similarly did the response of large onion flowers comprise mature embryo sac. Before culture, flowers are surface disinfestation usually with 96% ethanol for 1 min, followed by 5–10% sodium hypochlorite containing a few drops of Tween 20 for 15 min, and then rinsed thrice in sterilized water (Ponce et al. 2006).

8.6.1.3 Pre-treatment

It is hard to induce onion gametophytic alteration into sporophytic status without pretreatments (Chen et al. 2010). Pre-treating onion donor plants enhance the frequency of gynogenic embryogenesis from flower-bud cultures (Puddephat et al. 1999). The subjection of the explants to mild stress such as heat, cold and starvation is widely used pretreatments. As a general guideline, the efficiency of gynogenesis in onion improves when lowering the temperature of donor-plant during the final phase(s) of flower development. High light intensity may be beneficial in this context.

8.6.1.4 Medium Composition and Culture Incubation Conditions (Light and Temperature)

The three most-often used basal media have been reported as B5 (Gamborg et al. 1968), MS (Murashige and Skoog 1962) and BDS (Dunstan and Short 1977). Organic-nitrogen source and plant growth regulators (PGR) have been modified (Chen et al. 2010). Usually, different stages of gynogenesis development may require different supplies of nutrients (Jakse et al, 1996). Initial ovary explants is suggested to be cultured on medium with low concentrations of PGR. Then they are to be transferred to medium with higher PGR concentrations. However, Alan et al. (2004) compared single-step culture on BDS medium against two step culture (BDS/B1 medium) for induction of gynogenic. Single-step is suggested to be more suitable for large-scale haploids regeneration. Single-step culture is less time-consuming and more efficient than the two-step ovary culture (Bohanec and Jakse 1999). Cultures are usually maintained at 20–25 °C in Parafilm sealed petri dishes under 16-h light/8-h dark photoperiod and 80–100 $\mu mol\ m^{-2}\ s^{-1}$ light provided by fluorescent tubes (Bohanec and Jakse 1999).

8.6.1.5 Verification of Ploidy and Homozygosity Status and Chromosome Doubling

Ploidy status proof is essentially needed to discrete redundant heterozygous diploids which can be regenerated from the somatic tissue of flower-bud and ovary cultures, meanwhile to identify desirable regenerants (haploids and spontaneously doubled haploids). As a general rule, haploids are predominantly found in gynogenic regenerants. Ploidy status can be both directly and indirect accomplished. The regenerated plants are orphologically compared with the donor plants for traits as plant vigor and fertility and number of chloroplasts and their size in stomatal guard cells. This, however, is not precise as such traits are much affected by environmental conditions. Besides, haploid onion plants may be as vigor as donor plants. Ploidy determination can be assessed using more reliable methods comprising conventional cytological procedures (counting the chromosome number in cells from root tips) and amount of DNA content by flow cytometry. The flow cytometry is rapid and, consequently, it is suitable for large-scale ploidy determination. Flow cytometry further distinguishes mixoploid regenerants (cells with diverse ploidy) and their proportion.

Chromosome doubling is crucial for haploids, as they are missing a whole set of homologous chromosomes. Meiosis cell division is, therefore, interrupted and as a result, no seeds are produced. Spontaneously doubled plants will be preferred if the induced doubling is suspected to generate undesirable mutations. Frequently, anti-microtubule as colchicine is utilized which inhibits microtubule polymerization by binding to tubulin. Other treatments include pronamide, amiprophosmethyl (APM), oryzalin and trifluralin. All those are herbicides and are utilized at low concentrations. Chromosome duplication is applied after regeneration. Gynogenically derived embryos can be treated with colchine (Jakše et al. 2003) but later treatment is advantageous as being applied only to the already verified haploids.

8.6.2 Applications of Double Haploids in Plant Breeding

Double-haploid (DH) has been one of the most extensively and successfully utilized in vitro technology in advanced breeding programs in many crop species. Development of commercial vegetable crop varieties assisted by DH has been released in eggplant, melon, pepper and others. Exclusively, homozygous lines can be produced from heterozygous genotypes (comprising F1 or F2 of self-pollinated crops and populations of open-pollenated crops) in just one generation using DH technique. Differently, recurrent selfing can lead only to partial homozygous genotypes after several generations. DH may be directly used as new cultivar or parents to produce hybrid cv in self-pollinated crop species or in test-crosses of cross-pollinated species. DH can be considered as parents for crossing to produce the next cycle of certain recurrent selection cycle. It is expectable to get genetic advances due to the alternation of recombination and selection. Due to the lethal recessive alleles content, onion extremely suffers deadly selfing effects and shows serious inbreeding depression after only few generations; two on average. Thus, it is not feasible to produce completely homozygous lines by continuous selfing in onion. The lethal recessive alleles are not expressed in heterozygous genotypes. They are eliminated early thru DH application. DH, further, can aid to speed up the mutation breeding since it can fix the induced mutants directly following mutagenic treatment to the haploid cells in vitro. It may also speed up genetic transformation by fixing the induced haploid transformants directly after chromosome doubling treatment for the haploid cells in vitro.

8.7 Genetic Engineering

Classical breeding is indispensible method in spite of having several drawbacks. One of the drawbacks of conventional breeding is its slow process. Breeding of onions is slowdown or hampered due to its biennial growth habit besides its high heterozygosity as an open pollinated and severe inbreeding depression plus sterility or poor fertility of interspecific crosses. However, emerging tissue culture in vitro and molecular manipulations have provided potential tools that can enable faster and more precise development of improved onion cvs. Genetic transformation is one of such developing biotechnology techniques which have attracted the attention of the breeders. It is noticeable that developing in vitro protocols compatible with the genetic transformation system is a prerequisite for such application. In this context, in vitro methods have been developed in Allium species to accomplish basic micropropagation, in vitro plant regeneration and virus eradication. Besides, creating somaclonal variation for in vitro selection, germplasm preservation and secondary metabolic production have been investigated (Kahane et al. 1992). De novo meristems regeneration from totipotent cells may be directly from the explant tissues or thru callus phase. Regeneration is valued for creating somaclonal variation and conduct selection in vitro and for genetic transformation. Onion can be initiated in vitro using explants of almost all plant organ tissues with varying degrees of reaction. A comprehensive review on the in vitro regeneration of onion has been presented (Eady 1995; Marinangeli and Curvetto 2010).

8.7.1 Tissue Culture Methods and Its Applicability for Onion Transformation

Transgenic plants are obtained via integrating DNA into totipotent cell that can be regenerated to whole plant. The employed in vitro onion cultures media included MS (Murashige and Skoog 1962), B5 (Gamborg et al. 1968) and BDS (Dunstan and Short, 1977). Usually, auxins and cytokinins in different combinations are used. Various explant source and genotypes were assessed. Regeneration of onion plants from callus, however, has more specific necessities. Noticeable, onion as being highly heterozygous, variability within cultivar is habitually as large as those amongst cultivars or species. Darkness pretreatment of callus is favored for shoot regeneration when afterward transferred under light/dark conditions (Van Der Valk et al. 1992). Only limited assessments have been undertaken on Allium protoplast culture (Eady 1995). It may be possible to obtain embryonic suspension derived-protoplasts of onion cultivars. However, the efficacy and consistency of this system need substantial enhancement. Regeneration from protoplasts, on the other hand, is largely genotype-dependent.

To maximize the chance to attain transformants, there should be much quantity of callus cells that can both receive DNA and regenerate. Worth to mention that, field-grown onion bulbs are difficult to surface disinfestations and there is also doubt about the totipotency of their driver-callus. Generally, utmost callus has a low incidence of regeneration in onion (Eady et al. 2000). Many researchers (Zheng et al. 2001) used embryo and seedling-derived callus. It was shown that “high frequency” of regeneration can be attained from embryo-derived callus cultures. Such cultures would likely be a suitable tissue for onion transformation. However, it is a time consuming process to dissect embryos from onion seeds. Therefore, seedling derived “microbulbs” are used in transformation research. Microbulbs can be produced using seeds germinated on picloram supplemented-medium. The microbulbs can regenerate multiple-shoots from their actively dividing central dedifferentiated cells (Eady et al. 2000). Stable integration of plasmid DNA occurred following bombarded or Agrobacterium-infected fragmented bulb. Additionally, twin scale explants, split stems and shoot apex was studied for its possible beneficial use in transformation. In spite of the potential use of twin scale to regenerate adventitious shoots, delivery of the DNA into the totipotent cells may be difficult. This is due to the existence of competent cells at the base of the lower scale on the adaxial. Also, for incidence of regeneration, both scales and part of basal plate should exist. De novo plants via either organogenesis or embryogenesis can be regenerated from split stems derived from secondary shoots of twin scales (Dommisse 1993) but the frequency of regeneration was shown to be low. Embryogenic callus can be produced from onion shoot apex explants (Ramakrishnan et al. 2013). High frequency of somatic embryos forming plantlets successfully grew to maturity after hardening.

8.7.2 *Transformation System Choice and Factors Controlling It*

In onion, *Agrobacterium tumefaciens* and particle bombardment are dominantly used delivery vectors. Onion has a tremendously large genome and this is considered drawback regarding its genetic transformation (Arumuganathan and Earle 1991). This large genome may diminish the chance to attain correct expression (transcriptionally active region of the genome) for the introduced alien DNA. Host link of onion is restricted to few strains of *Agrobacterium*, as monocotyledonous plants, and merely in non-regenerative tissues (Eady 1995). Further, plant regeneration may be impaired while eliminating *Agrobacterium* after infection. As a result, relatively few attempts have been made to transform onion and it has been considered recalcitrant.

8.7.3 Optimization of Protocol for Transformation

Agrobacterium-mediated transformation of onion was reported employing two strains of *A. tumefaciens* (EHA105 and LBA4404). Immature embryos (Eady et al. 2000) and their derived callus (Zheng et al. 2001) were utilized. Toward optimizing protocol for transformation, expression of the *gus* (*Uid*A) reporter gene coding for *β*-glucuronidase is used as an indicator. The cultivars and species were among the most common influential factors. Using callus of mature embryos (Zheng et al. 2001) in onion transformation showed reduced callus size and enhanced the exposure to the selective agent (hygromycin). This reflected in elevated selection efficacy. For establishing stable transformation, callus induction media and co-cultivation duration were effective factors. Ineffective factors comprised callus age and *Agrobacterium* strains and sources and osmotic treatments during co-cultivation. Transformation methods using cell and protoplasts such as chemical and electroporation are not appropriate for onions currently. This is because reliable regeneration protocol for such material is not practicable yet. On the other hand, direct DNA delivery via a particle gun using intact embryogenic callus seems feasible. This system can be basically optimized (Eady et al. 1995) via transient assay. Puncture the cells with silicon fibers is another direct gene transfer (Kaeppler et al. 1992). The technique should provide comparable results to the biolistic particle gun.

8.7.4 Genetic Modified (GM) Traits/Cultivars

Crop plants that have been genetically transformed incorporated traits for superior resistance/tolerance to biotic and abiotic stresses and longer shelf-life. Moreover, useful products comprising biofuel or drugs have been offered by genetically modified organisms (GMO). Also, GMO capable to uptake toxins have been employed in pollution bioremediation. The benefits accessible by the genetically modified foods (GMF) (Shetty et al. 2018) includes: (1) improving crops production capacity in marginal soil, (2) rationalizing the utilization of nonrenewable resources, (3) rising crop use efficiency for water, nitrogen and other nutrients, (4) reducing applications of pesticides and herbicides, and (5) elevating crop nutritional products. The following are some possible potential risks (Sparrow 2010) associated with GM impact on human health, environment and biological diversity: (1) allergic reactions, (2) decrease biodiversity, (3) decreased antibiotic efficacy, (4) cross-pollination makes difficult to distinguish which crop fields are organic and which are not, (5) gene transfer where GMO altered genes may outflow to the wild, (6) GMF may be employed as a weapon, (7) the risk caused by widening gap of corporate sizes and consequently fewer competitors will control the prices, and food supply can be at risk as crop disaster can occur due to use of GMO seeds.

8.7.5 Cultivated Genetically Modified (GM) Crops

GM horticultural crops are principally far late compared to field crops. Noticeably, onion GM varieties have not been commercially released. The topmost ten GM crops are corn (field and sweet), alfalfa, apples, canola, cotton, papaya, soybeans, squash, potatoes, and sugar beet. A large scale commercial planting of GM crops was shown in 1996 in Argentina, Australia, Canada, China, Mexico, and US. Twenty six GM species have been accepted for commercial production by 2015. The majority of these species comprised genes for herbicide tolerance or resistance to insects. Other traits included virus resistance, delayed ripening, altered flower color and transformed composition. Twenty eight countries cultivated GM crops while 39 countries imported them in 2014 (https://en.wikipedia.org/wiki/List_of_genetically_modified_crops); accessed 14/05/2022; page was last edited on 14 May 2022, at 21:31 (UTC).

8.8 Mutations Breeding

8.8.1 The Main Role of Induced Mutations

Crop genetic improvement relies on availability of potential diversity. However, continuous depletion in genetic variation has eventually led to the interest in and the need of artificial mutation (Raina et al. 2016). Onion interspecific breeding is not that feasible (Chuda and Adamus 2009; Romanov et al. 2021) approach. Further, genetic variability for important traits such as disease resistance/tolerance is lacking. Thus induced mutation can provide source of genetic materials that may be difficult to get through cross breeding and other breeding approaches (Khan et al. 2009). Induced mutations have been, therefore, considered as a potential complementary approach for onion genetic improvement (Joshi et al. 2011; Singh et al. 2021). Mutation approach can enable generating improved crop varieties addressing the threats of climate changes and challenges of human food insecurity (Raina et al. 2016).

8.8.2 Mutagenic Agents and Advantages of Mutation Breeding

The concept of induced mutagenesis for crop improvement is dated back to the beginning of twentieth century (Kharkwal 2012) Mutation breeding is advantageous over other methodologies of breeding in some situations (Pathirana 2011; Raina et al. 2016). In cross breeding when well high adapted desirable cultivar is to be improved for one or few traits, a backcross strategy has to be employed to recover the elite genotype. However, mutation breeding has the advantage of improving a defect trait with no missing of the initial preexisting traits. Further, cross-breeding may be restricted by poor combining ability of the parents. Mutation breeding can assist in case when genes controlling unwanted trait are tightly linked to genes of desirable traits. In such a situation, mutation induction may lead to crossing-over occurrence or isolation of an independent mutant for the desired character. Mutation induction is the only straightforward alternative or the only feasible way for increasing genetic variability to develop new cultivars in crops that do not produce seeds (Raina et al. 2016). Mutation induction can be a valuable breeding strategy for cross-breeding of vegetative propagated perennial crop species.

Mutations induction has been studied using both physical and chemical mutagenic agents (Goyal and Khan 2010; Kirtane 2018). The chemical mutagens such as ethyl methane sulphonate (Kozgar et al. 2011), sodium azide (Mostafa 2011), diethyl sulphate (Aruna et al. 2010), N-nitroso methylurea (Alghamdi et al. 2010) have been widely used. More recently, plant mutations have been obtained using high-energy (220 MeV) ion beam implantations in Japan (Hase et al. 2010) and in China using low-energy (30 keV) ion beams (Feng et al. 2009; Raina et al. 2016). Molecular markers can be efficiently utilized to identify new variants/mutants (Mostafa and Alfrmawy 2011). Mutants can facilitate the isolation and cloning genes that utilized in designing crop plants improved for yield, stress tolerance, longer shelf-life and lessen input (Ahloowalia and Maluszynski 2001).

8.8.3 Mutagenesis Application in Onion

Onion (*Allium cepa* L.) is a biennial outcrossing and consequently highly heterozygous species. Therefore, improvement of this crop by conventional methods (hybridization, recombination and selection) is a time consuming process. In addition, onion suffers great inbreeding depression after only 2 cycles of selfing. It is hard to accomplish genetic linkage analysis (Cramer and Havey 1999) in onion due to the lack of highly inbred lines (Ahirwar and Verma 2016). To discover the chance of obtaining beneficial induced mutagenesis, researchers have tested the efficiency of various mutagens on onion (Asita and Matobole 2010; Joshi et al. 2011; Kirtane 2018). Most of the workers on mutagenesis of onion utilized radiations. As example for use of chemical mutagen, Joshi et al. (2011) studied sensitivity (suitable mutagen dose) in onion. The study was carried out using seeds of six variety treated with varied concentrations of sodium azide (SA, 0.1, 0.3, 0.5 and 0.7%) and ethyl methane sulphonate (EMS, 0.1, 0.15, 0.2 and 0.25%). Their results showed that low dose of both mutagens were suitable for the creation of variability in onion.

In general, the reported mutant cultivars and research on mutation induction in onion are very scanty (Kirtane 2014) relative to some other crop species. However,

Table 8.2 Number of officially released mutant cultivars with different types of mutagens

Mutagen used	Number of released mutant cultvars
Gama rays	910
X-rays	311
Gamma chronic	61
Fast neutrons	48
Thermal neutrons	22
Ethylemethane sulphonate	106
Sodium azide	11
N-ethyle-N-nitrosourea	57
N-Nitroso-N-methylurea	46

Source Raina et al. (2016)

onion has one of the most notable naturally occurred historic mutants; it is the "male-sterility" mutant found in onion variety Italian Red. The role of induced mutations is well recognized in improvement of onion. Four notable mutant varieties, namely, Compas, Brunette, KIK-11 and KIK-13 were released for the production in Netherlands and Russia (Maluszynski et al. 2000). Kirtane (2014) assessed the effect of sodium azide (0.1, 0.3, 0.5 and 0.7%) and gamma radiation (2, 4, 6, 8, 10 and 12 kR) alone and in combination for onion seeds. He used chlorophyll mutations to estimate the frequencies of induced mutations. Combination treatments were more effective while gamma radiation was more effective than the sodium azide to induce chlorophyll mutations. The utilization of mutations approach during the past five decades has played chief role in the developing onion cultivars worldwide. There has been an official release of 3222 plant mutant varieties. This was from 170 different plant species in more than 60 countries. (Feng et al. 2009). China has 810 mutants, Japan 481 and India 330. Based on the crop, the largest number of released mutant cultivars is in rice 815. The developed varieties increased the crop biodiversity and thus provided material for conventional plant breeding and directly donating to plant genetic resource. Number of officially released mutant cultivars with different types of mutagens is presented in Table 8.2.

8.8.4 Advances in Integrated Mutation Breeding

Advancements in mutation breeding techniques hold a promise (Raina et al. 2016) to enhance and facilitate the development of the crop varieties further. The advances such as in vitro mutagenesis provide convenient and more precise selection of new variants and their rapid multiplication. In vitro based methods have developed in the past two decades to help plant breeders (Oladosu et al. 2016). It is possible to work with large mutant populations in a limited space and the possibility of screening for some useful characters such as tolerance to salt toxic elements or fungal toxins in the laboratory environment. This allows the transfer of putative mutants only to the field for screening, and in the case of vegetative propagated crops, to remove chimaera through several repeated recultures.

Detection of mutants in haploids is faster where it can be detected in the first generation. Besides, it is easier and will be directly applicable in case if the breeding program is based on doubled haploids. It is possible for the breeders to employ molecular methodologies to efficiently select desired genotypes. Recently interest has renewed in mutation studies. Induced mutagenesis is of interest to plant molecular biology (Ahirwar and Verma 2016). Mutagenesis is utilized as tool to identify and isolate genes and assess their structure and function. Such these studies will certainly have a major influence on the crop species improvement programs. Mutation in association with the new biotechnologies of genetic engineering will be a routine tool of plant breeders in near future.

Most of the varieties released so far were obtained utilizing mutation and with the direct selection. Lately, in vitro culture and molecular tools have created new pattern in the utilization of mutation breeding. Mutation breeding has become increasingly widespread as an efficient tool for crop improvement (Basu et al. 2008). Small pieces of plant tissue can be exposed to mutagenesis in vitro (Xu et al. 2012). Callus or cell suspension may be used for in vitro selection as far as shoots can be regenerated from these cells/tissues. In vitro selection is potential method to breed for disease resistant. Further, irradiation of haploid cell cultures (anther and microspore or ovules) and regeneration of doubled haploid plants can be utilized. Induced mutation has been considered beneficial in the preparation of genetic maps. Consequently, it will facilitate molecular marker assisted-breeding. Use of mutation in developing molecular maps in structural and functional genomics could lead to rapid enhancement in plant yield and quality. The molecular techniques of DNA fingerprinting and molecular mappings such as RAPD (Random Amplified Polymorphic DNA,) AFLP (Amplified Fragment Length Polymorphisms) and STMS (Sequence-Tagged Microsatellite Sites) have contributed in the screening and analysis of mutants at the gene level. An example for employing molecular genotyping for identification of onion mutant lines is shown by Chakraborty et al. (2015). The authors used Simple Sequence Repeat (SSR) marker to characterize onion mutants of variety BARI Piaz-2 which possess resistance against purple leaf blotch (PLB) disease. All the four mutant lines showed band for the primer MatK-1RKIM-f/MatK-3FKIM-r indicating the existence of *PLB* gene for resistance against PLB. Band was also observed with marker rbcLa-F/rbcLa-R in all mutant lines except BP2-100/12 indicating absence of *PLB* gene in BP2 -100/12 which is susceptible to PLB of onion. An unknown allele was noticed and this may be either linked to *PLB* gene or a candidate gene triggers the *PLB* gene responsible for purple blotch of onion. Site directed insertion of transgenes based on chimeric RNA/DNA oligonucleotides and mutant tagging will be widely used in gene technology.

8.9 Conclusions and Prospects

The genus *Allium* is very large as it comprises about 750 species. Onion (*Allium cepa* L.) is preceded only by tomato (Solanum lycopersicum) on the list of valued cultivated vegetable species worldwide. This crop is valuable not only for its culinary uses but also as pharmaceutical/medicinal therapy. As far as onion production is concerned, its plants have relatively limited root system and slow growth especially during the early stages. Therefore, it cannot compete well with the weeds. Further, the plants are sensitive to salinity and drought and are severely attacked by many diseases and pests. Onion has a locality nature as it has to be adapted to certain level of day length with temperature interaction to set full dry bulbs. Introgression of genes of beneficial traits from onion relatives via traditional breeding has shown a very limited success while it is time consuming. Complete benefit of heterosis breeding (F1 hybrid cv) with high plant homogeneity has not been realized yet. The most current efforts are being concerned with population breeding following half-sib selection procedure. Onion is not attractive for breeders where they feel frustration due to its biennial life cycle, open pollination nature, severe depression after few number of selfing cycles and sterility of intraspecific crosses. New tools of in vitro tissue and cell cultures, molecular manipulations and genetic transformation have been available now and number of them can be employed alone or in combination with conventional methods to attain the breeding goals faster and more efficient. For instance, desired traits as resistance/tolerance to abiotic and biotic stresses may be obtained by in vitro mutation induction and selection. Further, molecular marker can aid selection of traits with low heritability in onion segregating populations. Also, ovary culture and regeneration of haploids and double haploids would help to fully attain heterosis and uniformity of F1 hybrid cultivars. In conjunction with molecular markers, most diverse haploids and double haploids can be selected to maximize F1 heterosis. Onion is of need of great attention to direct efforts and funds to exploit the potentials of emerging molecular and cell manipulations technologies toward improving this prominent crop.

Appendix 1: Research institutes

Institute name	Specialization research activities	Contact information and website
National Research Centre for Onion and Garlic	Onion and Garlic research	Rajgurunagar-410505, Pune, Maharastra, India
Department of Plant Breeding and Genetics	Onion Double Haploid and breeding	Cornell University, 303 Bradfield Hall, Ithaca, NY 14583 USA

(continued)

(continued)

Institute name	Specialization research activities	Contact information and website
National Horticultural Research and Development Foundation, Nashik-Aurangabad Road, Nashik, MS, India	Horticulture Res incl. Onion	Phone 01000298922 https://www.indiamart.com/nhrdf/aboutus.html
Hybrid Seed Company NZ Ltd Patumahoe, RD 3 Pukekohe, New Zealan	Hybrid Seed Company incl. onion	https://hybridseed.co.nz
Seed Company: CORA Seeds	Seed production Company incl. onion	+39 349 2477897 paolopagan@coraseeds.com www.coraseeds.com
Ethiopian Institute of Agricultural Research	Agricultural Research incl. onion	Wondo Genet *Agricultural Research Centre, Ethiopia* www.eiar.gov.et
Asian Vegetable Research and Development Center	Vegetable Research incl. onion	*P.O. Box 42 Shanhua, Tainan, Taiwan 741, Republic of China* AVRDC Website: http://www.avrdc.org.tw Institutional email: avrdcbox@netra.avrdc.org.tw
Department of Genetics, Plant Breeding and Seed Science	Plant Breeding Genetics incl. onion	University of Agriculture in Cracow, 29 Listopada 54, 31-425 Cracow, Poland a.chuda@ogr.ur.krakow.pl
New Zealand Institute for Plant and Food Research, Private Bag 4704, Lincoln 7608, New Zealand	Breeding Technologies and Plant Genomics	https://www.plantandfood.co.nz *Tel: +64 3 977 7340* *Fax: +64 3 325 2074*
Department of Vegetable Science	Vegetable research incl. onion	Punjab Agricultural University, Ludhiana, 141 004 Punjab, India
Plant Gene Resources of Canada, Centre 107 Science Place Saskatoon SK S7N 0X2	Gene Resources and Molecular breeding	Saskatoon Research Tel: (306) 385-9463 Fax: (306) 385-9465 Email: axel.diederichsen@canada.ca https://pgrc.agr.gc.ca
Kazusa DNA Research Institute, Kisarazu, Chiba, Japan	Marker and DNA sequencing	*2-6-7 Kazusa-kamatari, Kisarazu, Chiba 292-0818, JAPAN* qa-kdri@kazusa.or.jp www.kazusa.or.jp

(continued)

(continued)

Institute name	Specialization research activities	Contact information and website
Department of Plant Science, Plant Genomics and Breeding Institute, Vegetable Breeding Research Center, College of Agriculture and Life Sciences, Seoul National University, Seoul, South Korea	Vegetable Breeding Research Center; Vegetable Breeding Research Center Plant Genomics and Breeding	1 Gwanak-ro, Gwanak-gu, Seoul 08826, Republic of Korea Phone: +82-2-880-51141 1 Gwanak-ro, Gwanak-gu, Seoul 08826, Republic of Korea Phone: +82-2-880-51141 Email: calsweb@snu.ac.kr http://cals.snu.ac.kr/en/research/institute/institute?centidx=21
Nong Woo Bio Co. Ltd. researches and develops, 14-8 Central Town-ro Yeongtong-gu Suwon, 16506 Korea, Republic of (South)	Produce seeds (incl. onion) and agricultural products worldwide	www.nongwoobio.com Phone: 82-31–218-1116
Indian Institute of Horticultural Research, Bangalore, India	Molecular markers in the improvement of *Allium* crops	https://www.iihr.res.in/
IPGRI—the International Plant Genetic Resources Institute	Molecular crop breeding—markers for genebank management	www.bioversityinternational.org
Center of Molecular Biotechnology	Genomic in situ hybridization (GISH); *Allium cepa*; *A. roylei*; downy mildew resistance;	Russian State Agrarian University-Moscow Timiryazev Agricultural Academy, 49, Timiryazevskaya Str., 127550 Moscow, Russian Federation info@rgau-msha.ru khrustaleva@rgau-msha.ru
New Zealand Institute for Crop and Food Research Limited Private Bag 4704 Christchurch, New Zealand	HortResearch	http://www.plantandfood.co.nz
Instituto de Genética Ewald A. Favret CICVyA, INTA De los Reseros y Las Cabañas s/n Castelar, Argentina	Genetic transformation—onion	Tel/Fax: 54 11 4450-0805/1876 E-mail: rrios@cnia.inta.gov.ar http://www.criba.edu.ar/cerzos/
Departamento de Agronomía Universidad Nacional de Sur Bahía Blanca, Argentina	Genetic transformation—onion	Tel/Fax: 54 291 4862882

(continued)

(continued)

Institute name	Specialization research activities	Contact information and website
Vegetable Research Station, Department of Genetics and Plant Breeding, JAU Junagadh, Gujarat, Junagadh 362 001, Gujarat (INDIA)	Onion	+91-285-2672080-90 Telefax: +91-285-2672482 E-mail: dit@jau.in http://www.jau.in/index.php/contact-us
Nordic Genetic Resource Center, SE-230 53, Alnarp, Sweden	Genetic Resource	http://www.nordgen.org/ Phone +46 40-53 66 40
Department of Biotechnology and Bioinformatics, School of Life Sciences, Yogi Vemana University, Kadapa, Andhra Pradesh, India	Genetic diversity analysis—Onion (*Allium cepa* L.) germplasm	Phone: 08562-225429 Telefax: 08562-225419 Email: registraryvu@gmail.com www.yogivemanauniversity.ac.in
Faculty of Biotechnology and Biomolecular Science, Universiti Putra Malaysia (UPM), Serdang, 43400 Selangor, Malaysia	Plant mutagenesis in crop improvement	Ph: 0397691047 Fax: 0397697590 biotech_dean@upm.edu.my https://biotech.upm.edu.my/
The New Zealand Institute for Plant and Food Research Limited, Private Bag 11600, Palmerston North 4442, New Zealand	Plant mutation breeding in agriculture	https://www.plantandfood.co.nz *Tel: +64 6 953 7700 Fax: +64 6 351 7050 Email*: Ranjith.Pathirana@plantandfood.co.nz
Department of Vegetable Science, Punjab Agricultural University, Ludhiana -141 004, India	Production of doubled haploids in onion	E-mail: ajmerdhatt@gmail.co https://www.indcareer.com/punjab-agricultural-univer sity-ludhiana/Departments/department-vegetable-crops
Wageningen UR Plant Breeding, Wageningen University and Research Centre, PO Box 386, Wageningen 6700 AJ, The Netherlands	SNP-markers in Allium species to facilitate introgression breeding in	olga.scholten@wur.nl Droevendaalsesteeg 1 (Building no. 107) P.O. Box 16, 6700 AA Wageningen The Netherlands phone: +31 (0) 317 486 001 fax: +31 (0) 317
Agricultural Research Service, USDA, Department of Horticulture,	Genetic mapping—onion	University of Wisconsin, 1575 Linden Drive, Madison, WI 53706, USA
Laboratory of Plant Breeding, Department of Plant Sciences, Wageningen University and Research Centre, AJ, Wageningen, The Netherlands	Genetic mapping—onion	31 317 482836 secretariaat-plantbreeding.psg@wur.nl https://www.wur.nl/en/Res earch-Results/Research-Instit utes/plant-research/Plant-Bre eding/contact.htm

Appendix 2

Onion cultivars grown in top 10 countries.

Country	Cultivar
China	Fuzhou Red skin, Texas Early Grano, Red Skin, White Skin, Shanghai Red, Nanjing Yellow, Yangzhou Red, Yunnan Red, Kuming Purple, Minhou Purple
India	N-53, Nask Red, Pusa Red, Bellery Red, Punjab Selection, Pusa Ratner, Patna Red, Poona Red, White Patna, Red Globe, Large Red, Agrifound Light Red, Agrifound Dark Red, Agrifound White, Early Grano, Arka Niketane, Pusa Madhavi, Hisar-2, Arka Kalyan, Pusa White Round, Pusa Flat White, CO-3, CO-4, Agrifound Red, Punjab Red Round, Punjab Naroya, Arka bindu, Pusa White Flat, Pusa White Round
USA	Oie, Riverside, Ringmake, Big Mac, Valiant, Gringo, Daytona, Orogrande, Robin, Celebrity, Sweet Sandwich, Colorado No. 6, Sweet Spanish, Buffalo, Walla Walla, Southport White Globe, Red Wethersfield, Norstar 210, Rocket
Egypt	Giza 6, Beheri, Giza 20. Giza-6 Mohasan
Iran	Red Azarshahr, White Kashan, Texas Early Grano, Yellow Sweet Spanish, (Red) Tarom, Ray, Isfahan, (White) Qum
Turkey	Corum, Akgun, Kantartopu
Pakistan	Red Yunic, Texas Early Grano, Local White, Phulkara, Faisalabad Early, Desi Early Red, Swat, Chaltan, Sariab Red, Hazara
Brazil	Baia Performance Precoce, Grannex, Piroqra, Texas Grano, Pera Norte, Crioula, Texas Grano 502, Norte-14, Jubileu, IPA varieties, Granex 33, Granex 429, Alfa Tropical, Beta Cristal, Conquista, Roxa de Bareiro
Russia	Krasnodar, Ispanskij, Karatalskij, Chernomorskij, Andizhanj, Masallinskiij, Vertiuzhanskij, Dnestrovskij, Kaba, Skwirskij, Arzamasskij, Bessonovskij, Danilovskij, Kilichinskij, Mstersky, Rannij Zheltyi, Rostovskij repchatyi mestnyi, Strigunovskij, Pogrskij, Sibirskii, Skoropelyi, Tschernunchka
Republic of Korea	Cheongdangnang, Paechong-Joseng, Chenju-Whang, Kinkyu, Bonganghwang, Yongangwhang, Samda

Source Pareek et al. (2018)

Appendix 3: List of Allium Gene Banks in the World

Name of Institute	Web page	Number of accessions	Country
AVRDC: The world vegetable center	https://avodc.org	1100	Taiwan
ICAR-Directorate of onion and garlic research	http://www.doggr.res.in	2050	India

(continued)

(continued)

Name of Institute	Web page	Number of accessions	Country
European Cooperative Programme for Plant Genetic Resources	http://eurisco.ipk-gatersleben.de	14,400	Germany
N. I. Vavilov Research Institute	http://www.vir.nw.ru	1888	Russia
National Institute of Agrobiological Sciences	http://www.nias.affrc.go.jp	1352	Japan
Royal Botanical Gardens	http://kew.org	1100	UK
US Department of Agriculture	http://www.usda.gov	1304	USA
Warwick crop center	http://www.warwick.ac.uk	1755	UK

References

Abdelmotaal EG, Mohamed MF, Haridy AG, Abdalla RMA (2022) Variability, heritability, association and selection response to bulb weight and storability in Egyptian onion. Assiut J Agricult Sci 53(3):118–134. https://doi.org/10.21608/ajas.2022.134534.1136

Abdou R, Zakari M, Bakasso Y, Saaddou M, Bauoin J (2016) Analysis of the diversity of onion (*Allium cepa* L.) cultivars from Niger for their conservation and improvement. In: First young researchers overseas' day, Brussels, Belgium, 22 Mar 2016

Ahirwar R, Verma RC (2016) Ethyl methane sulphonate (EMS) induced translocation and inversion heterozygote in *Allium cepa* L. Cytologia 81(2):149–153

Ahloowalia BS, Maluszynski M (2001) Induced mutations—a new paradigm in plant breeding. Euphytica 118:167–173

Alan AR, Brants A, Cobb ED, Goldschmied PA, Mutschler MA, Earle ED (2004) Fecund gynogenic lines from onion (*Allium cepa* L.) breedin materials. Plant Sci 167:1055–1066

Alghamdi SS, Migdadi HM, Al-fifi SA, Ammar MH (2010) Evaluation of critical dose for mutagenic treatments of barley varieties with N-nitroso-N-methyl urea (NMU). Environ We Int J Sci Tech 5:13–25

Aoba T (1967) Effects of different temperatures on seed germination of garden ornamentals in *Allium*. J Jpn Soc Hort Sci 36:333–338

Arumuganathan K, Earle ED (1991) Nuclear DNA content of some important plant species. Plant Mol Biol Report 9(3):208–218

Aruna J, Prakash M, Kumar BS (2010) Studies on effect of physical and chemical mutagens on seedling characters in Brinjal (*Solanum melongena* L.). Int J Curr Res 3:38–41

Asita AO, Matobole RM (2010) Comparative study of the sensitivities of onion and broad bean root tip meristematic cells to genotoxins. Afr J Biotech 9:4465–4470

Astley D, Innes NL Van der Meer QP (1982) Genetic resources of *Allium* species. In: IBPGR secretariat, Rome, Italy, p 38

Austin RB (1972) Bulb formation in onions as affected by photoperiod and spectral quality of light. J Hort Sci 47:492–504

Baldwin SR, Revanna M Pither-Joyce M, Shaw K, Wright S, Thomson L, Moya R, Lee R, Macknight M, Callum J (2014) Genetic analyses of bolting in bulb onion (*Allium cepa* L.). Theor Appl Genet 127:535–547

Baldwin S, Revanna R, Thomson S, Pither-Joyce M, Wright K, Crowhurst R, Fiers M, Chen L, Macknight R, McCallum JA (2012) A toolkit for bulk PCR-based marker design from next-generation sequence data: application for development of a framework linkage map in bulb onion (*Allium cepa* L.). BMC Genom 13:637

Basu SK, Acharya SN, Thomas JE (2008) Genetic improvement of fenugreek (*Trigonella foenum-graecum* L.) through EMS induced mutation breeding for higher seed yield under western Canada prairie conditions. Euphytica 160(2):249–258

Bohanec B (2002) Doubled haploid onions. In: Rabinowich H, Currah L (eds) *Allium* crop science: recent advances. CABI Publishing House, Wallingford, UK, pp 145–148

Bohanec B, Jakse M (1999) Variations in gynogenic response among long-day onion (*Allium cepa* L.) accessions. Plant Cell Rep 18:737–742

Bohanec B, Jakse M, Ihan A, Javornik B (1995) Studies of gynogenesis in onion (*Allium cepa* L.): induction procedures and genetic analysis of regenerants. Plant Sci 104:215–224

Bohanec B, Jakse M, Havey MJ (2003) Genetic analyses of gynogenetic haploid production in onion. J Amer Soc Hortl Sci 128:571–574

Brahim A (2020) Exploring the genetic and phenotypic diversity within and between onion (*Allium cepa* L.) ecotypes in Morocco. In: Genetics and molecular biology, Webinar 2020, Paris, France, 24 Sept 2020

Brewster JL (2008) Crop production science in horticulture 15. In: Onion and other vegetable alliums, 2nd edn. CAB International

Chakraborty M, Quddus T, Rahmanl S, Azad MAK et al (2015) Molecular characterization of selected mutant lines of onion (*Allium cepa* L.) against purple leaf blotch disease using SSR markers. Am J Exp Agric 8(4):261–267

Chen JF, Cui L, Malik AA, Mbira KG (2010) In vitro haploid and dihaploid production via unfertilized ovule culture. Plant Cell Tiss Org Cult 68:51–59

Chuda A, Adamus A (2009) Aspects of interspecifc hybridization within dibble Alliaceae. Acta Physiol Plant 31:223–227

Collard BCY, Jahufer MZZ, Brouwer JB, Pang ECK (2005) An introduction to markers, quantitative trait loci (QTL) mapping and marker-assisted selection for crop improvement: the basic concepts. Euphytica 142:169–196

Cramer CS (2001) Comparison of open-pollinated and hybrid onion varieties for New Mexico. HortTechnology 11(1):119–123. https://doi.org/10.21273/HORTTECH.11.1.119

Cramer CS, Havey MJ (1999) Morphologial, biochemical and molecular markers in onion. Hort Sci 34:589–593

Damon S, Havey MJ (2014) Quantitative trait loci controlling amounts and types of epicuticular waxes in onion. J Am Soc Hortic Sci 139:597–602

Dhatt AS, Thakur P (2014) Production of doubled haploids in onion: a review. J Hort Sci 9(2):107–112

Do GS, Suzuki G, Mukai Y (2004) Genomic organization of a novel root alliinase gene, ALL1, in onion. Gene 325:17–24

Dommisse E (1993) Development of an *Agrobacterium* transformation system for onion (*Allium cepa* L.). Unpublished Ph.D. thesis. Lincoln University, New Zealand

Duangjit JB, Bohanec AP, Chan CD, Town MJ, Havey (2013) Transcriptome sequencing to produce SNP-based genetic maps of onion. Theor Appl Genet 126:2093–2101

Duangjit J, Welsh K, Wise ML, Bohanec B, Havey MJ (2014) Genetic analysis of anthocyanin concentrations and intensity of red bulb colour among segregating haploid progenies of onion. Mol Breed 34:75–85

Dunstan DI, Short KC (1977) Improved growth of tissue cultures of onion (*Allium cepa*). Physiol Plant 41:70–72

Eady CC (1995) Towards the transformation of onions (*Allium cepa*). N Z J Crop Hort Sci 23(3):239–250. https://doi.org/10.1080/01140671.195.9513895

Eady C, Lister CE, Schafer DS (1995) An onion (*Allium cepa* L.) in vitro culture system developed for particle-gun mediated transformation. In: 11th biennial conference of the New Zealand branch of the international association for plant tissue culture (Abstract)
Eady CC, Weld RJ, Lister CE (2000) *Agrobacterium tumefaciens*-mediated transformation and transgenic-plant regeneration of onion (*Allium cepa* L.). Plant Cell Rep 19:376–381
Fang H, Liu H, Ma R, Liu YL, Yu X et al (2020) Genome-wide assessment of population structure and genetic diversity of Chinese Lou onion using specific length amplified fragment (SLAF) sequencing. PLoS ONE 15(5):e0231753. https://doi.org/10.1371/journal.pone.0231753
FAOSTAT (2018) Onion (dried) production in 2017: crops/world regions/production quantity from pick lists. http://www.fao.org/faostat/en/#data/QC. Retrieved 18 June 2019
Feng HY, Yang G, Yu ZL (2009) Mutagenic mechanisms of ion implantation in plants. In: Induced plant mutations in the genomics era. In: Proceedings of an international joint FAO/IAEA symposium. International Atomic Energy Agency, Vienna, Austria, pp 220–222
Finkers R, van Workum W, van Kaauwen MPW, Huits H, Jungerius A, Vosman BJ, Scholten OE (2015) SEQUON—sequencing the onion genome source. In: Proceedings of the plant & animal genome XXIII—PAG, 2015 Event Plant & Animal Genome XXIII, San Diego, CA, USA
Fischer D, Bachmann K (2000) Onion microsatellites for germplasm analysis and their use in assessing intra-and interspecifc relatedness within the subgenus *Rhizirideum*. Theor Appl Genet 101:153–164
Fritsch RM, Friesen N (2002) Evolution, domestication and taxonomy. In: Rabinowitch HD, Currah L (ed) Allium crop science: recent advances. CABI Publishing, UK, pp 5–30
Galluzzi G, Eyzaguirre P, Negri V (2010) Home gardens: neglected hotspots of agro-biodiversity and cultural diversity. Biodivers Conserv 19(13):3635–3654
Gamborg OC, Miller RA Ojima K (1968) Nutrient requirements of suspension cultures of soybean root cells. Exp Cell Res 50:151–158
Geoffriau E, Kahane R Rancillac M (1997) Variation of gynogenesis ability in onion (*Allium cepa* L.). Euphytica 94:37–44
Ghodke, PH. Gijare UM, Andhale PS, Salunkhe VN, Mahajan V, Gupta AJ, Singh M (2018) Genetic diversity in onion by multivariate analysis under short day condition. Int J Curr Microbiol Appl Sci 7(1):2184–2194. https://doi.org/10.20546/ijcmas.2018.701.263
Glaszmann JC, Kilian B, Upadhyaya HD, Varshney RK (2010) Accessing genetic diversity for crop improvement. Curr Opin Plant Biol 13:1–7
Gopal J (2015) Onion research in India: status and challenges. Prog Hort 47(1):1–15
Goyal S, Khan S (2010) Induced mutagenesis in urdbean (*Vigna mungo* L. Hepper): a review. Int J Bot 6:194–206
Griffiths G, Trueman L, Crowther T, Thomas B, Smith B (2002) Onions: a global benefit to health. Phytother Res 16(7):603–615
Hai HD (2015) *Allium cepa* (bulb onion). The Worldwide Vegetables. http://theworldwidevegetables.weebly.com/allium-cepa-bulb-onion.html
Haile B, Babege T, Hailu A (2016) Constraints in production of onion (*Allium cepa* L.) in Masha District, Southwest Ethiopia. Gloal Sci Res J 4(2):314–321
Hamrick JL, Godt MJW (1996) Effects of life history traits on genetic diversity in plant species. Philos Trans R Soc Lond B Biol Sci 351:1291–1298
Hanci F, Gokce AF (2016) Genetic diversity evaluations in Turkish onion (*Allium cepa* L.) genotypes: principal component analyses (PCA) for breeding strategies. Acta Hortic 1143:227–234 https://doi.org/10.17660/ActaHortic.2016.1143.33
Hanelt P (1990) Taxonomy, evolution, and history. In: Rabinowitch HD, Brewster JL (eds) Onions and allied crops. Botany, physiology, and genetics, vol I. CRC Press, Boca Raton, pp 1–26
Hase Y, Okamura M, Takeshita D, Narumi I, Tanaka A (2010) Efficient induction of flower-color mutants by ion beam irradiation in petunia seedlings treated with high sucrose concentration. Plant Biotechnol 27(1):99–103
Heath OVS, Hollies MA (1965) Studies in the physiology of the onion plant. J Expt Bot 16:128–144

Huang X, Han B (2014) Natural variations and genome-wide association studies in crop plants. Annu Rev Plant Biol 65:531–551

Ibrahim AM, Kayat, FB, Susanto, Kashiani D, Arifullah MD (2016) Haploid induction in spring onion *Allium fistulosum* L. via gynogenesis. Biotechnology 15(1–2):10–16. https://doi.org/10.3923/biotech.2016.10.16

Jakše M, Bohanec B, Havey MJ (2003) Chromosome doubling procedures of onion (*Allium cepa* L.) gynogenic embryos. Plant Cell Rep 21:905–910

Jakse M, Bohanec B, Ihan A (1996) Effect of media components on the gynogenic regeneration of onion (*Allium cepa* L.) cultivars and analysis of regenerants. Plant Cell Rep 15:934–938

Janik J, Coors JG, Pandey S (1999) Exploitation of heterosis: uniformity and stability. The genetics and exploitation of heterosis in crops. In: Proceedings of an international symposium. CIMMYT, Mexico, 17–22 Aug 1999, 319–333

Jones HA Emsweller SL (1933) Methods of breeding onions. Hilgaria 7:625–642

Jones DF, Clarke AI (1943) Inheritance of male sterility in onion and the production of hybrid seed. Amer Soc Hort Sci Proc 43:189–193

Joshi N, Ravindran A, Mahajan V (2011) Investigations on chemical mutagen sensitivity in onion (*Allium cepa* L.). Int J Bot 7:243–248

Kaeppler HF, Somers DA, Rines HW et al (1992) Silicon carbide fiber-mediated stable transformation of plant cells. Theoret Appl Genet 84:560–566. https://doi.org/10.1007/BF00224152

Kahane R, De La Serve BT, Rancillac M (1992). Bulbing in long-day onion (*Allium cepa* L.) cultured in vitro: comparison between sugar feeding and light induction. Ann Bot 69(6):551–555. http://www.jstor.org/stable/42764440

Kale PN, Warade SD, Midmore DJ (1994) International symposium on alliums in the tropics, Bangkok, Thailand. 15-1993. Acta Hort 358:153–156

Kalia RK, Rai MK, Kalia S, Singh R, Dhawan AK (2011) Microsatellite markers: an overview of the recent progress in plants. Euphytica 177:309–334

Kamenetsky R, Faigenboim A, Mayer ES, Michael TBC et al (2015) Integrated transcriptome catalogue and organ-specific profiling of gene expression in fertile garlic (*Allium sativum* L.). BMC Genom 16(1):12. https://doi.org/10.1186/s12864-015-1212-2.16

Kato T (1964) Physiological studies on bulbing and dormancy of onion plant. III. Effects of external factors on the bulb formation and development. J Jpn Soc Hort Sci 33:53–61

Kavitha SJ, Rami Reddy PV (2018) Floral biology and pollination ecology of onion (*Allium cepa* L.). J Pharmacogn Phytochem 7(6):2081–2084

Keller J (1990) Culture of unpollinated ovules, ovaries and flower buds in some species of the genus *Allium* and haploid induction via gynogenesis in onion (*Allium cepa* L.). Euphytica 47:241–247

Keller ERJ, Korzum L (1996) Haploidy in onion (*Allium cepa* L.) and other *Allium* species. In: Jain SM, Sopory SK, Veilleux RE (eds) In vitro haploid production in higher plants, vol 3.. Springer Science and Business Media, B.V. Netherlands, pp 51–75

Khan S, Goyal S (2009) Improvement of mungbean variety through induced mutations. Afri J Plant Sci 3:174–180

Khan PSSV, Vijayalakshmi G, Raja MM, Naik ML, ermanà MA, Terry RG (2020) Haploid production in onion (*Allium cepa* L.) from gynogenesis to chromosome doubling. Plant Cell Tissue Organ Cult 142:1–22. https://doi.org/10.1007/s11240-020-01831-4

Khar K, Isiam S, KaliA P, Bhatia R, Kumar A (2019) Present status of haploidy research in onion (*Allium cepa*)—a review. Indian J Agricult Sci 89(3):396–405

Kharkwal MC (2012) A brief history of plant mutagenesis. In: Shu QY, Forster BP, Nakagawa H (eds) Plant mutation breeding and biotechnology. CABI, Wallingford, pp 21–30

Khosa JS, Negi KS, Dhatt AS, Khar A (2014) Utility of simple sequence repeat (SSR) markers to realize worth of germplasm in genus *Allium*. Indian J Plant Genet Resour 27:238–245

Khosa JS, McCallum J, Dhatt AS, Macknight TC (2016) Enhancing onion breeding using molecular tools. Plant Breed 135:9–20

Kim S, Kim MS, Kim YM, Yeom SI, Cheong K, Kim KT, Jeon J, Kim S, Kim DS, Sohn SH, Lee YH, Choi D (2015) Integrative structural annotation of de novo RNA-seq provides an accurate reference gene set of the enormous genome of the onion (*Allium cepa* L.). DNA Res 22:19–27
Kirtane SA (2014) Studies on induced mutations in onion: frequency and spectrum of chlorophyll mutations. Biol For Int J 6(2):141–144
Kirtane SA (2018) Comparative mutagenic effectiveness and efficiency of sodium azide and gamma radiation in onion (*Allium cepa* L.). Int J Theor Appl Sci 10(1):169–173
Kohn C, Kiełkowska A, Havey MJ (2013) Sequencing and annotation of the chloroplast DNAs and identification of polymorphisms distinguishing normal male-fertile and male-sterile cytoplasms of onion. Genome 56:737–742
Kononkov PF, Ustimenko GW, Perez AP (1969) Beitr trop subtrop landw. Trop Vet Med 183–192
Kozgar MI, Goyal S, Khan S (2011) EMS induced mutational variability in *Vigna radiata* and *Vigna mungo*. Res J Bot 6:31–37
Kuhl J, Cheung F, Yuan Q, Martin W, Zewdie Y, McCallum J, Catanach A, Rutherford P, Sink KC, Jenderek M, Prince JP, Town CD, Havey MJ (2004) A unique set of 11,008 expressed sequence tags (EST) reveals expressed sequence and genomic differences between monocot order asparagales and poales. Plant Cell 16:114–125
Lederhouse RC, Caron DM, Morse RA (1972) Onion pollination in New York. N Y Food Life Sci 1: 8–9
Long RF, Morand L (2011) Low hybrid onion seed yields relate to honey bee visits and insecticide use. UC Agricult Nat Resour Calif Agric 65(3). https://escholarship.org/uc/item/27x3c05d
Mallor C, Arnedo-Andres MS, Garces-Claver A (2014) Assessing the genetic diversity of Spanish *Allium cepa* landraces for onion breeding using microsatellite markers. Sci Hortic 170:24–31
Maluszynski M, Nichterlein K, Van-Zanten L, Ahloowalia BS (2000) Mutation breeding review. No. 12(dec). FAO/IAEA, Vienna, Austria
Marinangeli P, Curvetto N (2010) Improvement of in vitro plant regeneration for genetic transformation of Argentinean onion varieties. Amer J Plant Sci Biotechnol 3(special issue 1):31–37
Martin LB, Fei Z, Giovannoni JJ Rose, JK (2013) Catalyzing plant science research with RNA-seq. Front Plant Sci 4:66
McCallum J (2007) Onion. In: Kole C (ed) Genome mapping and molecular breeding in plants, vol 5, pp 331–342
McCallum JM, Pither-Joyce M, Shaw F, Kenel S, Davis R, Butler J, Scheffer J, Jakse M, Havey MJ (2007) Genetic mapping of sulphur assimilation genes reveals a QTL for onion bulb pungency. Theor Appl Genet 114:815–822
McCallum J, Baldwin S Shigyo M, Deng YS, Heusden V, Pither-Joyce M, Kenel F (2012) Allium map—a comparative genomics resource for cultivated Allium vegetables. BMC Genome 13:168
McCollum GD (1976) Onion and allies *Allium* (*Liliaceae*). In: Simmonds NW (eds) Evolution of crop plants. Longman, London, pp 186–190
Mostafa GG (2011) Effect of sodium azide on the growth and variability induction in *Helianthus annuus* L. Int J Plant Breed Genet 5:76–85
Mostafa GG, Alfrmawy AM (2011) RAPD analysis for detection and fingerprinting of sunflower mutants induced by sodium azide. Int J Plant Breed Genet 5:277–285
Mousa MAA, Mohamed MF (2009) Enhanced yield and quality of onion (*Allium cepa* L. cv Giza-6 produced using organic fertilization. Assiut Univ Bull Environ Res 12(1):9–17
Murashige T, Skoog F (1962) A revised medium for rapid growth and bioassays with tobacco tissue cultures. Physiol Plant 15:473–497
Muren RC (1989) Haploid plant induction from unpollinated ovaries in onion. Hort Sci 24:833–834
Murovec J, Bohanec B (2012) Haploids and doubled haploids in plant breeding. In: Abdurakhmonov I (ed) Plant breeding. In Tech, Europe, pp 87–106. Available from: https://www.intechopen.com/books/plant-breeding/haploids-and-doubled-haploids-in-plant-breeding
Musial K, Bohanec B, Przywara L (2001) Embryological study on gynogenesis in onion (*Allium cepa* L.). Sex Plant Repr 13:335–341

Oladosu Y, Mohd YR, Norhani A, Ghazali H, Asfaliza R, Harun AR, Gous M, Magaji U (2016) Principle and application of plant mutagenesis in crop improvement: a review. Biotechnol Biotechnol Equip 30(1):1–16. https://doi.org/10.1080/13102818.2015.108733

Pareek S, Sagar NA, Sharma S, Kumar V (2018) Onion (*Allium cepa* L) In: Yahia EM (ed) Fruit and vegetable phytochemicals: chemistry and human health, vol II, 2nd edn. Wiley & Sons Ltd.

Pathirana R (2011) Plant mutation breeding in agriculture. CAB Rev Perspect Agric Vet Sci Nutr Nat Resour 032. http://www.cabi.org/cabreviews

Peterson PM, Annable CR, Rieseberg LH (1908) Systematic relationships and nomenclatural changes in the 'Allium douglasii' complex. Syst Bot 13:207–214

Ponce M, Martinez L, Galmarini C (2006) Influence of CCC, putrescine and gellan gum concentration on gynogenic embryo induction in *Allium cepa*. Biol Plant 50:425–428

Puddephat IJ, Robinson HT, Smith BM, Lynn J (1999) Influence of stock plant pre-treatment on gynogenic embryo induction from flower buds of onion. Plant Cell Tiss Org Cult 57:145–148

Raina A, Laskar RA, Khursheed S, Ruhul R, Tantray YR, Parveen K, Khan S (2016) Role of mutation breeding in crop improvement past, present and future. Asian Res J Agric 2(2):1–13. Article no. ARJA.29334

Raines S, Henson C, Havey MJ (2009) Genetic analyses of soluble carbohydrate concentration in onion bulbs. J Am Soc Hortic Sci 134:618–623

Ramakrishnan M, Ceasar SA, Duraipandiyan V et al (2013) Efficacious somatic embryogenesis and fertile plant recovery from shoot apex explants of onion (*Allium cepa.* L.). Vitro Cell Dev Biol Plant 49(3):285–293

Rashid MH, Islam MAK, Mian MAK, Hossain T, Kabir ME (2013) Multivariate analysis in onion (*Allium cepa* L.). Bangladesh J Agric Res 37:573–582

Rivera A, Mallor C, Garcés-Claver A, García-Ulloa A, Pomar F, Silvar C (2016) Assessing the genetic diversity in onion (*Alliumcepa* L.) landraces from northwest Spain and comparison with the European variability. N Z J Crop Hortic Sci 44(2):103–120. https://doi.org/10.1080/01140671.2016.1150308

Romanov VS, Soldatenko AV, Pivovarov VF, Timin1 NI, Romanova OV (2021) IOP Conf Ser Earth Environ Sci 650:012064. https://doi.org/10.1088/1755-1315/650/1/012064

Scholten OE, Heusden AWV, Khrustaleva LI, Burger-Meijer K, Mank RA, Antonise RGC, Harrewijn JL, Van Haecke W, Oost EH, Peters RJ, Kik C (2007) The long and winding road leading to the successful introgression of downy mildew resistance into onion. Euphytica 156:345–353

Shetty MJ, Chandan K, Krishna HC et al (2018) Genetically modified crops: an overview. J Pharmacogn Phytochem 7(1):2405–2410

Sidhu AS, Bal SS, Rani M (2005) Current trends in onion breeding. J New Seeds 6(2):223–245

Singh BP, Rana RS (1994) Collection and conservation of *Allium* genetic resources: an Indian perspective. Acta Hort 358:181–190

Singh H, Khar A, Verma P (2021) Induced mutagenesis for genetic improvement of Allium genetic resources: a comprehensive review Genet Resour Crop Evol 68:2669–2690. https://doi.org/10.1007/s10722-021-01210-8

Sparrow PAC (2010) GM risk assessment. Mol Biotechnol 44(3):267–275

Sudha GS, Ramesh P, Sekhar AC, Krishna TS, Bramhachari PV Riazunnisa K (2019) Genetic diversity analysis of selected Onion (*Allium cepa* L.) germplasm using specifc RAPD and ISSR polymorphism markers. Biocatal Agric Biotechnol 17:110–118

Suzuki G, Do GS, Mukai Y (2002) Efficient storage and screening system for onion BAC clones. Breed Sci 52:157–159

Swarup V (1991) Breeding procedures for cross pollinated vegetable crops. ICAR, New Delhi, India, p 118

Takhtajan A (1997) Diversity and classification of flowering plants. Columbia University Press, New York, p 643

Valenzuela H, Shimabuku R, Kratky B, Hamasaki R (1999) Onions in Hawaii and around the world, In: Hamasaki R, Valenzuela H, Shimabuku R (eds) P1-8. Bulb onion production in Hawaii. College of Tropical Agriculture and Human Resources, University of Hawaii at Manoa

Van Der Valk P, Scholten O, Verstappen F, Jansen R, Dons J (1992) High frequency somatic embryogenesis and plant regeneration from zygotic embryo-derived callus cultures of three *Allium* species. Plant Cell Tissue Organ Cult 30:181–192

Vu HQY, Yoshimatsu LI, Khrustaleva N, Yamauchi SM (2012) Alien genes introgression and development of alien monosomic addition lines from a threatened species, *Allium roylei* Stearn, to *Allium cepa* L. Theor Appl Genet 124:1241–1257

Xu L, Najeeb U, Naeem MS, Wan GL, Jin ZL, Khan F, Zhou WJ (2012) In vitro mutagenesis and genetic improvement. In: Technological innovations in major world oil crops, vol 2. Springer, New York, pp 151–173

Zheng S-J, Khrustaleva L, Henken B, Sofiari E et al (2001) *Agrobacterium tumefaciens*-mediated transformation of *Allium cepa* L.: the production of transgenic onions and shallots. Mol Breed 7:101–115

Chapter 9
Overview on Oregano Breeding Strategies as Prince of Medicinal Plants

Gholam-Reza Sharifi-Sirchi

Abstract Oregano plant, a perennial herb from lamiaceae family is a high conception medicinal plant with valuable primary and secondary metabolites. Three constituents, carvacrol, thymol and rosmarinic acid, are the most important oregano bioactive compounds which makes its taste, spicy flavor and aroma. It is originated from the Irano-Turanian habitats, the Mediterranean countries and temperate western regions. Oregano plants conception is highly increasing in production of the food, drugs, perfume, and washing liquid. The medicinal effects of oregano plants constituents in a wide range of diseases, including; antioxidant, antitumor, antipathogens, and antiviral activity, as remedy against respiratory disorders, upset stomach, jaundice, childhood malnutrition, and rheumatoid arthritis, and its usage in the modern food industry, lead to the excessive harvest and conception of this valuable plant in its habitats which increased the oregano plants genetic erosion risk. The main goal of this paper is to make familiar the oregano farmer and breeders with the oregano agronomy and breeding characteristics. The conventional breeding methods accompany with the modern biotechnological tools can employed to develop the new oregano cultivars with enhanced productivity and adaptability. In this chapter, we present an overview of the origin, distribution, taxonomic position, genetic resource characterization and conservation, crop cultivation practices and recent developments on biotechnology, and molecular biology and their application for crop improvement in association with conventional breeding methods of the oregano plant.

Keywords Oregano · Breeding · Medicinal plant · Bioactive compounds · Molecular marker · Cell suspension

G.-R. Sharifi-Sirchi (✉)
Department of Agricultural Biotechnology, Faculty of Agriculture, Shahid Bahonar University of Kerman, Kerman, Iran
e-mail: sharifi@uk.ac.ir

J. M. Al-Khayri et al. (eds.), *Biodiversity and Genetic Improvement of Herbs and Spices*, Advances in Plant Breeding Strategies 8, https://doi.org/10.1007/978-3-031-81838-7_9

9.1 Introduction

Oregano, prince of herbs, is common name of *Origannum vulgare.* Hippocrates (460–370 B.C.), the ancient Greek physician scientist, named this plant as *Origanum,* which was originated from the Greek words, *oros* (mountain) and *ganos* (joy), meaning of "joy of the mountains" (Alekseeva et al. 2020).

Origanum is the genus name of more than 60 aroma and flavour mint family species which are rich in primary and secondary metabolites (Bedoya-Serna et al. 2018). The most well-known species of *Origanum* genus is *Origanum vulgare* L. with trade name of oregano, sometimes called wild marjoram. For thousands of years, it has been used in traditional medicine and as a culinary spice purposes around the world. Its flowers and leafs are rich in kind of bioactive compounds and contents. Fresh oregano has a strongly aromatic and camphoric aroma [reminiscent of clove and balsam, (a combination of mint, thyme, and rosemary)] and a slightly bitter and spicy flavor with a slightly astringent sensation. The dried herb has a pungent flavour with notes of peppermint, too. The warmth sensation in the mouth come from two constituents, carvacrol and thymol, due to their actions on temperature-sensitive structures which called ion channels (Xu et al. 2006; Lee et al. 2008).

Dried flowers and leafs are used as spice and flavors in beers, soups, yogurt, casseroles, teas, pizza, dough, eggs, stew, tomato-based dishes, sauces, Pasta, chilis, and several other foods (Meyers 2005). There are many different varieties of Oregano around the world which are used for culinary purposes. Some of Oregano varieties have quite different flavors from each other's. The oregano essential oil is used commercially as fragrances to scent perfumes, soaps, lotions (Meyers 2005), tea powders, baths, and gargles (Fleming 2000). Oregano increases appetite and energy of body.

Main varieties of Oregano which commonly used for culinary and medicine purposes are: Greek oregano (*Origanum vulgare* ssp. hirtum (Link) Ietswaart); Spanish origano (*Coridohymus capitatus* (L.)) (Oliver 1997), Turkish oregano (*Origanum onites* L.), Mexican oregano (*Lippia graveolens* HBK) (Lawrence 1984), Syrian oregano (*Origanum syriacum* or *Origanum maru*), Cuban oregano (*Plectranthus amboinicus*), Mexican bush oregano (*Poliomintha longiflora*), and Iranian Oregano (*Origanum vulgare* L.) (Pirigharnaei et al. 2011).

Medicinal uses of oregano date back to ancient civilizations, Persia, Greece, Roma, India and China which people used it as remedy against colds and an upset stomach, to treat skin sores, painful menstruation, urinary problems, vomiting, fever, dysentery, and jaundice, childhood malnutrition, and rheumatoid arthritis, to relieve aching muscles, for respiratory disorders (coughs, inflammation of the bronchial mucous membranes and as an expectorant) therapy, and as an emmenagogue, an antiseptic, and a diuretic (Alekseeva et al. 2020).

The essential oil from *O. vulgare* subsp. hirtum exhibits antimicrobial effects, including; antibacterial effects, suppressing the growth of gram-positive and gram-negative bacteria with the effective concentrations of 0.4–4000 μg/mL (Esen et al. 2007; Stefanakis et al. 2013; Grondona et al. 2014), antifungal and yeast effects

(Mancini et al. 2014; Spagnoletti et al. 2016). Two main components of oregano essential oil, carvacrol and thymol have antimicrobial effects with effective doses from 20 to 50 μg/mL (Kalemba and Kunicka 2003; Periago et al. 2004; Cristani et al. 2007). The water-soluble constituents of oregano extract have robust antioxidant properties (Milos et al. 2000; Sarikurkcu et al. 2015). The oregano extract constituents of ɣ-terpinene, carvacrol, ursolic acid, rosmarinic acid, protocatechuate-glycosides, and thymol are responsible for radical scavenging action, inhibition of nitric oxide activity, suppression of lipid peroxidation, and protection of DNA against H2O2-induced oxidant damage (Braga et al. 2006; Aherne et al. 2007; Tsai et al. 2007). The oregano essential oil due to potential AOX capacity can used in arthritis, Alzheimer's disease, chronic-inflammation, Parkinson's disease, diabetes, some types of cancer, and atherosclerosis therapy (Leyva-López et al. 2017).

The oregano extract constituents have antiparasitic effects (Skoufos et al. 2011). The rosmarinic acid constituent has an inhibitory effect on HIVE type 1 reverse transcriptase which reduced the mortality of mice infected with an encephalitis virus (Hooker and Harrich 2001; Swarup et al. 2007). The oregano extract and essential oil had antiparasitic effects against coccidian protozoa from the genus *Eimeria*, agent of coccidiosis disease in livestock. The oregano essential oil inhibited epimastigote growth of *Trypanosoma cruzi* (IC(50)/24 h = 175 μg/ml) and also induced trypomastigote lysis (IC(50)/24 h = 115 μg/ml) (Santoro et al. 2007). The oregano extract have shown to have antiparasitic activity against human intestinal enteric parasites; *Blastocystis hominis*, *Entamoeba hartmanni* and *Endolimax nana* (Force et al. 2000).The oregano essential oil has anticarcinogenic effects (Grondona et al. 2014). The *Origanum onites* essential oil was used for its potential anticancer effects in vitro and in vivo (Spyridopoulou et al. 2019). In vitro tests, showed a dose-dependent antiproliferative activity of oregano essential oil against the four human cancer cell lines; melanoma cells (A375), breast cancer cells (MCF-7), hepatocellular carcinoma cells (HepG2), and colon cancer cells (HT-29). The HT-29 colon cancer cell line was the highest sensitive line to the oregano essential oil antiproliferative activity (lowest IC_{50} value for 72 h, 0.35 ± 0.2 μg/mL). The oregano essential oil attenuated migration and induced apoptosis-related morphological changes in human (HT-29) and murine (CT26) colon cancer cells. Also, it had the antitumor activity on mouse model syngeneic CT26 colon tumors as oral administration, too.

The *O. vulgare* subsp. hirtum essential oil has been successfully used as an herbicide, inhibiting glutamate and aspartate metabolism by altering the photorespiratory pathway in *Arabidopsis thaliana* seedlings (Araniti et al. 2018). In in-vitro assays. Oregano essential oil had inhibitory effect on the seed germination of *Lolium perenne* L. and *Trifolium pratense* L. plants at concentrations ranging from 0.5–3.0 μL/mL. It seems, carvacrol-rich essential oils had a strong inhibitory effect on seed germination (Nikolova et al. 2021). Researcher showed thymol constituent of *Origanum vulgare* L. essential oil had a Potential selective bio-based herbicide on monocot plant species (Gruľová et al. 2020).

The antimicrobial and antioxidant activity of oregano essential oil has been used in the food industry. The carvacrol and thymol constituents are responsible for the characteristic odor, antioxidant activity (in retarding the process of lipid peroxidation in fatty foods, and scavenging free radicals), and as antibacterial agent (caused permeability of the bacterial cell membrane due to the hydrophobic domains impregnation) (Rodriguez-Garcia et al. 2016). Nagmetova et al. (2020) used biocellulose, impregnated with oregano essential oil against two acetic bacteria strains, *Gluconacetobacter hansenii* ATCC 23,769 and *Komagataeibacter* sp. GH1 and observed strong and moderate antimicrobial activity against them (Nagmetova et al. 2020). In in-vitro assays, the oregano essential oil showed high antibacterial activity against bacteria associated with food poisoning, *Staphylococcus aureus*, *Salmonella enterica*, *Erwinia rhapontici* and *Xanthomonas campestris* (Simirgiotis et al. 2020). In the broth microdilution method, the essential oil of *O. vulgare* presented antifungal effects against *C. albicans* and non-albicans Candida reference strains (Cid-Chevecich et al. 2022).

The oregano essential oil was extensively studied for its antimicrobial activity against *Staphyloccocus* spp. and *Pseudomonas aeruginosa.* It is used as a new therapeutic alternatives with high potential for skin disorders, such as acne, wounds or aging in the modern formulations, such as microemulsions and nanoemulsions (Bora et al. 2022).

Drying method of oregano plants effects on its essential oil quantity and quality. Caputo et al. (2022) evaluated the influence of some drying methods on the essential oil constituents of the oregano plants (Caputo et al. 2022). The plants aerial parts were dehydrated by convective drying techniques (shade, static oven), osmotic treatment, and microwave-assisted heating (three different treatments). The plants aerial parts dried by microwave and shade methods showed the highest essential oil yield based on the GC-FID and GC-MS analysis results (Caputo et al. 2022).

The oregano and its essential oil constituents have complex mixtures of strongly active compounds with many natural and safe properties which can introduced as an attractive substances for pharmaceutical (Table 9.1), cosmetics and food industries.

9.2 Taxonomy

Oregano, *O. vulgare*, is from planta kingdom, viridiplantae subkingdom, streptophyta infrakingdom, embryophyta super division, tracheophya division, spermatophytina subdivision, magnoliopsida class, asteranae super order, Lamiales order, Lamiaceae family, Origanum genus, and vulgare species based on Integrated Taxonomic Information System (ITIS) (https://www.itis.gov/servlet/SingleRpt/SingleRpt?search_topic=TSN&search_value=32632#null).

Table 9.1 The oregano essential oil constituents, biological activity and effects

Constituent name (Cid-Chevecich et al. 2022)	Biological activity	Effect	References
a-Thujene α-pinene Sabinene β-myrcene α-phellandrene α-terpinolene p-Cymene 3-carene α-terpinolene b-Myrcene Trans-allo-ocimene γ-Terpinene Thymol Carvacrol 4-thujanol Linalool Cis-sabinene hydrate Cis-β-terpineol Cis-2-menthenol Trans-2-menthenol Pinocarvone Endo-borneol 4-terpineol α-terpineol Trans-piperitol Cis-piperitol Methyl carvacro Linalyl acetate Geranyl acetate (E)-caryophyllene α-humulene Germacrene D Spathulenol Caryophyllene oxide Neryl acetate	Anti-inflammatory activity	Reduction on the levels of NO and ROS	Leyva-López et al. (2016)
	Antioxidant activity	Radical scavenging activity against DPPH	Arana-Sánchez et al. (2010)
	Antiproliferative activity	Inhibits colon carcinoma (LoVo) and hepatocarcinoma (HepG2) cell proliferation	Marrelli et al. (2016)
	Anti-genotoxic activity	Reduces the chromosomal aberration in bone marrow cells of rats	Mossa et al. (2013)
	Antifungal activity	Inhibits morphogenesis, adhesion, and proliferation of virulence patterns of *Candida* spp. strains	Cid-Chevecich et al. (2022)
	Antibacterial activity	Inhibits bacteria associated with food poisoning, *Staphylococcus aureus*, *Salmonella enterica*, *Erwinia rhapontici* and *Xanthomonas campestris*	Simirgiotis et al. (2020)
	Antiparasitic activity	Inhibitory effect on HIVE type 1 reverse transcriptase which reduced the mortality of mice infected with an encephalitis virus	Swarup et al. (2007)

9.3 Botany

Oregano is an herbaceous perennial plant, growing from 20 to 80 cm in height and spread from 30 to 60 cm with 1–4 cm long. It has narrow and pinnate, or arrow-shaped, soft and fuzzy texture and olive-green opposite leaves which grow in pair and have been covered with trichomes (fine hairs). Oregano plants have multiple branching stems (Fig. 9.1). The glaucous stems are become woody at the base at late mature step. The plants bloom in long erect spikes at the top of the stalks and their petals are from purple to pink to white in color which attract butterflies and bees (Chishti

Fig. 9.1 Oregano (*O. vulgre*) plant arrow-shaped, soft and olive-green opposite leaves and multiple branching stems (photographed by M.Sc. Nohi in Iran national botanical garden, Tehran 2022)

et al. 2013). The flowers are in 3–4 mm long. The fruits are small brown oval shape. The essential oil glands are present in the oregano leaves and flowers which produce fragrance volatile oils (Meyers 2005). Oregano is harvested just before or after the flowers bloom depending to People's taste. As usual, the leaves will be become so bitter when the plants bloomed. Oregano plants have a spreading root system which called rhizome, run underground and can grow their shoots many feet far from the mother plant. The root depth of plant is till 35 cm (Herb Society of America 2005). The mature seeds have small brown round shape and 1000 grain weight of 125 mg (https://www.botanicalinterests.com/product/Common-Oregano-Seeds).

9.4 Distribution

Oregano is native to the Irano-Turanian regions, the Mediterranean countries, temperate western and southwestern Eurasia regions including: Iran, Afghanistan, Tajikistan, Turkmenistan, Uzbekistan, north of Iraq, Syria, Lebanon, Palestine, Jordan, Turkey, Armenia, Azerbaijan, Greece, Albania, Italy, France, Monaco, Spain, Bosnia and Herzegovina, Montenegro, Croatia, Slovenia, Spain, Cyprus, Malta,

Morocco, Algeria, Libya, Tunisia, and Egypt (Lukas et al. 2015) (Fig. 9.2). Nowadays, oregano plants are widely distributed and naturalized every regions in the temperate northern hemisphere, from Britain, central Europa, and the Scandinavia countries to India and Taiwan; also, North America and Venezuela (Węglarz et al. 2020). According to *O. vulgare* plants taxonomy characteristics data including: the shape, color and length of root, stem, leaves, bracts, flower and seeds, the density of glandular hairs, and the number of oil glands on the leaves, there are six recognized subspecies of *O. vulgare* (Lukas et al. 2015; Leyva-López et al. 2016; Alekseeva et al. 2020). The most valuable sub-species is *O. vulgare* L. subsp. hirtum (Link) Ietswaart, Greek oregano, which is native to Mediterranean regions and neutralized and distributed almost all over the world, nowadays. The others are *O. v.* ssp. *gracile* (K. Koch) Ietsw. Which is originated from Tunisia, and Algeria, *O. v.* ssp. *gracile* (K. Koch) Ietsw. Which is Native from east Turkey to Afghanistan and South Siberia, *O. v.* ssp. *virens* (Hoffmanns. and Link) Ietsw. distributed from Azores, Madeira, Balearic Islands, Portugal, Spain, to Morocco, *O. v.* ssp. *viridulum* (Martrin-Donos) Nyman, Widespread from Corsica to Nepal, *O. v.* ssp. *Vulgare* distributed from Britain and Scandinavia to Taiwan and naturalized in North America and Venezuela (Alekseeva et al. 2020).

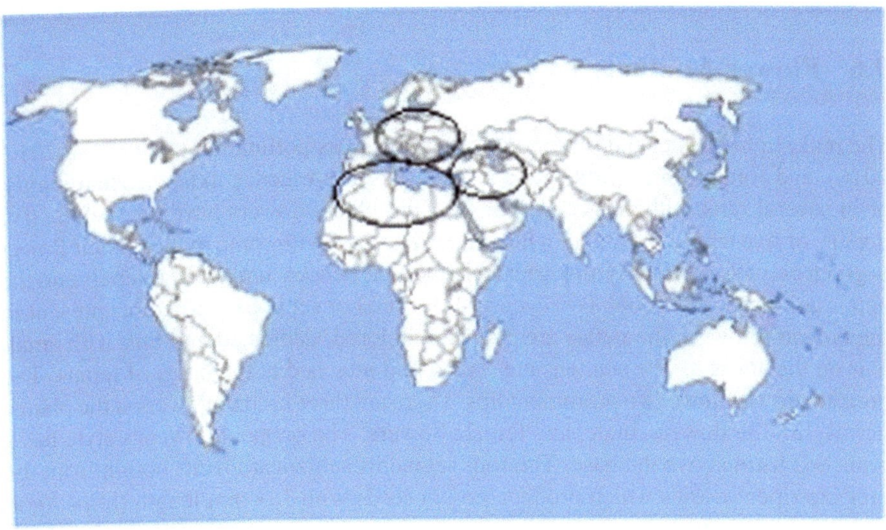

Fig. 9.2 The first main habitats of *O. vulgare* in the world including; the Irano-Turanian regions, the Mediterranean countries, temperate western and southwestern Eurasia regions

9.5 Propagation

Oregano plants are multiplied by seed, leaf and stem cuttings, root division, and by dividing existing plants (Pitzer 1999; Bonfanti et al. 2012). The seeds will dispersed on the soft prepared agriculture soil when daytime temperatures reach 20 °C. For better cultivation, the seeds will be lightly press into the soil surface. The seeds need 4 days light to germinate. Therefore, It should not be covered the tiny seeds and the maximum seed depth proposed by the seed company is 0.31–0.62 cm. The seeds cultivated in the rows with the 20 cm distance between the plants in the row and the 50–60 cm distance between the rows. The best time for the oregano root divisions, cuttings, and dividing existing plants is late spring as the plants are in the maximum growth. The stem cuttings size is about 7.6–12.7 cm and the stems cut at a 45° angle for better chance of rooting. The light potting mediums like: peat moss and vermiculite are used as the bed to keep moist until the cuttings rooted. Different kind of tissue culture techniques can be employ to propagate the oregano plants. Base on author experience, any kind of transplanting will be changed oregano plants phenotypic and biochemical characteristics, and the best way of oregano propagation is by the seeds.

9.6 Flower Anatomy

The pink, light purple, and plasticity flowers are cross pollination (Azizi et al. 2012, 2016), and formed in a spiral shape around the inflorescence axis, or occur singly, or in several rank with no nectar. The oregano plants flowers have five petals, five Sepals, or five tepals, 4 stamen, with a superior ovary with some ovules. Their flower 's green calyx is radially symmetrical and their carpels are united. The sepals and the petals are fused into a cup. Two cycles of the petal or sepal like structures present in the oregano flower. The petals are 5–7 mm, colored, dedicated and thin with small hair on the surface. The stamens are linked and attached at the bases of tepals. The oregano flowers have two prominent lips. There are three kind of flower on the plants, hermaphrodite flowers, male, and female flowers. The oregano flower 's style has a branched form above the base. The four segments schizocarp fruits obtain from the oregano ripen ovaries which produce brown seeds with 1–4 mm length (https://gobotany.nativeplanttrust.org/species/origanum/vulgare/).

9.7 Genome Size and Chromosomes Number

The genome size and chromosomes number of *O. vulgare* subsp. Virens and five other oregano species were studied in Morocco (Bakha et al. 2017). The genome size of *O. vulgare* was as small as 1.43 pg/2C base on flow cytometry data which was the smallest one in the four studied species. The all four oregano species had 2n = 2x = 30 chromosome number. In one other study, the total nuclear DNA amount *O. vulgare* L. was shown 2C = 1.36 pg (Mowforth 1986).

9.8 Breeding Strategies

Breeding of plants is a time consuming and expensive program. In average, it usually takes more than ten years to release a commercial cultivar. The first need of each breeding program is a gene pool or a plant germplasm pool with extensive genetic diversity. The plant germplasm consist of wild relative species (WR), land races (LR), breeding stocks, and modern cultivars. The WR and LR are used in back cross breeding program as donor parent to contribute novel target genes to modern cultivars. The successful plant breeding program is depend on three major factors; the extensive genetic diversity in the gene pool, the additive effect of the novel target gene, no or low the environment effect and the interaction effect of genotype and environment on the novel target gene. It means traits with high genetic response and heredity will have a successful breeding program. The cross-pollinated plants have a great degree of genetic variability and flexibility. There are several conventional breeding methods for cross pollinated plants as oregano including, recurrent selection (simple recurrent selection, recurrent selection for combining ability, and reciprocal recurrent selection), plants progeny selection, line breeding, synthetic and composite varieties, hybrid varieties, induced polyploidy.

Nowadays, biotechnology techniques are used in oregano plants quality and quantity traits improvement. The DNA sequencing technique presents the genetic codes information for the genes responsible for worthy economic, important traits of the oregano plants. Molecular marker, DNA and protein markers have the most significant effect on the oregano plants breeding program. Marker assisted selection is a noteworthy method to breed traits which closely-linked to molecular markers, and diminish breeding program time, which is a major limitation agent.

The shoot tip meristem culture, auxiliary bud culture, callus culture, embryo culture, pollen or microspore and anther culture, ovule and ovary culture, cell suspension, protoplast fusion are modern propagation techniques which can be used to multiply oregano plants to produce many health progeny plants rapidly. Haploid culture, pollen and ovule cultures, can be employed for the oregano pure line and double haploid plants production which are completely homozygote and can be used in hybridization programs as the pure line parents. Tissue culture techniques can be used to diminish the time and cost in the oregano resistance to biotic and abiotic stress breeding programs. The oregano callus culture can be produced somaclonal variation which can be used as base of the oregano breeding programs. Tissue culture techniques can be used in the oregano induced mutation breeding programs when produce proper explants, like callus, exposed to the chemical and physical mutagens. The DNA manipulation technique which is the main genetic engineering aspect of biotechnology, is employed in the plant modern breeding programs to improve the quality and quantity traits of plants. This technique can be used in the oregano plants to improve amount of essential oil components and the medicinal characteristics. The DNA manipulation and cell suspension techniques are strongly suggested in the *in-vitro* secondary metabolite production of the oregano plants. The induced gene expression and priming can be used to produce worthy components and molecular in the oregano plants.

The gene knockdown and knockout techniques like the RNAi, antisense RNA, and crisper technologies are used in the revers genetics and signaling pathway studies, can be recruited to recognize and improve of the medicinal characteristics of the oregano plants. Also, these techniques can be used to remove the undesirable genes in the oregano plants.

In-situ hybridization techniques such as FISH provide a powerful tool for identifying the novel target single gene locus and quantitative traits loci (QTL) locations on the metaphase chromosomes. These techniques can be used in the gene for gene relationships studies in the oregano plants. ELISA technology (enzyme-linked immunosorbent assay) is a modern technique to detect the protein with the foreign sources (bacteria, viruses, mycotoxins, fungi, and mycoplasma) and hormones in the oregano plants. This technique used antibodies which binds to specific protein molecules.

The term of quality in the oregano plants has different meanings according to the aim of the oregano consumption. It can mean the antioxidant, antiseptic, antimicrobial, medicinal, culinary, spicing and food tasting characteristics of the oregano leaves and flowers components. Also, the quality of oregano plants can refer to their characters which make it proper as additive in perfumes and mouthwashes. The major constituents of the oregano essential oil, the volatile compounds thymol and carvacrol along other bioactive compounds include; rosmarinic acid, phenolic acids (caffeic acid, p-coumaric acid), ursolic acid, caffeoyl derivatives, flavonoids, and carnosic acid, are the main agents of the oregano qualitative characteristics. Inside quality, breeders should work on the quantitative characteristics of the oregano plants include; number of stem and flower, number of leaves in the stem, size and form of the leaves, form of the flowers (single or spike), number of the glandular trichomes on the leaves, the plant height, width and biomass, essential oil output or percentage, and leaves and flowers color.

Breeders should notice that environmental factors can be lead to the extensive variation in the quality and quantity of the oregano plant product include; soil structure and texture, soil nutrients, humidity, light, temperature, height above mean sea level, precipitation and weather. Therefore, the environmental factors should be removed or calculated in the oregano plant breeding programs. Some researchers, make a big mistake and compare the oregano plants between different geographical locations without considering the environment factors effects and genotype × environment interaction effects. They should know that the environment factors effect on genetic gain. Other subject which effects on the results of the oregano research, is the age of the oregano herbs. The oregano plant is a perennial herbs, and the middle ages oregano plants are harvested for medicinal purposes. Therefore, if researcher studied on the oregano medicinal characteristics, the middle ages plants are the best material for the analysis.

Base on author experience, it would be better that the researchers prepare the oregano seeds from the different cultivars and cultivate, compare, and breed them in a high above mean sea level area in the temperate northern hemisphere.

9.9 Oregano Breeding Studies

In a conventional breeding program on Greek oregano (*O. vulgare* ssp. hirtum) for casting the elite self-plants from the wild initial population, a pedigree intra-selection method, based on widely spaced single-plant performance, was applied (Sarrou et al. 2017). The evaluated traits were biomass density, plant's growth, type of stems and inflorescences, dry weight of leaves and flowers, ratio of leaves and flowers per stem, the composition of essential oil and carvacrol content. The end-scope of this research was to obtain the high essential oil and carvacrol content genotypes and stabilization of valuable herbage components (Sarrou et al. 2017).

Shafiee-Hajiabad et al. (2014) compared the trichome morphology of Common oregano (*O. vulgare* L. subsp. *vulgare*) and Greek oregano (*O. vulgare* L. subsp. *hirtum*). The plant seeds were cultivated in pots under field conditions and the traits were investigated by light microscopy and scanning electron microscopy. They observed different density and size of trichomes between the two studied sub-species. The peltate trichomes size on both sides (abaxial and adaxial) of the subsp. *Hirtum* leaves was larger than subsp. *Vulgare*. Also, density of the peltate as well as capitate glandular trichomes on adaxial leaf side and essential oil content of subsp. *Hirtum* was higher than subsp. *Vulgare* (Shafiee-Hajiabad et al. 2014).

The morphological parameters and the raw materials quality of the two oregano subsp. Including; *O. vulgare* L. subsp. hirtum (Link) Ietswaart (Greek oregano) and *O. vulgare* L. subsp. Vulgare (common oregano) under cultivation condition in Central Europe with temperate climate were assayed and compared (Węglarz et al. 2020). The quality assays were performed on the content and the composition of their essential oils. The number of glandular trichomes on the leaves in Greek oregano (up to 4.85 per 1 mm^2) was visibly higher than common oregano cultivar. Subsequently, Greek oregano had higher content of essential oil (up to 3.36 g $\times$ 100 g^{-1} DW). Based on sesquiterpenes, Greek and common oregano were classified as mixed carvacrol/ -terpinene chemotype rich and mixed sabinyl/cymyl type rich, respectively. Greek oregano had the higher Rosmarinic acid and total phenolic acid contents than common oregano cultivar. The highest content of the essential oil in Greek and common oregano cultivars were at the full blooming stage and the beginning of blooming stage, respectively. Both subspecies shown the highest amount of the total phenolic acids and subsequently rosmarinic acid, at the beginning of seed-setting stage.

The sensory assays shown that Greek oregano had the higher levels of the odor and taste. Based on the results of this study, Greek oregano is quantitatively and qualitatively well adapted to Central Europe area within temperate climate (Węglarz et al. 2020).

Azizi et al. 2016 assayed the associations between the essential oil, carvacrol, and thymol contents from one side and the molecular markers including; AFLP (Amplified Fragment Length Polymorphism) and SAMPL (Selectively Amplified Microsatellite Polymorphic Loci) from the other side, in the forty-two *Origanum vulgare* L. accessions originated from Europe. They observed relatively high correlation between them and there were significant associations between 3 AFLP and 20 SAMPL markers with three key studied qualitative traits. The selected markers were suggested for the next maker association selection programs on improving oregano key essential oil components (Azizi et al. 2016).

The seed set formation of *O. syriacum* plants were assayed under three conditions including; the free insect visiting condition, the covered plant and hand cross pollination, and the covered plant (control) in Jordan with semiarid Mediterranean climate (Al-Ghzawi et al. 2009). The results shown that the plant in supplementary pollination condition produced the highest seed and seed set and the lowest seed and seed set formation was related to the covered plants (Al-Ghzawi et al. 2009). The *O. syriacum* flowers are so attractive for the *Apis mellifera* and also they are visited by twenty one bee species in Jordan and wind did not has significant role in the *O. syriacum* flowers cross pollination (Al-Ghzawi et al. 2009).

Genetic structure and differentiation among North Africa's *O. vulgare* subsp. Glandulosum plants were assayed by the ESTs developed thirteen SSR markers (Mechergui et al. 2017). The five Maghrebian natural populations from regions of Bargou, Sejnane, Nefza, Zaghouan and Krib and two other subspecies of *O. vulgare* including; hirtum (from USA) and vulgare (from Italy) as out-group were considered in this study. The results presented that the observed heterozygosity was lower than unbiased expected heterozygosity which indicated that a deficit in heterozygosity among the total populations (Mechergui et al. 2017).

In order to develop a core collection from the natural populations of Bulgarian *O. vulgare* subsp. *hirtum* (Link) Ietsw. two genetic and flower diversity experiments were designed by Alekseeva et al. 2021 (Alekseeva et al. 2021). They used two molecular markers; sequence-related amplified polymorphism (SRAP) and simple sequence repeat (SSR) with eight and eleven primer respectively, for genetic diversity experiment and gas chromatography/mass spectrometry (GC/MS) analysis for the flower volatiles diversity experiment on the natural plants of two Bulgarian regions of Kresna Gorge and Eastern Rhodopes. There were eight populations and 239 individual plants in this assay. In the genetic diversity experiment, SRAP markers were more informative than SSR markers, and together presented 0.0047–0.11 genetic differentiation between studied populations based on the pairwise fixation index (Fst) values and consisted 10 genetic clusters. In the flower volatile diversity experiment, the 63 compounds were identified. The most of plants were belonged to the carvacrol chemotype and just a single plant was belonged to the thymol chemotype. The 6 major and 23 minor metabolite clusters were identified based on an in-depth analysis. The overall data set and analysis were used to obtain a proper strategy for the selection of target plants for a core collection development with entire genetic and flower volatile diversity. The selected core collection represented better set of genetic cluster and minor metabolic cluster than a core set extracted by the PowerCore software (Alekseeva et al. 2021).

The *O. syriacum* L. and *O. ehrenbergii* nodal segments explants were cultured on supplemented Ms medium with the different concentrations of BAP (1, 1.5, 2 mg/L) for multiple shoot induction (El Beyrouthy et al. 2015). After four sub-cultures, a multiplication with mean rate of 2.0–3.7 (new shootlets per explant) was achieved. The shoots were completely rooted (100%) on MS basal medium. The species *O. ehrenbergii* explants were reacted better in entire stage of tissue culture. The overexploitation and habitat degradation of wild grown oregano plants lead to the destruction of the oregano plant resources. Therefore, authors suggested that the tissue culture techniques accompany with conventional propagation method reduce the extinction risk of oregano (El Beyrouthy et al. 2015).

The 17d-old seedling derived shoot apices explants were used to propagate *Origanum sipyleum* L. on modified MS medium containing 550 mg/l $CaCl_2$ (Oluk and Cakir 2009). On supplemented MS medium with 1 mg/L benzylaminopurine (BAP), a shoot multiplication with mean rate of 3.7 ± 0.3 (new shootlets per explant) was achieved and after subculture, it increased to 7.8 ± 0.4. 96% (Oluk and Cakir 2009). Indolebutyric acid (IBA) 0.5 mg/l used for root induction. The seventy six percentage of the tissue cultured plants were survived after acclimation in greenhouse (Oluk and Cakir 2009).

To produce polyphenols from *O. vulgare*, a cell suspension culture developed by Li et al. 2021and the salicylic acid elicitation and phenylalanine feeding were used as the stimulators (Li et al. 2021). The calli have produced from aseptic seedlings of oregano plants. MS medium supplemented with 0.5 mg/L 2, 4-dichlorophenoxy acetic acid and 3.0 mg/L Kinetin was used for both cells growth and polyphenols accumulation. The much better polyphenols [especially rosmarinic acid (RosA)] content and yield, 41.36 mg/g and 752.93 mg/L, respectively were obtained since used a combination of 200 μM SA and 100 μM Phe compared than individual use of each stimulators (Li et al. 2021). Also, the cell suspension stimulated with the combined elicitors produced the total polyphenols with higher antioxidant capacity than the controls. Authors suggested that a combination of SA and Phe elicitors can apply on natural polyphenols production in the *O. vulgare* cell suspension culture (Li et al. 2021).

9.10 Conclusion

Oregano plants with the plenty of worthy primary and secondary metabolites play an important and central role for various medicinal activities including; antitumore, antioxidant, antibacterial, antiviral, anti-inflammatory, antifungal, and insecticidal (Garcia-Beltran and Esteban 2016). Also, they are used as spice and flavors in the food and perfume industry because of strongly aromatic and camphoric aroma and bitter and spicy flavor with a slightly astringent sensation. The increasing human population, important role of oregano medicinal plant in human healthy and current and expected effects of climate changes have all led to threaten oregano natural habitats and loss of its distributional range. To overcome increasing consumption of oregano, need to employ plant conventional and modern breeding techniques. Today, utilizing modern sciences and techniques including cytogenetics, genetics, DNA sequencing, gene silencing, and transformation assist the oregano breeders to increase oregano yield and metabolite. This paper mentioned all strategies and techniques which should be employed by breeders for the global oregano security.

References

Aherne S, Kerry J, O'Brien N (2007) Effects of plant extracts on antioxidant status and oxidant-induced stress in $CaCo_2$ cells. Br J Nutr 97:321–328

Alekseeva M, Zagorcheva TZ, Atanassov I, et al (2020) *Origanum vulgare* L.—a review on genetic diversity, cultivation, biological activities and perspectives for molecular breeding. Bulg J Agric Sci 26(6):1183–1197

Alekseeva M, Zagorcheva T, Rusanova M et al (2021) Genetic and flower folatile diversity in natural populations of *Origanum vulgare* subsp. *hirtum* (Link) Ietsw. In Bulgaria: toward the development of a core collection. Front Plant Sci 12:1–13. https://doi.org/10.3389/fpls.2021.679063

Al-Ghzawi ABDAM, Zaitoun S, Freihat N et al (2009) Effect of pollination on seed set of *Origanum syriacum* under semiarid Mediterranean conditions. Acta Agric Scand B Soil Plant Sci 59:273–278. https://doi.org/10.1080/09064710802093862

Arana-Sánchez A, Estarrón-Espinosa M, Obledo-Vázquez EN et al (2010) Antimicrobial and antioxidant activities of Mexican oregano essential oils (*Lippia graveolens* H.B.K.) with different composition when microencapsulated in β-cyclodextrin. Lett Appl Microbiol 50:585–590

Araniti F, Landi M, Lupini A et al (2018) *Origanum vulgare* essential oils inhibit glutamate and aspartate metabolism altering the photorespiratory pathway in *Arabidopsis thaliana* seedlings. J Plant Physiol 231:297–309. https://doi.org/10.1016/j.jplph.2018.10.006

Azizi A, Hadian J, Gholami M et al (2012) Correlations between genetic, morphological, and chemical diversities in a germplasm collection of the medicinal plant *Origanum vulgare* L. Chem Biodivers 9:2784–2801

Azizi A, Ardalani H, Honermeier B (2016) Statistical analysis of the associations between phenolic monoterpenes and molecular markers, AFLPs and SAMPLs in the spice plant Oregano. Herba Polon 62:42–56

Bakha M, Faiz CA, Daoud M et al (2017) Genome size and chromosome number for six taxa of *Origanum* genus from Morocco. Botany Lett 164(4):361–370. https://doi.org/10.1080/23818107.2017.1395766

Bedoya-Serna CM, Dacanal GC, Fernandes AM et al (2018) Antifungal activity of nanoemulsions encapsulating oregano (*Origanum vulgare*) essential oil: in vitro study and application in Minas Padrao cheese. Braz J Microbiol 49(4):929–935

Bonfanti C, Iannì R, Mazzaglia A et al (2012) Emerging cultivation of oregano in Sicily: sensory evaluation of plants and chemical composition of essential oils. Ind Crop Prod 35:160–165

Bora L, Avram S, Zinuca Pavel L et al (2022) An Up-to-date review regarding cutaneous benefits of *Origanum vulgare* L. essential oil. Antibiotics 11:549–569

Braga P, DalSasso M, Culici M et al (2006) Antioxidant potential of thymol determined by chemiluminescence inhibition in human neutrophils and cell-free systems. Pharmacol 76:61–68

Caputo L, Amato G, Bartolomeis P et al (2022) Impact of drying methods on the yield and chemistry of *Origanum vulgare* L. essential oil. Sci Rep 12:3845–3856

Chishti S, Kaloo ZA, Sultan P (2013) Medicinal importance of genus *Origanum*: a review. J Pharmacognosy Phytother 5(10):170–177

Cid-Chevecich C, Müller-Sepúlveda A, Antonio Jara J et al (2022) *Origanum vulgare* L. essential oil inhibits virulence patterns of *Candida* spp. and potentiates the efects of fuconazole and nystatin in vitro Camila. BMC Complement Med Ther 22:39–52

Cristani M, D'Arrigo M, Mandalari G et al (2007) Interaction of four monoterpenes contained in essential oils with model membranes: implications for antibacterial activity. J Agric Food Chem 55:6300–6308

El Beyrouthy M, Elian G, Abou Jaoudeh C, et al (2015) In-vitro propagation of *Origanum syriacum* and *Origanum ehrenberg II*. Acta Hortic 1083:169–172. https://doi.org/10.17660/ActaHortic.2015.1083.19

Esen G, Azaz AD, Kurkcuoglu M et al (2007) Essential oil and antimicrobial activity of wild and cultivated *Origanum vulgare* L. subsp. hirtum (Link) letswaart from the Marmara region, Turkey. Flavour Frag J 22:371–376. https://doi.org/10.1002/ffj.1808

Fleming T (2000) PDR for herbal medicines, 2nd edn. Medical Economics Montvale (US). https://www.itis.gov/servlet/SingleRpt/SingleRpt?search_topic=TSN&search_value=32632#null

Force M, Sparks WS, Ronzio RA (2000) Inhibition of enteric parasites by emulsified oil of oregano in-vivo. Phytother Res 14:213–214

Garcia-Beltran JM, Esteban MA (2016) Properties and applications of plants of *Origanum* sp. genus. SM J Biol 2(1):1006–1015

Grondona E, Gatti G, Lopez AG et al (2014) Bio-efficacy of the essential oil of oregano (*Origanum vulgare* Lamiaceae. ssp. hirtum). Plant Foods Hum Nutr 69:351–357. https://doi.org/10.1007/s11130-014-0441-x

Gruľová D, Caputo I, Elshafie HS et al (2020) Thymol chemotype *Origanum vulgare* L. essential oil as a potential selective bio-based herbicide on monocot plant species. Molecules 25(3):595–605. https://doi.org/10.3390/molecules25030595

Herb Society of America (2005) Oregano & marjoram. An Herb Society of America fact sheet. Available at: http://www.herbsociety.org/factsheets/oregano.pdf

Hooker C, Lott W, Harrich D (2001) Inhibitors of human immunodeficiency virus type 1 reverse transcriptase target distinct phases of early reverse transcription. J Virol 75:3095–3104

https://www.botanicalinterests.com/product/Common-Oregano-Seeds.

Kalemba D, Kunicka A (2003) Antibacterial and antifungal properties of essential oils. Curr Med Chem 10:813–829

Lawrence B (1984) The botanicaland chemical aspects of oregano. Parfum Flovor 9:41–51

Lee S, Buber M, Yang Q et al (2008) Thymol and related alkyl phenols activate the TRPA1 channel. Br J Pharmacol 153:1739–1749

Leyva-López N, Nair V, Bang WY et al (2016) Protective role of terpenes and polyphenols from three species of oregano (*Lippia graveolens*, *Lippia palmeri* and *Hedeoma patens*) on the suppression of lipopolysaccharide-induced inflammation in RAW 264.7 macrophage cells. J Ethnopharmacol 187:302–312

Leyva-López N, Gutiérrez-Grijalva EP, Vazquez-Olivo G, Heredia B (2017) Essential oils of Oregano: biological activity beyond their antimicrobial properties. Molecules 22:989–1013. https://doi.org/10.3390/molecules22060989

Li YP, Tang DB, Wang XQ et al (2021) Development of *Origanum vulgare* cell suspension culture to produce polyphenols and the stimulation effect of salicylic acid elicitation and phenylalanine feeding. Biotechnol Bioprocess Eng 26:456–467. https://doi.org/10.1007/s12257-020-0193-4

Lukas B, Schmiderer C, Novak J (2015) Essential oil diversity of European *Origanum vulgare* L. (Lamiaceae). Phytochemistry 119:32–40

Mancini E, Camele I, Elshafie HS et al (2014) Chemical composition and biological activity of the essential oil of *Origanum vulgare* ssp. hirtum from different areas in the Southern Apennines (Italy). Chem Biodivers 11:639–651. https://doi.org/10.1002/cbdv.201300326

Marrelli M, Conforti F, Formisano C et al (2016) Composition, antibacterial, antioxidant and antiproliferative activities of essential oils from three *origanum* species growing wild in Lebanon and Greece. Nat Prod Res 30:735–739

Mechergui K, Jaouadi W, Bekele WA et al (2017) Genetic structure and differentiation among oregano [*Origanum vulgare* subsp. glandulosum (Desf.) Ietswaart] provenances from North Africa: bioinformatic approaches cause systematic bias. Genet Resour Crop Evol 64(4):717–732. https://doi.org/10.1007/s10722-016-0395-y

Meyers M (2005) Oregano and marjoram: an Herb Society of America guide to the genus *Origanum*. The Herb Society of America. https://www.herbsociety.org/

Milos M, Mastelic J, Jerkovic I (2000) Chemical composition and antioxidant effect of glycosidically bound volatile compounds from oregano (*Origanum vulgare* L. ssp. hirtum). Food Chem 71:79–83. https://doi.org/10.1016/S0308-8146(00)00144-8

Mossa ATH, Refaie AA, Ramadan A et al (2013) Antimutagenic effect of *Origanum majorana* L. essential oil against prallethrin-induced genotoxic damage in rat bone marrow cells. J Med Food 16:1101–1107

Mowforth M (1986) Variation in nuclear DNA amounts in flowering plants: an ecological analysis. Ph.D. dissertation. University of Sheffield, United Kingdom

Nagmetova G, Berthold-Pluta A, Garbowska M, Kurmanbayev A, Stasiak-Różańska L (2020) Antibacterial activity of biocellulose with oregano essential oil against Cronobacter strains. Polymers 12:1647–1657. https://doi.org/10.3390/polym12081647

Nikolova M, Traykova B, Yankova-Tsvetkova E et al (2021) Herbicide potential of selected essential oils from plants of Lamiaceae and Asteraceae families. Acta Agrobotanica 74:1–7. https://doi.org/10.5586/aa.7411

Oliver GW (1997) The world market of oregano. In: Pro-ceedings of the IPGRI international workshop on Oregano, Bari, 8–12 May 1996, pp 142–146

Oluk EA, Cakir A (2009) Micropropagation of *Origanum sipyleum* L., an endemic medicinal herb of Turkey. Afr J Biotech 8(21):5769–5772

Periago P, Delgado B, Fernandez P et al (2004) Use of carvacrol and cymene to control growth and viability of *Listeria onocytogenes* cells and predictions of survivors using frequency distribution functions. J Food Prot 67:1408–1416

Pirigharnaei M, Zare S, Heidary R et al (2011) The essential oils compositions of Iranian Oregano (*Origanum vulgare* L.) populations in field and provenance from Piranshahr district, West Azarbaijan province, Iran. Avicenna J Phytomed 1(2):106–114

Pitzer S (1999) Growing & using oregano. Storey Publishing, LLC. ISBN: 088266526X. https://gobotany.nativeplanttrust.org/species/origanum/vulgare/

Rodriguez-Garcia I, Silva-Espinoza BA, Ortega-Ramirez LA, Leyva JL, Siddiqui MW, Cruz-Valenzuela MR, Gonzalez-Aguilar GA, Ayala-Zavala JF (2016) Oregano essential oil as an antimicrobial and antioxidant additive in food products. Crit Rev Food Sci Nutr 56(10):1717–1727. https://doi.org/10.1080/10408398.2013.800832

Santoro GF, Cardoso MG, Guimarães LGL et al (2007) Effect of oregano (*Origanum vulgare* L.) and thyme (*Thymus vulgaris* L.) essential oils on *Trypanosoma cruzi* (Protozoa: Kinetoplastida) growth and ultrastructure. Parasitol Res 100(4):783–790. https://doi.org/10.1007/s00436-006-0326-5

Sarikurkcu C, Zengin G, Oskay M et al (2015) Composition, antioxidant, antimicrobial, and enzyme inhibition activities of two *Origanum vulgare* subspecies (subsp. vulgare and subsp. hirtum) essential oils. Ind Crop Prod 70:178–184. https://doi.org/10.1016/j.indcrop.2015.03.030

Sarrou E, Tsivelika T, Chatzopoulou P et al (2017) Conventional breeding of Greek oregano (*Origanum vulgare* ssp. hirtum) and development of improved cultivars for yield potential and essential oil quality. Euphytica 213:104–121. https://doi.org/10.1007/s10681-017-1889-1

Shafiee-Hajiabad M, Hardt M, Honermeier B (2014) Comparative investigation about the trichome morphology of Common oregano (*Origanum vulgare* L. subsp. *vulgare*) and Greek oregano (*Origanum vulgare* L. subsp. *hirtum*). J Appl Res Med Aromat Plants 1(2):50–58. https://doi.org/10.1016/j.jarmap.2014.04.001

Simirgiotis MJ, Burton D, Parra F, López J, Muñoz P, Escobar H, Parra C (2020) Antioxidant and antibacterial capacities of *Origanum vulgare* L. essential oil from the arid Andean region of Chile and its chemical characterization by GC-MS. Metabolites 10:414–426. https://doi.org/10.3390/metabo10100414

Skoufos I, Tsinas A, Giannenas I et al (2011) Effects of an oregano based dietary supplement on performance of broiler chickens experimentally infected with *Eimeria acervulina* and *Eimeria maxima*. Poult Sci 48: 194–200. https://doi.org/10.2141/jpsa.010123

Spagnoletti A, Guerrinia A, Tacchini M et al (2016) Chemical composition and bio-efficacy of essential oils from Italian aromatic plants: *Mentha suaveolens*, *Coridothymus capitatus*, *Origanum hirtum*, and *Rosmarinus officinalis*. Nat Prod Commun 11:1517–1520. https://doi.org/10.1177/1934578X1601101023

Spyridopoulou K, Fitsiou E, Bouloukosta E et al (2019) Extraction, chemical composition, and anticancer potential of *Origanum onites* L. Essent Oil Mol 24(14):2612. https://doi.org/10.3390/molecules24142612

Stefanakis MK, Touloupakis E, Anastasopoulos E et al (2013) Antibacterial activity of essential oils from plants of the genus *Origanum*. Food Control 34:539–546. https://doi.org/10.1016/j.foodcont.2013.05.024

Swarup V, Ghosh J, Ghosh S et al (2007) Antiviral and anti-inflammatory effects of rosmarinic acid in an experimental model of *Japanese encephalitis*. Antimicrob Agents Chemother 51:3367–3370

Tsai P, Tsai T, Yu C et al (2007) Evaluation of NO-suppressing activity of several Mediterranean culinary spices. Food Chem Toxicol 45:440–447

Węglarz Z, Kosakowska O, Przybył JL et al (2020) The quality of greek oregano (*O. vulgare* L. subsp. hirtum (Link) Ietswaart) and common oregano (*O. vulgare* L. subsp. vulgare) cultivated in the temperate climate of central Europe. Foods 9(11): 1671–1685 https://doi.org/10.3390/foods9111671

Xu H, Delling M, Jun J et al (2006) Oregano, thyme and clove-derived flavors and skin sensitizers activate specific TRP channels. Nat Neurosci 9:628–635

Chapter 10
Parsley: Breeding Approaches for Plant Improvement

Kalliopi I. Kadoglidou, Anastasia Boutsika, Ifigeneia Mellidou, Aliki Xanthopoulou, and Ioannis Ganopoulos

Abstract *Petroselinum crispum* (Mill.) Nyman (parsley) is among the most important vegetables belong to the Apiaceae (*Umbelliferae*) family with extensive use of its leaf, root and seeds, as herb and spice, in food processing, and in pharmaceutical industries. It is originated from the Mediterranean basin, whereas it can still be wild growing. In this review, we summarize comprehensive information concerning the origin, the distribution, the botanical classification, as well as the current state of the research on the phytochemistry, the cultivation practices and challenges and the germplasm biodiversity and conservation. We also present the traditional breeding and the molecular tools, like the molecular marker-assisted breeding, the transcriptomic analysis applied to elucidate the gene expression in parsley and a discussion with future perspectives. This review contributes as a guide reference for basic or applied research on parsley.

Keywords *Petroselinum crispum* · Essential oil · Germplasm · Traditional breeding · Molecular breeding · Trancriptomics

K. I. Kadoglidou (✉) · A. Boutsika · I. Mellidou · A. Xanthopoulou · I. Ganopoulos (✉)
Institute of Plant Breeding and Genetic Resources, Hellenic Agricultural Organization—Dimitra (ELGO-Dimitra), 570 01 Thermi, Thessaloniki, Greece
e-mail: pkadoglidou@elgo.gr

I. Ganopoulos
e-mail: iganopoulos@elgo.gr

A. Boutsika
e-mail: bouanastasia@outlook.com

I. Mellidou
e-mail: imellidou@elgo.gr

A. Xanthopoulou
e-mail: axanthopoulou@elgo.gr

J. M. Al-Khayri et al. (eds.), *Biodiversity and Genetic Improvement of Herbs and Spices*, Advances in Plant Breeding Strategies 8, https://doi.org/10.1007/978-3-031-81838-7_10

10.1 Introduction

10.1.1 Background, Chapter Objectives and Content Description

Apiaceae (*Umbelliferae*) family comprises from 434 genera and approximately 3780 species, among them several significant plants, including carrot (*Daucus carota*), dill (*Anethum graveolens*), coriander (*Coriandrum sativum*), celery (*Apium graveolens*) and parsley (*Petroselinum crispum*), primarily distributed in Northern Hemisphere (Wang et al. 2022). Among plants belong to the *Apiaceae* family is parsley, a popular herb originated on the Mediterranean basin, where it can still be wild growing. It is a vivid green biennial herbaceous plant in moderate environments, and an annual plant in subtropical latitudes. It has domesticated Europe and is extensively cultivated as a herb, and as a vegetable. The genus *Petroselinum* Hill was created with the goal of being used in the culinary, pharmaceutical, perfume, and cosmetic industries. Parsley is high in iron, vitamins A, B, and C, as well as in apiol, an oil extracted from the seeds, which is helpful to treat several health issues. Among other beneficial properties, it has been reported that parsley has antimicrobial, antianemic, menorrhagic, anticoagulant, antihyperlipidemic, antihepatotoxic, antihypertensive, hypoglycemic, antioxidative, estrogenic, and hyperuricemic characteristics, as well as hypoglycemic, antioxidative, estrogenic, and hyperuricemic. In some countries, it has been found to treat arterial hypertension, diabetes, kidney illness, and cardiovascular disease etc.

A comprehensive literature review revealed that the *Apiaceae* family is abundant in pharmaceutical secondary metabolites (Que et al. 2019a; Wang et al. 2022). Indeed, parsley, celery, carrot, (*P. crispum*), and fennel (*Foeniculum vulgare*) are, among others, *Apiaceae* vegetable crops which high carotenoids, flavonoids, vitamins, coumarin, and minerals content (Yan et al. 2014; Que et al. 2019a). Moreover, parsley and other Apiaceae members, such as caraway (*Carum carvi*) and cuminum (*Cuminum cyminum*), produce essential oils for use in food processing and as additives (Que et al. 2019a, b).

Furthermore, several uncommon species, such as *Dorema ammoniacum*, *Aegopodium podagraria*, *Peucedanum luxurians*, and *Seseli devenyense*, have been used as medicinal folk treatments in the long past (Tovchiga 2017; Widelski et al. 2018, 2021). Some common species like dill (*Anethum graveolens*), coriander, caraway (*Carum carvi*), and cumin (*Cuminum cyminum*), are produced for seasoning (Ferrie et al. 2011; Saleem et al. 2011).

However, fool's parsley (*Aethusa cynapium*) produces the toxin coniine (James et al. 2019). Generally, the toxins production of some perennial members of *Apiaceae* family is also referred in literature (like the toxin coniine from poison hemlock (*Conium maculatum*) and the toxin cicutoxin from water hemlock (*Cicuta maculata*) (Radulović et al. 2012; Hotti et al. 2015; Stonecipher et al. 2020).

Bioactive chemicals, pharmaceutical practices, omics, traditional and modern techniques for extraordinary compounds have all been researched in *Apiaceae* plants lately (Wang et al. 2022). Outlining these scientific advancements could help increase and improve the applications in Apiaceae plants.

In this review, we aim to summarize the complex origins of *P. crispum*, the botanical characterization and distribution of the plant, as well as its significance around the world, in terms of economy and health. Additionally, one section highlights the existing cultivation procedures and limitations, while the following sheds light to parsley's germplasm biodiversity and conservation. The traditional, and molecular breeding techniques, bioactive chemicals, and omics research will also be discussed. To improve *P. crispum*, numerous molecular markers, regulatory factors, and functional genes have been discovered, investigated, and utilized. As a result, recent trends in parsley application and research are briefly discussed, including the use of—omics research to discover new functional genes and metabolites. Moreover, the aspects of parsley in research and traditional/molecular breeding work are being covered in two different parts of this chapter. Breeding is severely ignored even though there is a plethora of available parsley literature. We hope that this chapter will enhance the knowledge for a better understanding of parsley and provide a reference for basic and applied research on *P. crispum* as a vegetable and a medicinal plant.

10.1.2 Classification and Distribution

Parsley is currently a member of the *Petroselinum* genus, with the name *P. crispum* (Mill.) Nyman, Consp. fl. eur. 2 [1879] 309 (nomen in syn.) ex A. W. Hill, Hand-list herb. pl. Kew ed. 3 (1925) 122 Airy-Shaw in Kew Bull. 256) (Airy Shaw 1938). *P. crispum* is an ancient crop that was well-known in ancient Greece and Rome (Agyare et al. 2017). Parsley belongs to the Apiaceae or *Umbelliferae* family and includes both leaf and root varieties. The *Apiaceae* family contains about 434 genera and nearly 3780 species, which is prevalent in sub-tropical, sub-temperate, and temperate climates (Wang et al. 2022). Figure 10.1 presented the geographical distribution of parsley in the world. It can be found in abundance in most Mediterranean and temperate areas, while it is considered to have originated in the Western Mediterranean. Species in this family are mostly herbaceous, and they're used mostly as aromatic medicinal plants and spices throughout the world.

The generic name '*Petroselinum*' comes from the Greek word 'petra', which means 'stone', and refers to the plant's preference for stony environments and the word 'selinin' means celery (Sarwar et al. 2016), while the name of the species '*crispum*' comes from the Latin crispus and means curled (Marthe 2020). In ancient antiquity, the Greek term for parsley was 'selinon'. It is also known as parsley or garden parsley in English, 'patraseli', 'patrasoli' or 'potrasoli' in Indonesia, 'phakchi-farang' in Thailand, 'vannsuy baraing' in Cambodia, 'paseri' in Japan, 'pietersielie' in Africa, 'persil' in France, and 'bagdouness' or 'maadnous' (Ipor and Oyen 1999; Quattrocchi 2012). *Apium crispum* Mill., *Apium petroselinum* L., *P. hortense* Hoffm., and *P. sativum* Hoffm. are all synonyms for *P. crispum* (Agyare et al. 2017).

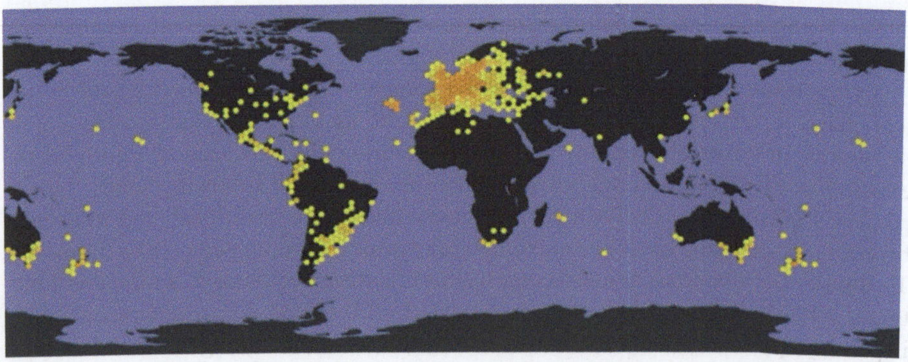

Fig. 10.1 Worldwide distribution of *P. crispum* with 9927 georeferenced records (GBIF Secretariat 2021)

In the European Union are cultivated approximately 5000 ha annually with parsley, whereas, France, Germany, Poland, and the Netherlands are major producers of dried parsley (Marthe 2020). Parsley's greens and chilled spiced food production are vital in France, Belgium, the Netherlands, the United Kingdom, Germany, and Poland. Moreover, some countries like The Netherlands, Germany, Belgium, France, the United Kingdom, and Poland all produce potted plants. Additionally, turnip-rooted parsley is grown and used in Central and Southeastern European countries such as Slovakia, Romania, and Bulgaria. These countries host a large number of gene bank accessions.

In America, Canada is a major supplier of dried parsley flakes to the United States (Marthe 2020). The latter is a substantial producer for raw and dried parsley, with production centered in California, New Jersey, Florida, and Texas, as well as, though in minor quantities, in Ohio, Massachusetts, and other states (Small 1997).

Concerning Africa, parsley is found as a relic or as an escape and it is increasingly being cultivated in East and West side. It is cultivated as a medicinal herb in Eritrea, Ethiopia, Mozambique, South Africa, Morocco, and Tunisia to treat cardiovascular disorders like hypertension (Gadi et al. 2009). It is known that Egypt's parsley cultivation is by far the most essential for international exports, including dried parsley, green bunches, and frozen spice (Marthe 2020).

According to Asia, Uzbekistan is a significant producer and exporter to other Asian countries; yet Japan, India, Malaysia, Indonesia, China, and the Philippines are all major parsley producers. In Turkey, parsley is extensively distributed and produced in gardens and fields (Yanardağ et al. 2003).

10.1.3 Botanical Description

P. crispum is mostly cultivated as an annual herb, although it grows as a biennial plant in temperate zones that flowers the year after it is sown. It makes a rosette of tripinnate leaves with several leaflets and a taproot that functions as a feed supplement over the cold season (Agyare et al. 2017). Indeed, parsley becomes a flowering plant in the second year, with sparser leaves and flat-topped diameter umbels having various yellow to yellowish-green flowers (Simon 1990).

There are three main types of parsley: the plain leaf type (ssp. *neapolitanum*, Danert), the curly leaf type (ssp. *crispum*), both are cultivated for their foliage, and the turnip-rooted or 'Hamburg' type *P. crispum* ssp. *tuberosum* primarily grown for its roots (Boutsika et al. 2021). These morphological differences of three types of parsley are presented in Fig. 10.2, whereas phenotype of some parsley cultivars and Greek landraces in Fig. 10.3.

It is a highly branched, aromatic, with silky texture herb, that can reach a height of 30–100 cm (Agyare et al. 2017). The stem is circular, slightly curved, and hollow. The leaflets are evenly oriented, dark green, shiny, flattened, or curved, and sheathed at the base. Older leaves have the biggest petiole. The pinnae have lengthy leaflets that are split into pointed parts and are elliptic to roughly linear. The higher the leaf gets, the less divided it gets, until the uppermost leaf is just a few acute segments.

In accordance with the morphology of *Apiaceae's* flowers, parsley's flower is bisexual, with five sepals and petals and an elongated disk at the bottom of the style that forms a prominent flat-topped umbel (Ajani et al. 2016). Parsley is a In addition, the cremocarp in *seeds* is split in two sections (Chomchalow 2002). The inflorescence is an apical or lateral composite umbel (Agyare et al. 2017). The 1–3 foliolate petals are only a few inches long. There are 3–15 subsidiary pedicels that range in length from 2 to 5 mm. The bisexual flowers are tiny and yellow green in color, while the sepal is not visible. The five petals are semi-spheric to egg-shaped and reach up to 1 mm × 0.5 mm in length. It is semi-marginate including one rigorous apical lobe. Five stamens are present. In parsley observed the phenomenon of protandry and

Fig. 10.2 The commonly three types of cultivated parsley: **a** the plain leaf type *P. crispum* ssp. *neapolitanum*, **b** the curly leaf type *P. crispum* ssp. *crispum*, and **c** the turnip-rooted or 'Hamburg' type *P. crispum* (ssp. *tuberosum*)

Fig. 10.3 Parsley seedlings (*P. crispum*) of different genotypes: **a** Champion Moss (curly-leafed), **b** Doble Rizado (curly-leafed), **c** Festival (flat-leafed), **d** Hermes (tournip-root), **e** HL-191/07 accession/Greek landrace (flat-leafed), **f** IS-44/07 accession/Greek landrace (flat-leafed)

for this reason anthers start the anthesis five days before the stigmata are receptive to pollen. Each carpel of the pistil has a dilated stylopodium, a style, and a round stigma, while the pistil has an inferior and 2-carpelled ovary. When ripe, the fruit is a schizocarp 2–3 mm long, ovoid, and divides into two mericarps having five short ribs each. As biennial plant, parsley is bloom in the second year, while the phenomenon of protandry occurs as parsley belong to the Apiaceae family. Figure 10.4 shows a parsley plant at blooming stage.

While younger, the roots are thin and fibroid, with a root system up to 1 m long, sometimes thicker, and with an entire arrangement of leaflets (Ipor and Oyen 1999). The seeds are oval, measuring 2–3 mm in length, with visible style remains at the apex. Following seed maturation, the plant usually dies (Huxley 1992). Seeds variability is lost quickly, consequently the sowing should be make as soon as possible after their harvest. In contrast, parsley germination usually is a slow process and

Fig. 10.4 Parsley plant (curly type) at blooming stage

seedlings appeared up to 8 weeks after the sowing since it dependent from factors such as soil temperature.

The morphological characterization in order to distinguish and to compare parsley genotypes in different growth stages are stated by UPOV (2006) in test guidelines, as well as in CPVO (2023).

10.1.4 Economic Importance

The *Apiaceae* family is perhaps the most important family inside the order Apiales, both economically and socially (Heywood 2014). The usable parts of parsley are stems, leaves and roots, utilized fresh, freeze or dried in cooking, as condiments, in confectionary, in beverages, in bakery etc. Moreover, the parsley essential oil is applied in the food industry and as a fragrance in perfumes and cosmetics (Feldheim 1999). However, the economic value of parsley's essential oil is relatively small in contrast to fresh and dried parsley (Marthe 2020). It has been estimated that the total global value of parsley essential oil from seed is about $600,000.00 (US), while the cost of "parsley herb" essential oil was $360,000.00 per year (Verlet 1993).

Except of the abovementioned uses of parsley, a plethora of studies have reported on its biological activities, i.e., antioxidant, antibacterial, antifungal, etc. Regarding its pharmaceutical importance, *P. crispum* is well-known for its therapeutic effects, including anticancer properties and the ability to manage diabetic and rheumatic joint pain. It also works to prevent osteoporosis and functions as a pain reliever with soothing properties. Furthermore, it provides relief from gastrointestinal ailments such as acid reflux, stomach disorders, swelling, and queasiness, while strengthening the immune system. It's also a nutrient-dense plant with plenty of nutrients and cell enhancers (Ajmera et al. 2019).

10.1.5 Essential oil Composition

Marthe (2020) presented a thorough review of the valuable plant secondary metabolites of parsley. Due to contained essential oil, the fresh above ground part of dill are used as flavor additive in cuisine and pastry (Aćimović and Milić 2017). Essential oil is extracted via distillation process of the parsley fruit (Punoševac et al. 2021). Even though the majority of the oil is found in the fruit, it could also be found in the leaves and roots. The color of the oil is described as pale yellow or colorless. Among other metabolites, this essential oil contains monoterpenes, phenylpropanoids, alcohols, sesquiterpenes, ketones, aldehydes, and aromatic chemicals (Teuscher et al. 2006; Farzaei et al. 2013). There have been various research comparing the essential oil content, antioxidant impact, and medicinal effect of various *P. crispum* varieties (Wong and Kitts 2006; Sabry et al. 2013; Soliman et al. 2015; Maodaa et al. 2016; Boutsika et al. 2021). Table 10.1 summarized the main constituents found either in essential oil or parsley leaf, root, and fruit tissues (Teuscher et al. 2006; Boutsika et al. 2021).

Based on the content of specific elements, parsley leaf essential oils could be classified into three groups: (1) a myristicin-rich oil (almost all the samples derived from curly leaf type); (2) a 1,3,8-p-menthatriene-rich oil (almost all the samples derived from flat leaf); and (3) an apiole-rich oil (almost all the samples derived from flat leaf). Significant quantities of phellandrene could be found in all three

Table 10.1 Chemical groups along with their main chemical constituents of parsley essential oil and plant tissue derived from leaf, root, and fruit

Chemical group	Compound	Leaf	Root	Fruit	Literature
Phenylpropanoids	Apiol	P	P	P	Farzaei et al. 2013, Craft and Setzer (2017), Jugreeta and Mahomoodally (2020), Fernandes et al. (2020) and Boutsika et al. (2021)
	Myristicin	P	P	P	
	Elemicin	P	A	P	
	1-allyl-2,3,4,5-tetramethoxybenzene	A	P	P	
	Eugenol	A	A	P	
Monoterpenes	*β*-phellandrene	P	P	P	
	p-mentha-1,3,8-triene	P	P	A	
	4-isopropenyl-1-methylbenzene	P	A	A	
	Myrcene	P	P	P	
	α-pinene	P	P	P	
	β-pinene	P	P	P	
	α-phellandrene	P	P	A	
	p-cymene	P	A	P	
	Terpinolene	P	P	A	
	α-terpinene	P	A	A	
	Myrtenyl acetate	P	A	A	
	Limonene	P	P	P	
	α, p-dimethylstyrene	A	P	A	
	Sabinene	P	A	P	
	γ-terpinene	P	A	P	
Sesquiterpenes	(E)-*β*-fernesene	P	N/A	N/A	
	Germacrene D	P	N/A	N/A	
	Angelic acid esters	P	N/A	N/A	
	α-copaene	P	N/A	N/A	
	α and *β*-cubebene	P	A	N/A	
	γ-elemene	P	N/A	A	
	α-humulene	P	A	N/A	
	γ-amorphene	P	N/A	N/A	
	α-, *γ*- and *δ*-cadinene	P	N/A	N/A	
	Germacrene A	A	P	N/A	
	β-bisabolene	P	P	N/A	
	Carotol	A	A	P	
Flavonoid	Apiin				
Furanocoumarins	Oxypeucedanin	P	P	P	
	Xanthotoxin	P	P	P	
	Bergaptene	P	P	P	

(continued)

Table 10.1 (continued)

Chemical group	Compound	Leaf	Root	Fruit	Literature
	Psoralen	P	A	P	
	Isopimpinellin	P	P	P	
	Imperatorin	P	P	P	
	Angelicin	P	P	A	
	Trioxsalen	P	P	A	

P present; *A* absent; *N/A* not available

groups (Craft and Setzer 2017). Indeed, in a thorough survey by Boutsika et al. (2021), the major compounds in leaf essential oil of twenty-four parsley commercial cultivars and Greek landraces (isolated by hydro-distillation), as determined by gas-chromatography (GC-FID), were *α*-pinene (0.6–11.8%), *β*-pinene (0.2–8.2%), myrcene (1.3–13.5%), *β*-phellandrene (9.6–39.1%), terpinolene (1.4–8.9%), 1, 3, 8-*p*-menthatriene (0.1–43.4%), *cis*-carveol (0.2–8.1%), iso-dihydro carveol acetate (0.2–6.9%), myristicin (0.51–44.4%) and apiole (0.01–35.2%).

Several parameters like climatic conditions, time of harvesting, parsley's growth of stage, and geographical region could all influence the chemical components of essential oils. For example, as a result of seasonal fluctuation, apiol levels were found to be greater in parsley cultivated in December and harvested in April than in parsley seeded in September and collected in January. Moreover, the concentrations of β-phellandrene and myristicin in flat leaf parsley decreased with age, while the levels of 1,3,8-p-menthatriene increased (Craft and Setzer 2017). The quantity of distinct elements greatly depends on the plant's location and variety (Farzaei et al. 2013).

Concerning essential oil of parsley derived roots from both *P. crispum* ssp. *foliosum* and *P. crispum* ssp. *tuberosum*, there are differences between them. Hence, the essential oil content of *P. crispum* ssp. *tuberosum* roots ranges from 0.1 to 0.3%, while the essential oil content of *P. crispum* ssp. *foliosum* roots ranges from 0.20 to 0.75% (Teuscher et al. 2006). The extraction method is determinant in the qualitative composition of essential oil. For example, when the oil derived by hydrodistillation method, the major elements are phenylpropanoids and monoterpenes; whereas, if oil is acquired via diethyl ether, the major constituents are polyacetylenes (falcarinol and falcarinidiol) (Wichtl 2002; Teuscher et al. 2006).

In the case that parsley's essential oil derived from seeds (fruits) differences were also observed. In detailed, it is referred that the oil yield extracted from of *P. crispum* ssp. *tuberosum* fruits was 1–4%, while that of *P. crispum* ssp. *foliosum* was 2–6%. Similarly, differences were recorded on the qualitative composition of the derived oil regarding the amount of phenylpropanoid: fruits of *P. crispum* ssp. *tuberosum* had the greatest quantity of apiole, whereas those of *P. crispum* ssp. *foliosum* contained mainly myrsticin (Punoševac et al. 2021).

10.1.6 Other Constituents

Parsley contains a considerable number of bioactive compounds like phenolic compounds, accompanied by polyacetylenes (Christensen 2011; Farzaei et al. 2013). Precisely, flavonoids, phenolic acids, coumarin and tannis, apiin and malonylapiin are responsible for the organoleptic characteristics of bitterness, astringency, color, flavor, and odor (Thiviya et al. 2021). Similarly, (Sousa et al. 2021) referred that the negative trait of bitterness taste of parsley was positive attributed to the existence of the chemical group of polyacetylenes in roots of these vegetable. Supplements such as vitamin A and B vitamin complex, vitamin K, tocopherols, and ergosterol are all found in parsley. Minerals (Fe, Zn, Ca, Mg, Na, K, P), fatty acids (such as linolenic and palmitic acid), carotenoids, and furanocoumarins are abundant too (Hozayen et al. 2016; Al-Yousofy et al. 2017; Nirumand et al. 2018; Fernandes et al. 2020). Table 1outlines the key furanocoumarins and flavonoids.

The main constituents of parsley leaves are flavonoids. Additional flavonoid chemicals identified exclusively in parsley leaves including glycosides like luteolin apigenin isorhamnetin 3,7-diglucoside, 7-glucoside, 6″-acetylapiin and 7-apiosylglucoside, as well as kaempferol quercetin and aglycones (Teuscher et al. 2006; Farzaei et al. 2013; Hozayen et al. 2016; Fernandes et al. 2020). In parsley liquid extracts have been found, such as naringenin, naringin, rutin, glycosides and hesperidin of luteolin and apigenin with arabinose, rhamnose, and glucose, as well as compounds, such as glycosides of isorhamnetin, kaempferol, and chrysoeriol (Farzaei et al. 2013; Hozayen et al. 2016; De Menezes et al. 2020). 0.2% of furanocoumarins are found in parsley leaves (Teuscher et al. 2006; Fernandes et al. 2020).

Vitamin C is also present in unprocessed parsley leaves in amounts ranging from 0.12–0.4% (Teuscher et al. 2006). The largest percentage of ascorbic acid is observed in the cultivated flat leaf type of *P. crispum* ssp. *foliosum,* while α- and γ- tocopherol are following. Flat leaf parsley type seems to have the highest levels of—tocopherol and the entire tocopherols. Differences in—tocopherol (the major type of vitamin E) composition could have a huge impact on this plant's antioxidant properties (Fernandes et al. 2020). Carbohydrates including apiose, that are often found in flavonoids glycosides, are as well present in parsley leaves (Farzaei et al. 2013). The liquid extracts of parsley leaf contained mannitol, glucose, mannose, arabinose, rhamnose, and sucrose (Hozayen et al. 2016). There are also carotenoids such as—carotene, neoxanthin, lutein, and violaxanthin (Farzaei et al. 2013; Fernandes et al. 2020), while polyacetylenes including falcarindiol, falcarinol, falcarinolone, and falcarinone exist (Christensen 2011).

According to a more recent study, targeted LC–MS/MS analysis was performed on extraction of aerial parts of different parsley cultivars and landraces, whereas sixteen polyphenolic compounds were quantitatively identified belonging to five main groups: coumarin derivatives, hydrocinnamic acids, flavanones, flavones and flavonols (Boutsika et al. 2021). More specifically, apigenin and luteolin derivatives were the identified flavones, and kaempferol, quercetin, galangin, and morin represented flavonols. Apiin (apigenin-7-*O*-apioglucoside) was the major flavone ranging from 1732.57 to 3676.43 mg 100 g $^{-1}$ dry weight. The concentration of apiin was in accordance with previous reports (Kaiser et al. 2013), that determined apiin content in two parsley cultivars ranging between 21 and 36 g kg^{-1} dry matter. Although (Justesen et al. 1998) showed a ratio of apigenin/luteolin of 185/1 in hydrolyzed extracts, in this research, luteolin, the 3′-hydroxylated apigenin, was only found in traces in the examined parsley extracts. Additionally, the total quantified apigenin derivatives are nearly 2000 times more numerous than the entire luteolin derivatives.

In regard to the roots, furanocoumarins (oxypeucetanin, bergaptene, xanthotoxin and others) are the primary chemical group that have been referred in literature (Teuscher et al. 2006; Fernandes et al. 2020). Phtalides including butylphtalide, (Z)-ligustilide, butylphtalide and senkyunolide are also detected in parsley root system at lesser quantities which have been shown to affect odor (Wichtl 2002; Teuscher et al. 2006). Most of the polyacetylenes (falcarinolone, falcarindiol, falcarinol, and falcarinone) have also been found in the roots (Wichtl 2002; Teuscher et al. 2006).

Apart from furanocoumarins, the major elements of parsley fruit are fatty oils, which represent the 25% (Wichtl 2002). More specifically, petroselinic acid (60–80%) is the most abundant fatty acid in the form of glycerides, while apiose was also discovered (Farzaei et al. 2013).

10.1.7 Domestication and Selection

Parsley has a lot of variety due to its long and varied history of use. It was domesticated and cultivated in many regions, whereas it is planted all over the world in the subtropical and temperate zones. Parsley is thought to have originated in the eastern Mediterranean region or Western Asia, where natural variability is greatest (Hanelt and Institute of Plant Genetics and Crop Research 2001). It also grows naturally in the western Mediterranean, where it overlaps with *P. segetum*. In Western Europe it grows wild from the British Isles to the Iberian Peninsula. North Africa likewise has a lot of variation and it provides seasonal accessions without specific cultivation requirements (Ipor and Oyen 1999).

Generally, the selection of genotypes for breeding purposes for the maintenance of an existent variety or for the development of a new one is based on several different criteria like resistance to pathogens, stress tolerance. In the parsley case, the knowledge of the existing variation of parsley genotypes based on the type of the primary use is significant.

Specifically, *P. crispum* var. *neapolitanum*, *P. crispum* var. *tuberosum*, and *P. crispum* var. *crispum* are the three primary variations of *P. crispum*, though there are also other cultivated varieties. *P. crispum* var. *crispum* is curled-leaf, while *P. crispum* var. *neapolitanum* is flat-leafed (Italian) parsley. Flat-leafed parsley is often more difficult to grow than curled-leaf parsley. *P. crispum* var. *tuberosum* is cultivated for its root. It's also referred as 'turnip-root parsley' or 'Hamburg parsley.' Root parsley is produced as a root vegetable, whilst curly and plain leaf types are produced for their leaves (Ticker and DeBaggio 2009). This kind of parsley creates significantly larger roots, in contrast to the parsley, which is being grown for its leaves, up to six times the size (Hanrahan and Frey 2005). A large number of cultivars include both the curled-leaf and flat-leaf types.

There are also, two types of convarieties (Marthe 2020). The convar. *radicosum* (Alef.) Danert in Mansf. Verzeichnis (1959) 322, having swelling taproots that can be eaten and convar. *crispum,* in which the roots cannot be consumed. The leaves of convar. *radicosum* are not crispy throughout populations, belonging to var. *tuberosum* (Bernh.) Crov. Gene banks owned germplasms of var. *erfurtense* Danert that were not crisp. Convar. *radicosum*'s parsley roots are used as an eaten vegetable. There are various leaf types among convar. *crispum*: germplasms with crispy leaves are ascribed to var. *crispum,* whereas germplasms with non-crispy leaves are referred to var. *vulgare* (Nois.) Danert and var. *neapolitanum.* With thick and longer petioles, Danert is also known as Neapolitan or Italian parsley. Some varieties are bred for these edible leaf stalks in the Mediterranean region (Small 1997). The wild type of parsley, convar. *crispum* var. *silvestre* (Alef.) Danert, was likewise unavailable from gene banks (Danert 1958).

10.2 Current Cultivation Practices and Challenges

10.2.1 Current Cultivation Practices

Parsley is a type of plant that can be grown in fields, greenhouses, plastic tunnels (Marthe 2020), and it can be grown as a potted plant also. It is referred that parsley can be successfully grown in floating and hydroponic system (Martins et al. 2019). *P. crispum* could also sprout semi-perennial in greenhouses or relatively warm regions. Seasonal accessions are also provided even without vernalization.

The parsley's growth peculiarities and production depend on the cultivated variety and of the growing method. Generally, sandy and sandy-clay soil types with a moderate to high level of compost from 1 to 5% and a pH of 6–7 are ideal for the growth of parsley (Dachler and Pelzmann 1999).

Parsley should be planted at the end of winter. That is from the beginning of March until the end of April through Central Europe, although less often, on moist, light humus soil it takes place in autumn (Punoševac et al. 2021). It is noteworthy that, due to auto-incompatibility, it cannot be sown for more than one season in the same location (Teuscher et al. 2006). The optimal temperature for germination is between 12 and 15 °C. As a member of *Apiaceae* family, parsley's seeds have an erratic and rather low germination rate, even under optimum conditions. Germination can take up to 26 days if kept at a cold temperature. According Marthe (2020), seeds pretreatment with 18 h soaking in warm water of 20 ± 2 °C and redrying at maximum

30 °C could enhance the germination. Moreover, a possible seed dormancy can be broken with briefly freeze at around 5 °C.

During primary growth stage, a rosette of leaves is formed after 10–12 weeks of cultivation, and the very first harvest follows. Within the species, the length of the tripinnate leaves with many leaflets differ, from approximately 8–25 cm. These of var. *crispum* are smaller, while those of var. *neapolitanum* are longer. Harvesting of the freshly grown leaflets might occur every month, depending on climatic conditions. Plants of the "leaf type" should attain a height of 20–25 cm before being harvested, while the cut must be 4–6 cm just above the surface (Hoppe et al. 2013).

In the second year and since parsley is a cool season plant, it blooms in temperatures within 7 and 16 °C, whereas the growth rate is significantly slow below the minimum limit (Small 1997).

Parsley leaves can withstand temperatures as low as 6–8 °C (Marthe 2020). Roots can tolerate subzero temperatures as − 10 °C during the wintertime. On the other hand, repeated freezing temperatures, may result in full loss due to decomposition. It may be essential to remove the root ends of the growing season and transplant them in the springtime for seed production in the following year.

A parsley-based crop rotation process should ensure a growing intermission of at least 5 years, preferably up to 7 years, between parsley and other plants in *Apiaceae* family (Marthe 2020). Optimal former crops include weed-deterrent species.

10.2.2 Current Agricultural Challenges

Vegetables are vital for human nourishment, and their role is growing all over the globe (Dursun et al. 2010). Installation of the crop is one of the most difficult challenges in crop production throughout the world, so producers and scientists understand its value (Murungu et al. 2004). Poor seedbed treatment, low-quality seeds, delayed seed sowing, and extreme temperatures following sowing, are all barriers to improved crop establishment (Harris 1996; Murungu et al. 2004).

In nurseries, parsley seed germination and appearance are crucial during the growth season, particularly in the early springtime, summer, and early winter. Parsley is among the most significant crops for which seed germination requires a lot of time and is hard to accomplish, especially in adverse weather conditions (Vural et al. 2000). In such circumstances, the germination rate of parsley seeds is less than 55–75% (Günay 2005). As a result of the sluggish appearance and reduced emergence rate, the seeds tend to be smaller (Cantliffe 2003). To maximize yield, durability, and profitability in agriculture, consistency and efficiency of seed germination and appearance are required (Cantliffe 2003; Duman 2005).

Seed priming strategies for several agricultural crops have been utilized to enhance crop stands and yields (Lee 2004). Seed priming is also referred to be an efficient strategy to reduce the amount of time to early crops and provide adaptability in managing plant schedules. In vegetable nurseries and consequently in parsley ones, are commonly applied techniques with priming agents in order to improve the seedlings production. Particularly, parsley seeds treat with water (hydropriming), polyethylene glycol (PEG, hosmopriming) or gibberellic acid (hormonal priming). Seed priming seems to be a physiologically crop improvement technique that involves the regulated moisture of seeds prior to sowing and consequently improve the uniformity of seed emergence and seed quality. da Silva et al. 2022 found that the best priming technique for parsley seeds was hydropriming under 24 h of water imbibition, followed by the osmopriming under 116 h of PEG imbibition at – 2.0 MPa. With the controlled hydration of seeds, pregerminated metabolic events occur, but the radicle protrusion through the seed coat cannot be carried out (Nascimento and de Aragão 2004). Briefly, priming agents allow a quick seed hydration at phase I of germination and accelerate the reserve substances' digestion at phase II (Chen et al. 2010). This approach has been utilized to improve seed vitality, germination speed, total germination rate, and seed homogeneity in several vegetable seeds, primarily under abiotic stress. In summary, priming techniques have the potential to improve crops and yield (Black and Bewley 2000; Duman 2005).

10.3 Germplasm Biodiversity and Conservation

10.3.1 Germplasm Diversity

P. crispum has 2n = 2x = 22 chromosomes (Darlington and Wylie 1955). By using flow cytometry, the overall degree of ploidy of 50 parsley accessions, which were collected from the national ex situ repository of agricultural and horticultural crops at the Leibniz Institute of Plant Genetics and Crop Plant Research in Gatersleben in Germany, revealed the diploid status (2n = 2x = 22), while the quantity of DNA was found to be 2C = 4.0 pg (Bennett and Leitch 1995).

In comparison to many other aromatic and pharmaceutical plants, parsley's biodiversity is well preserved in ex situ repositories around the globe (Marthe 2020). In a report of a network coordinating group of vegetables of the European Umbellifer database, a number of 314 parsley accessions is referred (Astley and Kotlińska 2001). Many germplasms and historic varieties can be found in these repositories, several of which significantly being used in various parts of the world. Recently, Marthe (2020), based on (EURISCO 2019) database, stated that '*the biggest parsley sets exist in the following countries: Spain, 351 accessions; Germany, Federal* ex situ *Gene Bank at Leibniz Institute of Plant Genetics and Crop Plant Research (IPK), 242 accessions; Poland, 229 accessions; Hungary, 163 accessions; Russian Federation, N.I. Vavilov All-Russian Research Institute of Plant Industry (VIR), 66 accessions; Ukraine, 163 accessions; Portugal, 113 accessions; Bulgaria, 82 accessions; United Kingdom, 67 accessions; Romania, 57 accessions; and Czech Republic, 40 accessions*'. Moreover, the U.S. national plant germplasm system of USDA presents 248 entries as parsley accessions (USDA 2019).

10.3.2 Phylogeny

Parsley species present an enormous variability, which could be partially attributed to the differentiation of its origin and to the high phylogenetically distance as well.

Numerous studies have identified taxonomic delimitation in the family *Apiaceae* using traditional criteria like morphology, particularly inflorescence pattern, seed, and fruit features, combined with molecular and biochemical markers. Previously, the notion of dividing the family into three subfamilies (*Hydrocotyloideae*, *Saniculoideae*, and *Apioideae*) and 12 tribes (Drude, 1897). *Apioideae* is the biggest and most significant economically of the three subfamilies. At many taxonomic levels, the family *Apiaceae* has undergone significant systematic reorganization. A heterogeneous set of morphologically different data was produced by research on the subfamily *Apioideae*, particularly tribe *Apieae* (Jiménez-Mejías and Vargas 2015). Several morphological traits confirmed the interconnections between subfamilies as shown by molecular data, although none of these morphological traits indicated the monophyly of subfamilies *Saniculoideae* or *Apioideae* (Calviño et al. 2008).

Many researchers have looked at the phylogeny of the *Apiaceae* subfamily *Apioideae* (Zhou et al. 2008; Winter et al. 2008; Tilney et al. 2009; Spalik et al. 2009; Downie et al. 2010a; Magee et al. 2010; Valiejo-Roman et al. 2012). Major clades' compositions as determined by phylogenetic studies using molecular data were highly inconsistent. It was also inconclusive to make a taxonomic distinction based on morphological characteristics. A modified categorization that represented phylogeny was published in 2001, in which earlier opinions were presented on the molecular systematics of the *Apioideae* (Downie et al. 2001). Intron (rpl 16, rps 16, rpo C1) sequences, cpDNA restriction sites, nuclear ribosomal DNA (nrDNA), and internal transcribed spacer sequences were used to identify the consistency among various phylogenetic analyses. Further groupings and tribes were discovered through further research (Downie et al. 2010b), and some of these clusters are now confirmed both by ITS and cpDNA analyses. Inside the subfamily *Apioideae*, the study has discovered 41 significant clades, 21 of which have previously received tribe or subtribal recognition. It has also shed light on the natural divisions that exist within the subfamily as a whole. (Sun and Downie 2010) used maximum parsimony and a Bayesian technique to analyze the morphological and molecular data sets from 123 endemic North American *Apioideae* species. After removing weak branches, phylogenetic trees based on morphological traits were consistent with those based on genomic information. There were found to be three main clades and a number of distinct subclades. A phylogenetic analysis of 100 species of the *Apiaceae* subfamily *Apioideae*, mostly found in China, was also conducted (Zhou et al. 2008). Some genera were found to be monophyletic, although the majority of them were not, showing wide generic borders and heterogeneous diversity.

The Internal Transcribed Spacer (ITS) region on nuclear ribosomal DNA is perhaps the most widely applied molecular marker in *Apioideae* (Baldwin et al. 1995). Downie and Katz-Downie (1996). combined into a single clade in the maximum likelihood tree the *A. graveolens*, *F. vulgare*, *P. crispum*, and *A. graveolens* (Celery), although *Trachyspermum ammi* (Ajwain) and *D. carota* were grouped in two different clades. Besides the identification of new main clades to account for expanded sampling, no significant alterations to the pre-existing phylogenetic categorization emerged. Diverse clades continue to exhibit inconsistent grouping patterns. The following conclusions are drawn from phylogenies derived using the parsimony, neighbor-joining, and maximum likelihood methods: *Apiaceae* subfamily *Apioideae* is monophyletic and is a synonym group to subfamily *Saniculoideae*; *Apiaceae* subfamily *Hydrocotyloideae* is not monophyletic; some members are vigorously aligned to *Araliaceae*, while some to *Apioideae* and *Saniculo* (Downie and Katz-Downie 1996).

Even though there was an attempt to determine the compatibility between them (Downie et al. 2010b), the hypothesized molecular systematics of the *Apiaceae* family by diverse scientists is erratically diverging from one another. Only a small number of genera, including *Pimpinella*, *Trachyspermum*, *Angelica*, etc., were divided into two or more main groupings. Only in a few instances did phylogenetic trees built from morphological characteristics, while those constructed from DNA data agree. In particular, the morphology of seedlings and the motifs of leaf margin insertion could be used to resolve taxonomic disagreements and ambiguities in determining phylogeny in the *Apiaceae* family. A recent study analyzed some carefully chosen *Apiaceae* members that varied in mature leaf and cotyledonary leaf morphology to determine the evolutionary lineages (Das 2017). In terms of the classification of taxa, the evolutionary groupings demonstrated a major divergence from the existing knowledge.

Saniculoideae frequently have simple leaves, either whole or with various levels of marginal incertation. All *Saniculeae* members have simple leaves, while *Steganotaenieae* members have compound leaves. In the tribe *Saniculeae* and subfamily *Saniculoideae*, simple leaves are plesiomorphic, while compound leaves are synapomorphic in the tribe *Steganotaenieae*. Subfamily *Apioideae* also has compound leaves. Subfamily *Apioideae* also has compound leaves. Nonpalmate leaves were a plesiomorphic trait shared by the progenitors of the tribes *Steganotaenieae* and *Saniculeae* as well as the subfamily *Saniculoideae*, which is the ancestor of the *Apioideae* (Shan and Constance 1951).

Simple intact leaves are thought to be the main formation with evolution progressing toward a steady growth in marginal insertion, resulting to pinnately divided, pinnatipartite, and decomposed leaves with small linear sections. Similar to this, longitudinal cotyledonary leaflets have evolved as a result of progressive shortening from ovate-lanceolate cotyledons, which may be thought of as the ancestral state. It makes sense to assume that the long coriander (*Eryngium foetidum*), subfamily *Saniculoideae*, indicates the ancestral type and that other species with pinnatifid, pinnatipartite to pinnatisect-decompound leaves with narrow linear segments may have derived from it. Coriander has undergone two distinct evolutionary tendencies, one of which produced Celery, Ajwain, and Radhuni (*Trachyspermum roxburghianum*), and the other of which produced Parsley, Carrot, Cumin, Dill, and Fennel.

The generated potential trend appears to be supported by the dendrogram constructed from morphological features. Long Coriander distinguished itself significantly from the others. Dill, Fennel, and Cumin were placed in a different group from Coriander, Celery, Parsley, Radhuni, and Ajwain. The clades corresponding to tribes or subtribes in molecular phylogeny (Downie et al. 2010b) display a diverse assembly of species with markedly divergent morphologies. The tribe *Apieae* includes genera like *Anethum, Foeniculum, Apium,* and *Petroselinum.* These genera differentiate from one another in terms of seedling and leaf edge insertion. Although cumin and carrot are members of the group that composes the subtribe *Daucinae*, and they are divided into two different groups in a following study (Das 2017).

10.3.3 Genetic Resources Conservation Approaches

A critical part of biodiversity is agrobiodiversity (Maxim et al. 2009). Plant genetic resources have the total hereditary material of species. The preservation of genetic diversity, which is expressed in local varieties, is necessary for the long-term viability of current agriculture (Chable 2005) and for the. These varieties often have distinct genes responsible for pests, diseases and abiotic stress resistance. Research focus on investigation of novel genetic combinations, some of which may confer in preventing future global challenges such as climatic change, parasite migration to the north and so on (Moldovan 2003). Meanwhile, due to their adaptability, these cultivars could be widely used throughout ecological crop culture systems.

There is no doubt that crop conservation 'in situ' is the most effective conservation strategy. This is because 'in situ' conservation permits the varieties to keep evolving at their point of origin, in the land, whilst seeds in gene banks (ex situ preservation) are not subjected to further natural selection or could experience harmful mutations (Hammer et al. 2003). Several regulations and protocols have been developed at the European and International level aiming to conserve biodiversity. Two of the most significant laws are, the Rio de Janeiro Convention on Biodiversity (1992) (http://orcp.hustoj.com/convention-on-biological-diversity-1992/), the 98/95 EEC Directive, and the International Convention on Genetic Resources of Plants for Alimentation and Agriculture (2001) (Zaharia and Kastler 2003; Maxim 2008).

Despite that preservation of genetic variability at crop plants is an essential precondition for sustainable agriculture, few studies have been performed according to parsley's biodiversity for the conservation of varieties (Maxim et al. 2009).

10.4 Traditional Breeding Methodology and Limitations

Two characteristics caused difficulties in parsley breeding currently are the availability of only open-pollinated crossbred genetic material in parsley cultivation and the lack of male sterility.

Several quantity and quality traits are considerable in parsley breeding. The International Union for the Protection of New Varieties of Plants (UPOV) offers precise descriptions for several of these traits under the UPOV code: PETRO CRI (UPOV 2005).

For disease resistance and stress tolerance, there are a lot of breeding goals. Research focus on the evaluation of parsley species for existence of pest resistance, especially in economically important pathogens and the possibility for transfer of this advantage in new cultivated varieties. This approach is also ecological and in agreement with the principles of sustainable agriculture with a cultivation system of reduced inputs management. Moreover, the current used pest control formulations, probably will not be available in the years ahead. In Sect. 4.4 of the present chapter, parsley breeding concerning disease resistance is presented.

The self-life time and the maintenance of green color of potted parsley are also included among the studied critical factors in breeding programs, since plant is used both as a spice and vegetable. The study of secondary metabolites that serve as antibiotics is another aim of parsley breeding, since different bacteria live on and inside its leaves.

Other significant breeding objectives are the compounds relating with taste and aroma. Regarding the resistant activity of some sensory beneficial compounds this breeding goal improvement should be attempted. The flavor and bioactive secondary metabolites of parsley are both exceptional and distinctive (Cazzola et al. 2011). Previous research has examined the sensory characteristics and volatile components of freshly collected and dried parsley, as well as of its essential oil (Masanetz and Grosch 1998). A total of about 80 volatiles have been found, and 17 odorants out of this smaller group exhibit a comparatively significant olfactory impact. P-mentha-1,3,8-triene, myrcene, 2-s-butyl-3-methoxypyrazine, myristicin, linalool, (Z)-6-decenal, and (Z)-3-hexenal are the major compounds responsible for the flavor of freshly collected parsley leaves but, after the chopping of parsley leaves, aroma changed rapidly (Masanetz and Grosch 1998). A technique for the assessment of sensory aspects and their variation by a screen of participants was established in 2007 (Hoberg et al. 2007). Previously, sensory characteristics in relation to the volatile profile and resistance to *Septoria petroselini* were investigated in parsley genotypes concluding that the disease resistant genotypes have several negative sensory characteristics such as bitter, grassy, herbaceous, pungent, chemical, and harsh and indicating that there is a strong correlation between disease resistance and volatile compounds (Ulrich et al. 2011). The aforementioned sensory traits across with the metabolites' determination could be an improvement tool for parsley breeding, useful for the evaluation of the preserved genetic material in gene banks. Nevertheless, there is currently insufficient knowledge about the interactions between fragrance chemicals and how they operate in resistance mechanisms.

The set of breeding objectives generated by pharmacopoeia criteria is not particularly interesting in terms of the practical use of the European Pharmacopoeia. Apart from the broad specifications of the monograph herbal medications, there is no monography for parsley in EDQM (2023).

10.4.1 Conventional Breeding

Landrace selections were used to initiate traditional parsley breeding (Marthe 2020). For instance, in the nineteenth century, the curled parsley 'Mooskrause' was subdivided into three main varieties: 'Mooskrause 1', 'Mooskrause 2', and 'Mooskrause 3'.

Several breeders-initiated projects to preserve the variety intact. Consequently, the selections 'Mooskrause 2' (Germany), 'Moskrul 2' (Netherlands), 'Moss Curled 2' (Great Britain), 'Nain Frisé mousse' (France), and 'Nano ricciuto 2' (Italy) varied (Italy). Numerous additional variations were created, by selection from this "umbrella" variety.

Further varieties include 'Amsderdamse Snij' (flat-leaf parsley), 'Einfache Schnitt 3 Typ Hamburger Schnitt' (turnip-rooted parsley), and 'Halblange–Berliner' (turnip-rooted parsley).

From the early start of the twentieth century, interspecific crosses between parsley and celery (*Apium graveolens* L.) have successfully been performed (Becker-Dillingen 1926; Becker 1943, 1962). Only celery as the female was effective within those sexual crossings. Following, reciprocal crossings had been performed too (Madjarova and Bubarova 1978; Honma and Lacy 1980). Bruznican et al. 2020 mentioned that repeated successive backcrossing enhances the resistance to Septoria disease. The goal of these crossings was to generate increased yields progeny, but the main goal was to transmit Septoria blight resistance into the progeny. *S. petroselini* has been resistant to celery, and *S. apiicola* to parsley. It has been found that some of the generated bastards were more resistant to *S. apiicola*, however the resistance was polygenetic and recessive (Honma and Lacy 1980). Developing resistant cultivars was impossible, due to the unstable resistances (Ochoa and Quiros 1989). The Bulgarian variety 'Festival 68' was created from a cross between parsley and celery (Madjarova and Bubarova 1978), however it is intolerant to *S. petroselini*. Interspecific crossings of parsley with celery with reciprocal crossings led to the production of intermediate phenotypes (Madjarova and Bubarova 1978).

10.4.2 Hybrid Breeding

Despite the huge variation of parsley genotypes originated globally and the consequent large phyllogentic distance, hybrid breeding is not possible. This is due to the lack of male sterility factors and to knowledge on the severity of the hybrid effect (Marthe 2020). Indeed, the male sterility is a necessary condition for hybrid variety breeding in members of *Apiaceae* family. It is well knowing that male sterility is not seldom in *Apiaceae* family, although in case of parsley no source for male sterility or cytoplasmatic male sterility was found (Marthe 2020).

10.4.3 Disease Resistance Breeding

For breeding projects, information about the biology of the pathogens is necessary. Moreover, the screening of accessions for diseases and the evaluation data for sources of resistance are essential.

To prevent and manage the incidence and spreading of plant diseases, three major strategies are usually used: (1) selecting disease-resistant cultivars, (2) enhancing crop control, and (3) applying fungicides (Wang et al. 2022). The most efficient approach of minimizing prevalence of the disease in crops, is the select disease-resistant germplasm resources (Nelson et al. 2017).

Literature concerning parsley diseases and the relative breeding for resistance varieties is limited. Powdery mildew (*Blumeria graminis* f. sp. hordei) affects *Apiaceae* vegetable crops and could prevent severe damage to leaves and petioles. A study regarding carrot and parsley plants in Washington of the United States was the very first record of powdery mildew in *Apiaceae* crops (Glawe et al. 2018). Particularly in parsley, the pathogen agent caused powedery mildew is the fungus *Erysiphe heraclei.* Marx and Gärber 2014 referred that twenty parsley varieties showed differences in their susceptibility against this fungus, whereas varieties Felicia and Grüne Perle showed a comparable susceptibility against fungal which were isolated with different origin but have the same aggressiveness. Generally, factors such as climate, infectiousness and parsley physiological stage affect this disease. Precisely, parsley plants were most sensitive and susceptible at four leaves stage than at cotyledon stage.

Furthermore, *Alternaria radicina*, which is a seed-borne fungus, can reduce the quality of the seed (Scott and Wenham 2012). Sclerotinia disease, could severely affect *Apiaceae* crops in storage (Saharan and Mehta 2008; Jensen et al. 2008). Although this disease can be found in celery and parsley, it has a minor influence on these plants.

Among 127 parsley accessions that were studied for their resistance against *Alternaria radicina, Erysiphe heraclei, Fusarium oxysporum,* and celery mosaic virus, parsley accessions of vars. *crispum* and *tuberosum* were weakly infected or immuned than those of vars. *vulgare* and *neapolitanicum* (Marthe et al. 2003). Moreover, sixteen of the var. *tuberosum* accessions were highly susceptible in *Erysiphe heraclei*, whereas a great variability was observed concerning the resistance against viruses: vars. *crispum, neapolitanicum* and *tuberosum* presented a high resistance and remained uninfected, although var. *vulgare* was sensitive with an infection rate about 15%. Parsley of var. *crispum* presented also the highest infection of *Fusarium oxysporum.* Marthe et al. (2013) stated that the center for disease resistance of parsley is probably Southern and Eastern Europe, since in their survey found that, among 220 accessions, most of the resistant parsley genotypes against *S. petroselini, Plasmopara petroselini* and *Erysiphe heraclei* originated from this region.

10.4.4 Insect Pest Resistance Breeding

Several *Apiaceae* plants—among them parsley—are affected by aphids (*Aphidoidea*), armyworms (*Mythimna separata*), and cutworms (*Agrotis* spp.) (Webb 2017). Leaf discoloration, necrotic spots, and stunted growth are symptoms caused by aphids, and could be effectively controlled by using resistant varieties and pesticides (Vanlerberghe-Masutti and Guillemau 2007). *Bacillus thuringiensis* has been effectively applied to prevent armyworms from spreading (Sanchis and Bourguet 2008), while field control is the most common method for preventing and controlling cutworm infestations. Previously, Webb (2005) referred that parsley and celery shared to a great degree the same insect pests, whereas among the most common insects attack these crops were American serpentine leafminer (*Liriomiza trifolii*), beet armyworm (*Feltia subterranean*), granulate cutworm (*Feltia subterranea*), black cutworm (*Agrotis ipsilon*) cabbage looper (*Trichoplusia ni*), click beetles (*Elateridae*), aphids and twospotted spider mite (*Tetranychus urticae*).

A small fly that attacks various *Apiaceae* plants, which include celery, parsnip, parsley, carrot, is the carrot fly (*Chamaepsila rosae*) (Wang et al. 2022). The roots of the plant that are being attacked by this fly are damaged, resulting in the death of the crops. The main cause, damaged roots are hard to store, is bacterial infections. In the case of carrot fly, breeding resistant cultivars has been shown to be an efficient approach of reducing the consequences of its infestations (Ellis and Hardman 1981; Guerin and Ryan 1984). Another vegetable pest that attacks parsley is the carrot weevil (*Listronotus oregonensis*), that also attacks carrots and celery (Graßmann 2005). This insect has a substantial impact on the yield and is difficult to handle, since no resistant variants are currently available (Boivin 1999).

10.4.5 Late-Bolting Breeding

Although parsley–being a biennial–flowers in the second year, it is possible in adverse conditions (e.g. high temperatures in summer) to bloom from its first year. This undesirable phenomenon is called 'early bolting' and reduces the value and production of parsley and other *Apiaceae* crops like carrot and celery (Yang et al. 2007; Ramin and Atherton 2015; Wang et al. 2022). Apium species' requirement for a cooling period is influenced both by habitats and genetics (Gratwick 1992). To the best of our knowledge, there are no studies dealing with parsley late-bolting breeding. However, a limited number of relevant studies have been published regarding celery and carrot, concerning the production of new slow-bolting cultivars by using traditional breeding methods, such as single selection and crossing (Wolf et al. 1993; WuTai et al. 2012).

10.5 Molecular Tools

10.5.1 Molecular Marker-Assisted Breeding

Breeding programs should consider genetic variation within species (Simmonds and Smartt 2000). Molecular markers are frequently designed to choose a variety of superior genotypes which could be applied in breeding programs (Said-Al Ahl et al. 2016). Furthermore, genetic markers are an effective tool for identifying genetic variation in natural communities and germplasm repositories (Sigrist et al. 2010).

However, molecular genetic research on parsley that determine the genetic connection between various parsley genotypes are limited (Maxim et al. 2009; Lohwasser 2010; Domblides et al. 2010; Ibrahim et al. 2017; Boutsika et al. 2021), despite the value and relevance of it as a medicinal herb and its applications. Nowadays, the exploitation of molecular breeding could be a great opportunity in minor crops such as parsley, due to the development of methods such as next-generation sequencing and the following reduction of its costs.

DNA markers provide a wide spectrum of applications in breeding programs, (1) including marker assisted analysis of breeding components, such as evaluating genetic diversity, parental selection, cultivar authenticity, and cultivar integrity (Winter and Kahl 1995), investigation of heterosis, and characterization of locations in the genome under selection, (2) marker assisted backcrossing, and (3) marker assisted pyramiding (Winter and Kahl 1995; Collard and Mackill 2008). Several of the constraints of morphological and biochemical approaches are eliminated by molecular marker methods, which can identify polymorphism at the DNA level (Verma et al. 2015). In addition, genetic markers based on DNA variations, such as random amplified polymorphic DNA (RAPD), amplified fragment length polymorphism (AFLP), and microsatellites or simple sequence repeat (SSR), have been generated and have become a frequent method applied for germplasm characterization, in particularly in *Apiaceae* members breeding like parsley, celery, carrot etc. (Yang and Quiros 1993; Mustafa and Abou-Ellail 2013; Liu et al. 2017; Wang et al. 2022).

The most used molecular marker in *Apioideae* subfamily and many other angiosperms has been the Internal Transcribed Spacer (ITS) region on nuclear ribosomal DNA (Baldwin et al. 1995). Downie et al. (1998) mentioned that maximum likelihood tree obtained from ITS 1 and ITS 2 sequences from *Apiaceae* crops showed that parsley, celery, dill and fennel included in a single clade, whereas carrot in two separate clades.

Recently, parsley genotypes were studied, in order to examine the genetic diversity between fifteen local landraces (flat-leafed), obtained from the Greek Gene Bank, and nine commercial cultivars (flat-leafed, curly-leafed, tournip-root) (Boutsika et al. 2021). Using molecular markers such as AFLP and ISSR, the genetic diversity of all landraces and commercial cultivars was examined. Partcularly, in the landrace population 63.83% of the loci were polymorphic, compared to 100% polymorphism for the cultivars, while landraces had significantly higher values of unbiased haploid gene diversity than the modern cultivars. Hence, the Greek germplasm collection may be a valuable source of genetic material with significant range in its genetic profiles.

Previously (Domblides et al. 2010) mentioned that the combined application of RAPD and ISSR markers are ideal for the detection of genetic variation in parsley. Specifically, they referred considerable phylogenetic distances among 39 parsley genotypes (populations and accessions of Russian gene bank) by using of RAPD and ISSR markers. The same authors concluded that ISSR analysis provides a sufficient number of microsatellite repeats at parsley phylogenetic studies.

10.5.2 Transcriptomics

Gene expression analysis, gene function identification, and development of molecular markers, all depend greatly on transcriptome data (Xie et al. 2013; Verma et al. 2015; Wang et al. 2022). Even though the *Apiaceae* family is composed of several members, only few *Apiaceae* vegetable crops have been subjected to transcriptome research (Iorizzo et al. 2011, 2016; Xu et al. 2014; Feng et al. 2018; Ma et al. 2018; Lu and Li 2019; Que et al. 2019b; Song et al. 2020; Liu et al. 2021).

To begin with, crops generally, are subjected to different stressors throughout their development, due to worldwide issues such as global warming, ecological degradation, and hazardous waste (Li et al. 2014). Drought, salt stress, and harsh temperatures are the primary constraints to increased vegetative growth and yield. According to these abiotic stressors, several genes in plant species are expressed (Holmberg and Bülow 1998). Genes typically deal with stress in one of two ways: directly or through regulating the expression of certain genes (Hasegawa et al. 2000; Figueiredo et al. 2012). Stress tolerance in plants is linked to several transcription factors including AP2/ERF, NAC, bZIP, and WRKY (Hsieh et al. 2010; Rushton et al. 2010; Zhuang et al. 2010). These transcription factors play a role in controlling gene expression and constructing intricate gene regulatory networks (Shinozaki et al. 2003; Chen and Zhu 2004; Shinozaki and Yamaguchi-Shinozaki 2007). In the case of parsley, transcriptome sequencing analysis used and identified the relation between stress and 88 AP2/ERF transcription factors (Li et al. 2014). AP2/ERF factors had important functions in development, hormonal regulation and abiotic responses in salt, drought and extreme temperature) (Li et al. 2014). The same researchers concluded that stress tolerance was controlled by multiple genes, and further studies are required to identify the complex regulatory networks of these AP2/ERF genes in parsley. Later, Li et al. (2016) found that, among eight candidate genes in parsley, EF-1α and TUB were the most stable genes for heat, cold, salt, and drought stresses, whereas EF-1α, GAPDH, and TUB were suitable genes for normalizing under GA, SA, MeJA, and ABA hormone stimulation procedures.

Essential oil content (Zhang et al. 2006; Petropoulos et al. 2008) and flavonoid compounds (Boldizsár et al. 2013) have been the priority of parsley studies, however knowledge on parsley's ability to withstand abiotic stress, molecular biology, and gene function is limited. The lack of genome sequenced *Apiaceae* species has narrowed scientific work. Precisely three transcriptome sequences have been acquired from celery (Fu et al. 2013; Yan et al. 2014) and carrot (Iorizzo et al. 2011) inside the *Apiaceae* family. Parsley can hardly be studied any further, due to the lack of available resources (Li et al. 2014).

In the last decades, the yeast one-hybrid method was applied. Two WRKY transcription factors from parsley were identified (Cormack et al. 2002), while three bZIP genes was cloned as well (Weisshaar et al. 1991), presenting that these genes are implicated as parsley response to abiotic and biotic stress, such as environmental changes and disease invasion, respectively. On the other hand, nearly no AP2/ERF members have been discovered on this vegetable. Firstly, Rushton et al. (1996) studied on cultured parsley cell the effect of a defined pathogen-derived elicitor (Pep 25) on WRKY1, WRKY3 mRNA levels and they concluded that WRKY 1, 2 and 3 proteins are transcriptors factors with significant role as a signalling in transduction pathway and in the gene activation as regulation of defense mechanism. Indeed, they showed that an extremely rapid and transient increase in WRKY1 and -3 mRNA levels occurred after a specifically fungal pathogen-derived elicitor (Pep25) in cultured parsley cells.

10.6 Conclusion, Prospects and Recommendations for Future Research

Parsley has a considerable economic impact due to its broad utilization as a spice, leafy or root vegetable for human consumption, its biological activity, and its association with food and pharmaceutical industries. All these perspectives render parsley an interesting agriculture product item for further investigation. Until nowadays, notable research has already been carried out worldwide regarding diverse sections related to the improvement of its cultivation.

In the current chapter, we provide a comprehensive presentation of various aspects, such as classification, distribution, botanical description, economic importance, applications, bioactive compounds, cultivation practices, germplasm biodiversity, as well as a thorough review of the research and development in traditional and molecular parsley breeding.

Two disadvantages concerning the sexual reproduction of the available genetic material, which become source of difficulties for parsley breeding, are the open-pollination and the lack male sterility. Due to those characteristics, hybridization is not possible, despite the huge variation of parsley genotypes worldwide.

For most crops and in the case of parsley also, yield components (at leaf, root, or seed level) are included among the major objectives for their breeding. However, in parsley breeding other principal objectives are the improvement of essential oil yield, of flavonol content, of flavor and aroma and of the resistance against diseases such as septoria, powdery mildew and alternaria, or against pests. The self-lifetime and the maintenance of green color are also included among the critical breeding objectives.

Regarding the traditional breeding in the past, interspecific crosses between parsley and celery, with celery only as female have been successfully realized, whereas the method of repeated successive backcrossing generate progeny with enhanced yield and resistance to Septoria disease. Despite the fact that the molecular genetic markers are a valuable tool for the identification of genetic variation of parsley germplasm collections, they have been used to a limited extent, thus far. Indeed, the Internal Transcribed Spacer (ITS) region on nuclear ribosomal DNA is a frequently used molecular marker, whilst random RAPD, AFLP, and SSR, have been developed and applied for members of Apiaceae family and for parsley germplasm exploration, particularly.

Based on the so far gain knowledge concerning the cultivation and breeding of parsley, as well as the current impetus in genomic tools which could facilitate the breeding, further challenges arise for the research and development of new parsley varieties with improve yield and qualitative traits.

References

Aćimović MG, Milić NB (2017) Perspectives of the Apiaceae hepatoprotective effects—a review. NPC 12:309–317

Agyare C, Appiah T, Boakye YD, Apenteng JA (2017) Petroselinum crispum: a review. In: Medicinal spices and vegetables from Africa: therapeutic potential against metabolic, inflammatory, infectious and systemic diseases. Elsevier Inc., pp 527–547

Airy Shaw HK (1938) The correct name of the common parsley. Bull Misc Inf (Royal Botanic Gardens, Kew)

Ajani Y, Bull-Hereñu K, Claßen-Bockhoff R (2016) Patterns of flower development in Apiaceae-Apioideae. Flora 221:38–45. https://doi.org/10.1016/J.FLORA.2016.02.004

Ajmera P, Kalani S, Sharma L (2019) Parsley-benefits & side effects on health. Int J Physiol 4:1236–1242

Al-Yousofy F, Gumaih H, Ibrahim H, Alasbahy A (2017) Parsley! Mechanism as antiurolithiasis remedy. AJCEU 5:62

Astley D, Kotlińska T (2001) Report of a network coordinating group on vegetables. IPGRI 14

Baldwin BG, Sanderson MJ, Porter JM et al (1995) The ITS region of nuclear ribosomal DNA: a valuable source of evidence on angiosperm phylogeny. Ann Missouri Bot Gard 82:247. https://doi.org/10.2307/2399880

Becker T (1943) Blütenbiologische Studien an Zwiebeln, Möhren, Sellerie und Petersilie. Kühn-Archiv 60:466–492

Becker G (1962) Knollensellerie (*Apium graveolens* L. var. rapaceum (Miller) D.C.). In: Kappert H, Rudorf W (eds) Handbuch der Pflanzenzüchtung, Bd. VI: Züchtung von Gemüse, Obst, Reben und Forstpflanzen. Berlin, pp 104–130

Becker-Dillingen J (1926) Ein neuer Gattungsbastard, *Apium graveolens* (Sellerie) × *Petroselinum sativum* (Wurzelpetersilie). Z Pflanzenz 11:199–204

Bennett MD, Leitch IJ (1995) Nuclear DNA amounts in angiosperms. Ann Bot 76:113–176. https://doi.org/10.1006/ANBO.1995.1085

Black M, Bewley DJ (2000) Seed technology and its biological basis. Sheffield Academic Press; CRC Press, Sheffield, England, Boca Raton, FL

Boivin G (1999) Integrated management for Carrot Weevil. J Integr Pest Manag 4:21–37. https://doi.org/10.1023/A:1009624714172
Boldizsár I, Füzfai Z, Molnár-Perl I (2013) Characterization of the endogenous enzymatic hydrolyses of Petroselinum crispum glycosides: determined by chromatography upon their sugar and flavonoid products. J Chromatogr A 1293:100–106. https://doi.org/10.1016/J.CHROMA.2013.03.037
Boutsika A, Sarrou E, Cook CM et al (2021) Evaluation of parsley (*Petroselinum crispum*) germplasm diversity from the Greek Gene Bank using morphological, molecular and metabolic markers. Ind Crops Prod 170:113767. https://doi.org/10.1016/J.INDCROP.2021.113767
Bruznican S, de Clercq H, Eeckhaut T et al (2020) Celery and celeriac: a critical view on present and future breeding. Front Plant Sci 10:1699. https://doi.org/10.3389/FPLS.2019.01699/BIBTEX
Calviño CI, Martínez SG, Downie SR (2008) Morphology and biogeography of Apiaceae subfamily Saniculoideae as inferred by phylogenetic analysis of molecular data. Am J Bot 95:196–214. https://doi.org/10.3732/AJB.95.2.196
Cantliffe DJ (2003) Seed enhancements. Acta Hortic 607:53–59. https://doi.org/10.17660/ACTAHORTIC.2003.607.8
Cazzola R, Camerotto C, Cestaro B (2011) Anti-oxidant, anti-glycant, and inhibitory activity against α-amylase and α-glucosidase of selected spices and culinary herbs. Int J Food Sci Nutr 62:175–184. https://doi.org/10.3109/09637486.2010.529068
Chable V (2005) Conserving and developing crop biodiversity. In: Biodiversity and local ecological knowledge in France
Chen WJ, Zhu T (2004) Networks of transcription factors with roles in environmental stress response. Trends Plant Sci 9:591–596. https://doi.org/10.1016/J.TPLANTS.2004.10.007
Chen K, Arora R, Arora U (2010) Osmopriming of spinach (*Spinacia oleracea* L. cv. Bloomsdale) seeds and germination performance under temperature and water stress. SST 38:36–48. https://doi.org/10.15258/SST.2010.38.1.04
Chomchalow N (2002) Production of herbs in Asia: an overview. AU JT 6:95–108
Christensen LP (2011) Aliphatic C17-polyacetylenes of the falcarinol type as potential health promoting compounds in food plants of the Apiaceae family. Recent Pat Food Nutr Agric 3:64–77
Collard BCY, Mackill DJ (2008) Marker-assisted selection: an approach for precision plant breeding in the twenty-first century. Philos Trans R Soc Lond B Biol Sci 363:557–572. https://doi.org/10.1098/rstb.2007.2170
Cormack RS, Eulgem T, Rushton PJ et al (2002) Leucine zipper-containing WRKY proteins widen the spectrum of immediate early elicitor-induced WRKY transcription factors in parsley. Biochim Biophys Acta Gene Regul Mech 1576:92–100. https://doi.org/10.1016/S0167-4781(02)00298-1
CPVO Technical Protocols | CPVO. https://cpvo.europa.eu/en/applications-and-examinations/technical-examinations/technical-protocols/cpvo-technical-protocols?t=parsley&field_crop_sector_tid=All. Accessed 24 Feb 2023
Craft JD, Setzer WN (2017) The volatile components of parsley, *Petroselinum crispum* (Mill.) Fuss. Am J Essent 5:27–32. https://doi.org/10.5513/JCEA01/23.1.3293
da Silva ALBR, de Barros MFL, Foshee W et al (2022) Priming strategies for parsley seedling production. HortScience 57:1086–1091. https://doi.org/10.21273/HORTSCI16675-22
Dachler M, Pelzmann H (1999) Arznei- und Gewürzpflanzen. Österr Agrarverlag 253–261
Danert S (1958) Zur Systematik von Papaver somniferum L. Kulturpflanze 6:61–88
Darlington C, Wylie AP (1955) Chromosome atlas of flowering plants, 2d edn. Allen & Unwin, London
Das S (2017) Leaf and seedling morphology: the two prospective parameters to trace phylogeny in apiaceae subfamily apioideae. Ann Plant Sci 6:1698. https://doi.org/10.21746/APS.2017.6.10.3
De Menezes M, Epifanio N, Rykiel Iglesias Cavalcanti L, Falcão Dos Santos K et al (2020) Chemical characterization and in vivo antioxidant activity of parsley (*Petroselinum crispum*) aqueous extract. Food Funct 11:5346–5356. https://doi.org/10.1039/D0FO00484G

Domblides AS, Domblides EA, Kharchenko VA, Potekhin GA (2010) Study of genetic variation among parsley (*Petroselinum crispum* (Mill.) Nym.) samples using RAPD and ISSR markers. Mosc Univ Biol Sci Bull 65:152–154. https://doi.org/10.3103/S0096392510040073

Downie SR, Katz-Downie DS (1996) A molecular phylogeny of apiaceae subfamily Apioideae: evidence from nuclear ribosomal DNA internal transcribed spacer sequences. Am J Bot 83:234–251. https://doi.org/10.1002/J.1537-2197.1996.TB12701.X

Downie SR, Ramanath S, Katzdownie DS, Llanas E (1998) Molecular systematics of Apiaceae subfamily Apioideae: phylogenetic analyses of nuclear ribosomal DNA internal transcribed spacer and Plastid rpoC1 intron sequences. AJB 85:563–591. https://doi.org/10.2307/2446441

Downie SR, Katz-Downie DS, Lee BY et al (2001) Tribes and clades within apiaceae subfamily apioideae: the contribution of molecular data. Edinb J Bot 58:301–330. https://doi.org/10.1017/S0960428601000658

Downie SR, Spalik K, Katz-Downie DS, Reduron J (2010a) Major clades within Apiaceae subfamily Apioideae as inferred by phylogenetic analysis of nrDNA ITS sequence. PDE 111–136

Downie SR, Spalik K, Katz-Downie DS, Reduron J-P (2010b) Major clades within Apiaceae subfamily Apioideae as inferred by phylogenetic analysis of nrDNA ITS sequences. PDE 128:111–136. https://doi.org/10.1127/1869-6155/2010/0128-0005

Drude CGO (1897) Umbelliferae. Die natürlichen Pflanzenfamilien, vol. 3, Wilhelm Engelmann, Leipzig, Germany, pp 63–250. In: Engler A, Prantl K (eds) Wilhelm Engelmann, Leipzig, Germany, pp 63–250

Duman İ (2005) Olumsuz çevre koşullarında tohum-ların çimlenme ve çıkış performansı nasıl artırılabilir? Hasad Dergisi 21:72–78

Dursun A, Ekinci M, Dönmez MF (2010) Effects of foliar application of plant growth promoting bacterium on chemical contents, yiled and growth of tomato (*Lycopersicon esculentum* L.) and cucumber (*Cucumis sativus* L.). Pak J Bot 42:3349–3356

EDQM European Directorate for the Quality of Medicines and Healthcare—European Directorate for the Quality of Medicines & HealthCare. https://www.edqm.eu/en/. Accessed 24 Feb 2023

Ellis PR, Hardman JA (1981) The consistency of the resistance of eight carrot cultivars to carrot fly attack at several centres in Europe. Ann Appl Biol 98:491–497. https://doi.org/10.1111/J.1744-7348.1981.TB00781.X

EURISCO (2019) EURISCO. https://eurisco.ipk-gatersleben.de/apex/eurisco_ws/r/eurisco/taxon-search-results?p26_genus=PETROSELINUM&p26_species=CRISPUM. Accessed 24 Feb 2023

Farzaei MH, Abbasabadi Z, Ardekani MRS et al (2013) Parsley: a review of ethnopharmacology, phytochemistry and biological activities. J Tradit Chin Med 33:815–826. https://doi.org/10.1016/S0254-6272(14)60018-2

Feldheim W (1999) Petersilie als Desodorans? Petersilienöl mindert Knoblauchgeruch. Deut Apotheker Ztg 139:1552–1557

Feng K, Hou XL, Li MY et al (2018) CeleryDB: a genomic database for celery. Database-Oxford 2018. https://doi.org/10.1093/DATABASE/BAY070

Fernandes Â, Polyzos N, Petropoulos SA et al (2020) Phytochemical composition and nutritional value of pot-grown turnip-rooted and plain and curly-leafed parsley cultivars. Agronomy 10:1416. https://doi.org/10.3390/AGRONOMY10091416

Ferrie AMR, Bethune TD, Arganosa GC, Waterer D (2011) Field evaluation of doubled haploid plants in the Apiaceae: Dill (*Anethum graveolens* L.), caraway (*Carum carvi* L.), and fennel (*Foeniculum vulgare* Mill.). PCTOC 104:407–413. https://doi.org/10.1007/S11240-010-9821-6

Figueiredo DD, Barros PM, Cordeiro AM et al (2012) Seven zinc-finger transcription factors are novel regulators of the stress responsive gene OsDREB1B. J Exp Bot 63:3643–3656. https://doi.org/10.1093/JXB/ERS035

Fu N, Wang Q, Shen HL (2013) De Novo assembly, gene annotation and marker development using illumina paired-end transcriptome sequences in Celery (*Apium graveolens* L.). PLoS ONE 8:e57686. https://doi.org/10.1371/JOURNAL.PONE.0057686

Gadi D, Bnouham M, Aziz M et al (2009) Parsley extract inhibits in vitro and ex vivo platelet aggregation and prolongs bleeding time in rats. J Ethnopharmacol 125:170–174. https://doi.org/10.1016/J.JEP.2009.05.014

GBIF Secretariat (2021) GBIF backbone taxonomy. GBIF Backbone Taxonomy. https://doi.org/10.15468/39OMEI

Glawe DA, Pelter GQ, du Toit LJ (2018) First report of powdery mildew of carrot and parsley caused by *Erysiphe heraclei* in Washington State. APS 6. https://doi.org/10.1094/PHP-2005-0114-01-HN

Graßmann J (2005) Terpenoids as plant antioxidants. Vitam Horm 72:505–535. https://doi.org/10.1016/S0083-6729(05)72015-X

Gratwick M (1992) Crop pests in the UK. In: Gratwick M (ed) CPP. Springer, Netherlands, Dordrecht, pp 252–254

Guerin PM, Ryan MF (1984) Relationship between root volatiles of some carrot cultivars and their resistance to the carrot fly, Psila rosae. Entomol Exp Appl 36:217–224. https://doi.org/10.1111/J.1570-7458.1984.TB03431.X

Günay A (2005) Sebze yetiştiriciliği cilt II. İzmir

Hammer K, Gladis T, Diederichsen A (2003) In situ and on-farm management of plant genetic resources. Eur J Agron 19:509–517. https://doi.org/10.1016/S1161-0301(02)00184-3

Hanelt P, Institute of Plant Genetics and Crop Research (2001) Mansfeld's encyclopedia of agricultural and horticultural crops (except ornamentals). Springer, Berlin

Hanrahan C, Frey RJ (2005) Parsley. In: Longe JL (ed) The Gale encyclopedia of alternative medicine. Farmington Hills : Thomson/Gale

Harris D (1996) The effects of manure, genotype, seed priming, depth and date of sowing on the emergence and early growth of *Sorghum bicolor* (L.) Moench in semi-arid Botswana. Soil Tillage Res 40:73–88. https://doi.org/10.1016/S0167-1987(96)80007-9

Hasegawa PM, Bressan RA, Zhu JK, Bohnert HJ (2000) Plant cellular and molecular responses to high salinity. Annu Rev Plant Physiol Plant Mol Biol 51:463–499. https://doi.org/10.1146/ANNUREV.ARPLANT.51.1.463

Heywood VH (2014) The socio-economic importance of the Apiales. J Pharm 44:113–130

Hoberg E, Marthe F, Ulrich D (2007) Sensory quality of parsley (*Petroselinum crispum* [Mill.] Nym.). Z Arznei Gewürzpfla 12:180

Holmberg N, Bülow L (1998) Improving stress tolerance in plants by gene transfer. Trends Plant Sci 3:61–66. https://doi.org/10.1016/S1360-1385(97)01163-1

Honma S, Lacy ML (1980) Hybridization between pascal celery and parsley. Euphytica 29:801–805. https://doi.org/10.1007/BF00023227

Hoppe B, Marthe F, Böhme M et al (2013) Petersilie (*Petroselinum crispum* [Mill.] Nyman ex A. W. Hill). In: Hoppe B (ed) Handbuch des Arznei- und Gewürzpflanzenbaus, Band 5, Arznei-und Gewürzpflanzen von L – Z, Verein für Arznei- und Gewürzpflanzen. Saluplanta e.V., Bernburg, Deutschland, pp 276–309

Hotti H, Seppänen-Laakso T, Arvas M et al (2015) Polyketide synthases from poison hemlock (*Conium maculatum* L.). FEBS J 282:4141–4156. https://doi.org/10.1111/FEBS.13410

Hozayen WG, El-Desouky MA, Soliman HA et al (2016) Antiosteoporotic effect of *Petroselinum crispum, Ocimum basilicum* and *Cichorium intybus* L. in glucocorticoid-induced osteoporosis in rats. BMC Complement Altern Med 16:1–11. https://doi.org/10.1186/S12906-016-1140-Y/FIGURES/3

Hsieh TH, Li CW, Su RC et al (2010) A tomato bZIP transcription factor, SlAREB, is involved in water deficit and salt stress response. Planta 231:1459–1473. https://doi.org/10.1007/S00425-010-1147-4

Huxley A (1992) The new RHS dictionary of gardening. MacMillan Press, London

Ibrahim HMM, Abou El-Leel OF, Emam KA (2017) Molecular profiling for genetic variability in *Petroselinum crispum* based on ISSR and RAPD markers. Middle East J Agric Res 6:67–75

Iorizzo M, Senalik DA, Grzebelus D et al (2011) De novo assembly and characterization of the carrot transcriptome reveals novel genes, new markers, and genetic diversity. BMC Genom 12:1–14. https://doi.org/10.1186/1471-2164-12-389/TABLES/4

Iorizzo M, Ellison S, Senalik D et al (2016) A high-quality carrot genome assembly provides new insights into carotenoid accumulation and asterid genome evolution. Nat Genet 48:657–666. https://doi.org/10.1038/NG.3565

Ipor IB, Oyen LPA (1999) Nigella sativa L. In: de Guzman CC, Siemonsma JS (eds) Plant resources of South-East Asia 13: spices. Backhuys Publishers, Leiden, pp 148–151

James LF, Ralphs MH, Nielsen DB (2019) The ecology and economic impact of poisonous plants on livestock production, 1st edn. CRC Press

Jensen BD, Finckh MR, Munk L, Hauser TP (2008) Susceptibility of wild carrot (*Daucus carota* ssp. carota) to *Sclerotinia sclerotiorum*. Eur J Plant Pathol 122:359–367. https://doi.org/10.1007/S10658-008-9300-7/TABLES/3

Jiménez-Mejías P, Vargas P (2015) Taxonomy of the tribe Apieae (Apiaceae) revisited as revealed by molecular phylogenies and morphological characters. Phytotaxa 212:057–079. https://doi.org/10.11646/phytotaxa.212.1.2

Jugreeta BS, Mahomoodally MF (2020) Essential oils from 9 exotic and endemic medicinal plants from Mauritius shows in vitro antibacterial and antibiotic potentiating activities. S Afr J Bot 132:355–362. https://doi.org/10.1016/j.sajb.2020.05.001

Justesen U, Knuthsen P, Leth T (1998) Quantitative analysis of flavonols, flavones, and flavanones in fruits, vegetables and beverages by high-performance liquid chromatography with photo-diode array and mass spectrometric detection. J Chromatogr A 799:101–110. https://doi.org/10.1016/S0021-9673(97)01061-3

Kaiser A, Carle R, Kammerer DR (2013) Effects of blanching on polyphenol stability of innovative paste-like parsley (*Petroselinum crispum* (Mill.) Nym ex A. W. Hill) and marjoram (*Origanum majorana* L.) products. Food Chem 138:1648–1656. https://doi.org/10.1016/J.FOODCHEM.2012.11.063

Lee JM (2004) Advances in seed treatments for horti-cultural crops. Chron Horticult 44:11–20

Li MY, Tan HW, Wang F et al (2014) De Novo transcriptome sequence assembly and identification of AP2/ERF transcription factor related to abiotic stress in Parsley (*Petroselinum crispum*). PLoS ONE 9. https://doi.org/10.1371/JOURNAL.PONE.0108977

Li MY, Song X, Wang F, Xiong AS (2016) Suitable reference genes for accurate gene expression analysis in Parsley (*Petroselinum crispum*) for Abiotic stresses and hormone stimuli. Front Plant Sci 7. https://doi.org/10.3389/FPLS.2016.01481

Liu PY, Zhang B, Zhao H (2017) The study of based on RAPD molecular markers in celery cultivars relationship. Sci Discov 5:312–316

Liu JX, Jiang Q, Tao JP et al (2021) Integrative genome, transcriptome, microRNA, and degradome analysis of water dropwort (*Oenanthe javanica*) in response to water stress. Hortic Res 8:1–17. https://doi.org/10.1038/s41438-021-00707-8

Lohwasser U (2010) Taxonomy of plant genetic resources—use of morphological, molecular and phytochemical data in order to verify existing classifications. Agric Conspec Sci 75:175–178

Lu CL, Li XF (2019) A review of *Oenanthe javanica* (Blume) DC as traditional medicinal plant and its therapeutic potential. Evid Based Complement Altern Med. https://doi.org/10.1155/2019/6495819

Ma J, Li J, Xu Z et al (2018) Transcriptome profiling of genes involving in carotenoid biosynthesis and accumulation between leaf and root of carrot (*Daucus carota* L.). ABBS 50:481–490. https://doi.org/10.1093/ABBS/GMY027

Madjarova DJ, Bubarova MG (1978) New forms obtained by hybridization of *Apium graveolens* and *Petroselinum hortense*. Acta Hortic 65–72. https://doi.org/10.17660/ACTAHORTIC.1978.73.7

Magee AR, Calviño CI, Liu MR et al (2010) New tribal delimitations for the early diverging lineages of Apiaceae subfamily Apioideae. Taxon 59:567–580. https://doi.org/10.1002/TAX.592021

Maodaa SN, Allam AA, Ajarem J et al (2016) Effect of parsley (*Petroselinum crispum*, Apiaceae) juice against cadmium neurotoxicity in albino mice (Mus Musculus). Behav Brain Funct 12:1–16. https://doi.org/10.1186/S12993-016-0090-3/FIGURES/8

Marthe F (2020) *Petroselinum crispum* (Mill.) Nyman (Parsley). J Appl Res Med Aromat Plants 435–466

Marthe F, Scholze P, Krämer R et al (2003) Evaluation of parsley for resistance to the pathogens *Alternaria radicina*, *Erysiphe heraclei*, *Fusarium oxysporum*, and celery mosaic virus (CeMV). Plant Breed 122:248–255. https://doi.org/10.1046/J.1439-0523.2003.00840.X

Marthe F, Bruchmüller T, Börner A, Lohwasser U (2013) Variability in parsley (*Petroselinum crispum* [Mill.] Nyman) for reaction to *Septoria petroselini* Desm., *Plasmopara petroselini* Săvul. et O. Săvul. and *Erysiphe heraclei* DC. ex Saint-Aman causing Septoria blight, downy mildew and powdery mildew. Genet Resour Crop Evol 60:1007–1020. https://doi.org/10.1007/S10722-012-9897-4/FIGURES/8

Martins JB, Santos Júnior JA, da Silva Júnior FJ et al (2019) Production of parsley in hydroponic conditions under isosmotic brackish nutrient solutions. Cienc Agrotec 43:23418. https://doi.org/10.1590/1413-7054201943023418

Marx P, Gärber U (2014) Powdery mildew on parsley—developing a screening method for resistance breeding. Z Arznei Gewürzpfla 19:162–169

Masanetz C, Grosch W (1998) Key odorants of parsley leaves (*Petroselinum crispum* [Mill.] Nym. ssp. crispum) by odour ± activity values. Flavour Fragr J 13:115–124. https://doi.org/10.1002/(SICI)1099-1026(199803/04)13:2

Maxim A (2008) General and applied ecology. Cluj-Napoca

Maxim A, Sandor M, Jidavu M et al (2009) Research concerning the conservation of genetic diversity of parsley (*Petroselinum crispum* Mill.). Bull UASVM Agric 66:144–148

Moldovan F (2003) Phenomena of climate risk. Equinox, Cluj-Napoca

Murungu FS, Chiduza C, Nyamugafata P et al (2004) Effects of 'on-farm seed priming' on consecutive daily sowing occasions on the emergence and growth of maize in semi-arid Zimbabwe. Field Crops Res 89:49–57. https://doi.org/10.1016/J.FCR.2004.01.020

Mustafa NSA, Abou-Ellail M (2013) Genetic discrimination of Egyptian fig cultivars revealed by RAPD fingerprints. Int J Agric Res 8:17–25

Nascimento WM, de Aragão FAS (2004) Muskmelon seed priming in relation to seed vigor. Sci Agric 114–117

Nelson R, Wiesner-Hanks T, Wisser R, Balint-Kurti P (2017) Navigating complexity to breed disease-resistant crops. Nat Rev Genet 19:21–33. https://doi.org/10.1038/nrg.2017.82

Nirumand MC, Hajialyani M, Rahimi R et al (2018) Dietary plants for the prevention and management of kidney stones: preclinical and clinical evidence and molecular mechanisms. Int J Mol Sci 19:765. https://doi.org/10.3390/IJMS19030765

Ochoa O, Quiros CF (1989) Apium wild species: novel sources for resistance to late blight in celery. Plant Breed 102:317–321. https://doi.org/10.1111/J.1439-0523.1989.TB01262.X

Petropoulos SA, Daferera D, Polissiou MG, Passam HC (2008) The effect of water deficit stress on the growth, yield and composition of essential oils of parsley. Sci Hortic 115:393–397. https://doi.org/10.1016/J.SCIENTA.2007.10.008

Punoševac M, Radović J, Leković A, Kundaković-Vasović T (2021) A review of botanical characteristics, chemical composition, pharmacological activity and use of parsley. Arh Za Farm 71:177–196. https://doi.org/10.5937/ARHFARM71-30800

Quattrocchi U (2012) CRC world dictionary of medicinal and poisonous plants common names, scientific names, eponyms, synonyms, and etymology, vol 5, 1st edn. CRC Press

Que F, Hou XL, Wang GL et al (2019a) Advances in research on the carrot, an important root vegetable in the Apiaceae family. Hortic Res 6:1–15. https://doi.org/10.1038/s41438-019-0150-6

Que F, Wang YH, Xu ZS, Xiong AS (2019b) DcBAS1, a carrot brassinosteroid catabolism gene, modulates cellulose synthesis. J Agric Food Chem 67:13526–13533. https://doi.org/10.1021/ACS.JAFC.9B05241/ASSET/IMAGES/MEDIUM/JF9B05241_0010.GIF

Radulović N, DorDević N, Denić M et al (2012) A novel toxic alkaloid from poison hemlock (*Conium maculatum* L., Apiaceae): identification, synthesis and antinociceptive activity. FCT 50:274–279. https://doi.org/10.1016/J.FCT.2011.10.060

Ramin AA, Atherton JG (2015) Manipulation of bolting and flowering in celery (*Apium graveolens* L. var. dulce). III. Effects of photoperiod and irradiance. Int J Hortic Sci 69:861–868. https://doi.org/10.1080/14620316.1994.11516522

Rushton PJ, Torres JT, Parniske M et al (1996) Interaction of elicitor-induced DNA-binding proteins with elicitor response elements in the promoters of parsley PR1 genes. EMBO J 15:5690–5700

Rushton PJ, Somssich IE, Ringler P, Shen QJ (2010) WRKY transcription factors. Trends Plant Sci 15:247–258. https://doi.org/10.1016/J.TPLANTS.2010.02.006

Sabry RM, Kandil MAM, Ahmed SS (2013) Comparative study of some parsley cultivars grown in Egypt for some potent compounds. Res J Appl Sci 9:6419–6424

Saharan GS, Mehta N (2008) Sclerotinia diseases of crop plants: biology, ecology and disease management. Springer Science & Business Media

Said-Al Ahl HAH, Abou-Ellail M, Omer EA (2016) Harvest date and genotype influences growth characters and essential oil production and composition of *Petroselinum crispum* plants. J Chem Pharm 8:992–1003

Saleem F, Eid AH, Shetty K (2011) Potato–Herb synergies as food designs for hyperglycemia and hypertension management. Wiley

Sanchis V, Bourguet D (2008) Bacillus thuringiensis: applications in agriculture and insect resistance management. A review. Agron Sustain Dev 28:11–20. https://doi.org/10.1051/agro:2007054

Sarwar S, Ayyub MA, Rezgui M et al (2016) Parsley: a review of habitat, phytochemistry, ethnopharmacology and biological activities. IJCBS 9:49–55

Scott DJ, Wenham HT (2012) Occurrence of two seed-borne pathogens, *Alternaria radicina* and *Alternaria dauci*, on imported carrot seed in New Zealand. N Z J Agric Res 16:247–250. https://doi.org/10.1080/00288233.1973.10421142

Shan RH, Constance L (1951) The genus sanicula: umbelliferae, in the old world and the new. University of California Publications in Botany

Shinozaki K, Yamaguchi-Shinozaki K, Seki M (2003) Regulatory network of gene expression in the drought and cold stress responses. Curr Opin Plant Biol 6:410–417. https://doi.org/10.1016/S1369-5266(03)00092-X

Shinozaki K, Yamaguchi-Shinozaki K (2007) Gene networks involved in drought stress response and tolerance. J Exp Bot 58:221–227. https://doi.org/10.1093/JXB/ERL164

Sigrist MS, Pinheiro JB, Azevedo-filho JA et al (2010) Development and characterization of microsatellite markers for turmeric (*Curcuma longa*). Plant Breed 129:570–573

Simmonds NW, Smartt J (2000) Principles of crop improvement, 2nd edn. Blackwell Science

Simon JE (1990) Essential oils and culinary herbs. In: Janick J, Simon JE (eds) Advances in new crops. Timber Press, Portland, OR, pp 427–483

Small E (1997) Petroselinum. Culinary herbs. NRC Reasearc Press, Ottawa, pp 475–480

Soliman HA, Eltablawy NA, Hamed MS (2015) The ameliorative effect of *Petroselinum crispum* (parsley) on some diabetes complications. J Med Plants Stud 3:92–100

Song X, Wang J, Li N et al (2020) Deciphering the high-quality genome sequence of coriander that causes controversial feelings. Plant Biotechnol J 18:1444–1456. https://doi.org/10.1111/PBI.13310

Sousa RMOF, Cunha AC, Fernandes-Ferreira M (2021) The potential of Apiaceae species as sources of singular phytochemicals and plant-based pesticides. Phytochemistry 187:112714. https://doi.org/10.1016/J.PHYTOCHEM.2021.112714

Spalik K, Downie SR, Watson MF (2009) Generic delimitations within the Sium alliance (Apiaceae tribe Oenantheae) inferred from cpDNA rps16-5′trnK(UUU) and nrDNA ITS sequences. Taxon 58:735–748. https://doi.org/10.1002/TAX.583004

Stonecipher CA, Welch KD, Lee ST et al (2020) Geographical and seasonal variation in water hemlock (*Cicuta maculata*) toxins. Biochem Syst Ecol 89:104012. https://doi.org/10.1016/j.bse.2020.104012

Sun FJ, Downie SR (2010) Phylogenetic relationships among the perennial, endemic Apiaceae subfamily Apioideae of western North America: additional data from the cpDNA trnF-trnL-trnT region continue to support a highly polyphyletic cymopterus. PDE 128(1–2):151–172. https://doi.org/10.1127/1869-6155/2010/0128-0007

Teuscher E, Bauermann U, Werner M (2006) Medicinal spices: a handbook of culinary herbs, spices, spice mixtures, and their essential oils. Medpharm Scientific Publishers, Germany

Thiviya P, Gamage A, Piumali D et al (2021) Apiaceae as an important source of antioxidants and their applications. Cosmetics 8(4):111. https://doi.org/10.3390/cosmetics8040111

Ticker AO, DeBaggio T (2009) The encyclopedia of herbs: a comprehensive reference to herbs of flavor and fragrance. Timber Press, Portland, London

Tilney PM, van Wyk BE, Downie SR, Calviño CI (2009) Phylogenetic relationships in the genus Lichtensteinia (Apiaceae) based on morphological, anatomical and DNA sequence data. S Afr J Bot 75:64–82. https://doi.org/10.1016/J.SAJB.2008.07.011

Tovchiga OV (2017) Metabolic effects of goutweed (*Aegopodium podagraria* L.) tincture and metformin in dexamethasone-treated rats. JDMP 2:117. https://doi.org/10.11648/J.JDMP.20160206.17

Ulrich D, Bruchmüller T, Krüger H, Marthe F (2011) Sensory characteristics and volatile profiles of parsley (*Petroselinum crispum* [Mill.] Nym.) in correlation to resistance properties against septoria blight (*Septoria petroselini*). J Agric Food Chem 59:10651–10656. https://doi.org/10.1021/JF202120D

UPOV (2005) International union for the protection of new varieties of plants, Geneva (2005) TG/136/5, parsley, UPOV code: PETRO CRI

UPOV (2006) International union for the protection of new varieties of plants: hop. Guidelines for the Conduct of Tests for Distinctness, Uniformity and Stability. http://www.upov.int/portal/index.html.en

USDA (2019) U.S. National Plant Germplasm System. https://npgsweb.ars-grin.gov/gringlobal/search.aspx. Accessed 24 Feb 2023

Valiejo-Roman CM, Terentieva EI, Pimenov MG et al (2012) Broad polyphyly in Pleurospermum S. l. (umbelliferae-apioideae) as inferred from nrDNA ITS and chloroplast sequences. Syst Bot 37:573–581. https://doi.org/10.1600/036364412X635593

Vanlerberghe-Masutti F, Guillemau TT (2007) La résistance des pucerons aux insecticides . Biofutur 27–30

Verlet N (1993) Essential oils: suppy, demand and price determination. Acta Hortic 9–16. https://doi.org/10.17660/ACTAHORTIC.1993.344.2

Verma S, Singh S, Sharma S et al (2015) Assessment of genetic diversity in indigenous turmeric (*Curcuma longa*) germplasm from India using molecular markers. Physiol Mol Biol Plants 21:233–242. https://doi.org/10.1007/S12298-015-0286-2

Vural H, Eşiyok D, Duman İ (2000) Kültür sebzeleri (sebze yetiştirme) 440

Wang X-J, Luo Q, Li T et al (2022) Origin, evolution, breeding, and omics of Apiaceae: a family of vegetables and medicinal plants. Hortic Res 9. https://doi.org/10.1093/HR/UHAC076

Webb S (2005) Insect management for celery and parsley: ENY-463/IG149, 2002–2017. EDIS 14

Webb SE (2017) Insect management for celery and parsley. ENY 463, Entomology and Nematology Department, Florida Cooperative Extension Service, Institute of Food and Agricultural Sciences, University of Florida

Weisshaar B, Armstrong GA, Block A et al (1991) Light-inducible and constitutively expressed DNA-binding proteins recognizing a plant promoter element with functional relevance in light responsiveness. EMBO J 10:1777–1786. https://doi.org/10.1002/J.1460-2075.1991.TB07702.X

Wichtl M (2002) Teedrogen und phytopharmaka. Wissenschaftliche Verlagsgesellschaft mbH, pp 587–589

Widelski J, Luca SV, Skiba A et al (2018) Isolation and antimicrobial activity of coumarin derivatives from fruits of *Peucedanum luxurians* Tamamsch. Molecules 23. https://doi.org/10.3390/MOLECULES23051222

Widelski J, Luca SV, Skiba A et al (2021) Coumarins from Seseli devenyense Simonk.: isolation by liquid-liquid chromatography and potential anxiolytic activity using an in vivo Zebrafish Larvae model. Int J Mol Sci 22:1–21. https://doi.org/10.3390/IJMS22041829

Winter P, Kahl G (1995) Molecular marker technologies for plant improvement. World J Microbiol Biotechnol 11:438–448. https://doi.org/10.1007/bf00364619

Winter PJD, Magee AR, Phephu N et al (2008) A new generic classification for African peucedanoid species (Apiaceae). Taxon 57:347–364. https://doi.org/10.2307/25066009

Wolf E, White J, Stubblefield R (1993) Florida Slobolt M68': a spring celery cultivar for Florida. HortScience 28:754–755

Wong PYY, Kitts DD (2006) Studies on the dual antioxidant and antibacterial properties of parsley (*Petroselinum crispum*) and cilantro (*Coriandrum sativum*) extracts. Food Chem 97:505–515. https://doi.org/10.1016/J.FOODCHEM.2005.05.031

WuTai W, Feng W, GuoXun G et al (2012) A new slow-bolting celery cultivar "Juventus." Acta Hortic Sin 39:1007–1008

Xie C, Li B, Xu Y et al (2013) Characterization of the global transcriptome for *Pyropia haitanensis* (Bangiales, Rhodophyta) and development of cSSR markers. BMC Genom 14:1–13. https://doi.org/10.1186/1471-2164-14-107/TABLES/4

Xu ZS, Tan HW, Wang F et al (2014) CarrotDB: a genomic and transcriptomic database for carrot. Database-Oxford 2014. https://doi.org/10.1093/DATABASE/BAU096

Yan J, Yu L, Xu S et al (2014) Apigenin accumulation and expression analysis of apigenin biosynthesis relative genes in celery. Sci Hortic 165:218–224. https://doi.org/10.1016/J.SCIENTA.2013.11.018

Yanardağ R, Bolkent Ş, Tabakoğlu-Oğuz A, Özsoy-Saçan Ö (2003) Effects of *Petroselinum crispum* extract on Pancreatic B cells and blood glucose of streptozotocin-induced diabetic rats. Biol Pharm Bull 26:1206–1210. https://doi.org/10.1248/BPB.26.1206

Yang X, Quiros C (1993) Identification and classification of celery cultivars with RAPD markers. TAG 86:205–212. https://doi.org/10.1007/BF00222080

Yang X, Yu YJ, Zhang FL et al (2007) Linkage map construction and quantitative trait loci analysis for bolting based on a double haploid population of *Brassica rapa*. J Integr Plant Biol 49:664–671. https://doi.org/10.1111/J.1744-7909.2007.00447.X

Zaharia H, Kastler G (2003) European Union: seeds and plants—legislative progress. Citizen Earth 6

Zhang H, Chen F, Wang X, Yao HY (2006) Evaluation of antioxidant activity of parsley (*Petroselinum crispum*) essential oil and identification of its antioxidant constituents. Int Food Res J 39:833–839. https://doi.org/10.1016/J.FOODRES.2006.03.007

Zhou J, Peng H, Downie SR (2008) A molecular phylogeny of Chinese Apiaceae subfamily Apioideae inferred from nuclear ribosomal DNA internal transcribed spacer sequences. Taxon 57:402–416. https://doi.org/10.2307/25066012

Zhuang J, Xiong AS, Peng RH et al (2010) Analysis of Brassica rapa ESTs: gene discovery and expression patterns of AP2/ERF family genes. Mol Biol Rep 37:2485–2492. https://doi.org/10.1007/S11033-009-9763-4

Chapter 11
Peppers (*Capsicum* spp.): Domestication, Breeding and Epigenetics

Ponnam Naresh, Patcharaporn Suwor, Nattika Kumsee, Roshni Pydi, and Sanjeet Kumar

Abstract The genus *Capsicum* ($2n = 24$, 3.5 Gb), a member of the nightshade family, comprises five cultivated species. These pepper species were domesticated in the arid regions of the Andes Mountains (Peru and Bolivia) and later spread and acclimatized to the temperate and tropical regions. Pepper plants produce both pungent (chili pepper) and non-pungent (sweet pepper) fruits valued for food (spices and vegetables) and non-food uses. We briefly described the diverse uses of peppers and their products, domestication syndromes, the recent discovery of super-hot pepper landraces, market types, and breeding approaches. The breeding for generic traits like resistance to pests and abiotic stresses are listed. The application of genomics and post-genomics tools to improve quality traits, the exploration of epigenetic potentials, and the enhancement of public and private collaborations have advanced pepper breeding strategies.

Keywords Capsaicinoids · Carotenoids · Chili pepper · Genomics · Sweet pepper

P. Naresh · R. Pydi
Division of Vegetable Crops, ICAR-Indian Institute of Horticultural Research (ICAR-IIHR), Bengaluru, India
e-mail: nashhorti@gmail.com

R. Pydi
e-mail: pydiroshni64@gmail.com

P. Suwor · N. Kumsee
School of Agricultural Technology, King Mongkut's Institute of Technology (KMITL), Ladkrabang, Bangkok, Thailand
e-mail: patcharaporn.su@kmitl.ac.th

N. Kumsee
e-mail: nattika.chpu@gmail.com

S. Kumar (✉)
College of Agriculture, Acharya Narendra Deva University of Agriculture & Technology (ANDUA&T), Ayodhya, India
e-mail: sanjeet.kumar@nduat.org; sanjeetk1@gmail.com

J. M. Al-Khayri et al. (eds.), *Biodiversity and Genetic Improvement of Herbs and Spices*, Advances in Plant Breeding Strategies 8, https://doi.org/10.1007/978-3-031-81838-7_11

11.1 Introduction

The genus *Capsicum* belongs to the Solanaceae/nightshade family which includes 2500 species of flowering plants. The *Capsicum* (peppers) is native to tropical Central and South America and consists of 30 wild and five domesticated species (Khoury et al. 2020) cultivated as commercial crop and in household gardens worldwide. The fruits of five cultivated species are indispensably valued for various pungency levels (sweet pepper; zero pungent, chili pepper: mild to highly pungent) and flavours that they provide in local and global cuisines and on dining plates. The *C. annuum* peppers are most widely cultivated across the continents. *C. frutescens* and *C. chinense* are cultivated in Central America, West and Central Africa, and Oceania, while *C. baccatum* and *C. pubescens* are currently limited to Central and South America and the Caribbean. Pepper production has grown considerably over the past 20 years, from 2000 to 2020 (FAO, 2020), increasing from 2.46 to about 4.15 million tons for dry chilli and from 20.87 to 36.13 million tons for fresh types. India is the largest producer of dry chilli with an annual estimated 1.87 million tons followed by China with 0.25 million tons (FAOSTAT, 2022). The world's largest producer of fresh pepper is China with an estimated 12.87 million tons annually, followed by Mexico with about 2.07 million tons (FAOSTAT 2022).

In this chapter, we describe domestication and breeding progress on economically important traits and emphasize future research opportunities in epigenetics.

11.2 Versatile and Innovative Uses

Pepper fruits contain more than 30 carotenoids including capsanthin, capsorubin, zeaxanthin, beta-carotene (the precursor of vitamin A), ascorbic acid (vitamin C) etc. Over 15 capsaicinoids (alkaloids) are uniquely produced in chili pepper fruits. These compounds are secondary metabolites synthesized in the placenta of the fruits and are responsible for the pungency (the heat sensation). The capsaicin and dihydrocapsaicin constitute about 70% of total capsaicinoids. Food and non-food uses of peppers include (a) fresh use of immature green fruits, mature red fruits, leave; (b) fresh processing, for sauces, pastes, pickles, beer, etc.; (c) dried spices, from mature whole fruits and powder; (d) colouring and flavouring agents from oleoresin (carotenoids) extracts or powder; (e) ethnobotanical/ traditional medicine from fruit extracts and powder; (f) modern medicine/pharmaceuticals from extracts of capsaicinoids and carotenoids; (g) insecticides/repellents from capsaicinoids extracts; (h) spiritual, use of whole fruits, e.g. 'ristras'; (i) ornamental, use of whole plant or fruits; and (j) defense/punishment, use of capsaicin extracts or powder (Kumar et al. 2006). The biochemical compounds carotenoids, capsaicinoids, and capsinoids have wide applications in the spice, phytofeed, cosmeceutical, and pharmaceutical industries (Barik et al., 2022). Paprika powders are categorized as a food additive, specifically as a natural dye (EFSA, 2015). Chilli Spent Residue (CHSR), an oleoresin by-product, can be used in the bread manufacturing industry to enhance the physical and sensory properties of bread (Sowbhagya et al., 2015). Chilli seeds contain 16–25% oil, which is rich in linoleic acid (68–72%) and palmitic acid (13–15%) (Reddy and Sarojini, 1987), making them valuable compounds in the healthcare industry. Capsaicin has broad applications in the pharmaceutical industry (Hayman and Kam, 2008). As Qutenza®, capsaicin is used effectively to treat neuropathic pain (Aitken et al., 2017). Additionally, these compounds have applications as cosmetic colorants in products such as eye makeup, lipsticks, shampoos, and bath gels (Baenas et al., 2019).

11.3 Domestication of *Capsicum* Spp.

Understanding the domestication of a crop species provides valuable insights that can benefit in genetic improvement, conservation of genetic diversity, aid in crop management practices and is significant for economic and cultural prosperity. Pasquale et al. (2021) introduced a method called ReMIXTURE, which utilizes genetic data to assess the similarity between pepper collections from various regions. Their analysis revealed that Mesoamerica, East/South/Southeast Asia, and Africa are significant centres of diversification, each displaying large proportions of region-specific peppers. The results from genome-wide association studies and selection scans corroborate that pungency was a major factor in the initial appeal of peppers.

The telomere-to-telomere (T2T) gap-free genomes of pungent *C. annuum* and non-pungent relative *C. rhomboideum* revealed placenta-specific accessible chromatin regions that regulate capsaicinoid accumulation. Tandem duplications of key genes were identified between pungent and non-pungent species. Phylogenetic analysis showed *Capsicum* more closely related to *Physalis* than to *Solanum.* Capsaicinoids biosynthesis, unique to *Capsicum,* likely evolved after its split from *Physalis* around 17 million years ago. *C. baccatum* diverged from *C. annuum* and *C. chinense* about 5 million years ago (Weikai et al., 2024).

Sequencing, de novo assembly and annotation of cultivated *C. annuum* (Zunla-1) and its wild progenitor *C. annuum* var. *glabriusculum* suggested about 81% of repetitive sequences are due to retrotransposon expansion (LTR expansion). Comparative analyses also revealed molecular footprints of artificial selection and potential domestication genes along with tandem duplications likely contributing to pepper's pungency (Qin et al., 2014a, b). The pan-genome variation map derived from 500 accessions of the five domesticated *Capsicum* spp. and their wild relatives depicted phenotypic variations among the species arising from selections in separate genomic regions. The introgressions from *C. baccatum* into *C. chinense* and *C. frutescens* were identified as contributing to enhanced stress tolerance (Feng et al., 2023).

Fig. 11.1 Small-round, erect, soft-flashed, and deciduous fruits of wild *C. annuum*

11.4 Domestication Syndromes

Domestication is defined as the selection of wild species for cultivation and human use and the domestication syndrome refers to a set of desirable/favorable traits that were or are, being selected by the domesticators (Gepts 2014). After acquiring such desirable traits, domesticated plants lose their ability to grow in the wild and need human care to complete their reproductive cycles. The wild *C. annuum* chilli (bird's-eye chilli) typically has small-round, erect, pungent, deciduous, and soft-fleshed fruits (Fig. 11.1). These traits facilitate easy seed dispersal by birds. The major domestication syndromes of peppers are (i) non-deciduous fruits (remain on the plant until harvested manually; (ii) pendent fruits (associated with size increase, better protection from sun exposure, and predation by birds); and (iii) fruit appearance and varied degree of pungency (associated with consumer preference) (Kumar et al. 2018). The cost-share for manual harvesting is considerably high (>20%) in total chili pepper production costs. Therefore, in countries with a shortage of manual labor, breeders are manipulating primitive traits like deciduous and erect fruits, for easy machine harvesting.

11.5 Discovery of Super-Hot Landraces

Naga Jolokia, a landrace from the Assam state of India, was found to be India's hottest chilli (Mathur et al. 2000). This triggered considerable interest leading to the discovery of highly pungent (super-hot pepper) genotypes in the past two decades. Trinidad Scorpion, a landrace from Trinidad and Tobago, was found to be hotter than Bhut Jolokia, followed by Butch T Scorpion (from Australia), and the hottest known Trinidad Moruga Scorpion (Bosland and Votava, 2012). Several variants of Trinidad Scorpion are cultivated in Trinidad and Tobago, e.g., Trinidad 7-Pot Jonah and Douglah Trinidad Chocolate, which are genetically distinct from Bhut Jolokia (Bosland et al. 2012). Bhut Jolokia widely grown in the Northeast regions of India is believed to have originated from sympatric domesticated species (Rai et al. 2013). The natural variability within and between super-hot landraces is maintained *in-situ* by local communities and warrants systematic *ex-situ* conservation. "Carolina Reaper with high pungency of 2.2 million Scoville units broke earlier records of high pungency and later "Pepper X" was bred by Ed Currie with the highest 3.18 million heat units. Notably, capsaicinoids are normally produced in secreting vesicles found on the placenta of pepper fruits. However, super-hot landraces developed accessorial vesicles on the pericarp in addition to the placenta leading to more than 1 million Scoville heat units (Bosland et al. 2015).

11.6 Market Types and breeding Approach

Although in international trade paprika refers to non-pungent pepper genotypes, in many regions (e.g., Eastern Europe) paprika is synonymous with peppers and includes both sweet and chilli varieties.

Tremendous variability with plant characteristics exists in peppers. Based on fruit size, shape, and pungency levels, more than 20 commercial types (e.g., cayenne, ancho, bell, jalapeño, pasilla, yellow wax, jwala, etc.) are recognized and cultivated (Fig. 11.2). The pepper breeding objectives are broadly grouped into five market types: (i) fresh market (green, red, multicolored whole fruits); (ii) fresh processing (sauce, paste, canning, pickling); (iii) dried spice (whole fruits and powder); (iv) oleoresin and capsaicin extraction; and (v) ornamental (plants and/or fruits) (Poulos 1994). The common breeding objectives for industrial extraction are increased fruit yield and capsaicin (pungency) and oleoresin (colour value) contents. The breeding effort, however, for higher capsinoids has been scanty. Capsinoids are a non-pungent capsaicin analog that possesses medicinal properties for natural drug formulation against obesity, an alarming chronic disease mankind faces today due to unbalanced food and lifestyle.

Modern pepper improvement programs are in the hands of public and private breeding groups operating at local, regional, and global scales. Depending on the organizational capacity and targeted markets for the seed, breeding objectives and approaches vary from simple selection to gene deployment using emerging tools including multi-omics (Lozada et al. 2022). Quantitative trait loci (QTLs) for important fruit traits such as size, yield, shape, colour, chlorophyll content, metabolite content, pungency, and post-harvest water loss have been identified in both intraspecific and interspecific crosses (Mohan and Paran 2019). Most breeding programs focus on developing and commercializing hybrid seeds to generate continuous revenue. Additionally, hybrids make it easier to combine several desirable traits, and intellectual property rights are protected as the parent lines are kept as trade secrets. Nonetheless, the area under unimproved (landraces) and improved open-pollinated types still dominate in several regions.

Fig. 11.2 Globally recognized commercial fruit types of chili (spice) and sweet (vegetable) peppers. *Photo credit* Susan Lin, World Vegetable Center, Taiwan

11.7 Important Traits

Modern pepper breeding, like other crops, focuses on several biotic and abiotic stresses of economic importance (Barchenger et al., 2019; Table 11.1). The discovery of resistant germplasm, interspecific hybridization, and genomic and post-genomic tools are employed to accelerate product development. Breeding for biotic stress resistance and combining resistance with desirable fruit market types is a long-term process.

11.7.1 Anthracnose

Among the fungal pathogens, Anthracnose fruit rot is major devastating pathogen, to date, 24 species of *Colletotrichum* infecting chilli have been identified, of which most commonly pathogenic are: *C. truncatum* (syn. *C. capsici*), *C. scovillei* (previously identified as *C. acutatum*) and *C. siamense* (previously identified as *C. gloeosporioides* (Ridzuan et al. 2018). The application of pesticides at fruit ripening stage to control this disease raises food safety concerns and pesticide residues thereby limiting the export value. Exploring host plant resistance is the best advocated strategy and to develop resistant varieties/ F_1 hybrids to combat the losses due the disease. Globally, *C. baccatum* (PBC80 & PBC81) and *C. chinense* (PBC932) (AVRDC Report, 1998) have been identified as potential donors for the resistant genes for anthracnose fruit

Table 11.1 Major biotic and abiotic stresses, symptoms, host plant resistance sources, and genetics in *Capsicum*

S. no	Stress	Symptoms	Resistant sources	Genetics	References
Biotic stresses					
1	Root-knot nematode *Meloidogyne* spp (*M. javanica, M. arenaria, M. incognita, M. hapla*)	Plants show chlorosis, stunted growth, wilting and reduced leaf, and fruit size, gall formation in roots	S343, Surjamukhi, Punjab Tej, Japani Longi, Sel 15, JCA288, Sel 6, CM334	Monogenic dominant; and *Me* gene cluster on chromosome (Chr) 9	Fazari et al. (2012), Naresh et al. (2019)
2	Phytophthora root rot (*Phytophthora capsici*)	Causes root rot, stem blight, and fruit rot; wilting of plants and water-soaked lesions on leaves and fruits	CM334, PI201232, PI201234, KTC144, KTC148, chili local, KTPL19, Pant Chilli, and PBC631	Cross-specific: recessive, dominant, polygenic with additive and epistatic loci (*5.1*, *5.2*, and *5.3*) for broad-spectrum resistance	Thabuis et al. (2004), Naresh et al. (2019) Siddique et al. (2019), Katoch and Singh (2021)
3	Bacterial wilt (*Ralstonia solanacearum*)	Wilting of lower leaves followed by sudden death of the plans	IIHR-B-HP-130, LS2341, Manganji, PM687, White Kandari, Ujwala, Anugraha, BVRC 1, KC352	Complementary genes, polygenic and monogenic recessive; *pBWR-1* QTL on Chr 1, *qRRs-10.1* on Chr 10, partial dominance	Mimura et al. (2009), Du et al. (2019), Naveena et al. (2020), Deepu (2020), Thakur et al. (2014), Chae et al. (2022)
4	Powdery Mildew (*Leveillula taurica*)	Yellowing and premature defoliation, with powdery growth on the bottom of leaves	H3, H-V-12, 4638, K515R, PBC167 (*C. annuum*), CNPH 36, 38, 50, 52, 279, 288 (*C. baccatum*)	MonogenicDominant and polygenic; *PMR1* mapped on Chr 4	Jo et al. (2017), Gabor et al. (2017)
5	Anthracnose fruit rot (*Colletotrichum capsaici* and *C. gleosporides*)	Dark sunken spots sporangia scattered or arranged in concentric rings on fruits	PBC80, PBC81, PI594137 (*C. baccatum*), PBC932 (*C. chinense*)	Monogenic recessive, duplicate gene, polygenic, quantitative trait loci (QTLs) identified	Gniffke et. al. (2013), Lee et al. (2010), Ridzuan et al. (2018)

(continued)

Table 11.1 (continued)

S. no	Stress	Symptoms	Resistant sources	Genetics	References
	Anthracnose fruit rot (*Colletotrichum scovillei*)	Upward curling and distorted leaves	PBC932 (*C. chinense*)	Complementary two dominant genes (green fruit), two recessive genes (red fruit); *$AnR_{GO}5$ locus*; mapped on Chr 5	Zhao et. al. (2020)
6	*Cucumber mosaic virus* (CMV)	Mosaic, leaf distortion, and lesions on fruits	Bukang, BJ0747, IHR2451, IHR4503 (*C. annuum*) Pen 3 and PI439381 (*C. baccatum*), BG2814-6, S1839-2–4, PBC688 (*C. frutescens*)	Monogenic recessive, partially dominant, single dominant, and polygenic	Kang et al. (2010) Naresh et al. (2016), Yao et al. (2013), Guo et al. (2017)
7	Poty viruses	Vein banding, mosaic and necrosis symptoms	IHR2451, IHR4503 (ChiVMV); PI 159,236, PI 152,225) (*Tobacco etch virus, Pepper mottle virus*); Yolo Y (*Potato virus* Y)	Monogenic recessive (*pvr* genes), dominant Pvr4 (*Pvr7*) mapped on Chr 10	Naresh et al., (2016), Venkatesh et al. (2018)
8	*Tospoviruses Tomato spotted wilt virus* (TSWV)	Concentric zones on leaves and fruits	PI152225, PI159236, PI159234, AC09207 (*C. chinense*), CNPH 275, C00943, ECU973, PIM26-1 (*C. baccatum*)	Dominant resistance (*Tsw* mapped on Chr *10*, monogenic dominant and non-allelic gene actions	Soler et al. (2015), Hoang et al. (2013)
9	Groundnut bud necrosis virus (GBNV)	Mosaic and ring spots and necrosis	EC631810, IHR4360, IIHR4577, IHR4578, IIHR4582, IHR4585, IIHR4587, IHR4588	–	Pavitra et al. (2020)
10	Begomoviruses	Chlorosis, curling and reduction in leaf size; no fruits set or tiny fruits	Bhut Jolokia, DLS-Sel-10, WBC-Sel-5, PBC145, PBC143, Kalyanpur Chanchal, HR4517, IHR4615, IHR4630	Single recessive and monogenic dominant	Adluri et al. (2017), Srivastava et al. (2017), Singh et al. (2016), Yadav et al. (2022),Thakur et al. (2019), Maurya et al. (2019)

(continued)

Table 11.1 (continued)

S. no	Stress	Symptoms	Resistant sources	Genetics	References
11	Thrips (*Frankliniella occidentalis*)	Deformation and wrinkling of leaves	CGN16975, AC1979, CM331, Aji Blanco Christal (*C. baccatum*)	Major QTL on Chr 6	Maharijaya et al. (2011, 2015), Haperen et al. (2021)
12	Mites (*Polyphagotarsonemus latus*)	Leaf mottling and downward curling, bronzing of young tips and leaves	IC342390, IC572492, IC337281, IC344366, Kalyanpur Red, Punjab Lal, Goli Kalyanpur, X1068, X204	–	Rameash et al. (2015), Dhall (2015)
13	Whitefly (*Bemisia tabaci*)	Leaf deformation and yellowing	GCh-3, GCh-1, IAC1544, IAC1545, IAC1579 P2, P4, ACC1, ACC12	–	Rajput et al. (2017), Pantoja et al. (2018), Jeevanandham et al (2018)
Abiotic stresses					
1	Heat stress	Stunted growth, flower abortion and poor fruit set	Sy-2, Pepsi-17–2, Punjab Guchhedar, Ajeet 1	Genomic regions for fruit set and pollen germination at high temperatures \mapped on Chr 3 and 6	Yamazaki et. al. (2020), Kaur et al. (2016), Mishra et al. (2020)
2	Salt stress	Stunted growth	CO1, K1, Jayanthi, Arka Suphal, EC497636	–	Balasankar et al. 2017
3	Drought stress	Stunted growth, flower abortion, and poor fruit set	IHR4502 (*C. chinense*)	Additive and non-additive, predominantly by dominance effects for root traits	Naresh et al. 2017

rot. *C. baccatum* is cross-incompatible with *C. annuum*, which is the main hindrance in practical breeding.

Recently, *C. annuum* chilli pepper derived from different *Capsicum* spp. lines (PBC80 and PBC932) were developed using marker-assisted selection and evaluated against two aggressive isolates of *Colletotrichum acutatum* (*Ca*) and *C. capsici* (*Cc*) (Suwor et al., 2015). In the second step, they used introgression lines derived from PBC932 and PBC80 and a susceptible parent to develop three-way populations. MAS facilitated time and cost-effective generation advancement to combine resistance from two non-*annuum* (*baccatum* and *chinense*) sources (Suwor et al. 2017). Currently, these introgression lines are being used to improve fruit quality (Saengarun et al., 2025).

Some of these lines are now being evaluated at the on-farm trail in Thailand (Fig. 11.3).

Fig. 11.3 On-farm evaluation of anthracnose-resistant lines in Nakhon Ratchasima Province, Thailand

11.7.2 *Phytophthora*

Phytophthora capsici Leonian is a major devastating pathogen causing global yield losses. Criollo de Morelos-334 (CM334) was identified as the best donor for resistance against wide virulent strains (Sy and Bosland 2005). Meta QTL analysis indicated three QTLs, viz. *Meta Pc5.1*, *Meta Pc5.2*, and *Meta Pc5.3*, on chromosome 5 for Phytophthora root rot resistance (Mallard et al., 2013). Recently, Kumar et al. (2012) screened public domain molecular markers, namely CAMS 405, Hpms 1–62, Hpms MADS, SSR 9, and CAMS 839, linked to the PcRR resistance loci, and validated them in IHR3575 (derivative of Criollo de Morelos 334; CM334) and IHR3476 (highly susceptible) derived populations. Further, new SSR markers, HpmsE034 and HpmsE062 (explaining PVE of 9%), linked to PcRR resistance against the Indian isolate were identified.

11.7.3 *Begomoviruses*

Among the viruses, Begomoviruses have become a major threat to chilli pepper production worldwide. Exploiting host plant resistance is an efficient, environmentally friendly, and sustainable approach. Globally, various resistant sources have been identified, such as BG-3821 (*C. chinense*) (Anaya-Lopez et al., 2003), BAPep-5 (*C. annuum)* (Koeda et al., 2021), GR1 (*C. chinense*) (Mori et al., 2022), and LP97 (*C. annuum)* (Siddique et al., 2022), mainly against the Pepper leaf curl Indonesia virus (PepYLCVIV). Accessions 9852–123 (*C. annuum*) (Barchenger et al., 2019) and PSP11-7 & PSP11-10–1 (*C. annuum*) (Suwor et al., 2021) are resistant against the *Pepper yellow leaf curl* Thailand virus (PepYLCTHV). The single recessive gene locus *pepy-1* was mapped on chromosome 5 in BAPep-5 (Koeda et al., 2021). Field studies of *pepy-1* carrying accession BApep-5 and plants in the F_2 population of BApep-5 × BApep-4 indicated that *pepy-1* conferred resistance against multiple begomoviruses (PepYLCIV, TYLCKaV, AYVV, and PepYLCAV) (Pohan et al., 2022). Through genotyping by sequencing and QTL mapping, three QTLs—*peplcv-1*, *peplcv-7*, and *peplcv-12*—on chromosomes 1, 7, and 12, respectively, explained 9.9%, 31.7%, and 6.3% of the phenotypic variation for resistance in the LP97 x ECW30R F_2 population against PepYLCV (Siddique et al., 2022). Recently, a new resistant source GR1 (*C. chinense*) has been identified against Pepper yellow leaf curl Indonesia virus. Inheritance studies showed that resistance is a dominant trait controlled by multiple genes. Two QTLs on chromosomes 11 and 4, explaining 19.7% and 31.6% of the phenotypic variation, respectively, have been reported in the GR1 x Habanero F_2 population (Mori et al., 2022). Another resistant source "PG1-1" (*C. annuum*) for Pepper yellow leaf curl Aceh virus (PepYLCAV) and Pepper yellow leaf curl Indonesia virus (PepYLCIV) was identified. Inheritance of resistance was the monogenic dominant and the *Pepy-2* locus was mapped on chromosome 7 in the PG1-1 × SCM334 F_2 and F_3 populations. The candidate gene CaRDR3 (RNA-dependent RNA polymerase 3a) was also identified in the region associated with resistance (Koeda et al., 2022). In India, DLS-Sel-10, WBC-Sel-5, and PBC 142 were identified as resistant sources against chilli leaf curl disease (Srivastava et al., 2017). An accession S-343 with resistance to ChiLCD, mainly caused by Tomato leaf curl Joydebpur virus (ToLCJoV), was identified and inheritance was monogenic dominant (Thakur et al., 2019). Using this accession, a GMS-based F_1 hybrid CH-27 highly resistant to ChiLCVD was developed (Dhaliwal et al., 2015). The resistance was mapped on chromosome 6, and two SSR markers (PAU-LC-343–1 and CA 516,044) were reported as linked markers for chilli leaf curl disease resistance, at genetic distances of 8.9 and 6.8 cM, respectively (Thakur et al., 2020). The resistant sources, IHR4615 and IHR4517 for Chilli leaf curl virus-Raichur isolates were identified with monogenic dominant inheritance (Yadav et al., 2022).

11.7.4 Male Sterility

Considering the importance of male-sterility traits in hybrid development, both nuclear male sterility (genic male sterility, GMS) and cytoplasmic male sterility (CMS) systems are utilized in peppers. The CMS chilli pepper lines are used for commercial hybrids, however, the use of the CMS system in sweet peppers is limited. This is because of the unstable expression of male sterility at low temperatures and the lack of strong restorer sweet pepper genotypes (Lin et al. 2015). Two independently isolated and commercially utilized S-cytoplasms of pepper are genetically similar (Kumar et al., 2009). Thus, diversification of S (CMS)-cytoplasm is desirable to avoid risks associated with the vulnerability of commercial chilli pepper hybrids (Yeh et al. 2016). More than 20 monogenic recessive nuclear male sterile (*ms*) genes in pepper are known and molecular markers linked to several *ms* genes facilitate marker-assisted breeding (Naresh et al. 2018). Similarly, restorer of fertility (*Rf*) was fine mapped to a region of 821 kb in length on chr 6 and a PPR gene (CaPPR6) was identified as a strong candidate for *Rf* based on expression pattern and characteristics of encoding sequence (Jo et al., 2016). A single dominant gene *CaRfHZ* was mapped between P06gInDel-66 and P06gInDel-89 on chromosome 6 and fine-mapped to 533.81 kb region and *Capana06g002968* was proposed as the strongest candidate for the *CaRfHZ* gene (Nie et al., 2022).

11.7.5 Transgenic Peppers

Peppers, unlike tomatoes, tobacco, and petunia, have an inherent problem with in vitro regeneration. The severe recalcitrant morphogenic nature, formation of rosette shoots, ill-defined shoot buds, and genotypic-dependent tissue culture response in peppers hampered progress in genetic transformation. Genetically engineered transgenic pepper cultivars have not been commercialized. *The agrobacterium*-mediated transformation technique is mostly used for developing experimental transgenic peppers (Table 11.2). These protocols show promise for advancing the progress of pepper gene-editing research (Park et al. 2021).

11.8 Nuclear and Organelle Genomes

Ever since the first full genome sequences of cultivated *C. annuum* CM334 (Kim et al. 2014), whole-genome sequences of Zunla-1 and their wild progenitor *C. annuum* var. *Glabriusculum* chiltepin (Qin et al. 2014a, b), *C. baccatum* PBC81 and *C. chinense* PI159236 (Kim et al. 2017), and F_1 hybrid (CM334 × blocky pepper; Hulse-Kemp et al. 2018) have been reported. In addition to six full draft sequences, a pangenome of 383 pepper cultivar's databases and several online tools are available on various platforms (Li et al. 2019). Reduced complexity sequencing approach making use of whole-genome pepper sequences as references is underway to produce high-density genetic maps for tracking disease resistance loci. Comparative analyses between the whole genome sequences of pepper and other species of Solanaceae have pinpointed key genes involved in the evolution and domestication of pepper (Qin et al. 2014a, b). The chloroplast genome of several *Capsicum* species and the mitochondrial genome of at least one accession have been sequenced (Kumar et al. 2019).

These sequences, databases, and tools have conferred unprecedented insight into the pepper gene contents, enabling the rapid identification of genes implicated with metabolic pathways, disease resistance, and other traits and could also facilitate strategic information generation in epigenetics.

11.9 Epigenetics

Like many crop plants, the field of epigenetics in pepper breeding remains largely unexplored. This is partly because plant breeding and protection researchers have been more passionate about accelerating crop improvement based on the principles of central dogma (DNA-RNA–Protein). Historically trained with the same principles, the approach is also reflected in the mindset of the young researchers creating a self-evolving bias that limits progress in epigenetics.

The application of genomics and post-genomics tools is vital for efficient pepper breeding operations. This has been successfully demonstrated for desirable gene discovery to deployment, pathogen diagnosis, and analyzing host–pathogen interactions. Since new virulent pathogen races keep emerging, identifying novel resistant sources and marker developments is a continuous process. Contrastingly, markers once developed for biotic stress and quality traits could be used continuously—a one-time technological tool. In our experience, marker application in peppers is

Table 11.2 Examples of experimental transgenic peppers developed in the past decade for various purposes

Genotype-cultivar; Transformation method	Constructs	Salient results	Reference
Chili pepper—Habanero; *Agrobacterium- mediated*	GUS + β-glucuronidas:: pCAMBIA2301 and pCAMex + cDNAs encoding	Highest (55%) transformation efficacy found in leave explants	Arcos-Ortega et al. (2010)
Chili pepper—Aiswarya 2103; *Agrobacterium- mediated* (EHA105)	Agrobacterium EHA105 + pBINASCOSM:: CaMV 35S + *npt* II	Salt resistance transgenic plants with stable integration and expression of osmotin gene	Subramanyam et al. (2011)
Chili pepper—Pusa Jwala; *Agrobacterium- mediated* (EHA 105)	pBin AR βC1::*nptII* + *βC1* (kanamycin and rifampicin)	Transgenic plants with *βC1* gene (*satellite DNA β* molecule of *Chilli leaf curl virus,* useful in understanding *βC1*-mediated pathogenesis	Kumar et al. (2012)
Sweet pepper—California Wonder; *Agrobacterium- mediated* (LBA 4404)	pBI121::*npt-II* + gus reporters (cefotaxime and kanamyein)	Standardization of pre-culture and co-cultivation time	Verma et al. (2013)
Sweet pepper—Arka Gaurv; *In planta; Agrobacterium- mediated*	pCAMBIA-35sCaNPR1 Pathogenesis-related gene (*NPRI*) for resistance to powdery mildew	Transgenic plants overexpressing *NPR1* showed 76.7% reduced incidence of powdery mildew	Arthikala et al. (2014)
Chili pepper—G4; *Agrobacterium-mediated* (LBA4404)	pCAMBIA 2301 + *rd29:dreb1A* and neomycin phosphotransferase II (*npt II*) under CaMV 35S promoter and *nos* terminator	Rooting and shooting highest in MS medium with 2 mg L-1 indole butyric acid (IBA); transgene integration confirmed by PCR and Southern blotting	Maligeppagol et al. (2015)
Chili pepper—Balady, Topepo Rosso; *Agrobacterium-mediated*	Agrobacterium GV3101 + *β-Lcy*::*nptII*	Increased β-carotene content from 5.2 mg/100 g dw in the wild plants to 36.39 mg/100 g dw in the transgenic plants	Nagar and Mekawi (2018)
Chili pepper—Pusa Sadabahar, Pusa Jwala; *Agrobacterium-mediated* (LBA444)	LBA4404 + pCAMBIA2301:: GUS + *NPT-II*	Hypocotyl explants were more amenable with high transformation (30%) efficacy in both cultivars. The developed protocol was reproducible	Mahto et al. (2018)

(continued)

Table 11.2 (continued)

Genotype-cultivar; Transformation method	Constructs	Salient results	Reference
Chili pepper—G4; *Agrobacterium*-mediated (LBA4404	Agrobacterium LBA4404 + pBin438 (*TaNHX2*):: CaMV 35 promoter + *NPT II* gene	Salt-tolerant transgenic plants showed enhanced levels of proline, chlorophyll, superoxide dismutase, ascorbate peroxidase, and reduced levels of hydrogen peroxide (H_2O_2)	Bulle et al. (2016)
Chili pepper—NuMex 6–4, CM-334; *Agrobacterium*-mediated	RB gene construct pCLD04541, along with its native promoter	The RB gene from *S. bulbocastanum* confers resistance to *P. capsici*, demonstrating that the RB gene has a broader host range	Bagga et al. (2019)
Chili pepper—AA3; *Agrobacterium*-mediated (GV3101	VIGS + *CaSABP12::* TRV2 vector digested with *BamHi* and *KpnI* enzymes	The CaSBP12 negatively regulates salt stress tolerance that may relate to ROS signaling cascades	Zhang et al. (2020)

case-specific even if the marker is from a coding sequence (Yeh et al. 2016; Naresh et al. 2018). This is because, unlike tomato, pepper has a larger pepper genome size (3.5 Gb) with the presence of pseudogenes, homeotic mutants, repetitive elements (REs), genotype x environment interactions, and the undeciphered role of epigenetics (Madhavi et al., 2023). In a series of experiments, mostly conducted in Japan, researchers have demonstrated cell sap transmission of genetic materials and graft-induced inherited variation has been demonstrated in peppers (Miyashita and Takaki 1980; Kollárová et al., 2018; Taller et al., 1998).

11.10 Conclusions and Future Prospects

The abundance of REs in *Capsicum* affects the intra- and inter-specific genome size variations and the genome generally has dense cytosine methylation (Rawoof et al. 2019). DNA methylation is an important epigenetic mark, and therefore, it is imperative to conduct impact-focused trans-disciplinary experiments in peppers to get more insights into genome complexity and possibly unlock the potential of epigenetics. The researchers could typically study natural single-gene knockout mutants and their marker systems, for example, the capsaicin synthase (CS) gene. Contrasting genotypes are available from California Wounder (0 Scoville heat units) to Trinidad Moruga Scorpion (> 2 million Scoville heat units) (Aamir et al., 2022). The advancement and regulatory acceptance of gene-editing techniques in pepper (Park et al., 2021) provide new opportunities for impactful research. Approaches like De novo domestication require well-characterized wild, semi-wild species and cultivated landraces as basic raw materials. Hence, acquiring, exchanging, and utilizing gene bank pepper germplasm should get renewed attention in globally high-performing pepper breeding programs.

The improved seeds available to growers are the final products of pepper breeding. Typically, these enhanced seeds are hybrids developed and marketed by private seed companies and their stakeholders. The public institutions (national agricultural research systems and universities) continuously train critical human resources, supply enhanced germplasm, and provide an amenable policy for private seed businesses. Thus, pepper breeding strategies in a positive-sum competitive era would need enhanced private and public sector's collaborative platforms to reach farmers and consumers.

References

Aamir S, Sadaruddin C, Muzafaruddin C, Nazir A, Changfei G, Pingxian Z (2022) Recent Advances in DNA Methylation and Their Potential Breeding Applications in Plants. Horticulturae 8(7):562–562

Adluri PK, Baldoldiya GM, Nath PD (2017) Screening of Bhut Jolokia (*Capsicum chinense* Jacq.) germplasm of Northeast India against chili leaf curl virus. Int J Pure Appl Biosci 5:1189–1196

Aitken E, McColl G, Kingsmore D (2017) The Role of Qutenza® (Topical Capsaicin 8%) in treating neuropathic pain from critical ischemia in patients with end-stage renal disease: an observational cohort study. Pain Med 18(2):330–340. https://doi.org/10.1093/pm/pnw139

Anaya-Loopez JL, Torres-Pacheco I, Gonzalez CMJA, Garzon T, Pons HJL, Guevara GRG, Munoz SCI, Guevara OL, Rivera BRF, Hernaandez VS (2003) Resistance to geminivirus mixed infections in Mexican wild peppers. HortSci 38(2):251–255

Arcos-Ortega GF, Chan-Kuuk RA, Gonzalez KWA, Souza RP, Nakazawa UYE, Aviles BE, Godoy HG, Lawton M, Aguilar J (2010) *Agrobacterium tumefaciens*-transient genetic transformation of Habanero pepper (*Capsicum chinense* Jacq.) leaf explants. J Biotechnol 3:13

Arthikala MK, Nanjareddy K, Lara M, Rohini S (2014) Utility of a tissue culture-independent Agrobacterium-mediated in planta transformation strategy in bell pepper to develop fungal disease-resistant plants. Sci Hortic 170:61–69

Asian Vegetable Research and Development Center (AVRDC) (1998) Report, pdf.usaid.gov

Baenas N, Belovic M, Ilic N, Moreno DA, GarcíanVC (2019) Industrial use of pepper (*Capsicum annum* L.) derived products: technological benefits and biological advantages. Food Chem 274:872–885

Bagga S, Lucero Y, Apodaca Wathsala R K, Phillip L, Jose LO, Champa SG (2019) Chile (*Capsicum annuum*) plants transformed with the RB gene from *Solanum bulbocastanum* are resistant to Phytophthora capsici. PLoS One 14: e0223213. https://doi.org/10.1371/journal.pone.0223213

Balasankar D, Praneetha S, Arumugam T, Jeyakumar P, Manivannan N, Arulmozhiselvan K (2017) Genotypic response of chilli (*Capsicum annuum* L.) on germination and seedling characters to different salinity levels. Int J Curr Microbiol App Sci 6:2197–2205

Barchenger DW, Naresh P, Kumar S (2019) Genetic resources of *Capsicum*. In: Kole C, Ramchiary N (eds) The Capsicum genome. Springer Nature, Switzerland, pp 9–23

Barik S, Naresh P, Anand CR, Lakshman RDC, Koushik S, Acharya GC, Madhavi RK (2022) Breeding peppers for industrial uses: progress and prospects. Ind Crop Prod 178. https://doi.org/10.1016/j.indcrop.2022.114626

Bosland PW, Votava EJ (2012) Peppers, vegetable and spice *Capsicum*, 2nd edn. CABI, Wallingford, UK

Bosland PW, Coon D, Reeves G (2012) Trinidad Moruga Scorpion pepper is the world's hottest measured chile pepper at more than two million Scoville heat units. HortTechnology 22:535–538

Bosland P, Coon D, Cooke P (2015) Novel formation of ectopic (nonplacental) capsaicinoid secreting vesicles on fruit walls explains the morphological mechanism for super-hot chile peppers. J Amer Soc Hotic Sci 140:253–256

Bulle M, Yarra R, Abbagani S (2016) Enhanced salinity stress tolerance in transgenic chilli pepper (*Capsicum annuum* L.) plants overexpressing the wheat antiporter (TaNHX2) gene. Mol Breed 36 https://doi.org/10.1007/s1103201604515

Chae SY, Lee K, Do JW, Sun CH, Kang HL, Myeong CC, Eun YY, Jae BY (2022) QTL Mapping of resistance to bacterial wilt in pepper plants (*Capsicum annuum*) using genotyping-by-sequencing (GBS). Horticulturae 8:115

Deepu M (2020) Analysis of *QTL Bw1* and marker CAMS451 associated with the bacterial wilt resistance in hot pepper (*Capsicum annuum* L.). Plant Gene 24:2352–4073

Dhaliwal MS, Jindal SK, Cheema DS (2015) CH-27: A multiple disease resistant chilli hybrid. Agric Res J 52(4):127–129

Dhall RK (2015) Breeding for biotic stresses resistance in vegetable crops: a review. Res Rev: J Crop Sci Technol 4:13–27

Du H, Wen C, Zhang X, Xiulan X, Jingjing Y, Bin C, Sansheng G (2019) Identification of a Major QTL (qRRs-10.1) That confers resistance to Ralstonia solanacearum in pepper (*Capsicum annuum*) using SLAF-BSA and QTL mapping. Intel J Mol Sci 20:5887

EFSA (2015) Scientific opinion on the re-evaluation of paprika extract (E 160c) as a food additive. EFSA J 13:4320–4371

FAO (2020) www.fao.org/faostat

FAOSTAT (2022) www.fao.org/faostat

Fazari A, Palloix A, Wang LH, Hua MY, Sage PAM, Zhang BX, Djian CC (2012) The root-knot nematode resistance N-gene co-localizes in the Me-genes cluster on the pepper (*Capsicum annuum* L.) P9 chromosome. Plant Breed 131:665–673

Feng L, Jiantao Z, Honghe S, Cheng X, Xuepeng S, Xin W, Zhongyi W, Robert J, Jin W, Bingqian T, Hao X, Bowen H, Huan S, Bozhi Y, Lijun O, Xuefeng L, Shudong Z, Sha Y, Zhoubing L, Fang Y, Zhenming P, Yanqing M, Xiongze D, Shan W, Zhangjun F, Xuexiao Z (2023) Genomes of cultivated and wild Capsicum species provide insights into pepper domestication and population differentiation. Nat Commun 14:5487

Gabor BK, Just BJ, Huang C, Carl MJ, Dirk V, Joel MK, Pablo AQ, Terry GB, Anton PA, Wenwen X (2017) Methods and compositions for producing Capsicum plants with powdery mildew resistance. US Patent No 9,689,045

Gepts P (2014) The contribution of genetic and genomic approaches to plant domestication. Curr Opinion Pl Biol 18:51–59

Gniffke PA, Shieh SC, Lin SW, Sheu ZM, Chen JR, Ho FI, Tsai WS, Chou YY, Wang JF, Cho MC, Schafleitner, Kenyon LR, Ebert AW, Srinivasan R, Kumar S (2013) Pepper research and breeding at AVRDC–The World Vegetable Center. In: XV EUCARPIA meeting on genetics and breeding of Capsicum and eggplant, 2–4 Sept, Turin, Italy, 305–311

Guo G, Wang S, Liu J, Baogui P, Weiping D, Wei G, Changzhou G, John CS (2017) Rapid identification of QTLs underlying resistance to Cucumber mosaic virus in pepper (*Capsicum frutescens*). Theor Appl Genet 130:41–52

Haperen V P, Voorrips RE, V Kaauwen M, Henriette DLM, Van E, Ric CHV, Joop JAV, Ben V (2021) Fine mapping of a thrips resistance QTL in Capsicum and the role of diterpene glycosides in the underlying mechanism. Theor Appl Gent 134:1557–1573

Hayman M, Kam Peter CA (2008) Capsaicin: a review of its pharmacology and clinical applications. Curr Anaesth Crit Care 19(5–6):338–343

Hoang N, Yang HB, Kang BC (2013) Identification and inheritance of a new source of resistance against tomato spotted wilt virus (TSWV) in *Capsicum*. Sci Hortic 161:8–14

Hulse-Kemp AM, Maheshwari S, Stoffel K, Kevin S, Theresa AH, David J, Stephen RW, Neil W Srividya R, Vijay K, Preyas S, Michael CS, Deanna MC, Allen VD (2018) Reference quality assembly of the 3.5-Gb genome of *Capsicum annuum* from a single linked-read library. Hortic Res 5(4) https://doi.org/10.1038/s41438-017-0011-0

Jeevanandham N, Marimuthu M, Natesan S, Shanthi M, Sathiyamurthy A (2018) Levels of plant resistance in *Capsicum* spp against whitefly, *Bemisia tabaci*. Int J Curr Microbiol App Sci 7:1441–1449

Jo J, Venkatesh J, Han K, Lee HY, Choi G, Lee H, Choi D, Kang BC (2017) Molecular mapping of *PMR1*, a novel locus conferring resistance to powdery mildew in pepper (*Capsicum annuum*). Front Plant Sci 8:2090

Jo YD, Ha Y, Lee JH, Ziyan Z, Fang H, Jichi D, Weili C, Junjie C, Zhiming W, Kailin H (2016) Fine mapping of Restorer-of-fertility in pepper (*Capsicum annuum* L.) identified a candidate gene encoding a pentatricopeptide repeat (PPR)-containing protein. Theor Appl Genet 129(10):2003–2017 https://doi.org/10.1007/s00122-016-2755-6

Kang WH, Hoang NH, Yang HB, Jin KK, Sung HJ, Jang KS, Kook HK, Doil C, Byoung CK (2010) Molecular mapping and characterization of a single dominant gene controlling CMV resistance in peppers (*Capsicum annuum* L.). Theo Appl Gen 120:1587–1596

Katoch S, Singh A (2021) Screening of capsicum germplasm for resistance against *Phytophthora capsici* causing leaf blight and root rot. J Entomol Zool Stud 9(2):627–630. https://doi.org/10.22271/j.ento.2021.v9.i2i.8634

Kaur N, Dhaliwal MS, Jindal S, Parjeet S (2016) Evaluation of hot pepper (*Capsicum annuum* L.) genotypes for heat tolerance during reproductive phase. Intl J Bio-Resour Stress Manag 7:126–129

Khoury CK, Carver D, Barchenger DW, Barboza GE, van Zonneveld M, Jarret R, Lynn B, Michael K, Mark U, Kristin M, Gary PN, Paul WB, Stephanie LG (2020) Modelled distributions and conservation status of the wild relatives of chile peppers (*Capsicum* L). Divers Distrib 26:209–225

Kim S, Park J, Yeom SI, Yong MK, Eunyoung S, Ki TK, Myung SK, Je ML, Kyeongchae C, Ho SS, Saet BK, Koeun H, Jundae L, Eunbi C, So EL, Jongbum J, Hyunbin K, Cho D (2017) New reference genome sequences of hot pepper reveal the massive evolution of plant disease-resistance genes by retroduplication. Genome Biol 118(1):210. https://doi.org/10.1186/s13059-017-1341-9

Kim S, Park M, Yeom SI Yong MK, Je ML, Hyun AL, Eunyoung S, Jaeyoung C, Kyeongchae C , Ki TK, Kyongyong J, Gir WL, Sang KO, Chungyun B, Saet BK, HyeYL, Shin YK, Myung SK, Byoung CK, Yeong DJ, Hee BY, Hee JJ, Won HK, Jin KK, Chanseok S, Jae YL, June HP, Paul WB, Gregory R, Sung HJ, Hyung TC, Hee SC, Min SL, Yeisoo Y, Yang D C, Beom SP, Allen D, Hamid A, Theresa H, Woo T K, Hyun-S P, Hee KA, Inhwa Y, James J G, Jocelyn KCR, Iben S, Sang JL, Ryan W K, Ik YC, Beom SC, Jong SL, Yong HL, Doil C (2014) Genome sequence of the hot pepper provides insights into the evolution of pungency in *Capsicum* species. Nat Genet 46:270–278

Koeda S, Onouchi M, Mori N, Nadya SP, Atsushi JN, Elly K (2021) A recessive gene pepy-1 encoding Pelota confers resistance to begomovirus isolates of PepYLCIV and PepYLCAV in *Capsicum annuum*. Theor Appl Genet 134:2947–2964

Koeda S, Mori N, Horiuchi R, Chiho W, Atsushi JN, Hayato S (2022) PepYLCIV and PepYLCAV resistance gene Pepy-2 encodes DFDGD-Class RNA-dependent RNA polymerase in Capsicum. https://doi.org/10.21203/rs.3.rs-1396048/v1

Kollárová K, Závodská L, Kočeková L (2018) Transgenerational changes in sweet pepper (*Capsicum annuum* L.) after grafting. Hortic Sci 45(2):55–60. https://doi.org/10.17221/101/2017-HORTSCI

Kumar R, Kumar S, Devedi N, Sanjeet K, Ashutosh R, Major S, Dasrath SY, Mathura R (2009) Validation of SCAR markers, diversity analysis of male-sterile cytoplasms (S-) and isolation of an alloplasmic CMS in *Capsicum*. Sci Hortic 120:167–172

Kumar RV, Sharma VK, Chattopadhyay B, Chakraborty S (2012) An improved plant regeneration and Agrobacterium—mediated transformation of red pepper (*Capsicum annuum* L.). Physiol Mol Biol Plants 18:357–364

Kumar S, Kumar R, Singh J (2006) Cayenne/American pepper (*Capsicum* species). In: Peter KV (ed) Handbook of herbs and spices, vol 3. Woodhead Publishing, Cambridge, pp 299–3121

Kumar S, Shieh HC, Lin SW, Schafleitner R, Kenyon L, Srinivasan R, Wang JF, Ebert A, Chou YY (2018) Peppers (*Capsicum* spp.): domestication and breeding for global use. In: Mandal D, Shukla AC, Siddiqui MW (eds) Sustainable horticulture, vol 1: diversity, production, and crop improvement, Part III. Crop improvement and biotechnology. Apple Academic/CRC Press, Waretown, pp 387–400

Kumar N, Islam K, Ramchiary N (2019) Sequencing of *Capsicum* organellar genomes. In: Kole C, Ramchiary N (eds) The *Capsicum* genome. Springer Nature, Switzerland, pp 153–171

Lee J, Hong JH, Do J, Jae BY (2010) Identification of QTLs for resistance to anthracnose to two *Colletotrichum* species in pepper. J Crop Sci Biotechnol 13:227–233. https://doi.org/10.1007/s12892-010-0081-0

Li F, Liu M, Zhou Y (2019) Sequence databases and online tools for Capsicum Research. In: Kole C, Ramchiary N (eds) The Capsicum genome. Springer Nature, Switzerland, pp 221–232

Lin SW, Shieh HC, Wang YW, Chee WT, Roland S, Wen JY, Sanjeet K (2015) Restorer breeding in sweet pepper: introgressing Rf allele from hot pepper through marker-assisted backcrossing. Sci Hortic 197:170–175

Lozada DN, Bosland P, Barchenger D (2022) Chile pepper (*Capsicum*) breeding in the multi-omics era. Front Plant Sci. https://doi.org/10.3389/fpls.2022.879182

Madhavi RK, Rajesh K, Naresh P, Indivar P, Satya PB, Roshni P, Sai T, Pavani N, Mahebub S, Karishma P(2023) Chili: breeding and genomics. Veg Sci 50(Special Issue):177–18

Maharijaya A, Vosman B, Steenhuis B, Asep HG, Agus P, Richard GF, Roeland EV (2011) Screening of pepper accessions for resistance against two thrips species (*Frankliniella occidentalis* and *Thrips parvispinus*). Euphytica 177:401–410

Maharijaya A, Vosman B, Steenhuis BG, Koen P, Agus P, Richard GF, Roeland EV (2015) QTL mapping of thrips resistance in pepper. Theor Appl Genet 128:1945–1956

Mahto BK, Sharma P, Rajam MV, Pallavolu MR, Swatismita DR (2018) An efficient method for Agrobacterium-mediated genetic transformation of chili pepper (*Capsicum annuum* L.). Indian J Physiol 23:573–581

Maligeppagol M, Manjula R, Navale PM, Babu KP, Bhimanagoud K, Laxman R (2015) Genetic transformation of chilli (*Capsicum annuum* L.) with Dreb 1A transcription factor known to impart drought tolerance. Indian J Biotechnol 15:17–24

Mallard S, Cantet M, Massire A, Alexandre B, Sophie E, Véronique L (2013) A key QTL cluster is conserved among accessions and exhibits broad-spectrum resistance to *Phytophthora capsici*: a valuable locus for pepper breeding. Mol Breed 32:349–364

Mathur R, Dangi RS, Dass SC, Malhotra RC (2000) The hottest chilli variety in India. Curr Sci 79:287–288

Maurya P, Srivatstava A, Mangal M, Akshay T, Bikash M, Vikas S, Anil K, Pritam K, (2019). Genetic analysis for resistance to leaf curl disease in Chilli Peppers (*Capsicum annuum* L.) under specific situations. Indian J Genet 79(4):741–748

Mimura Y, Kageyama T, Minamiyama Y, Hirai M (2009) QTL analysis for resistance to *Ralstonia solanacearum* in *Capsicum* accession 'LS2341' J. Japan. Soc. Hortic. Sci., 78 :(3)307–313 https://doi.org/10.2503/jjshs1.78.307.

Mishra S, Laxman RH, Venugopalan RMR (2020) TIR approach and stress tolerance indices to identify donor for high temperature stress tolerance in pepper (*Capsicum annuum* L.). Plant Genet Resour: Charact Util 18:19–27

Miyashita Y, Takaki I (1980) Graft-induced heritable variations in *Capsicum annuum*. Jpn J Breeding 30(4):388–389

Mohan V, Paran I (2019) Molecular mapping and identification of QTLs and genes for economically important traits in the *Capsicum* genome. In: Kole C, Ramchiary N (eds) The Capsicum genome. Springer Nature, Switzerland, pp 105–120

Mori N, Hasegawa S, Takimoto R, Jianying C, Qingliang N, Weimin Z (2022) Identification of QTLs conferring resistance to begomovirus isolate of PepYLCIV in *Capsicum chinense*. Euphytica 218(20). https://doi.org/10.1007/s10681-022-02970-9

Nagar MME, Mekawi M (2018) Genetic engineering to improve β-carotene content in pepper. In: 4th International conference on biotechnology applications in agriculture (ICBAA), Benha University, Moshtohor and Hurghada, 4–7 April. Egypt Plant Biotechnol, pp 119–126

Naresh P, Krishna RM, Hema C, Madhavi KR (2016) Screening chilli (*Capsicum* spp.) germplasm against Cucumber mosaic virus and Chilli veinal mottle virus and inheritance of resistance. Eur J of Plant Pathol 146:451–464

Naresh P, Bhatt RM, Venkatachalapathi V, Gangadhararao P, Madhavi KR (2017) Inheritance of root traits in an interspecific cross of *Capsicum annuum* × *C. chinense* in the presence of low moisture. Int J of Veg Sci 23:575–583

Naresh P, Meenu K, Acharya GC, Anand CR, Lakshmana RDC (2019) Genetics and molecular markers for resistance to major soil-borne pathogens in chilli (*Capsicum annuum* L.). Res J Biotechnol 14:101–105

Naresh P, Lin SW, Lin CY, Yen WW, Roland S, Andrzej K, Sanjeet K (2018) Molecular markers associated with two non-allelic genic male sterility genes in peppers (*Capsicum annuum* L.). Front Plant Sci https://doi.org/10.3389/fpls.2018.01343

Naveena GS, Sandeep V, Ponnam N, Meenu K, Acharya GC, Srinivas P, Sangeetha G, Sahu GS (2020) Genetic analysis of bacterial wilt resistance in hot pepper (*Capsicum annuum* L.). Veg Sci 47(1):153–156

Nie Z, Song Y, Wang H et al (2022) Fine mapping and gene analysis of restorer-of-fertility gene CaRfHZ in pepper (*Capsicum annuum* L.). Int J Mol Sci 23(14):7633. https://doi.org/10.3390/ijms23147633

Pantoja KF, Rocha KC, Melo AM, Julio MM, Edson LLB, João PFB, Ricardo G, Rômulo FK, Marcelo AP, Renate KS (2018) Identification of *Capsicum* accessions tolerant to Tomato severe rugose virus and resistant to *Bemisia tabaci* Middle East-Asia Minor 1 (MEAM1). Trop Plant Pathol. 43:138–145

Park SI, Kim HB, Jeon HJ, Hyeran K (2021) Agrobacterium-mediated *Capsicum annuum* gene editing in two cultivars, hot pepper CM334, and bell pepper Dempsey. Int J Mol Sci 22:3921

Pasquale T, Mark TRW, Lorenzo B, Sandip K, Salvatore E, Alberto A, Roland S, Maarten Z, Jaime P, Maria JD, Bernard C, Arnaud B, Ezio P, Richard F, Sergio L, Ilan P, Veronique L, Giovanni G, Nils S (2021) Global range expansion history of pepper (*Capsicum* spp.) revealed by over 10,000 genebank access. Proc Natl Acad Sci 118(34):e2104315118

Pavithra BS, Reddy KM, Kedarnath G, Krishna MR (2020) Identification of resistant sources in chilli (*Capsicum* spp.) genotypes to Groundnut bud necrosis virus (GBNV). Australas Plant Patho 49:15–23

Pohan N, Alfan G, Khalil M, Putra B, Rayhan H, Yusuf H, Nurul H, Mika O, Ryosuke S, Yoshinori K, Kanami H, Sabaruddin Z, Elly K, Sota K (2022). Pepper (*Capsicum annuum*) plants harboring the begomovirus resistance Gene pepy-1 show delayed symptom progress and high productivity under natural field conditions. Hortic J 92. https://doi.org/10.2503/hortj.QH-015

Poulos JM (1994) Pepper breeding (*Capsicum* spp.): achievements, challenges, and possibilities. Plant Breed Abstr 64:143–155

Qin C, Changshui Y, Yaou S, Xiaodong F, Lang C, Jiumeng M, Jiaowen C, Shancen Z, Meng X, Yong L, Yulan Y, Zhiming W, Likai M, Haiyang W, Changying LH, Huangkai Z, Haijian L, Sandra GM, Diana LTS, Hao T, Xin T, Maojun Z, Zhiyong H, Anwei Z, Xiaoming Y, Junjie C, Wenqi L, Zhe C, Yongqiang F, Yongchao N, Shimin B, Xiuwei Y, Weipeng L, Huimin C, Yonggang Z, Shuang Y, Xiaodian W, Jiaohui X, Xueqin L, Shuaicheng L, Junyi W, Alain P, Paul WB, Yingrui L, Anders K, Rafael FRB, Luis HE, Ye Y, Jiping Y, Kailin H, Zhiming Z (2014a) Whole-genome sequencing of cultivated and wild peppers provides insights into Capsicum domestication and specialization. Proc Natl Acad Sci 111(14):5135–5140

Qin C, Yu C, Shen Y, Xiaodong F, Lang C, Jiumeng M, Jiaowen C, Shancen Z, Meng X, Yong L, Yulan Y, Zhiming W, Likai M, Haiyang W, Changying LH, Huangkai Z, Haijian L, Sandra GM, Diana LTS, Hao T, Xin T, Maojun Z, Zhiyong H, Anwei Z, Xiaoming Y, Yan Z, Min W, Xiaodan L, Bo W, Hongjun L, Hemi L, Yonggang Z, Shuang Y, Xiaodian W, Jiaohui X, Xueqin L, Shuaicheng L, Junyi W, Alain P, Paul WB, Yingrui L, Anders K, Rafael FRB, Luis HE, Ye Y, Jiping Y, Kailin H, Zhiming Z (2014b) Whole-genome sequencing of cultivated and wild peppers provides insights into Capsicum domestication and specialization. Proc Natl Acad Sci 111:5135–5140

Rai VP, Kumar R, Kumar S, Ashutosh R, Sanjeet K, Major S, Sheo PS, Awadesh BR, Rajneesh P (2013) Genetic diversity in Capsicum germplasm based on microsatellite and random amplified microsatellite polymorphism markers. Physiol Mol Biol Pl 19:575–586

Rajput VS, Prajapati BG, Rebari GN (2017) Screening of different genotypes against major insect pests of chilli (*Capsicum annum* L.). J Entomol Zool Stud 5:1552–1554

Rameash KS, Pandravada SR, Sivaraj N, Babu B, Chakrabarty S (2015) Screening chilli (*Capsicum annuum* L.) genotypes for resistance to broad mite (*Polyphagotarsonemus latus* Banks) and analysing the geographic distribution of resistance. Elec J Plant Breed 6:928–937

Rawoof A, Ahmad I, Ramchiary N (2019) Epigenome landscape in *Capsicum* genome. In: Kole C, Ramchiary N (eds) The Capsicum genome. Springer Nature, Switzerland, pp 187–199

Reddy BS, Sarojini G (1987) Chemical and nutritional evaluation of chili (*Capsicum annuum*) seed oil. J Am Oil Chem Soc 64:1419–1422

Ridzuan R, Rafii MY, Ismail SI, Martini MY, Gous M, Magaji U (2018) Breeding for anthracnose disease resistance in chilli: progress and prospects. Int J Mol Sci 19:3122

Saengarun P, Suwor P, Inprasert H, Kumsee N, Tsai W-S, Techawongstien S, Tainta T, Kumar S, Jeeatid N, Chatchawankanphanich O, Kramchote S (2025). Combining ability analysis in anthracnose-resistant introgression inbreds and developing improved resistant lines of pepper (*Capsicum annuum* L.). Scientia Hortic 339. https://doi.org/10.1016/j.scienta.2024.113899

Siddique H, Lee N, Ro KKH, Jelli V, Abate MS, Abhinandan SP, Amornrat C, Jin KK, Byoung CK (2019) Identifying candidate genes for *Phytophthora capsici* resistance in pepper (*Capsicum annuum*) via genotyping-by-sequencing-based QTL mapping and genome-wide association study. Sci Rep 9:9962

Siddique MI, Lee JH, Ahn JH, Meirina KK, Ramadhani S, Asep H, Jin KK, Byoung CK (2022) Genotyping-by-sequencing-based QTL mapping reveals novel loci for Pepper yellow leaf curl virus (PepYLCV) resistance in *Capsicum annuum*. PLoS ONE 17(2):e0264026. https://doi.org/10.1371/journal.pone.0264026

Singh AK, Kushwaha N, Chakraborty S (2016) Synergistic interaction among begomoviruses leads to suppression of host defense-related gene expression and breakdown of resistance in chilli. Appl Micorbiol Biotechnol 100:4035–4049

Soler S, Debreczeni DE, Vidal E, Aramburu J, Lopez C, Galipienso L, Rubio L (2015) A new *Capsicum baccatum* accession shows tolerance to wild-type and resistance-breaking isolates of Tomato spotted wilt virus. Ann Appl Biol 167:343–353

Sowbhagya HB, Chandrashekhar S, Dasappa I, Srinivas P (2015) Physico-chemical characteristics of chilli spent residue and its effect on the rheological, microstructural and nutritional qualities of bread. J Food Sci Technol 52. https://doi.org/10.1007/s13197-015-1824-1

Srivastava A, Mangal M, Saritha RK, Pritam K (2017) Screening of chilli pepper (*Capscium* spp.) lines for resistance to the begomovirus causing chili leaf curl disease in India. Crop Prot 100:177–185

Subramanyam K, Sailaja KV, Lakshmidevi K (2011) Ectopic expression of an osmotin gene leads to enhanced salt tolerance in transgenic chilli pepper (*Capsicum annum* L.). Plant Cell Tissue Organ Cult 105:181–192

Suwor P, Thummabenjapone P, Sanitchon J, Sanjeet K, Suchila T (2015) Phenotypic and genotypic responses of chili (*Capsicum annuum* L.) progressive lines with different resistant genes against anthracnose pathogen (*Colletotrichum* spp.). Eur J Pl Pathol 143:725–736

Suwor P, Sanitchon J, Thummabenjapone P, Sanjeet K, Suchila T (2017) Inheritance analysis of anthracnose resistance and marker-assisted selection in introgression populations of chilli (*Capsicum annuum* L.). Sci Hortic 220:20–26

Suwor P, Masirayanan T, Khingkumpungk H, Wen ST, Kanjana S, Suchila T, Sanjeet K, Somsak K (2021) Chili (*Capsicum annuum* L.) genotypes resistant to Pepper yellow leaf curl Thailand virus (PepYLCTHV). J Plant Prot Res 61(4):377–383

Sy O, Bosland PW (2005) Inheritance of *Phytophthora* stem blight resistance as compared to Phytophthora root rot and Phytophthora foliar blight resistance in *Capsicum annuum* L. J Amer Sct 130(1):75–78

Taller J, Yagishita HY, N, Kita M, Ogata S (1998) Graft-induced genetic changes and the inheritance of several characteristics in pepper (*Capsicum annuum* L.). Theor Appl Genet 97:705–713

Thabuis A, Palloix A, Servin B, Daubeze AM, Signoret P, Hospital F, Lefebvre V (2004) Marker-assisted introgression of 4 *Phytophthora capsici* resistance QTL alleles into a bell pepper line: validation of additive and epistatic effects. Mol Breed 14:9–20

Thakur PP, Mathew D, Nazeem PA, Abida PS, Indira P, Girija D, Shylaja MR, Valsala PA (2014) Identification of allele specific AFLP markers linked with bacterial wilt [*Ralstonia solanacearum* (Smith)] resistance in hot peppers (*Capsicum annuum* L.). Physiol Mol Plant Pathol 87:19–24

Thakur H, Jindal SK, Abhishek S, Dhaliwal MS (2019) A monogenic dominant resistance for leaf curl virus disease in chilli pepper (*Capsicum annuum* L.). Crop Prot 116:115–120

Thakur H, Jindal SK, Sharma A, Major SD (2020) Molecular mapping of dominant gene responsible for leaf curl virus resistance in chilli pepper (*Capsicum annuum* L.). 3 Biotech 10(182) https://doi.org/10.1007/s13205-020-02168-7

Venkatesh J, An J, Kang WH, Molly J, Byoung CK (2018) Fine mapping of the dominant potyvirus resistance gene Pvr7 reveals a relationship with Pvr4 in *Capsicum annuum*. Phytopathology 108:142–148

Verma S, Dhiman K, Srivastava D (2013) Agrobacterium-mediated genetic transformation of bell pepper (*Capsicum annuum* L. cv. California Wonder) with gus and npt-II genes. Int J Adv Biotechnol Res 4(3):397–403

Weikai C, Xiangfeng W, Jie S, Xinrui W, Zhangsheng Z, Dilay HA, Shu Y, Ming Y, Lili Z, Tan M, Yu M, Jun L, Dian M, Jianxin B, Ke W, Lu W, Shaoying C, Ruidong C, Jingyun J, Bosheng L, Xingping Z, Xing WD, Hang H, Li G (2024) Nat Commun 15:4295

Yadav RK, Krishna RM, Ashwathappa KV, Manish K, Naresh P, Madhavi KR (2022) Screening of Capsicum germplasm and inheritance of resistance to chilli leaf curl virus. Indian Phytopathol 75:1129–1136

Yamazaki A, Shirasawa K, Hosokawa M (2020) Transgressive segregation and gene regions controlling thermotolerance of fruit set and pollen germination in *Capsicum chinense*. Euphytica 216:1–15

Yao M, Li N, Wang F, Zhibiao Y (2013) Genetic analysis and identification of QTLs for resistance to Cucumber mosaic virus in chili pepper (*Capsicum annuum* L). Euphytica 193:135–145

Yeh T, Lin S, Teoh YTY, Sanjeet K, Naqib K (2016) Markers for cytoplasmic male sterility (cms) traits in chili peppers (*Capsicum annuum* l.). I: Multiplex PCR and validation. SABRAO J Breed Genet 48:465–473

Zhang HX, Zhu WC, Feng XH, Jing HJ, Ai MW, Zhen HG (2020) Transcription factor CaSBP12 negatively regulates salt stress tolerance in pepper (*Capsicum annuum* L.). Intl J Mol Sci 21. https://doi.org/10.3390/ijms21020444

Zhao Y, Liu Y, Zhang Z, Cao Y, Yu H, Ma W, Wang L (2020) Fine mapping of the major anthracnose resistance QTL AnRGO5 in *Capsicum chinense* 'PBC932.' BMC Plant Bio 20:1–8

Chapter 12
Advances in Saffron (*Crocus Sativus* L.) Breeding Strategies

Amira A. Ibrahim, Sawsan Abd-Ellatif, El-Sayed S. Abdel Razik, and Khaled F. M. Salem

Abstract This chapter provides an overview of the recent advancements in breeding strategies for saffron (*Crocus sativus* L.), a highly valuable spice crop known for its unique and vibrant color, distinctive flavor, and various medicinal properties. Saffron production is predominantly dependent on the cultivation of its genetically diverse and economically valuable source, *Crocus sativus*. The chapter highlights the significance of saffron breeding to address challenges such as low and inconsistent yields, susceptibility to diseases, and changing climatic conditions. It discusses various traditional and modern breeding techniques employed to improve saffron cultivation, focusing on both genetic and agronomic aspects. Traditional breeding methods, including selection, hybridization, and mutation breeding, have played a crucial role in saffron improvement. The chapter outlines how these methods, with advancements in phenotypic and genotypic selection approaches, have contributed to

A. A. Ibrahim (✉)
Botany and Microbiology Department, Faculty of Science, Arish University, Al-Arish 45511, Egypt
e-mail: amiranasreldeen@sci.aru.edu.eg

S. Abd-Ellatif
Bioprocess Development Department, Genetic Engineering and Biotechnology Research Institute (GEBRI), City of Scientific Research and Technology Applications (SRTA-City), Borg EL-Arab 21934, Alexandria, Egypt
e-mail: sabdellatif@srtacity.sci.eg

E.-S. S. A. Razik
Livestock Research Department, Arid Lands Cultivation Research Institute (ALCRI), City of Scientific Research and Technology Applications (SRTA-City), Borg EL-Arab 21934, Alexandria, Egypt
e-mail: eshabaan@srtacity.sci.eg

K. F. M. Salem
Department of Biology, College of Science and Humanitarian Studies, Shaqra University, Qwaieah, Saudi Arabia

Department of Plant Biotechnology, Genetic Engineering and Biotechnology Research Institute (GEBRI), University of Sadat City, Sadat City, Egypt

K. F. M. Salem
e-mail: khaled.salem@gebri.usc.edu.eg; khaledfathi@su.edu.sa

J. M. Al-Khayri et al. (eds.), *Biodiversity and Genetic Improvement of Herbs and Spices*, Advances in Plant Breeding Strategies 8, https://doi.org/10.1007/978-3-031-81838-7_12

enhancing desirable traits such as improved yield, disease resistance, and increased adaptability to adverse conditions. Furthermore, the chapter delves into the emerging fields of molecular breeding, genomics, and biotechnology, which have opened new avenues for saffron improvement. It explores the utilization of marker-assisted selection (MAS), genotyping-by-sequencing (GBS), and other genomic tools for efficient identification and introgression of beneficial traits. The potential of genetic engineering and transgenic approaches in saffron breeding is also discussed. In addition to genetics, the chapter sheds light on the significance of agronomic practices and cultural management techniques that can contribute to sustainable and enhanced saffron production. It discusses the importance of optimizing irrigation, nutrient management, and crop protection strategies to ensure healthy plant growth and maximum yield. Overall, this chapter provides a comprehensive overview of the recent advances in saffron breeding strategies, encompassing traditional and molecular approaches. It emphasizes the importance of integrating genetic studies with agronomic practices to develop high-yielding and resilient saffron varieties that can meet the increasing global demands sustainably.

Keywords Agronomy · Breeding strategies · Genetic improvement · Genomics · Molecular breeding · Yield enhancement Saffron · (*Crocus sativus* L.)

12.1 Introduction

Saffron, also known as Crocus sativus L., is a botanical species that is highly regarded for its aromatic and medicinal properties. The stigmas of this plant are quite remarkable due to their ability to produce Saffron, a highly prized commodity often known as red gold. Saffron has a rich botanical history that spans thousands of years, with connections to various cultures and regions. Saffron has been widely used in the cultural traditions of ancient Greco-Roman, Egyptian, and Persian civilizations. The historical spread of Saffron to the regions of India and China is believed to have originated from Persia. During the early phases of the medieval period, specifically in the ninth century, there was a noticeable increase in the cultivation of Saffron in North Africa. This coincided with the Arab-Muslim invasions. The growth of saffron farming in Europe, especially in Spain, is believed to be a direct outcome of the Moors' conquest of the region. The historical connection between the introduction of Saffron in France and the Crusades era is quite fascinating, spanning from the 11th to the thirteenth centuries. According to the research conducted by El Midaoui et al. (2022), it was observed that saffron was grown in Germany, Switzerland, and Italy during the mentioned period. This observation provides valuable insights into the distribution and cultivation of saffron in various regions. The topic at hand revolves around the source of nutrients that support maintaining good health and preventing age-related neurological disorders. Saffron, also known as red gold, is a beautiful flowering plant species called Crocus sativus L. It is widely recognized as the most valuable spice in the world. According to a study by Gresta et al. (2008), the dried stigmas of Saffron have valuable properties that make them beneficial in various applications such as food, perfumes, and cosmetics. In earlier times, saffron was not given much importance and its full potential was not realized. The cultivation of this plant was primarily aimed at diversifying agriculture in less productive lands, with the goal of exploring the potential benefits of sustainable farming techniques (Gresta et al. 2008). It has been observed that the use of Plant growth-promoting bacteria (PGPB) has a positive effect on the growth and quality of Saffron. The implementation of this approach has shown promising results in improving the overall productivity and quality of saffron cultivations. Saffron, scientifically known as C. sativus L., is a geophyte from the Iridaceae family. Flowering usually occurs in the autumn. According to historical records, the cultivation of this plant in the Mediterranean Basin dates back to the Late Bronze Age. This is supported by studies conducted by Negbi (1999) and Gresta et al. (2008). The dried red stigmas of this particular specimen are highly regarded for their high value and are used to add flavor and seasoning in culinary settings, as well as for coloring and in traditional healing practices with a long history. This data has been collected from various sources, suggesting a thorough analysis of the subject matter. In modern times, it serves a particular market and is primarily used to add color, taste, and scent to food and drinks. The topic being discussed involves the implementation of crop rotation, specifically within the field of agricultural science. Iran is the leading producer of saffron globally, contributing 160 tonnes out of 205 tonnes, making up around 80% of the total production. On the other hand, the Jammu and Kashmir region makes a 5% contribution with a yield of 8–10 tonnes, while Greece contributes 3% with a yield of 4–6 tonnes. Both Morocco and Spain play a significant role in the overall biomass production. Morocco's contribution accounts for 0.5% of the total, falling within the range of 0.8–1 tonne. Meanwhile, Spain's contribution represents 0.25% of the total, with a range of 0.3–0.5 tonne. Plant biologists have been captivated by the Crocus genus due to its enigmatic origin, intricate classification, and the challenge of differentiating between superior species. As a result, scientists have focused their research on examining the physical, cellular, chemical (including the study of genes, proteins, and metabolites), and genetic characteristics of this specific group of plants. According to the research conducted by Ahmad et al. (2011), Beiki et al. (2010), Caiola and Canini (2010), and Poma et al. (2012), it is believed that the origins of C. sativus can be traced back to either C. catwrightianus or C. thomasii. The aim of this study is to evaluate the genetic diversity of Saffron by examining its genetic makeup. Even though there is a high demand for Saffron as a spice and colorant, its cultivation still relies on traditional knowledge and has not been extensively industrialized. On the other hand, C. sativus can undergo sexual reproduction to produce offspring with new and desirable traits. However, this process requires meticulous selection of parental genotypes, as pointed out by Rashed-Mohassel (2020). Developing a comprehensive understanding of the genetic and geographic origins of Saffron is essential for effectively carrying out the required selection procedures. In order to address this knowledge gap, numerous studies have been conducted in the past decade to gain insights into the domestication of Saffron and determine whether it is an auto or allopolyploid species. There are two distinct categories that images can be classified into based on their characteristics.

When it comes to the world of plant biology, it's crucial to differentiate between artworks that use pigments from the Crocus plant and those that portray the saffron crocus flower. The collection of visual representations has the potential to offer valuable insights into the potential timeline of saffron usage. Alternatively, examining the morphological traits of the crocus species used during that period can shed light on its origin (Kazemi-Shahandashti et al. 2022).

12.1.1 Origin and Geographical Distribution

According to recent reports by Palomares (2022) and Dubois (2022), Iran, Greece, Morocco, Spain, and India are currently the leading saffron producers worldwide. The region stretching from the Mediterranean basin to India (Fig. 12.1) boasts the highest global production level. The geographic subtlety mentioned earlier was referenced in the Encyclopaedia of Diderot and d'Alembert. France, Switzerland, Italy, Turkey, Azerbaijan, Pakistan, China, Japan, and the USA are also included, although to a lesser extent.

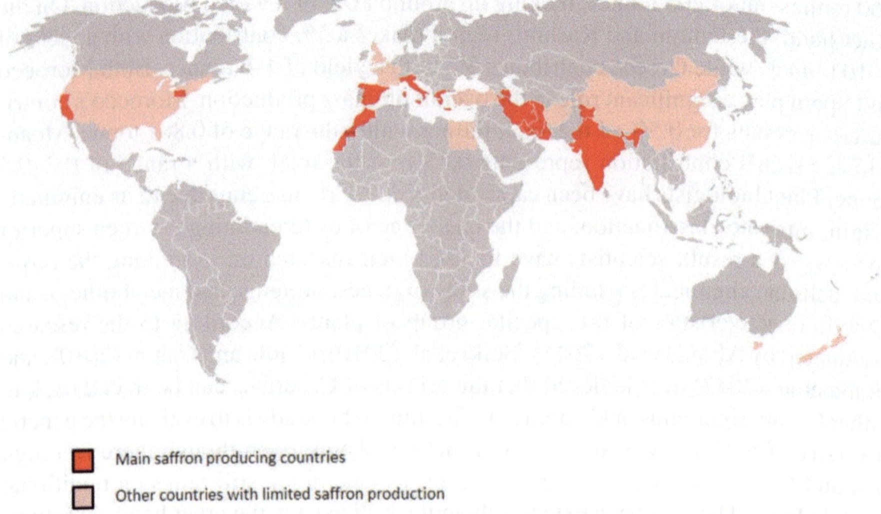

Fig. 12.1 The primary region of saffron cultivation worldwide.. *Source* Adil El Midaoui et al. (2022)

12.1.2 Botany and Cultivation

According to Crozet et al. (2012), Saffron is an extensively studied botanical species. Numerous internet search engines offer significant agricultural and botanical data that provides a comprehensive overview of Saffron. Its classification is as follows, as per the works of Claire Palomares (2022) and Al-Snafi (2016).

Kingdom: Plant

Phylum: Spermatophyte

Subphylum: Angiosperms (Magnoliophyta)

Class: Monocotyledons (Liliopsida)

Subclass: Liliidae

Order: Liliales

Family: Iridaceae

Subfamily: Crocoidae;

Genus: *Crocus*

Species: *C. Sativus* L.

Common name: Saffron

The Plant known as Saffron is frequently depicted in botanical illustrations (Fig. 12.2). It is a type of perennial, bulbous Plant with a bulb referred to as a corm. The Plant in question can attain a height of 30 cm and possesses elongated and slender leaves. Its flowers are cup-shaped and exhibit a hue of parma/purple, as depicted in Fig. 12.2.

12.1.3 Importance of Health Benefits

There has been a great deal of attention given to the medicinal properties of saffron, which has sparked a wave of research, cultivation, and improvement efforts among experts in the field (Ashktorab et al. 2019; Hosseinzadeh and Nassiri-Asl 2013; Ibrahim et al. 2023a; b). It has been noted that the use of certain bacteria can significantly improve the growth and quality of Saffron. This approach has shown great potential in enhancing the productivity and overall quality of saffron harvests. In modern times, Saffron has once again captured the interest of stakeholders and the scientific community. There is a renewed focus on its biological characteristics, including its potential anti-cancer properties, antioxidant effects, and ability to enhance memory (Ghadrdoost et al. 2011; Talaei et al. 2015). The subject under consideration possesses intricate and distinctive characteristics that can be attributed to its remarkable metabolic reserve. This reserve contains certain compounds that

Fig. 12.2 Aspects of Saffron. The illustration depicts a cormus with a bulbous shape and flowers that exhibit visible stigmas. *Source* Adil El Midaoui et al. (2022)

have been scientifically verified to be responsible for its vivid red hue and acrid flavor. As a plant researcher, I can confirm the importance of these compounds in the overall health and lifespan of the plant. Crop rotation is a method used in the agricultural industry that involves strategically planting different crops in the same area over multiple cultivation cycles.

12.1.4 Phytochemical Profile of Saffron

The plant commonly referred to as saffron has been used for various purposes, including medicine and cooking, for a very long time. This is evident from its rich historical documentation (José Bagur et al. 2017). This plant has a wide range of chemical constituents, with over 150 different compounds, both volatile and non-volatile. In this specific context, there are certain chemical components that are of importance. These include carotenoids, a specific group of polyphenols and flavonoids, and terpenes. The composition of these elements can vary depending on the geographic location of the specimen, as shown in several studies (Ramadan and Ibrahim, 2021). Prior research has extensively outlined the approximate composition of the substance (Al-Snafi 2016; Hosseini et al. 2018; Palomares 2022). The composition mentioned above consists of a combination of different components. It includes water (10%), proteins and amino acids (12%), lipids (5%), minerals (5%), fibers (5%), various sugars (63%), and vitamins such as B1 (riboflavin) and B2 (thiamin). Saffron consists of four primary biologically active compounds. These include crocin and crocetin, which are carotenoids derived from zeaxanthin. There is also picrocrocin, an apocarotenoid that gives saffron its unique flavor, and safranal, a terpene with an aldehyde function that contributes to the distinctive aroma of saffron. Various studies have documented the compounds responsible for the yellow color, flavor, and odor of Saffron (Nanda and Madan 2021; Song et al. 2021). Recent studies have revealed the existence of various compounds in saffron, as shown in research conducted by Abu-Izneid et al. (2022) and Koocheki and Khajeh Hosseini (2020). The compounds mentioned in the text include a variety of polyphenols, amino acids, and phytosterols, as reported in several studies. In addition, Daoud et al. (2021) have reported the identification of quinones in these compounds. Studies conducted by Bukhari et al. (2018) have reported the presence of vitamins, while terpenoids have been identified in research by Rigi et al. (2021) and Pandita (2021).

12.1.5 Biological Properties of Saffron

In recent years, there has been a renewed interest in Saffron for its possible use in the cosmetics industry. Throughout history, people have used saffron for cosmetic purposes, either by infusing it or applying it topically. Because of its well-known reputation for its anti-aging properties, it has been combined with fat or infused in donkey milk. It was included in Cleopatra's cosmetic formulations. Based on the principles of traditional Iranian medicine, it is believed that Saffron has the ability to improve the complexion and could be used as a possible treatment for erysipelas. Based on ancient Greek medicine, the application of this substance is believed to have rejuvenating effects on facial skin. It is also used to balance the liver's production of bile and treat various conditions like acne, skin ailments, and wounds.

In addition, according to research conducted by Zargari (1997) and Mir (2004), it has been noted that the body can display a more youthful and radiant appearance. It has been reported that saffron has the potential to reduce the production of melanin, which is responsible for skin coloration. As a result, it demonstrates great effectiveness in lightening the skin. Das et al. (2010) found that a particular formulation had significant depigmentation and anti-arrhythmic effects on human skin when it included an extract from C. sativus. The main way antioxidant activity is shown in C. sativus is through the presence of monoterpenoids, crocin, quercetin, kaempferol, and other phenolic compounds. De Leeuw et al. (2001) discovered that these compounds have the ability to reduce skin melanin by inhibiting tyrosinase activity. Using a combination of natural ingredients like saffron, coconut or olive oil, and unpasteurized milk can be a great way to exfoliate your face and improve blood circulation to your skin. It has been reported that saffron has properties that can help alleviate the symptoms of erythema, a dermatological condition characterized by inflammation of the skin, red lesions, or a rash. Saffron is a plant-based source that is rich in antioxidants. It is thought to have the potential to reduce the presence of inflammation markers, such as tumor necrosis factor (TNF) and interleukin. A formulation containing a 3% extract of C. sativus could potentially offer therapeutic benefits for the treatment of melanoma in humans when applied topically. The clinical trials have confirmed the effectiveness of Saffron in relieving itching and promoting healthy skin in skincare products. In a study conducted by Xing et al. in 2008, it was found that the effectiveness of Saffron is higher than that of a placebo. Saffron has a long-standing reputation as a mood booster, spanning across different eras and cultures. Depression is a highly prevalent illness that affects millions of people worldwide. This condition affects a significant portion of the global population, with an estimated prevalence of around 11.6%. Throughout various ancient civilizations, saffron and other spices were believed to have aphrodisiac properties. According to Dwyer et al. (2011), Saffron has historically been used as a sexual stimulant by various cultures.

Carotenoids, like crocin and crocetin, have been recognized for their valuable impact on human health because of their powerful antioxidant properties (Hosseinzadeh et al. 2009). When combined with certain synthetic compounds like sodium selenite or sodium arsenite, saffron extract has the potential to display a powerful synergistic effect. This discovery implies that the combination could play a crucial role in the prevention of cancer chemotherapy. Riverón (2002) suggests that the effectiveness of Saffron in inhibiting malignant cells depends on the dosage given (Botsoglou et al. 2005). As per Eghdami et al. (2013), the administration of a high dose of crocetin at a rate of 40 mg/kg has been found to be effective in preventing the development of insulin resistance. This is achieved by preventing compensatory hyperinsulinemia. A study conducted by Mashmoul et al. (2014) compared the effectiveness of saffron extract cream and silver sulfadiazine (SSD) in treating heat-induced burns in rats. Throughout history, different countries have utilized Saffron to address a range of eye conditions such as corneal disease, sore eyes, cataracts, and purulent eye infections. These uses have been well-documented by Khorasani et al. (2008). A study conducted by Ghahghaei et al. (2013) discovered that Saffron and crocins have the ability to inhibit beta-amyloid aggregation, a key step in the progression of Alzheimer's disease.

12.1.6 Economic Importance

The economic significance of Saffron lies in its elevated market value, which confers a substantial increase in worth. Apart from its economic significance, the importance of the subject matter also lies in the domains of agronomy, environment, and society. The cultivation of Saffron entails the mobilization of a significant labor force, with a notable proportion comprising female workers, particularly during the harvesting and pruning stages. The cultivation of Saffron necessitates a substantial amount of resources, as evidenced by the fact that a single kilogram of the spice necessitates the utilization of 150,000 to 200,000 flowers and approximately 400 h of labor. The present discourse endeavors to explicate the contemporary comprehension of the therapeutic attributes of Saffron, its correlation with diverse phytochemicals frequently present in this valuable spice, and its distinct applications that have earned it a notable reputation, particularly in the realm of cosmetics. Notably, there is a dearth of published literature on using this spice in the cosmetics industry, as highlighted by Ibtissam Mzabri et al. (2019). The recognition of Saffron 's significant medicinal value and antioxidant properties resulted in an increase in its commercial value as a spice over subsequent eras, which facilitated its dissemination across the Mediterranean region (Abrishami 2004; Caiola and Canini 2010). Despite the high demand for Saffron as a spice and colorant, its cultivation relies heavily on traditional knowledge and has thus far remained largely unaffected by industrialization.

12.2 Current Agricultural Problems and Challenges

From an agronomic perspective, it has been observed that Saffron propagates vegetatively through corms. However, some degree of variability is observed among saffron plants originating from different geographical regions (Siracusa et al. 2013). Saffron, due to its well-suited adaptation to the Mediterranean region, has the potential to serve as a lucrative alternative crop in this particular environment (Gresta et al. 2008). The fragility of saffron flowers and corms poses a challenge to their mechanization, necessitating manual management practices. The management of weed growth constitutes a significant challenge in cultivating Saffron. Saffron exhibits a low-growth habit and lacks competitive ability against neighboring weed species. According to Parizad et al. (2017), the competition among weeds commences during the flowering period in autumn, particularly in spring, when the foliage retains its green coloration. The challenge of mechanizing saffron production creates a favorable environment for its cultivation in countries with lower manual labor costs, such as Iran, India, and Afghanistan. In these regions, the absence of mechanization is not perceived as a limiting factor. Implementing optimal agricultural practices is crucial for achieving maximum production by efficiently utilizing all available resources. Guidelines for such practices can offer effective solutions to address current challenges in saffron production, such as weed management, through adopting strategies such as increasing plant density and implementing suitable crop rotation techniques. According to Shokrpour (2019), implementing virus-free corms and a cropping period of fewer than five years in the perennial system can decrease the likelihood of infections. Saffron is susceptible to damage from specific fauna, such as wild boars and rodents, as well as particular types of fungi, which can be considered its primary adversaries. Culturing a saffron plantation necessitates a high degree of attentiveness and substantial exertion (El Midaoui et al. 2022).

12.3 Genetic Improvement Objectives

The phenomenon mentioned above originates in the genetic makeup of Saffron, which is characterized by being a triploid male-sterile plant that can solely be propagated through vegetative means. According to Caiola and Canini (2010), any modification in the production of Saffron necessitates corresponding adjustments in its vegetative propagation and quality control measures. On the contrary, the sexual propagation of Saffron would yield novel and favorable characteristics, albeit necessitating the curation of parental genotypes. Acquiring knowledge about Saffron 's genetic and geographic origin is imperative to facilitate the requisite selection process. In recent years, numerous studies have endeavored to address the knowledge gap regarding the domestication of Saffron and to elucidate its auto- or allopolyploid ancestry, as reported by Kazemi-Shahandashti et al. (2022). The precise genetic ancestry of *C. sativus* remains uncertain. One theory posits that it may have arisen via auto triploidy from a wild *Crocus*, potentially resulting from the fertilization of a diploid unreduced egg cell by a haploid sperm cell or a haploid egg cell by two haploid sperms. Alternatively, *C. sativus* may be an allopolyploid that emerged through the hybridization of *C. cartwrightianus* and *C. hadriaticus* (Castillo et al. 2005). The available information regarding the ancestry of Saffron is not conclusive. In a karyological investigation, Brighton (1977) proposed that potential progenitors of *C. sativus* could be *C. cartwrightianus* or *C. thomasii*. According to Zubor et al. (2004), the compatibility of DNA 's quantitative and qualitative traits with *C. sativus* was verified through recent AFLP analysis (Amplified Fragment Length Polymorphisms). Several authors suggest that *C. cartwrightianus* is the most likely progenitor (Grilli Caiola. 2004).

Furthermore, the process of flowering in *Crocus cartwrightianus* exhibits significant resemblances to that of *C. sativus*. According to Brighton (1977), Saffron displays relatively consistent and enduring biological characteristics across the globe, with only slight variations in morphological and biochemical traits, such as certain morphometric features (Tammaro 1990). The assertion mentioned above was partially substantiated by a recent inquiry that employed RAPD (Random Amplified Polymorphic DNA) methodology to examine the DNA of Saffron from five distinct regions (Europe and Israel). The investigation conducted by Caiola (2004), did not detect any genomic disparities. Nevertheless, discernible morphological distinctions were observed among the samples collected from various nations, thus indicating that the screening technique employed may have been inadequate in identifying genetic variations.

Numerous scholars have researched phenotypic variations, encompassing accessions exhibiting enlarged styles, stamens, augmented stigmas, intensified tepal coloration, larger corm dimensions, increased flower count, and enhanced viability. Some reports have indicated the existence of phenotypic variations in Saffron that exhibit a higher number of stigmas. However, these modifications may not always be inheritable due to their aseptic characteristics. The accumulation of genetic erosion resulting from the urbanization of cultivable saffron land and the absence of genomic variation through corm propagation contribute to the restricted genetic diversity, as reported by various studies (Ali et al. 2013; Abd El-Moneim et al. 2021; EL-Mansy et al. 2021; Keify and Beiki 2012; El-Sheikh et al. 2020; Al-Shamrani et al. 2022; Mourad et al. 2019; Safhi et al. 2022; Salem et al. 2008; 2010; 2021d, f).

12.4 Germplasm Biodiversity and Conservation

12.4.1 Germplasm Diversity

Exploring the genetic variability and characterizing the germplasm are crucial aspects in the field of plant biology, as they provide valuable insights for plant breeding programs and aid in making informed decisions related to germplasm conservation and management (Abdelsalam et al. 2021; Alsharari et al. 2023; Alqahtani et al. 2024; Mattar et al. 2009; Salem et al. 2024). As per the findings of Shokrpour (2019), the germplasm constitutes a wide array of assortments, encompassing cultivated cultivars, landraces, mutants, and wild species. The scientific literature presents a perplexing question regarding the monomorphism or polymorphism of *C. sativus*. Despite its esteemed status, numerous studies have shown limited genetic diversity in this particular crop, providing a chance for genetic improvement (El Rabey et al. 2013; Essa et al. 2023; 2024; Ibrahim et al. 2024a, b; Namayandeh et al. 2013; Salem and Ibrahim 2024; Zheng et al. 2013). The interrelation between morphological characteristics and genetic variation is not the sole correlation, as their expression is also impacted by environmental and geographical factors (Fernández 2004).

12.4.2 Cultivars Characterization and Phylogeny

Many studies have been conducted on the Crocus series, driven by the immense fascination surrounding the origin of Saffron. According to Nemati et al. (2018), there is still uncertainty regarding the taxonomic relationships within the series Crocus. This is because there has not been a comprehensive investigation of all known taxa in a single analysis, and the analyzed sequence regions have not shown enough variability. However, various investigations have revealed inconsistencies among different accessions. The study conducted by Izadpanah et al. (2015) utilized morphological markers to evaluate phenotypic variation. These markers included vegetative attributes such as leaf number and length, leaf area, spathe number, and fresh and dry weights of the shoot, as well as flower attributes like style weight and length, and petal and calyx dry weight.

12.4.3 Genetic Resources Conservation Approaches

The Convention on Biological Diversity has identified the objective of mitigating the current rate of biodiversity loss as a crucial goal. The conservation efforts involve preserving plant taxa that are linked to species of socioeconomic significance, including those used for culinary, medicinal, aromatic, and decorative purposes

(Börner et al. 2010; Maxted et al. 2007). Ensuring the adequate and diverse representation of plant genetic resources is crucial for future purposes, such as breeding programs. Several studies have emphasized the importance of improving the systematic conservation of these resources (Abdel-Hamid and Salem 2021; Aboshama et al. 2015; Abu-Ellail et al. 2021; Aldahak et al. 2021; Emara et al. 2011; 2013; Hassan et al. 2021; Ibrahim et al. 2023a, b; Mattar et al. 2016; Nawara et al. 2016; 2017; Salem and Sallam 2016; Salem 1997; Saleh et al. 2020a, b; Saleh et al. 2021; Salem et al. 2021a, b, c, d, e; 2023; Badr et al. 2024). Enhancing the systematic preservation of plant genetic resources is crucial to ensure a wide range of genetic resources for future use, such as in breeding programs. The subfamily Crocoideae, to which Crocus belongs, is the largest among the four subfamilies identified in the Iridaceae family (Goldblatt et al. 2006). As someone who studies plants, it is crucial to consider the taxonomic classification of a plant species, which includes its subfamily and family. This group consists of 88 species of perennial plants that are known for their small size and possession of corms. These species can be found across a wide area, stretching from Southwest Asia to Western China. These plants are well-known for their colorful flowers, carefully bred varieties, practical applications, and unique specimens that collectors eagerly seek. Their aesthetic appeal has made them a highly sought-after plant in the world of horticulture.

These findings have been documented by multiple sources such as Petersen et al. (2008) and Rashed-Mohassel (2007). Most taxa, including species and subspecies, have a limited geographical range that is found only in Turkey and the Balkan Peninsula. According to the research conducted by Tsoktouridis et al. (2009), it has been discovered that Greece is the primary location for almost 40% of the global population of wild Crocus variations. On the other hand, Turkey's plant life consists of a total of 32 different species, with 18 of them being unique to the region, according to Arslan et al. (2007). Various countries have a wide range of Crocus species, with Italy having a total of 10 species, Spain with 6 species, and Hungary with 6 species, among other nations.

The decline in the land area allocated to saffron cultivation in various regions may have led to genetic erosion, further worsening the already limited genetic diversity in C. sativus caused by its sterility. Establishing a germplasm repository for this particular species is a significant accomplishment in the realm of plant biology. It has been noted that the inclusion of wild species can greatly enhance the qualities of Saffron by introducing a wide range of beneficial traits. According to Khoury et al. (2010), the uncultivated portion of a crop's gene pool typically contains a significantly higher level of genetic diversity compared to the cultivated varieties.

12.4.3.1 Ex Situ Conservation and Multiplication

The genetic material of Saffron and its wild relatives is conserved in the form of corms, while in the case of wild species, it is preserved in seeds. Engels and Visser (2003) have established international genebank standards that prescribe general conservation and multiplication strategies for collections. Furthermore, formulating particular methodologies and developing bank architecture for ex-situ conservation were primarily predicated on data about the species origin, reproductive physiology, propagation mode, sample classification, and collection goals. The design encompasses three primary assemblages, namely the reserve vegetative collection (comprising Saffron and related species), the exchange vegetative collection (comprising Saffron and related species), and the seed collection (comprising wild *Crocus*).

Cryopreservation presents a compelling mechanism for the enduring maintenance of sterile plants. The technique of cryopreservation was utilized for the preservation of shoot tips. The corms were formerly stored at a temperature of 2 °C in a refrigerated environment for 14 days. Subsequently, the fibrous layer was eliminated, and the denuded corms were sterilized using a 2% sodium hypochlorite solution. The study conducted by Zadeh et al. (2009) involved the cultivation of bare, aseptic corms in a medium known as LS, which was supplemented with 0.3 M sucrose. The cultivation process was carried out at 2 °C for 2 days.

12.4.4 Cytogenetics

Aghamohammadi (1977) has extensively studied the range of morphotypes of C. sativus that exhibit elongated stigmas. The observed variability may be attributed to developmental anomalies, chromosomal aberrations, or an uncommon genetic mutation. Through careful examination of the root apices, the researcher successfully determined the chromosome number of four distinct plant species to be 2n = 24, aligning with the chromosome number observed in typical saffron plants. According to a study by Ghaffari and Bagheri (2009), these flowers are formed when multiple floral buds merge on a single corm. Through cytological and morphological investigations, it has been discovered that this trait lacks genetic control and exhibits a certain level of instability. The study conducted by Behdani and Izanloo (2019) did not discover any significant correlation between microsatellite marker alleles and the number of stigmas. This study explores the various stages of cell division and meiotic behavior, as well as the assessment of pollen stainability. The ongoing research is centered around a study of C. sativus corms that were collected from various agricultural regions in Iran, including Karaj to Shahdasht and four other areas in Khorasan, namely Birjand, Qaen, Gonabad, and Bajestan. The corms were grown for this investigation. An examination was conducted on the somatic chromosomes found in the meristematic cells situated at the tips of roots. A procedure was performed to remove the apical meristems, which were then treated with a saturated solution of α-bromonaphthalene for 7 h at a temperature of 4 °C. After collection, the plant specimens were immersed in Pienaar's solution, which is a mixture of absolute alcohol, chloroform, and propionic acid, for a duration of 24 h. Following this, the plant samples were preserved in a solution containing 70% ethanol. The staining procedure was performed using the Feulgen reaction, which was improved by gently pressing in a two percent acetocarmine solution. Observations were made on the meiotic chromosomes found in the pollen mother cells. The young flower buds were immersed in Pienaar's solution for a day and then stored in 70% ethanol for preservation. A staining technique was used to assess the pollen viability, cotton blue was used for this purpose. The results of the mitotic analyses were consistent across all sites, as indicated by the findings of the research study. Figure 12.3a provides a visual representation of the different stages of the mitotic process.

During the prometaphase stage, an interesting observation was made regarding the chromosomes. They displayed a distinct morphology, with two chromatids and a tightly constricted centromere where they were joined together. (Fig. 12.3b). According to Fig. 12.3c, there are 24 chromosomes present at the metaphase plate, which suggests a diploid number of 2n. According to research conducted by Ghaffari in (1986) and Ghaffari and Bagheri (2009), it was observed that the chromosomes are arranged in the form of eight triplets. During anaphase, the sister chromatids naturally separated, as shown in Fig. 12.3d. The research findings presented here provide evidence for the auto triploidy of Saffron, as documented in various studies conducted by different researchers over the years.

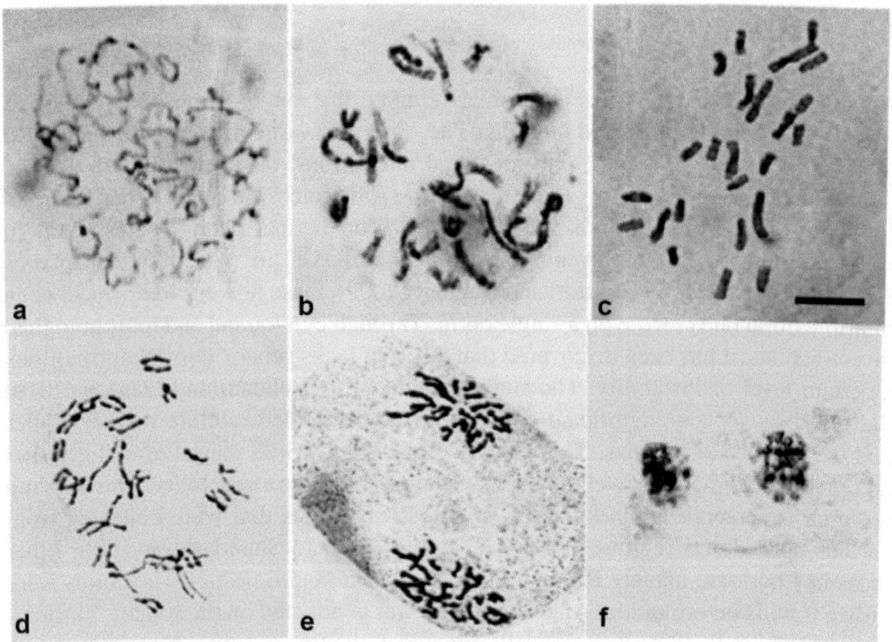

Fig. 12.3 Mitotic phases of Saffron (A-F). **a** prophase; **b** prometaphase; **c** metaphase $2n = 24$; **d** the start of anaphase (chromatids begin to separate); **e** anaphase; **f** telophase. Scale bar = 5 μm. *Source* Ghaffari (2022)

12.5 Traditional Breeding Methodologies and Limitations

The use of traditional breeding techniques to improve the genetics and create new combinations is limited by the sterility of Saffron. This sterility is due to its triploid genome and the use of corms for vegetative propagation (Shokrpour 2019). Many farmers are of the opinion that by using large quantities of chemical fertilizers, they can promote the growth of plants and increase the yield of Saffron. On the other hand, the overuse of chemical fertilizers by farmers leads to the degradation of the soil and the surrounding ecosystem. As per the findings of Chamkhi et al. (2022), the inclusion of bacterial inoculants aligns with the organic agricultural management practices necessary for attaining exceptional saffron quality.

Using corms for the vegetative propagation of Saffron suggests that the plant has remained largely unchanged over time. This means that a single cultivar could potentially represent Saffron globally. Considering the situation, it seems that using clonal selection may not be a practical approach to improve the genetic composition of Saffron. The historical evidence surrounding the cultivation of Saffron in ancient times suggests that certain factors can result in different types of mutation. Although the organism has the ability to reproduce asexually, it is still vulnerable to genetic changes caused by various processes such as chromosomal recombination, deletions, inversions, translocations, polyploidy, incomplete segregation, or segregation distortion (Soliman et al. 2017; 2023). Plants that have experienced genetic modifications through natural mutations create new clones that exist together in a diverse population. However, the sterility of these plants ensures that there is no genetic mixing, as evidenced by studies conducted by Abd El-Moneim et al. (2021), Al Shamrani et al. (2022), EL-Mansy et al. (2021), and Safhi et al. (2022).

Hybridization is a widely used technique in plant breeding that helps separate populations and create new combinations of genes. Hybrid variety development is often accomplished by crossing selfed lines. It is impossible to achieve this outcome due to the sterility of Saffron. Irregularities in meiosis can occur when a triploid genome is present, leading to abnormalities in both sporogenesis and gametogenesis. This can result in the production of abnormal pollen, which can ultimately lead to self-sterility. This phenomenon has been observed and documented by Chichiriccò (1984) and Grilli Caiola (2005).

12.6 Mutation Breeding

The induced mutation process is focused on altering specific traits in plants and can occur at either the chromosomal or genetic level. Exploring the potential of mutation in plant breeding offers a hopeful alternative for generating genetic variability, especially in crops like Saffron that do not undergo sexual reproduction. According to Soliman et al. (2023), the ability to detect, select, and conserve M1 generation mutants is made possible by their clonal nature. Through mutation breeding, various traits in Saffron can be modified, resulting in changes such as an enhanced concentration of valuable metabolites in the stigma, a higher abundance of flowers, increased multiplication of corms, a greater number of stigmas, and an elevated dry weight.

In the study conducted by Khan (2004), a range of gamma-ray doses were given to saffron corms in order to trigger mutations that could lead to the creation of high-yielding mutants. Individuals with mutations displaying five-branched stigmas and increased dry weight and length were chosen for this purpose. In a study conducted by Khan et al. (2011), an experiment was carried out to improve the production of daughter corms per mother corm. This was achieved by using physical (gamma ray) and chemical (EMS, colchicine, and ethidium bromide) mutagenic agents. As previously mentioned, one of the treatments, specifically gamma radiation at a dosage of 0.2 KR, had a positive impact on the traits. The authors suggested that by adopting a standardized technique, there could be a notable rise in the quantity of saffron corms, thereby allowing for the expansion of cultivation areas. Through the duplication of chromosome number and elevation of ploidy level, the process of mutagenesis has the potential to overcome barriers to sterility. It is often accomplished through colchicine treatment, as mentioned by Zaffar et al. (2004).

12.7 Molecular Breeding

The genetic information of Saffron 's ancestors remains uncertain from a botanical perspective. Based on traditional botanical research that focused on plant structure, C. cartwrightianus is considered to be the most closely related species to C. sativus. It is possible that saffron originates from C. cartwrightianus. One notable distinction between C. sativus and C. cartwrightianus is their size. In a study conducted by Gresta et al. (2008), it was discovered that the floral morphology of C. sativus was 100% larger than that of C. cartwrightianus. In their study, Caiola and Canini (2010) utilized RADP analysis to uncover the ancestors of C. sativus. C. cartwrightianus shares a strong genetic resemblance with C. sativus, while C. thomasii is considered to be closely related. In a study conducted by Zubor et al. (2004), AFLP analysis was employed to validate the DNA characteristics, both quantitative and qualitative, of C. cartwrightianus, C. thomasii, and C. sativus.

The utilization of DNA markers in Crocus spp. has facilitated the characterization of germplasm collections, thereby providing valuable assistance to breeding projects. It is highly skilled in the field of genetic diversity and taxonomy research (Mesfer ALshamrani et al., 2022). DNA marker methods have made significant advancements in the past decade, capturing the interest of experts in plant biology. Various methods have been employed in plant genetics research. This list includes various molecular markers used in plant research (El-Absawy et al. 2015; Eltaher et al. 2018; 2021; 2022; El-Zanaty et al. 2011; Ghazy et al. 2020; Ismail et al. 2013; Mattar et al. 2016; Salem 2009, 2015; Salem et al. 2007; 2010; 2015; 2021c; 2022; Salem and Mattar 2014a, b; Varshney et al. 2008). Multiple scholars have demonstrated the effectiveness of these methods, as evidenced by the studies conducted by Beiki et al.

(2010), Rubio Moraga et al. (2010), Erol et al. (2011), and Safhi et al. (2022). Despite using RAPD, AFLP/SSR, and IRAP techniques, the relationships between distantly related species in the series remained elusive. Saffron, also known as C. sativus, is a fascinating enigma in the world of plants. In a study conducted by Nemati et al. (2018), it was proposed that certain species, including C. cartwrightianus, C. hadriaticus, C. pallasii, and/or C. thomasii, could potentially be the wild progenitors. Molecular markers have been used to evaluate the genetic diversity of saffron across different regions. This study utilizes various molecular markers, including RAPD, ISSR, SSR, AFLP, Retero-transposons, SNP, and EST (Ali et al. 2013; Beiki et al. 2011; Caiola and Canini 2010; Imran et al. 2010; Keify and Beiki 2012; Namayandeh et al. 2013; Rubio-Moraga et al. 2009; Zheng et al. 2013).

12.8 Genetic Engineering and Gene Editing

Due to the economic and medicinal value of Saffron, many biotechnological projects have been initiated to improve its genetics. However, the results of these endeavors have been quite limited. In order to expand CRISPR-cas mediated genome editing to Saffron, it is crucial to establish a highly effective transformation system. The objective of this study was to create a reliable transformation platform for Saffron by developing a highly effective in vitro protocol. Experiments were carried out to enhance the production of callus and somatic embryos with maximum efficiency. The enhanced development of these plant tissues led to a more frequent production of cormlets. After establishing the necessary protocols, we developed methodologies to achieve a stable transformation system using the Cas9 gene. The findings presented in the study by Chib et al. (2020) are anticipated to have significant implications for CRISPRcas9-mediated gene editing in Saffron, with the potential to enhance the apocarotenoid content.

12.9 Conclusion and Prospects

The final remarks of the section discussing the developments in breeding approaches for *Crocus sativus* L. (Saffron) emphasize the noteworthy advancements achieved in enhancing Saffron 's growth through breeding methodologies. The field of saffron biology has observed novel methodologies that strive to improve diverse facets of saffron cultivation, such as productivity, characteristics of excellence, immunity against diseases, and acclimatization to varying habitats. The chapter highlights the efficacy of conventional breeding techniques, including selection and hybridization, in enhancing saffron cultivars. Recent developments in molecular biology and biotechnology have presented novel opportunities for expediting the process of plant breeding.

Utilizing molecular markers, genetic mapping, and genomic selection has enabled the detection of crucial characteristics and the early-stage selection of favorable specimens. The implementation of this technique has resulted in a notable decrease in the duration of saffron cultivar development aimed at achieving desired traits. The application of tissue culture methodologies and in vitro regeneration protocols has facilitated the large-scale propagation of high-quality saffron plants, guaranteeing a steady provision of exceptional planting resources. The significance of integrating genomic information and bioinformatics techniques in saffron breeding approaches is emphasized in the chapter. The recent technological advancements have facilitated a deeper understanding of the genetic underpinnings of diverse traits, enabling breeders to make judicious choices while selecting parental lines for crossing experiments. The chapter's discourse on saffron breeding techniques has resulted in superior saffron cultivars that exhibit improved productivity and quality. This advancement is crucial in fulfilling the growing need for Saffron and guaranteeing sustainable saffron cultivation in various global regions.

References

Abd El-Moneim D, ELsarag EIS, Aloufi S et al (2021) Quinoa (*Chenopodium quinoa* Willd.): genetic diversity according to ISSR and SCoT markers, relative gene expression, and morpho-physiological variation under salinity stress. Plants 10(12):2802. https://doi.org/10.3390/plants10122802

Abdel-Hamid AME, Salem KFM (2021) Breeding strategies of garden pea (*Pisum sativum* L.). In: Al-Khayri JM, Jain SM, Johnson DV (eds) Advances in plant breeding strategies: vegetable crops, Springer, Cham. https://doi.org/10.1007/978-3-030-66969-0_9

Abdelsalam NR, Hasan ME, Rabie SMA, El-wakeel HE, Zaitoun AF, Ghareeb RY, Abdelsalam AZ, Aly HM, Ibrahim AA, Hemeida AA (2021) Endorsement and phylogenetic analysis of some fabaceae plants based on DNA barcoding gene MatK. https://doi.org/10.1101/2021.07.27.454001

Aboshama HM, Salem KFM, Hassan HMF, Tahoun MH (2015) Estimating heterosis and combining ability for *in vivo* and *in vitro* traits using diallel cross on *tomato* (*Solanum lycopersicum* L.). Minufiya J Agric Res 40(3):757–770

Abrishami MH (2004) Saffron, from yesterday till today, an encyclopaedia of its production, trade and use. Amirkabir, Tehran

Abu-Ellail FFB, Salem KFM, Saleh MM et al (2021) Molecular breeding strategies of beetroot (*Beta vulgaris* ssp. vulgaris var. conditiva Alefeld). In: Al-Khayri JM, Jain SM, Johnson DV (eds) Advances in plant breeding strategies: vegetable crops. Springer, Cham. https://doi.org/10.1007/978-3-030-66965-2_4

Abu-Izneid T, Rauf A, Khalil AA et al (2022) Nutritional and health beneficial properties of saffron (*Crocus sativus* L.): a comprehensive review. Food Sci Nutr 62:2683–2706

Aghamohammadi Z (1977) An investigation of asexual reproduction, induction of variability of mutagens and sexual sterility of Saffron (*Crocus sativus*) as measured by pollen stainability and pollen germination in relation to chromosome pairing. M Sc Thesis, University of Tehran, Tehran, Iran

Ahmad M, Zaffar G, Mir SD, Razvi SM, Rather MA, Mir MR (2011) Saffron (*Crocus sativus* L.) strategies for enhancing productivity. Res J Med Plant 5(6):630–649

Aldahak L, Salem KFM, Al-Salim SHF, Al-Khayri JM (2021) Advances in chicory (*Cichorium intybus* L.) breeding strategies. In: Al-Khayri JM, Jain SM, Johnson DV (eds) Advances plant

breeding strategies: vegetable crops. Springer, Cham. https://doi.org/10.1007/978-3-030-66969-0_1
Ali G, Iqbal AM, Nehvi F et al (2013) Prospects of clonal selection for enhancing productivity in saffron (*Crocus sativus* L.). Afr J Agric Res 8:460–467
Alqahtani MM, Saleh MM, Alwutayd KM, Safhi FA, Okasha SA, Abdelsatar MA, Ali MS, Saif MI, Ibrahim AA, Salem KF (2024) Performance and genotypic variability in diverse date palm (Phoenix dactylifera L.) cultivars for fruit characteristics. Genet Resour Crop Evol. https://doi.org/10.1007/s10722-023-01719-0
ALshamrani MS, Safhi FA, Alshaya DS, Ibrahim AA, Mansour H, Abd El Moneim D (2022) Genetic diversity using biochemical, physiological, karyological and molecular markers of *Sesamum indicum* L. Front Genet 13:1035977. https://doi.org/10.3389/fgene.2022.1035977
Alsharari SF, Ibrahim AA, Okasha SA (2023) Combining ability for yield, oil content, and physio-biochemical characters of canola (*Brassica napus* L.) under salt stress conditions. SABRAO J Breed Genet 55(4):1003–1024. https://doi.org/10.54910/sabrao2023.55.4.1
Al-Snafi AE (2016) The pharmacology of *Crocus sativus*- A review. IOSR J Pharm 6:8–38
Arslan N, Ozer AS, Akdemir R (2007) Cultivation of saffron (*Crocus sativus* L.) and effects of organic fertilizers to the flower yield. Acta Hort 826:237–240
Ashktorab H, Soleimani A, Singh G et al (2019) Saffron: the golden spice with therapeutic properties on digestive diseases. Nutrients 11:943. https://doi.org/10.3390/nu11050943
Badr A, Basuoni MM, Ibrahim M et al (2024) Ameliorative impacts of gamma-aminobutyric acid (GABA) on seedling growth, physiological biomarkers, and gene expression in eight wheat (*Triticum aestivum* L.) cultivars under salt stress. BMC Plant Biol 24:605. https://doi.org/10.1186/s12870-024-05264-5
Behdani MA, Izanloo A (2019) Evaluation of genetic variations in Saffron with more than three stigmas using SSR and ISSR molecular markers. Saffron Agron Technol 7:347–357
Beiki A, Keify F, Mozafari J (2011) Rapid genomic DNA isolation from corm of Crocus species for genetic diversity analysis. J Med Plants Res 5:4596–4600
Beiki AH, Keifi F, Mozafari J (2010) Genetic differentiation of *Crocus* species by random amplified polymorphic DNA. Genet Eng Biotechnol J 18:1–10
Börner A, Landjeva S, Salem KFM et al (2010) Plant genetic resources- a prerequisite for drought tolerance breeding in wheat. Tagungsband 60. Tagung der Vereinigung Pflanzenzüchter und Saatgutkaufleute Österreichs, Gumpenstein, Österreich, 24–26 Nov 2009
Botsoglou NA, Florou PP, Nikolakakis I et al (2005) Effect of dietary saffron (*Crocus sativus* L.) on the oxidative stability of egg yolk. Br Poult Sci 46:701–707
Brighton CA (1977) Cytology of *Crocus sativus* L. and its allies (Iridaceae). Plant Syst Evol 128:137–157
Bukhari SI, Manzoor M, Dhar MK (2018) A comprehensive review of the pharmacological potential of *Crocus sativus* and its bioactive apocarotenoids. Biomed Pharm 98:733–745
Caiola MG, Canini A (2010) Looking for saffron's (*Crocus sativus* L.) parents. In: Husaini AM (ed) Saffron. Global Science Books, UK, pp 1–14
Castillo R, Fernandez JA, Gomez-Gomez L (2005) Implications of carotenoid biosynthetic genes in apocarotenoid formation during the stigma development of *Crocus sativus* and its closer relatives. Plant Physiol 139:674–689
Chamkhi I, El Omari N, Balahbib A et al (2022) Is the rhizosphere a source of applicable multi-beneficial microorganisms for plant enhancement? Saudi J Biol Sci 29:1246–1259. https://doi.org/10.1016/j.sjbs.2021.09.032
Chib S, Thangaraj A, Kaul S et al (2020) Development of a system for efficient callus production, somatic embryogenesis and gene editing using CRISPR/Cas9 in Saffron (*Crocus sativus* L.). 16:47. https://doi.org/10.1186/s13007-020-00589-2
Chichiriccò G (1984) Karyotype and meiotic behaviour of the triploid *Crocus sativus* L. Caryologia 37:233–239
Crozet A, de Sus-Rousset H, de Durfort SJ (2012) *Crocus sativus* L. (Iridaceae), le safran (I). Phytothérapie 10:121–125

Daoud M, Loukidi B, Taibi W et al (2021) Phenotypic and phytochemical diversity of saffron (*Crocus Sativus* L.). Genet Biodivers J 4:71–80

Das I, Das S, Saha T (2010) Saffron suppresses oxidative stress in DMBA-induced skin carcinoma: a histopathological study. Acta Histochem 112:317–327

De Leeuw SM, Smit NP, Van VM et al (2001) Melanin content of cultured human melanocytes and UV-induced cytotoxicity. J Photochem Photobiol 61:106–113

Dubois A (2022) Analyse de la Filière Safran du MAROC: Quelles Perspectives Pour la Mise en Place D'Une Indication Géographique? http://bionaturehouse.com/files/StudieSafranMarokkoCIHEAM.pdf

Dwyer A, Hawrelak J, Whitten D (2011) Herbal Medicines, other than St. John's Wort, in the treatment of depression: a systematic review. Sci Rev Altern Med 16:40–49

Eghdami K, Erfanparast A, Farshid AA et al (2013) Comparison of the effects of crocin, safranal and diclofenac on local inflammation and inflammatory pain responses induced by carrageenan in rats. Pharmacol Rep 65:1272–1280

El Midaoui A, Ghzaiel I, Vervandier-Fasseur D et al (2022) Saffron (*Crocus sativus* L.): a source of nutrients for health and for the treatment of neuropsychiatric and age-related diseases. Nutrients 14:597. https://doi.org/10.3390/nu14030597

El Rabey H, Salem KFM, Mattar MZ (2013) The genetic diversity and relatedness of eight rice (*Oryza sativa* L.) cultivars as revealed by AFLP and SSRs markers. Life Sci J 10(1):1471–1479

El-Absawy EA, Nowar AA, Salem KFM et al· (2015) Estimation of genetic diversity among some bread wheat (*Triticum aestivum* L.) genotypes using molecular markers and agronomic traits. Minufiya J Agric Res 40(3):223–233

EL-Mansy AB, Abd El-Moneim D, ALshamrani SM et al (2021) Genetic diversity analysis of tomato (*Solanum lycopersicum* L.) with morphological, cytological, and molecular markers under heat stress. Horticulturae 7:65. https://doi.org/10.3390/horticulturae7040065

El-Sheikh MH, Ghareeb RY, Ibrahim AA, Hafez EE (2020) Genetic diversity among *Juglans regia* and *Prunus dulcis* genotypes using RAPD and ISSR molecular markers. Plant Archives, 20(2)

Eltaher S, Baenziger S, Belamkar V et al (2021) GWAS revealed effect of genotype × environment interactions for grain yield of Nebraska winter wheat. BMC Genomics 22:1–14

Eltaher S, Sallam A, Belamkar V et al (2018) Genetic diversity and population structure of $F_{3:6}$ Nebraska winter wheat genotypes using genotyping-by-sequencing. Front Genet 9:76. https://doi.org/10.3389/fgene.2018.00076

Eltaher S, Sallam A, Emara HA et al (2022) Genome-wide association mapping revealed SNP alleles associated with spike traits in wheat. Agronomy 12:1469. https://doi.org/10.3390/agronomy12061469

El-Zanaty AM, Salem KFM, Esmail RM (2011) Detection of genetic diversity in Egyptian cotton (*Gossypium barbadense* L.) varieties using RAPD markers and morphological traits. J American Sci 7(12):1107–1115

Emara HA, Nower AA, Salem KFM, Eltaher S (2011) Estimating heterosis and combining ability for *in vivo* and *in vitro* traits using diallel cross in Maize (*Zea mays* L). J Prod Dev 16:101–120

Emara HA, Salem KFM, Abdel-Halim AA, Abd El-Kowy RNF (2013) An *in vivo* and *in vitro* traits analysis of a diallel cross in rice (*Oryza sativa L.*). Egyptian J Plant Breed 17:591–610

Engels JMM, Visser L (eds) (2003) A guide to effective management of germplasm collections. IPGRI Handbooks for Genebanks No 6. IPGRI, Rome, Italy

Erol O, Şık L, Kaya HB et al (2011) Genetic diversity of *Crocus antalyensis* B. Mathew (Iridaceae) and a new subspecies from southern Anatolia. Plant System Evolution 294:281–287

Essa NM, Ibrahim AA, Soliman MI (2023) Comparative studies among some cultivated species and their wild relatives in Egypt using phytochemical screening, nutritive value antioxidant activity and GC-MS analysis. Catrina 27(1):1–19

Essa NM, Ibrahim AA, Soliman MI (2024) Association studies between some cultivated species and their wild relatives from Apiaceae, Asteraceae and Brassicaceae families based on molecular and DNA barcoding in Egypt. Genet Resour Crop Evol https://doi.org/10.1007/s10722-023-01681-x

Fernández JA (2004) Biology, biotechnology and biomedicine of saffron. Recent 250 Res. Dev Plant Sci 2:127–159

Ghadrdoost B, Vafaei AA, Rashidy-Pour A et al (2011) Protective effects of saffron extract and its active constituent 253 crocin against oxidative stress and spatial learning and memory deficits induced by 254 chronic stress in rats. Eur J Pharmacol 667:222–229

Ghaffari SM (2022) Cytogenetics of saffron (*Crocus sativus*), a natural autotriploid species; an investigation of the reasons for its sterility. Iranian J Bot 28(1):57–63

Ghaffari SM (1986) Cytogenetic studies of cultivated *Crocus sativus* (Iridaceae). Plant Syst Evol 153:199–204

Ghaffari SM, Bagheri A (2009) Stigma variability in saffron (*Crocus sativus* L.). Afr J Biotech 8:601–604

Ghahghaei A, Bathaie SZ, Kheirkhah H, Bahraminejad E (2013) The protective effect of crocin on the amyloid fibril formation of A42 peptide in vitro. Cell Mol Biol Lett 18:328–339

Ghazy MI, Salem KFM, Sallam A (2020) Utilization of genetic diversity and marker-trait to improve drought tolerance in rice (*Oryza sativa* L.). Mol Biol Rep. https://doi.org/10.1007/s11033-020-06029-7

Goldblatt P, Davies TJ, Manning JC et al (2006) Phylogeny of Iridaceae subfamily Crocoideae based on a combined multigene plastid DNA analysis. Aliso 22:399–411

Gresta F, Lombardo GM, Siracusa L, Ruberto G (2008a) Saffron, an alternative 258 crop for sustainable agricultural systems. A review. In: Lichtfouse E, Navarrete M, Debaeke P et al (eds) Sustainable agriculture. Springer, Dordrecht. https://doi.org/10.1007/978-90-481-2666-8_23

Grilli Caiola M (2004) Saffron reproductive biology. Acta Hort 650:25–37

Grilli Caiola M (2005) Embryo origin and development in *Crocus sativus* L. (Iridacee). Plant Biosyst 139:335–343

Hassan MN, Mekkawy SA, Mahdy M et al (2021) Recent molecular and breeding strategies in lettuce (*Lactuca* spp.). Genet Resour Crop Evol 68:3055–3079. https://doi.org/10.1007/s10722-021-01246-w

Hosseini A, Razavi BM, Hosseinzadeh H (2018) Pharmacokinetic properties of saffron and its active components. Eur J Drug Metab Pharmacokinet 43:383–390

Hosseinzadeh H, Molaei M, Shakeri M et al (2009) Evaluation of *Crocus sativus* L. (saffron) on male erectile dysfunction: a pilot study. Phytomedicine 16:690–693

Hosseinzadeh H, Nassiri-Asl M (2013) Avicenna's (Ibn Sina) the canon of medicine and saffron (*Crocus sativus*): a review. Phytother Res 27:475–483. https://doi.org/10.1002/ptr.4784

Ibrahim AA, Abdel Razik ES, Abd-Ellatif S, Ramadan S (2023a) *Nigella sativa* and stressful conditions. In: Latef AAHA (ed) Medicinal plant responses to stressful conditions. CRC Press. https://doi.org/10.1201/9781003242963

Ibrahim AA, Abd-Ellatif S, Abdel Razik ESS, Salem KFM (2024) Biodiversity of cereal crops and utilization in food and nutritional security. In: Al-Khayri JM, Jain, SM, Penna S (eds) Sustainable utilization and conservation of plant genetic diversity. sustainable development and biodiversity, vol 35, Springer, Singapore. https://doi.org/10.1007/978-981-99-5245-8_2

Ibrahim AA, Alwutayd KM, Safhi FA et al (2024b) Characterization and comparative genomic analyses of complete chloroplast genome on *Trema orientalis* L. Genet Resour Crop Evol 71:1085–1099. https://doi.org/10.1007/s10722-023-01678-6

Ibrahim AA, Salem KFM, Abdein MA, Ramadan SM (2023) Advances in summer squash (*Cucurbita pepo* L.) molecular breeding strategies. In: Singh S, Sharma D, Sharma SK, Singh R (eds) Smart plant breeding for vegetable crops in post-genomics era, Springer, Singapore. https://doi.org/10.1007/978-981-19-5367-5_8

Ibrahim AA, Salem KFM, Ramadan S (2023b) *Allium sativum* and stressful conditions. In: Latef AAHA (ed) Medicinal plant responses to stressful conditions. CRC Press. https://doi.org/10.1201/9781003242963

Imran S, Nehvi F, Wani S, Zaffar G, Khan MA (2010) Studies in relation to molecular variability in Saffron. In: III International symposium on saffron: forthcoming challenges in cultivation, research and economics, vol 850, pp 75–78

Ismail M, Salem KFM, El-Zanaty AM, Esmail RM (2013) Association mapping of fiber and agronomic characters in Egyptian cotton (*Gossypium barbadense* L.) genotypes. J Agric Chem Biotech Mansoura Univ 4:205–216

Izadpanah F, Kalantari S, Hasani N et al (2015) Molecular and morphological variation in some Iranian saffron. Genetika 47(2):711–722

José Bagur M, Alonso Salinas GL, Jiménez-Monreal AM et al (2017) Saffron: an old medicinal plant and a potential novel functional food. Molecules 23:30

Kazemi-Shahandashti SS, Mann L, El-Nagish A et al (2022) Ancient artworks and crocus genetics both support saffron's origin in early Greece. Front Plant Sci 13:834416. https://doi.org/10.3389/fpls.2022.834416

Keify F, Beiki AH (2012) Exploitation of random amplified polymorphic DNA (RAPD) and sequence-related amplified polymorphism (SRAP) markers for genetic diversity of saffron collection. J Med Plants Res 6:2761–2768

Khan IA (2004) Induced mutagenic variability in Saffron (*Crocus sativus* L.). Acta Hortic 650:281–283

Khan MA, Nagoo S, Naseer S et al (2011) Induced mutation as a tool for improving corm multiplication in Saffron (*Crocus sativus* L.). J Phyt 3(7):8–10

Khorasani G, Hosseinimehr SJ, Zamani P et al (2008) The effect of saffron (*Crocus sativus*) extract for healing of second-degree burn wounds in rats. Keio J Med 57:190–195

Khoury C, Laliberte´ B, Guariano L (2010) Trends in ex-situ conservation of plant genetic resources: a review of global crop and regional conservation strategies. Genet Resour Crop Evol 57:625–639

Koocheki A, Khajeh-Hosseini M (2020) Saffron: science, technology and health. Woodhead Publishing, Oxford, UK

Mashmoul M, Azlan A, Yusof BN et al (2014) Effects of saffron extract and crocin on anthropometrical, nutritional and lipid profile parameters of rats fed a high-fat diet. J Funct Foods 8:180–187

Mattar MZ, Eissa OA, Salem KFM (2009) Genotypic variability in some wheat genotypes (Triticum aestivum L.) exposed to salinity at germination stage. Egyptian J Exp Biol (Bot) 5:143–148

Mattar MZ, Salem KFM, El-Abed AB (2016) Identification of candidate microsatellite markers associated with agronomic traits in rice (*Oryza sativa* L.). J Bot 15(5):213–232

Maxted N, Scholten M, Codd R, Ford-Lloyd B (2007) Creation and use of national inventory of crop wild relatives. Biol Conserv 140:142–159

Mir H (2004) Herbal Knowledge: usage of herbs in prevention and treatment of diseases, with latest research around the world. Daftare Nashre Farhange Eslami, Tehran, Iran

Mesfer ALshamrani S, Safhi FA, Alshaya DS, Ibrahim AA, Mansour H, Abd El Moneim D (2022) Genetic diversity using biochemical, physiological, karyological and molecular markers of Sesamum indicum L. Front. Genet. 13:1035977. https://doi.org/10.3389/fgene.2022.1035977

Mourad AMI, Alomari DZ, Alqudah AM, Sallam A, Salem KF (2019) Recent advances in wheat (*Triticum* spp.) breeding. In: Al-Khayri J, Jain S, Johnson D (eds) Advances in plant breeding strategies: cereals. Springer, Cham. https://doi.org/10.1007/978-3-030-23108-8_15

Mzabri I, Addi M, Berrichi A (2019) Traditional and modern uses of saffron (*Crocus Sativus*). Cosmetics 6:63. https://doi.org/10.3390/cosmetics6040063

Namayandeh A, Nemati Z, Kamelmanesh MM et al (2013) Genetic relationships among species of Iranian Crocus (*Crocus* spp.). Crop Breed J 3:61–67

Nanda S, Madan K (2021) The role of Safranal and saffron stigma extracts in oxidative stress, diseases and photoaging: a systematic review. Heliyon 7:e06117

Nawara HM, Mattar MZ, Salem KFM, Eissa OA (2016) Estimating heterosis and combining ability for agronomic and tissue culture traits using diallel cross on bread wheat (*Triticum aestivum* L.). Egyptian Journal of Botany, special issue 6 International Conference on Biotechnology and its applications in Botany and Microbiology 11–12 May 2016, pp 177–200

Nawara HM, Mattar MZ, Salem KFM, Eissa OA (2017) Diallel study on some in vitro traits of bread wheat (*Triticum aestivum* L.) under salt stress. Inter J Agric Environ Res 3 (1):1988–2006

Negbi M (1999) Saffron cultivation: past, present and future prospects. In: Negbi M (ed) Saffron: *Crocus sativus* L. Harwood Academic Publishers, Australia, pp 1–18

Nemati Z, Blattner FR, Kerndorff H (2018) Phylogeny of the saffron-crocus species group, *Crocus* series *Crocus* (Iridaceae). Mol Phylogenet Evol 127:891–897. https://doi.org/10.1016/j.ympev.2018.06.036

Palomares C (2022) Le Safran, Précieuse Épice Ou Précieux Médicament? Available online: https://hal.univ-lorraine.fr/hal-01732922/document

Pandita D (2021) Saffron (*Crocus sativus* L.): phytochemistry, therapeutic significance and omics-based biology. In: Aftab T, Hakeem KR (eds) Medicinal and aromatic plants. Woodhead Publishing, Sawston, UK, pp 325–396

Parizad S, Dizadji A, Koohi Habibi M et al (2017) Prevalence of saffron latent virus (SaLV), a new potyvirus species, in saffron fields of Iran. J Plant Path 99(3):802

Petersen G, Seberg O, Thorsoe S et al (2008) A phylogeny of the genus Crocus (Iridaceae) based on sequence data from five plastid regions. Taxon 57:487–499

Poma A, Fontecchio G, Carlucci G, Chichiricco G (2012) Anti-inflammatory properties of drugs from saffron crocus. Anti-Inflamm Anti -Allergy Agents Med Chem 11:37–51

Ramadan SM, Ibrahim AA (2021). Fruits and vegetables as sources of functional phytochemicals for the management of obesity, diabetes, and cancer. Springer. Springer Nature Switzerland AG 2021 C. Egbuna, S. Hassan (eds.), Dietary Phytochemicals. https://doi.org/10.1007/978-3-030-72999-8_8

Rashed-Mohassel MH (2007) Saffron from the wild to the field. Acta Hort (ISHS) 739:187–193

Rashed-Mohassel MH (2020) Evolution and botany of saffron (*Crocus sativus* L.) and allied species. In: Koocheki A, Khajeh-Hosseini M (eds) Saffron: science, technology and health. Woodhead Publishing, Sawston, United Kingdom, pp 37–57

Rigi H, Mohtashami L, Asnaashari M, Emami SA, Tayarani-Najaran Z (2021) Dermoprotective effects of saffron: a mini-review. Curr Pharm des 27:4693–4698

Riverón NL (2002) The combination of natural and synthetic agents: a new pharmacological approach in cancer chemoprevention. Proc West Pharmacol Soc 45:74–75

Rubio Moraga A, Traper-Mozos A, Gómez-Gómez L, Ahrazen O (2010) Intersimple sequence repeat markers for molecular characterization of *Crocus cartwrightianus* cv albus. Ind Crops Prod 32:147–151

Rubio-Moraga A, Castillo-López R, Gómez-Gómez L, Ahrazem O (2009) Saffron is a monomorphic species as revealed by RAPD, ISSR and microsatellite analyses. BMC Res Notes 2(1):189. https://doi.org/10.1186/1756-0500-2-189

Safhi FA, ALshamrani SM, Jalal AS, El-Moneim DA, Alyamani AA, Ibrahim AA (2022) Genetic characterization of some Saudi Arabia's accessions from *Commiphora gileadensis* using physio-biochemical parameters, molecular markers, DNA barcoding analysis and relative gene expression. Genes 13:2099

Saleh MM, Ali WA, Alwan MH et al (2020a) Multi-environmental evaluation of wheat tetraploid genotypes for agronomic traits under rainfed conditions in Syria. J Aridland Agric 6:1–5

Saleh MM, Alnaddaf LM, Almuhammady AK, Elabsawy EA, Mierah HHSalem KF, Alloosh MT, Al-Khayri JM (2021) Applications of plant-derived nanomaterials in mitigation of crop abiotic stress. In: Al-Khayri JM, Ansari MI, Singh AK (eds) Nanobiotechnology, Springer, Cham. https://doi.org/10.1007/978-3-030-73606-4_9

Saleh MM, Salem KFM, Elabd AB (2020b) Definition of selection criterion using correlation and path coefficient analysis in rice (*Oryza sativa* L.) genotypes. Bull Natl Res Cent 44(1):1–6

Salem KFM (1997) Breeding studies on rice (*Oryza sativa* L.). M Sc Thesis, Faculty of Agriculture, Minufiya University, Shibin El-Kom, Egypt

Salem KFM (2009) Relationship between genetic diversity based on SSRs markers with heterosis and combining ability in diallel cross of bread wheat (*Triticum aestivum* L.). Minufiya J Agric Res 6(34):2159–2178

Salem KFM (2015) Allelic state at the microsatellite locus *Xgwm261* marking the dwarfing gene *Rht8* in Egyptian bread wheat (*Triticum aestivum* L.) genotypes released from 1947 to 2004. Genetika 47:741–750

Salem KFM, Alghuthaymi MA, Elabd AB, Elabsawy EA, Mierah HH (2022) Prediction of heterosis for agronomic traits in half-diallel cross of rice (*Oryza sativa* L.) under drought stress using microsatellite markers. Plants 11:1532. https://doi.org/10.3390/plants11121532

Salem KFM, Alloosh MT, Saleh MM, Alnaddaf LM, Almuhammady AK, Al-Khayri JM (2021a) Utilization of nanofertilizers in crop tolerance to abiotic stress. In: Al-Khayri JM, Ansari MI, Singh AK (eds) Nanobiotechnology, Springer, Cham. https://doi.org/10.1007/978-3-030-73606-4_11

Salem KFM, Elabsawy A, Aldahak L, Elshamy H (2021b) Evaluation of stem reserve mobilization in Egyptian bread wheat (*Triticum aestivum* L.) genotypes and F_1 hybrids under post-anthesis chemical desiccation stress. J Genet Environ Resour Conserv 9(1):176–182

Salem KFM, El-Absawy EA, Khidr YA, El-Shamy H (2021c) Assessment genetic diversity among some Egyptian Bread wheat (*Triticum aestivum* L.) cultivars using RAPD markers. J Genet Environ Resour Conserv 9(1):168–175

Salem KFM, El-Zanaty AM, Esmail RM (2008) Assessing wheat (*Triticum aestivum* L.) genetic diversity using morphological characters and microsatellite markers. World J Agric Sci 4(5):538–544

Salem KFM, Ibrahim AA (2024) Plant biodiversity in the context of food security under climate change. In: Al-Khayri JM, Jain, SM, Penna S (eds) Sustainable utilization and conservation of plant genetic diversity. sustainable development and biodiversity, vol 35, Springer, Singapore. https://doi.org/10.1007/978-981-99-5245-8_1

Salem KFM, Mattar MZ (2014) Genetic diversity in old and modern Egyptian bread wheat (*Triticum aestivum* L.) varieties revealed by simple sequence repeats. Egypt J Genet Cytol 43:143–156

Salem KFM, Matter MZ (2014b) Identification of microsatellite alleles for salt tolerance at seedling stage in wheat (*Triticum aestivum* L.). Life Sci J 11(12s):1064–1073

Salem KFM, Röder MS, Börner A (2007) Identification and mapping quantitative trait loci for stem reserve mobilisation in wheat (*Triticum aestivum* L.). Cereal Res Common 35:1367–1374

Salem KFM, Röder MS, Börner A (2015) Assessing genetic diversity of Egyptian hexaploid wheat (*Triticum aestivum* L.) using microsatellite markers. Genet Resour Crop Evol 62:377–385

Salem KFM, Safhi FA, Alwutayd KM, Abozahra MS, Almohisen IA, Alsharari SF, Gangwar P, Rady AM, Hendawy MF, Ibrahim AA (2024) Analysis of genetic diversity, population structure and phylogenetic relationships of rice (*Oryza sativa* L.) cultivars using simple sequence repeat (SSR) markers. Genet Resour Crop Evol. https://doi.org/10.1007/s10722-023-01789-0

Salem KFM, Saleh MM, Abu-Ellail FFB, Abbas HS, Mahmoud AS (2021d) Role of quantum dots, polymeric NPS and dendrimers in emphasizing crops tolerate biotic and abiotic stresses. In: Faizan M, Hayat S, Yu F (eds) Sustainable agriculture reviews, vol 53, Springer, Cham. https://doi.org/10.1007/978-3-030-86876-5_1

Salem KFM, Saleh MM, Abu-Ellail FFB, Aldahak L, Alkuddsi YA. (2021e) The Role of salicylic acid in crops to tolerate abiotic stresses. In: Hayat S, Siddiqui H, Damalas, CA (eds) Salicylic Acid- A Versatile Plant Growth Regulator. Springer, Cham. https://doi.org/10.1007/978-3-030-79229-9_7

Salem KFM, Saleh MM, Aldahak L, Elabd AB (2021f) Assessment phenotypic diversity of rice (*Oryza sativa* L.) genotypes by multivariate analysis. J Arid Agric 7:52–59. https://doi.org/10.25081/jaa.2021.v7.6516

Salem KFM, Sallam A (2016) Analysis of population structure and genetic diversity of Egyptian and exotic rice (*Oryza sativa* L.) genotypes using SSR markers. Comptes Rendus Biol 339:1–9. https://doi.org/10.1016/j.crvi.2015.11.003

Salem KFM, Tawfik E, Mahdy M, Alghuthaymi MA, Mekkawy SA, Hassan MN, Ibrahim AA (2023) Advances in lettuce (*Lactuca* spp.) molecular breeding strategies. In: Singh S, Sharma D, Sharma SK, Singh R (eds) Smart plant breeding for vegetable crops in post-genomics era, Springer, Singapore. https://doi.org/10.1007/978-981-19-5367-5_11

Salem KFM, Varshney RK, Röder MS, Börner A (2010) EST-SSR based estimates on functional genetic variation in a barley (*Hordeum vulgare* L.) collection from Egypt. Genet Resour Crop Evol 57(4):515–521

Shokrpour M (2019) Saffron (*Crocus sativus* L.) breeding: opportunities and challenges. Springer Nature Switzerland AG. https://doi.org/10.1007/978-3-030-23265-8

Siracusa L, Gresta F, Avola G et al (2013) Agronomic, chemical and genetic variability of saffron (*Crocus sativus* L.) of different origin by LC-UV–vis-DAD and AFLP analyses. Genet Res Crop Evol 60:711–721

Soliman MI, Ibrahim AA, El-Metwally MA, Eshak DA (2017) Cytotoxicity effects of biological control and antioxidants using *Vicia faba* chromosomal aberration assay. J Environ Sci Technol 10:230–237. https://doi.org/10.3923/jest.2017.230.237

Soliman MI, Mohammed NS, EL-Sherbeny G, Safhi FA, ALshamrani SM, Alyamani AA, Alharthi B, Qahl SH, Al Kashgry NA, Abd-Ellatif S, Ibrahim AA (2023) Antibacterial, antioxidant activities, GC-Mass characterization, and cyto/genotoxicity effect of green synthesis of silver nanoparticles using latex of *Cynanchum acutum* L. Plants 12:172. https://doi.org/10.3390/plants12010172

Song YN, Wang Y, Zheng YH et al (2021) Crocins: a comprehensive review of structural characteristics, pharmacokinetics and therapeutic effects. Fitoterapia 153:104969

Talaei A, Moghadam MH, Tabassi SAS, Mohajeri SA (2015) Crocin, the main active saffron constituent, as an adjunctive treatment in major depressive disorder: a randomized, double-blind, placebo-controlled, pilot clinical trial. J Affect Disord 174:51–56

Tammaro F (1990) *Crocus sativus* L. cv. Piano di Navelli (L'Aquila saffron): environment, cultivation, morphometric characteristics, active principles, uses. In: Tammaro F, Marra L (eds) Proceedings of the international conference on saffron (*Crocus sativus* L.). L'Aquila, pp 47–57

Tsoktouridis G, Krigas N, Karamplianis T, Constantinidis T, Maloupa E (2009) Genetic differences among wild Greek Crocus taxa and cultivated saffron (*Crocus sativus* L.). In: 3rd International symposium on saffron Forthcoming challenges in cultivation research and economics. Greece Book of Abstracts, Krokos, Kozani, p 37

Varshney RK, Salem KFM, Röder MS et al (2008) Assessment and comparison of genetic diversity in barley germplasm collection using gSSR, eSSR and SNP markers. Plant Genet Res 6:167–174

Zadeh SM, Khosrowshahli M, Taeb M (2009) Cryopreservation of the axial meristem of *Crocus sativus* L. Cryobiology 59:412

Zaffar G, Wani SA, Anjum T, Zeerak NA (2004) Colchicine induced variability in Saffron. Acta Hortic 650:277–280

Zargari A (1997) Medicinal plants. Tehran University Publication, Tehran, Iran

Zheng H, MengHua L, CuiPing Y (2013) Construction of DNA fingerprinting for dry saffron. Afr J Pharm Pharmacol 7:2807–2812

Zubor AA, Suranyi G, Gyori Z, Borbély G, Prokisch J (2004) Molecular biological approach of the systematics of *Crocus sativus* L. and its allies. In: Abdullaev F (ed) Proceedings of the first international symposium on saffron biology and biotechnology, pp 85–93

Chapter 13
Advances in Breeding Strategies of Common Sage (*Salvia officinalis* L.)

Rashmi Ramakrishnan, Akshatha Banadka, Wudali Narasimha Sudheer, Chole Pranjali Bajrang, Jameel M. Al-Khayri, and Praveen Nagella

Abstract Common sage, *Salvia officinalis*, is a fragrant, medicinal herb with a broad range of pharmacological and culinary applications. In traditional medicine, *S. officinalis* is highly prized because it has been used for centuries to treat conditions like inflammation, infection, and antioxidants. Furthermore, because of its fragrant leaves, it is a mainstay in cuisines all over the world. Sage (*Salvia officinalis*) cultivation requires particular soil types, environmental conditions, and management strategies to maximize yield and quality; however, sage cultivation is still difficult due to its susceptibility to a range of biotic and abiotic stresses. Diverse germplasm ensures the availability of the traits needed for environmental adaptation and the development of better cultivars. Enhancement of yield, quality, and stress tolerance have been the primary objectives of genetic development programs for *S. officinalis*. These efforts aim to speed up the development of superior cultivars through a combination of conventional and molecular breeding techniques. Breeding strategies are designed to tackle particular issues like environmental stress, disease resistance, and market demands. The integration of genomic tools, marker-assisted selection, hybridization

R. Ramakrishnan · A. Banadka · W. N. Sudheer · C. P. Bajrang · P. Nagella (✉)
Department of Life Sciences, School of Sciences, Christ University, Bengaluru, Karnataka 560 029, India
e-mail: praveen.n@christuniversity.in

R. Ramakrishnan
e-mail: rashmir@res.christuniversity.in

A. Banadka
e-mail: akshatha.b@res.christuniversity.in

W. N. Sudheer
e-mail: wudali.sudheer@res.christuniversity.in

C. P. Bajrang
e-mail: cholepranjali.bajrang@res.christuniversity.in

J. M. Al-Khayri
Department of Agricultural Biotechnology, College of Agriculture and Food Sciences, King Faisal University, Al-Ahsa, Hofuf 31982, Saudi Arabia
e-mail: jkhayri@kfu.edu.sa

J. M. Al-Khayri et al. (eds.), *Biodiversity and Genetic Improvement of Herbs and Spices*, Advances in Plant Breeding Strategies 8, https://doi.org/10.1007/978-3-031-81838-7_13

techniques, tissue culture techniques and traditional breeding methods are promising avenues for achieving desired traits in *S. officinalis* cultivars. The chapter demonstrates the medicinal, culinary, and economic significance of *S. officinalis* in a variety of contexts. The chapter also offers significant insights for researchers, breeders, and stakeholders engaged in the cultivation and use of this adaptable plant species by clarifying cultivation methods, germplasm resources, genetic improvement initiatives, and breeding strategies.

Keywords *Salvia officinalis* · Sage · Germplasm · Hybridization · Essential oil · Molecular marker · Bioinformatics · Terpene synthase

13.1 Introduction

13.1.1 Background and Chapter Objectives

Salvia officinalis, also known as sage, common sage, or Dalmatian sage are among the most significant medicinal and fragrant plants in terms of nutrition and medicine due to its variety of key active components. It is mainly attracted by people for its strong aroma and aesthetic beauty of leaves and flowers. Apart from that, sage is known for its essential oils which are identified as potent antioxidants and are widely explored in the cosmetics and medicinal fields. The expansion of the sage essential oil market is likely to be fueled by a shift in consumer demand for natural and organic cosmetic products, particularly in developed countries. To address the expanding global demand for sage essential oils, a comprehensive assessment of sage breeding studies is required. Currently, majority of sage research is focused on its biochemical and medicinal effects. Only a few studies have been carried on sage breeding. In this context, the current conventional and modern breeding strategies in the production of novel cultivars and their large-scale propagation, as well as the application of biotechnological and bioinformatic tools in sage breeding research to meet worldwide requirements, are discussed in this chapter.

13.1.2 Botanical Classification and Distribution

Kingdom	Plantae
Division	Tracheophyta
Class	Magnoliopsida
Order	Lamiales
Family	Lamiaceae

(continued)

(continued)

Kingdom	Plantae
Subfamily	Nepetoideae
Genus	*Salvia*
Species	*officinalis*

Salvia officinalis is one among the 900 species of the genus, which belongs to the Lamiaceae family with a chromosome number 2n = 14 (Papafotiou et al. 2021). It is a small perennial plant that will reach 40–80 cm, with simple, opposite leaves with greenish-grey or greenish-white hairs on the upper surface and white hairs on the lower leaf surface. The leaves are slender and petiolate, with a toothed edge, a rugose texture, and basal lobes rarely. Blooms consist of five to ten violet-blue flowers in pseudoverticillasters. Their blooming period varies from March to July, contingent upon the climate and habitat (Jakovljević et al. 2019).

As per the records of Walker and team (Walker et al. 2004), the genus Salvia, which has over 900 species, with a polyphyletic origin, in which *Salvia officinalis* belonging to the clade 1, seems to be a monophyletic origin and thorough studies confirm it (Stojanović et al. 2015; Will and Claßen-Bockhoff 2014). The Middle East and Mediterranean region are the species' native habitat. Its cultivation is extensive in Yugoslavia, Italy, Albania, Turkey, Portugal, Spain, Cyprus, England, Canada, Switzerland, Germany, and the United States because of its pharmacological, cultural, and aesthetic interests worldwide. Jammu is the only place in India where it is grown sparingly (Ghorbani and Esmaeilizadeh 2017).

13.1.3 Importance (Economic and Health Benefits) of Sage

Sage is regarded as the "Salvation Plant" because its name comes from the Latin word "salvarem," which means "to save or cure" (Jakovljević et al. 2019). From prehistoric days, *S. officinalis* has been employed as a flavoring agent in meals, hydroalcoholic medication, and tea in folkloric medicine to treat a variety of illnesses (Salomón et al. 2022) and has also proven its application in the perfumery and cosmetic industry (Arslan 2022). Sage is a safe medicine for many ailments, including heart disease, cancer, and gastrointestinal issues, based on the reports of Food and Drug Administration. It can also be used to cure a sore throat, mouth inflammations, gingivitis, and toothache when used as a mouthwash (Sharifi-Rad et al. 2018). Recent reports also support the application of sage in the medications of oral diseases (Elchaghaby et al. 2022).

Several bioactive molecules in sage are thought to be responsible for its biological effects, including 1,8-cineole, borneol, camphor, thujone, ursolic acid, humulene, trans-caryophyllene, etc., (Martini 2022). The presence of phenolic acids (caffeic, vanillic, ferulic, and rosmarinic acids) and flavonoids (lutein, apigenin, and quercetin), and primary diterpenes such as carnosic acid, carnosol, and methyl carnosate also added the therapeutic value of sage (Pavić et al. 2019). The chemical makeup of *S. officinalis* essential oils varies based on a variety of parameters including genetic background, location, climate, season, plant components taken for essential oil extraction, and method of extraction (El Jery et al. 2020). Sage has a variety of pharmacological qualities, including antimicrobial, anti-inflammatory, antidiabetic, and anticancer actions, all of which are linked to its active components (D'Aquila et al. 2022; Mourabiti et al. 2024).

Common sage has long been praised for its alleged potential to aid memory and cognitive function (Dinel et al. 2020; McEwen Bradley 2020). Diverse memory functions have been shown to be influenced considerably and positively in various clinical and preclinical studies (Dinel et al. 2020; McEwen Bradley 2020). The capacity of sage extracts to inhibit AChE backs up its historic use in the ailment of cognitive illnesses such as Alzheimer's (Sharma et al. 2020).

Apart from that, nano emulsification of sage essential oil with cyclodextrin increased the antifungal activity (Nait Bachir et al. 2022). The efficiency of sage extracts in curing female reproductive tract infections was also reported (Sokolik and Prozorova 2022). The antimicrobial properties of sage extracts against a variety of human diseases like HSV1 also point to its potential in the treatment of lethal covid infections (Trivedi et al. 2022). Previous studies by Babaeekhou et al. (2021) also provide evidence to the capability of sage extracts in covid treatment.

The anti-inflammatory and antidiabetic capacity of sage is also supported by studies conducted by Ben Khedher et al. (2018). The role of carnosic acid in the anti-inflammatory properties of sage extracts was elucidated by Wang et al. (2018) via a joint process of bioinformatics and proteomics. The sage extract was found to have anti-inflammatory properties in various human cell lines, suggesting that it may help to lower peripheral inflammation and improve cognitive function (Margetts et al. 2022). The hepatoprotective and anticancer activity was well explained by Mohammed et al. (2021) and Jažo et al. (2023).

S. officinalis exhibits promising results in the treatment of various menstrual problems in females (Moradi et al. 2023). Wilfried et al. (2021) revealed the usefullness of menosan- a drug prepared from sage extracts that cured the physical and psychological symptoms of menopausal women. Recent studies show that the administration of tablets containing sage extracts enhances sexual function in women after menopause (Zeidabadi et al. 2022).

Recently various studies have reported the use of sage in feed formulation for different organisms. In the broiler chicken, dietary supplementation of sage (0.2%) could increase antioxidant status (enzymatic activity), regulate pulmonary hypertensive response, intestinal morphometry, and absorptive surface (Bahadoran et al. 2022). The effectiveness of the usage of sage as fish feed was also reported (Dadras et al. 2020; Salomón et al. 2022).

Apart from its medicinal properties, sage is commonly used for flavoring foods and in cosmetics because of its high antioxidant and antimicrobial activity. In the cosmetic sector, sage essential oil is quite important. Sage essential oil is rich in antioxidant that is frequently employed in a variety of skin treatments as anti-aging substance. It is used for wounds, surgical incisions, post-natal traumas, ulcers, and lesions in the pharmaceutical sector as an antiseptic. Sage essential oil is a tonic and can be used to condition and darken hair in the realm of the trilogy. India Essential Oils, The Essential Oil Company, Taytonn PTE LTD, Augustus Oils Ltd, Bontoux S.A, Lansdowne Chemicals, Lluch Essence S.L., M&U International LLC, and O'Laughlin Industries Inc. are the global competitors in the sage essential oil market (https://www.transparencymarketresearch.com/sage-essential-oil-market.html).

13.1.4 Domestication, Selection, and Early Improvements

Herbs like sage were first used as medicine before being introduced into the meals as a food, demonstrating the various benefits of food plants to our wellbeing beyond providing basic nutrition. The need to improve population health led to the evolution of the practice of flavoring food in many cultures. Furthermore, the great cultural relevance of therapeutic foods is suggested by the close association between medical and ritual applications in ancient and current times (Meyer et al. 2012).

The ethnobotanical significance of sage was well described by Sharma et al. (2019) who reported that *S. officinalis* has been utilized as a panacea in Greece since ancient times. The Egyptians employed sage to enhance fertility, and the Romans were aware of sage tea as a stimulating beverage (Mehta 2012). To cure infected wounds, lesions, and torpid ulcers, Hippocrates prescribed sage poultices. It was also once widely favoured by Chinese folks, who likened it to traditional tea in the winter (Hamidpour et al. 2014). In Brazilian folk medicine, sage leaves and their infused decoctions are utilized to treat toothaches, ulcers in mouth, bleeding of gum, and foul breath, while fresh leaves are consumed as a mouth refresher (Ghorbani and Esmaeilzadeh 2017). Sage, known as Karpooravalli in traditional Indian Siddha medicine, has been used to treat asthma, bronchitis, migraines, and rhinitis (Sharma 2017). The inhaling of steam from hot infusion of sage leaves has been practiced to cure respiratory problems such as cough and cold, according to folkloric medicine. Sage leaves were employed as a carminative, a light sleep-inducing drug, painkiller, tonic, and spasm reliever in Iranian folk medicine (Miraj and Kiani 2016). Oropharyngeal and gastrointestinal diseases are treated with leaf decoction. It has been used as antiseptic, anti-scabies, antisyphilitic, anti-inflammatory, and antiseptic in Jordanian traditional medicine for several skin problems (Abu-Darwish et al. 2013). The ethnobotanical studies show the evidence of the domestication of sage as medicine. With the limited information available, it can be assumed that the key selection characteristics for herbs like sage, might be high dry-leaf yield, high leaf/stem ratio, good pigmentation after drying, mechanical harvestability, and so on (Dudai 2012).

13.2 Current Cultivation Practices and Challenges

13.2.1 Current Cultivation Practices

Salvia officinalis, a medicinal herb that is predominantly grown in mediterranean, tropic and temperate regions, is an annual or a perennial shrub (Fern 2014). This plant adapts to various habitats ranging from cliffs, fallows, lawns, moors, mountains, plains, pastures and slopes (dry, arid and rocky) (Warbletoncouncil 2022). Common sage preferably grows well under sunlight with a temperature ranging from 15–26 °C and can tolerate up to a temperature of 5–30 °C. However, some plants prefer shade under extreme sunlight especially in hot weather. The plant requires dry sandy loamy calcareous soil with a pH ranging from 5–6.5 and can bear the pH of 4.2–8.3 (Fern 2014). The plants are prone to fungal attack when the soil is soggy or under high humid conditions which can be controlled by good air flow around the plants. The plants have to be watered when the top layer of soil (1- 2 inches) dries out (Iannotti 2022).

The best season to grow sage is the summer and spring season following the last frost. The plant grows well in autumn and winter in warm climatic conditions. According to Sharma et al. (2019), seeds, stem cuttings, layering, and division are the usual methods of propagation. The propagation by seeds can be done by sowing the seeds at a depth of 3–6 mm, 2–3 weeks prior to the last frost or in the beginning of spring following the last frost. The soil should be moist and has to be watered during summer until the germination starts. Maintaining a distance of 30–90 cm between the seeds is recommended. The sown seeds take a minimum of 2–3 weeks to germinate (Ravenscroft 2022). Despite seed propagation being one of the best and most convenient methods for multiplication, seed dormancy is a major incumbrance for establishing seedling through seed propagation (Mossi et al. 2011).

Seed priming and seedling production of sage have received very little attention, particularly in organic farming approaches. Priming approaches include the use of chemical, biochemical, biological, and environmental techniques (Sharma et al. 2019). Different priming methods are reported to be successful in sage germination so far. According to Ghorbanpour and Hatami 2014, when sage seeds were exposed to a plant growth-promoting rhizobacteria strain (*Pseudomonas putida*), seed germination was found to be significantly higher (78.5%) than when they were not (41.25%). As per Dasanpoor et al. (2013), hydropriming *S. officinalis* seeds resulted in increased germination, root length, branch length, and seedling vigor as compared to plain seeds. The seed priming with biochemicals like gibberellins, and biophysical agents like laser irradiation and magnetic field are also helping to increase the seed vigor and germination in sage (Abdani Nasiri et al. 2018).

Thus, vegetative propagation serves as an alternative when propagation through seed germination fails (Nicola et al. 2003). Cutting, grafting and budding are the most common vegetative propagation methods. Cutting propagation is the simplest method which involves the rooting of the stem cuttings. The best season for rooting stem cuttings is in spring to late autumn and in winter, the plants become dormant,

Fig. 13.1 A **a** *Saliva officinalis* habit, **b** Sage cultivation

and rooting will be difficult. Softwood stem cuttings given with auxin-containing rooting hormones in a sand-peat combination proved to be optimal for establishing a strong rooting system and ensuring cutting development in the summer (Nicola et al. 2005). Stem cuttings of sage treated with auxins like IAA, IBA, and NAA also showed improved root formation (Gudeva et al. 2017). The air layering method can also be used by placing the soil with a wire around a long sage stem leaving around 4 inches of the tip free (Dubey 2017). The better quality of the plant can be maintained by pruning the bloom stems and dead leaves and stems. Additionally, pruning prevents the stem from becoming woody with time (Jeanne 2016) (Fig. 13.1).

The plants can be harvested 10–12 weeks after transplanting. It can be harvested all around the year as an evergreen herb. While harvesting, the plant is cut at a height of 10–15 cm above the ground for regeneration of sage (Zutic et al. 2004), or the top 20–25 cm could be harvested in temperate regions (Sharma et al. 2019). However, harvesting in the summer season yields better quality leaves with high essential oil yield (Verma et al. 2015). In the study conducted by Zutic et al. (2004), the period between milk maturity to seed maturity and a period of 70–100 days after that are proposed to be the best harvesting periods. Harvesting during the flowering season yields high essential oils (Ben Farhat et al. 2016). Once after the harvest, sage leaves are dried either naturally under sunlight or artificially. There have been reports of artificial drying techniques like freeze drying at -50 °C and oven drying at 30 and 60 °C (Venskutonis 1997). The highest essential oil yields were reportedly obtained by infrared drying and oven drying at 45 °C (Sellami et al. 2012). According to Delfine et al. (2024), the quality of essential oil was more in rainfed plants that watered plants.

13.2.2 Current Agricultural Challenges

Sage plants are susceptible to many diseases, such as leaf spot, powdery mildew, and, grey mold (*Botrytis cinerea Pers.*) as well as weeds and pests such as aphids, and mealy bugs hinder sage cultivation (Szpyrka and Słowik-Borowiec 2019). *Phomopsis sclarea, Alternaria alternata, Colletotrichum dematium, Rhizoctonia solani, Ascochyta sclarea, Phytophthora cryptogea, Phodosphaera inequalis, Sclerotinia sclerotiorum* and *Erysiphae polygoni* are also found to be attacking on sage and leading to root rot and damping off, which questions the agricultural input (Zimowska 2008; Çakir et al. 2017). Apart from these winters hardiness is also a challenge for sage cultivation (Schmiderer and Novak 2020).

According to the assessment of medicinal and aromatic plants of Albania, even though Albania is the key exporter of sage in European countries, they predominantly depend on the wild populations. Since processed products like dried leaves or essential oils are getting more profit than fresh leaves, the cost of processing hinders the farmers from sage cultivation (Schmiderer and Novak 2020).

13.2.3 Genetic Improvement Objectives

Genetic improvement enables the plants to overcome the challenges of cultivation to some extent which can be achieved through traditional breeding, hybridization, genetic engineering, and so on. The main objectives of genetic improvement of common sage are,

- To produce high-yielding varieties
- Disease resistant cultivars
- High essential oil content
- Climatic adaptability

13.3 Germplasm Biodiversity and Conservation

13.3.1 Germplasm Diversity

Germplasm is a valuable resource for developing new plant varieties with interesting characteristics and increasing agricultural output. Unfortunately, there is very little published information on *Salvia officinalis* germplasm. However, recent study (Schmiderer and Novak 2020) has reported some commonly using cultivars of *S. officinalis* including.

- 'Newe Ya'ar No. 4','Moran', 'Nazareth', 'Silverline' and 'Caspit' from Israel
- 'Regula' from Switzerland

- 'Krajova' from Czech Republic.
- 'Bona' from Poland
- 'Extrakta' and 'Gute Selektion Herkunft PHASA' from Germany
- 'De Resniresti' from Romania
- 'Dessislava' and 'Freitaler Herkunft' from Bulgaria
- 'Foggia' and 'AT11' from Austria

Apart from this, Woodcote Farm, Tricolor, 'Rubriflora', Berggarten, 'Icterina', 'Purpurascens', Albiflora, and Compacta are some of the ornamental varieties of sage (Tucker and Maciarello 1990).

13.3.2 Genetic Resources Conservation Approaches

Since sage is recognized with the highest amount of diversity in both the plastid (Gao et al. 2020) and nuclear genome (Liber et al. 2014; Radosavljević et al. 2012), it can't be considered as an endangered species, and it is hoped that it will retain its reproductive fitness as well as the ability to evolve and adapt to changes in the environment over time (Frankham et al. 2002). However, global demand for *S. officinalis* and its diverse byproducts is growing, and it appears that commercial production alone will not be able to meet it. Because wild populations have been considered as a readily accessible indigenous reserve of these medicinal herbs, this poses a serious threat to the species and has been highly susceptible to over-exploitation. So, it seems to be important to conserve the natural populations either through in situ or ex-situ conservation methods according to the vulnerability of the population which is necessary for future research on genetic variabilities, phylogenetic studies, and breeding programs. According to Stojanović et al. (2015), the natural populations which are isolated and with small sizes should be conserved by ex-situ methods including cryopreservation and those having enough size should be under in situ conservation and it is important to avoid the mixing of populations to keep the variability as such (Stojanović et al. 2015).

Nowadays, with the aid of gene banks and in vitro cultures, it is possible to store plant materials for a long period of time. Somatic seeds, that possess any regenerative plant part in a protective bead, are one of the storage forms in gene banks. A simplified in vitro somatic seed production method for sage was developed by Deja (2021) using apical and axillary buds from multi-shoot cultures that were two weeks old. These buds are submerged in sodium alginate solutions, and the resulting hard polymer-coated beads containing solitary explants are then drip-coated in calcium chloride solutions. Following a month of storage at 4 °C, somatic seeds are cultivated for shoot formation on MS medium containing BAP, and subsequently rooted on MS medium containing IAA to produce plantlets that are acclimated. This method produced somewhat lower axillary buds' conversion rates (62.5%) and high apical buds conversion rates (85%) indicating its efficiency in long term storage of valuable sage genotypes.

As reported by Schmiderer and Novak (2020), there are notable collections of sage species in Croatia, with 116 *S. officinalis* accessions and, in the Czech Republic, with 24 *S. officinalis* accessions, in Germany, with 21 *S. officinalis* accessions and 1 hybrid of *S. officinalis x S. fruticosa* accessions, and in the United States, with 11 *S. officinalis* metabolites accessions. Mader et al. (2010) reported nineteen accessions of *S. officinalis L.* in the gene bank collection of Gatersleben (Germany). Exploration of these genebank accessions' genetic structures can be very beneficial for conservation efforts and hold great promise for upcoming breeding initiatives.

13.4 Traditional Breeding

13.4.1 Improvement Strategies

The demand for sage essential oil increased when consumers began to focus on cosmetics and medications made with natural ingredients. But still sage is considered a minor crop. So, it is important to expand the current cultivation practices to meet the economic demand for sage through breeding strategies. As previously mentioned by Schmiderer and Novak (2020), the main objectives of sage breeding are,

- Increased number of leaves with wide leaf area
- High leaf to stem ratio with lateral branches and bushy growth
- Enhanced essential oil productivity
- Desired composition of essential oil (less thujone content)
- Improved rosmarinic acid and carnosol content
- High climatic adaptability
- Biotic and abiotic stress resistance
- Increased tolerance to fungal pathogens

13.4.2 Traditional Breeding Methodologies

Traditional breeding also referred to as conventional or classical breeding, involves the development of new varieties of plants by employing natural sexual and asexual reproduction methods. It produces offspring with desired traits by crossing plants with relevant traits. Knowledge of floral biology is important in plant breeding. As described by Schmiderer and Novak (2020) vernalization is necessary to flower sage plants. However, *S. officinalis* is a bisexual plant with a protandrous nature, both self-pollination and cross-pollination can be applied for breeding (Schmiderer and Novak 2020). The special flower structure enables cross-pollination in sage by attracting pollinators like *Apis mellifera* and *Bombus terrestris* (Koch et al. 2017).

13.4.3 Mutation Breeding

Mutation breeding is a useful method for increasing desired traits in plants and adding genetic diversity. It has the advantage of allowing for the modification of particular traits without altering the genetic composition of the variety as a whole (Ke et al. 2019). The pretreatment of chemical mutagen, hydrazine hydrate to sage seeds led to the upregulation of the terpene synthase genes SoGPS, SoMYRS, SoNEOD, SoCINS, SoSABS, SoLINS, SoFPPS, SoHUMS, SoTPS6, SoSQUS, SoGGPS, and SoGA2, and it also had concentration-dependent effects on the essential oil composition and percentage of major terpenoids like 1–8 cineole and caryophyllene (Ali et al. 2023).

13.4.4 Polyploidy Breeding

Polyploidy in medicinal and aromatic plants has been exploited to increase the quantity and quality of specialized metabolites in recent decades. Polyploidy induction has been carried out in *S. miltiorrhiza Bunge* (Chen et al. 2018), with the goal of boosting growth and increasing the content of specialized metabolites. Recently, Tavan et al. (2021) found that polyploid induction in *S. officinalis* can modify the breeding traits in terms of terpenoids. Further research is needed to expand the scope of polyploids in sage breeding.

13.4.5 Hybridization

There are limited studies on hybridization in *S. officinalis*. The cold and drought-resistant ability of *S. officinalis* is pointing to its use in the inter-specific hybridization with other species of Salvia, for agronomic, economic as well as aesthetic values. In 1990, Putievsky et al. conducted hybridization experiments with three salvia species, *S. officinalis, S. tomentosa* and *S. fruticosa* using *S. tomentosa* as a female plant. They have carried out conventionally by emasculation followed by manual crossing and bagging. Pollen viability, crossability, and essential oil content were analyzed in F1 populations and both the hybrids *S. officinalis x S. fruticosa* and *S. officinalis x S. tomentos*a showed a less amount of pollen fertility as well as seed viability compared to parent populations. A medium level of essential oil components (1,8-cineole, alpha and beta thujone, camphor, beta-caryophyllene and alpha and β-pinene) was observed in *S. officinalis x S. fruticosa* hybrids and *S. officinalis x S. tomentosa* hybrids, the composition was similar to *S. officinalis* except for β-thujone. This study gives clues for the efficacy of *S. tomentosa* and *S. fruticosa* in sage breeding (Putievsky et al. 1990).

Later, Dudai et al. in 1999 produced a unique sage hybrid named Newe Ya'ar No. 4. using clones of *S. officinalis* from Dalmatian coast and *S. fruticosa* selection from Mount Carmel (Israel). This hybrid shows intermediate morphology and an essential oil composition similar to that of *S. officinalis*. It showed high productivity during winter under greenhouse conditions compared to others and became the leading cultivar in Israel (Dudai et al. 1999). In addition to this, Herraiz-Peñalver et al. examined a new sage hybrid *S. lavandulifolia x S. officinalis* produced by spontaneous hybridization between parents. The predominant components of this hybrid's essential oil are 1,8-cineole and alpha thujone (the key components of their respective parents) (Herraiz-Peñalver et al. 2015).

In 2019, extensive research on the morphological and epigenetic, and genetic analysis of hybrid sage (*S. officinalis x S. fruticosa*) was done by Radosavljević et al. using molecular markers. The results support the intermediate morphological characters of hybrids. But the genetic analysis showed high allelic richness more similar to *S. officinalis* and the homoploid hybridization was also exposed by the SSR results since the hybrids showed diploid chromosomes (Radosavljević et al. 2019).

The toxicity of camphor and thujone in sage poses a threat. *S. officinalis* is recognized for its strong thujone and camphor content. *S. fruticosa* and *S. aramiensis*, on the other hand, are recognised for their high eucalyptol concentration. Unlike *S. aramiensis*, both *S. officinalis* and *S. fruticosa* have substantial essential oil content. Based on these characteristics, Bahadirli and Ayanoğlu recently conducted hybridization events to create novel chemotypes (low camphor and thujone and high eucalyptol). The results show that the hybrids possess unique essential oil compositions diverse from parents and based on the oil composition of interest, superior ones among the hybrid populations (*S. aramiensis x S. fruticosa, S. officinalis x S. aramiensis, S. fruticosa x S. aramiensis* and *S. officinalis x S. fruticosa*) were selected for further ongoing studies (Bahadirli and Ayanoğlu 2019).

Lately, *S. fruticosa* and *S. officinalis* clones were hybridized with *S. pomifera, S. ringens* and *S. tomentosa* by Papafotiou et al. (2021) and found that interspecific crosses of *S. officinalis* with *S. pomifera ssp. pomifera, S. ringens,* and *S. tomentosa* were effective, but only when *S. officinalis* was employed as the female parent. The crosses of *S. fruticosa* with *S. pomifera, S. ringens* and *S. tomentosa* were successful with less than a 50% rate when used as the female parent. Only the cross *S. fruticosa* with *S. tomentos*a was fruitful when used as a male parent. Based on selection characteristics like rapid growth, early and vigorous flowering, small plant size, fast regeneration by cuttings and nurturing inside a greenhouse, and ease of greenhouse maintenance, four hybrids (*S. officinalis × S. pomifera, S. officinalis × S. tomentosa, S. officinalis × S. ringens, and S. fruticosa × S. ringens*) were identified for their ornamental value and practicality (Papafotiou et al. 2021). The low crossability of *S. officinalis* with other species indicates that they are reproductively isolated mainly because of environmental factors and not by genetic reasons (Papafotiou et al. 2021; Putievsky et al. 1990).

13.4.6 Hybrid Cultivars

Hybrids	Distinguishing characters	Reference
S. officinalis × *S. pomifera*	Intermediary characteristics of parents in leaf shape and color, height etc. have unique aroma more like male parent	(Papafotiou et al. 2021)
S. officinalis × *S. ringens*	Segmented leaves and light aroma with more flowers, taller than parents	(Papafotiou et al. 2021)
S. officinalis × *S. tomentosa*	Increased no. of lateral branches with small-sized leaves. Flowers more and have a light pink color, light aroma	(Papafotiou et al. 2021)
S. fruticosa × *S. ringens*	Hairy segmented leaf, sparsely arranged flowers with long stem, light aroma, purple flowers	(Papafotiou et al. 2021)
S. aramiensis × *S. fruticosa*	High essential oil ratio	(Bahadirli and Ayanoğlu 2019)
S. officinalis × *S. aramiensis*	High essential oil ratio	(Bahadirli and Ayanoğlu 2019)
S. fruticosa × *S. aramiensis*	Low essential oil ratio	(Bahadirli and Ayanoğlu 2019)
S.officinalis × *S.fruticosa*	Medium level of essential oil components like thujone	(Radosavljević et al. 2019)
Newe Ya'ar No. 4	High essential oil yield during winter season	(Dudai et al. 1999)
S. lavandulifolia × *S. officinalis*	Increased plant height and size of leaves	(Herraiz-Peñalver et al. 2015)
S. officinalis × *S. tomentosa*	Essential oil without β-thujone	(Putievsky et al. 1990)

13.5 Molecular Breeding

The plastid genome and nuclear DNA were extensively employed in molecular research on the genus Salvia, which focused mostly on phylogeny and evolution (Wang et al. 2013; Cao et al. 2020; Tursun et al. 2021). In order to focus on genetic affinity, population structure, and even inter-specific hybridization, research has also targeted various parts of DNA (RAPD, AFLP, SSR, EST, etc.) (Mader et al. 2010; Radosavljević et al. 2011, 2012; Liber et al. 2014; Bahadirli and Ayanoglu 2021). SRAP markers are also useful in the examination of genetic and population structure among sage species (Aghaei et al. 2017). Identification of six primary genotypes from 43 examined genotypes in Northern Albania using RFLP PCR markers of Sage cpDNA has been reported (Papa and Bacu 2016).

Numerous molecular marker approaches have been created and widely employed in analyzing genetic diversity, population structure, and phylogenetic relationships as a result of advances in plant molecular biology. Innovations in genomic tools have enabled a plethora of new marker techniques in recent years, including functional and gene-targeted markers, as well as the synthesis of numerous novel DNA-based marker systems (Etminan et al. 2018).

13.6 Molecular Marker-Assisted Breeding

Marker-assisted breeding employs near-isogenic lines, recombinant inbred lines, doubled haploids, F2 and back-cross populations, and recombinant inbred lines to enable the simultaneous selection of multiple desirable traits (Jiang 2013). A molecular marker, identified through southern blotting and PCR, is a brief segment of DNA sequence exhibiting polymorphism (base substitution, insertion, and deletion) among distinct individuals (Nadeem et al. 2018). They have long been regarded as an effective tool for determining plant genetic variation (Sunar et al. 2020). The most often used plant molecular markers are RFLP, AFLP, SNP, RAPD, and SSR.

There are several studies that describe the implementation of molecular markers in sage molecular breeding. Mader et al. (2010) used a method designed to examine co-dominant markers like SNPs and SSRs by high-resolution melting analysis to examine the genetic makeup of nineteen genebank accessions of *S. officinalis*. Nine microsatellite primers rich in ACA and AGA repeats were discovered by Radosavljević et al. in 2011 and reported to be useful for molecular studies of sage populations, both wild and cultivated, including phylogeny, genetic diversity, population structure, and the identification of closely related groups (Radosavljević et al. 2011). The genetic diversity of sage populations was analyzed by Etminan et al. (2018) using ISSR and SCoT markers and concluded that the novel SCoT markers were successful in evaluating genetic relationships among sage populations.

SSR markers have been employed for the evaluation of genetic variation and population genetics (Radosavljević et al. 2012). The variation between *S. officinalis* and *S. x sylvestris* was analyzed using the ITS region of nuclear ribosomal DNA sequence and concluded that the ITS region alone can't be used for the identification of species (Tursun et al. 2021). Recently, Cao et al. identified *Salvia yangii* as a sister species of sage by chloroplast genome analysis (GenBank accession No. MT537168) (Cao et al. 2020). Thus, the identification of related species through molecular markers will be helpful for future hybridization programs.

The identification of metabolic biomarkers, which are becoming more and more common in plant breeding, especially in aromatic and medicinal plants where metabolites are the main target traits for selection and betterment, is made possible by the recent development of targeted metabolite analysis. Since *S. officinalis* shows a high level of variance at the genomic level and metabolic profile, Sarrou et al. investigated the effectiveness of ISSR markers in seven sage populations of N. W Greece with metabolic profiling for the identification of parent populations in the implementation of breeding methodologies, by considering maximum polyphenol content and appropriate essential oil composition as main selection criteria. They found that since there is no correlation between ISSR markers and metabolites, molecular markers themselves cannot be used for breeding purposes, but coupling with metabolic data is advisable for the parent lines (Sarrou et al. 2017).

13.6.1 Functional Genomics

Functional genomics is the study of transcriptional and translational levels of gene expression, function, and regulation. The invention of several sequencing tools resulted in the explosion of gene sequencing of several plants and there is still a large no. of sequences whose function remains unexplained. The importance of functional genomics is to explore the role of certain genes with reference to their expression in growth, metabolite production, stress response, etc. (Jha et al. 2015). In *S. officinalis*, most of the functional genomics studies were performed to analyze the terpene biosynthesis and rosmarinic acid (RA) production.

Previous studies reported the genes SoHPPR, PAL, C4H, 4CL, TAT, HPPR, RAS, and CYP98A14, which are overexpressed with enhanced rosmarinic acid production in sage, which helped in the elucidation of genes engaged in the rosmarinic acid production (Barberini et al. 2013; Swamy et al. 2018). The role of the PAL gene in RA production with the application of salicylic acid was also revealed (Ejtahed et al. 2015). In 2017, Ali et al. designed a transcriptome database for *S. officinalis* leaves by employing NGS technology in order to uncover and describe genes related to the terpenoid biosynthesis pathway (Ali et al. 2017). Recent studies show that two genes SoBDH1 and SoBDH2 from sage are involved in the production of borneol dehydrogenases (Drienovská et al. 2020). According to Aslani et al. (2023) the sabine syanthene (ss) gene was overexpressed and geranyl diphosphate synthase (gpps) and borneal diphosphate synthase (bpps) has been down regulated under drought stress. Similarly, Schmiderer et al. (2023) revealed that the sage genome has 12 open reading frames encoding monoterpene synthase, one of which is a pseudogene.

The specific functions of the genes SoNEOD, SoCINS, SoSABS, SoLINS, and SoTPS6, related to the biosynthesis of terpenes in sage were explored by expressing these genes in recombinant *N. tabacum* and affirming that they can turn a prenyl diphosphate substrate into various products (Ali et al. 2017). In addition, the seasonal expression of terpene biosynthesis genes was studied by comparing the metabolic profile and mRNA abundance and noticed a positive correlation between terpene synthase mRNA with 1,8-cineole and camphor, and a negative correlation with thujone content (Grausgruber-Gröger et al. 2012). A recent genome-wide study of common sage revealed the gene repertoire involved in terpenoid biosynthesis as well as 188 variable single-nucleotide loci within these genes. These markers will be helpful for upcoming research endeavors aimed at analyzing and modifying terpenoid composition traits (Ceylan et al. 2024).

Apart from this, Marchica et al. conducted an experiment to analyze the role of signaling molecules under O_3 stress in sage and noticed that the genes WRKY TFs are involved in regulating oxidative protection and thus help in tolerating plants from O_3 damage (Marchica et al. 2019). Recently, the functional role of the NAD + -dependent farnesol dehydrogenase (SoFLDH) in terpenoid biosynthesis was confirmed using overexpression of SoFLDH in *Arabidopsis thaliana* plants (Ali et al. 2022). Hereby the functional analysis of terpene synthase genes will be a foundation for the improvement of plant breeding programs and also economic production of important metabolites through novel strategies like genetic engineering, gene silencing, as well as genome editing. Recently, *Salvia officinalis*'s mitogenome has been explored by Yang et al. (2023) and found that, it was unusually organized, with two circular chromosomes, suggesting a different genomic architecture within the Lamiales order. Moreover, significant RNA editing events in important genes that code for proteins highlight how dynamically the sage mitochondrial genome regulates gene expression.

13.6.2 Bioinformatics

Bioinformatics tools can be used to analyze genes having high protein sequence similarity to that of known enzymes. Two genes for aldehyde dehydrogenases, SoBDH1 and SoBDH2 from *S. officinalis* were discovered by Drienovská et al. (2020) through BLAST (using PBLAST) against known genes for borneol dehydrogenase available in a gene bank, which were inserted into plasmids and utilized to transform *E. coli* for the production of borneol dehydrogenases, which are required for the enzymatic oxidation of racemic borneol, a concise way for the production of enantiomerically pure borneol and camphor. The phylogeny of these genes is also done by the neighbor-joining method by the ClustalO program and reveals the divergent evolution of these terpene biosynthesis genes (Drienovská et al. 2020).

Recently, Tursun et al. also checked the homology of the ITS region of sequenced 19 nrDNA fragments from *S. officinalis* with the aid of the NCBI BLASTn program and constructed a phylogenetic tree (Tursun et al. 2021). With the aid of bioinformatics, Craft et al. (2017) studied the chemotaxonomy of sage essential oil and noticed major chemotypes having essential oil with the composition of α-thujone > camphor > 1,8-cineole.

In addition, Gao et al. (2020) have done the similarity analysis via BLAST of the chloroplast genome of *Salvia spp* and constructed a circular genome map for *S. officinalis* along with others. Ali et al. (2017) generated and sequenced a sage cDNA library, which they compared to public databases like NCBI NR and SWISSPORT. The unigenes were then classified into three categories based on homology analysis using the BLAST GO program: biological process (BP), cellular component (CC), and molecular function (MF) (Gao et al. 2020). The unigenes were also used as query sequences in the KEGG (Kyoto Encyclopedia of Genes and Genomes) pathways database (Ali et al. 2017).

Apart from the use of bioinformatics tools in understanding phylogeny, evolution, and breeding purposes, recent bioinformatic studies are mainly focusing on pharmacognostic properties and drug designing (Babaeekhou et al. 2021). The efficacy of the use of the key phenolic compounds, pinocembrin, and luteolin retrieved from the sage extract in phytotherapy since it exhibits molecular properties like synthetic drugs supported by bioinformatic analysis using the tool SwissTargetPrediction (Georgiana et al. 2022).

13.7 Tissue Culture Applications

Tissue culture approaches help in the regeneration of clones of salvia plants that have desired traits. Plant in-vitro regeneration strategies help to meet the need for raw materials for medicinal products. Modern biotechnological approaches like gene transformation and optimized elicitation strategies will also aid in increasing the biomass and desired metabolites which play a major role in the pharmacological property of the plant.

Callus-mediated indirect organogenesis is reported in *S. officinalis* where shoot apex explants were observed to develop callus when cultivated on MS medium supplemented with 1.0 mg/L TDZ, and shoot differentiation was aided by subsequent culturing on medium supplemented with 1.0 mg/L BA. MS medium supplemented with 1.0 mg/L IBA helped in rooting of the in-vitro shoots with a 70% acclimatization rate in ex-vivo conditions (Tawfik and Mohamed 2006). Shoot apex explants when inoculated on MS medium supplemented with 4.5 μM TDZ induced callus and maintained on the same medium supplemented with 0.45 mM ascorbic acid. Further shoot differentiation was seen when callus was sub-cultured onto the MS medium supplemented with 4.4 or 8.8 μM BA along with 0.45 mM ascorbic acid. Developed shoots were later transferred onto a rooting medium supplemented with 4.9 μM IBA, among which 75% acclimatization efficiency was recorded in ex-vivo conditions (Tawfik and Mohamed 2007). MS medium supplemented with 2.0 mg/L NAA along with 0.5 mg/L BAP helped the leaf and internode explants in the induction of callus making the internode explant the best with 95% initiation efficiency. It is seen that MS medium supplemented with 0.5 mg/L TDZ showed the best regeneration efficiency with respect to shoots and rooting was established in ½ strength MS medium supplemented with 1.0 mg/L IBA (Jafari et al. 2017).

In vitro shoots of salvia were induced from nodal segments when cultured on the MS medium supplemented with 0.05 mg/L of 2,4-D along with varied concentrations of KN (i.e., 1.5, 2.0, 4.0 mg/L) showed production of antioxidant rich phenolic compounds (Santos-Gomes et al. 2002). MS medium supplemented with 1.5 mg/L TDZ along with 0.1 or 0.5 mg/L IAA induced the highest number of shoots from nodal explants, similarly when leaf explants inoculated on MS medium supplemented with 1.0 mg/L TDZ along with 0.1 mg/L 2,4-D helped in the induction of adventitious shoots (Mohamed et al. 2019). Supplementation of additives like polyamines like putrescine at 50 mg/L concentration along with TDZ at 0.5 mg/L concentration helped in the production of adventitious shoots when internode explants were inoculated. Rooting of the in-vitro shoots was established when IBA was supplemented at 1.0 mg/L concentration (Jafari et al. 2017).

Somatic embryogenesis is a successful approach for the in-vitro production of plantlets. Heart and torpedo-shaped embryos from callus were developed when cultured on MS medium supplemented with 10.5 mM NAA along with 10.5 mM BAP. Similarly, globular embryos were induced on a medium fortified with 1.8 mM 2,4-D along with 1.8 mM KN. Once the embryos matured and the developed adventitious shoots were cultured on MS medium supplemented with IBA at 4.92 μM concentration for the root development (Ioja-Boldura et al. 2010). Development of globular embryos from callus induced from leaf explants was seen when cultured on MS medium supplemented with 4.5 μM 2,4-D along with 4.5 μM KN (Kintzios et al. 1999). The preparation of synthetic seeds or artificial seeds will help in the conservation of desired explants and make them viable for a long time. Shoot tips of salvia were one of the best explants used for regeneration. These were encapsulated using 2–3% sodium alginate and 50 mM calcium chloride. It was observed that in salvia the 24 weeks of stored synthetic seeds made with 3% sodium alginate supplemented with MS medium and GA_3 at 0.25 mg/L concentration maintained the viability and helped in regeneration. Similarly, synthetic seeds that were not stored for a long time showed increased efficiency when encapsulated using 2% sodium alginate supplemented with 0.5 mg/L GA_3 (Grzegorczyk and Wysokińska 2011).

13.8 Genetic Engineering and Gene Editing

Sage plants can be manipulated by integrating genetic engineering technologies with in vitro culture techniques for biomedical goals such as alternative systems for the production of pharmacologically important secondary metabolites. But because of its high antimicrobial activity against a spectrum of organisms including *Agrobacterium tumefaciens*, the possibilities of genetic modifications by vector-based methods in sage are very limited (Luwańska et al. 2017).

However, there are some studies in which rDNA technology has been used for the examination of the expression of several genes from *S. officinalis* for the production of some commercially important proteins. For example, transgenic *E. coli* for the production of borneol dehydrogenases (Drienovská et al. 2020), and recombinant *N. tabaccum* plants for terpene biosynthesis (Ali et al. 2017). Similarly, by overexpressing the transcription factor SmERF6 (Salvia miltiorrhiza Ethylene Responsive Factor 6), transgenic lines of *S. officinalis* were created via an in-planta *Agrobacterium*-mediated genetic transformation method. This resulted in significantly higher levels of carnosol (which increased four-fold) and carnosic acid (which increased three-fold) when compared to non-transgenic counterparts (Bharadwaj et al. 2023).

13.9 Conclusions and Future Prospects

Salvia species are utilized in a variety of industries, including traditional and mainstream medicine, spices, and perfumes. Salvia has been the subject of about 12,000 investigations, the majority of which are concerned with its chemical composition and bioactive properties. The rising market demand for sage essential oil demonstrates the crop's economic potential. However, sage cultivation alone is insufficient to supply worldwide demand. Breeding research with *S. officinalis* is still in its early stages. Molecular marker-assisted breeding is suitable for the identification of suitable parents in the breeding programs and is yet to be explored. Tissue culture can be used to produce homogenous plant materials on a large scale. Genetic engineering tools can be used for in vitro synthesis of essential oils via transgenic organisms to overcome agricultural obstacles such as processing costs and labor costs.

13.9.1 Current Research Initiatives to Combat Global Climate Change

The climatic changes like global warming with an increasing temperature and reducing water levels leading the scientific world to focus more on the production of drought-resistant plants. Breeding programs for drought resistance are important for agricultural and medicinal plants to meet global needs. Since the global demand for sage is increasing for its medicinal, culinary, and ornamental use, the research on resistant breeding has already started.

Because some sage species like *S. officinalis* and *S. fruticosa* have a high tolerance for water shortages, several scientists have focused on breeding drought-resistant plants through hybridization. To introduce new drought-tolerant plants into the urban floriculture sector in arid and semi-arid regions, Papafotiou et al. (2021) explored the potential of crossbreeding five Greek sage species known for their appealing floral characteristics and drought resilience. They were picked in order to create interspecific hybrids with a variety of features, including growth pattern, floral color, duration and timing of flowering, scent from the leaves, resistance to cold and drought, and more. The results show that interspecific hybrids (*S. officinalis* × *S. pomifera*, *S. officinalis* × *S. tomentosa*, *S. officinalis* × *S. ringens*, and *S. fruticosa* × *S. ringens*) have increased ability to survive in a water-deficient environment (Papafotiou et al. 2021).

The reports of Herraiz-Peñalver et al., suggests that, the drought-resistant species *S. lavandulifolia* was also successful in hybridization with *S. officinalis*, but the hybrid *S. lavandulifolia x S. officinalis* was only identified by essential oil composition. Further research is necessary to determine whether this hybrid can be used as a promising horticultural crop for semi-arid and rainfed soils, given its exceptional water-use efficiency, drought tolerance, and high essential oil production (Herraiz-Peñalver et al. 2015).

Sage cultivation faces significant challenges due to its sensitivity to heat, particularly in the warm season. Additionally, high temperatures pose a significant risk to the output of fresh-market salvia. Recent studies have shown that *S. officinalis* plant appearance and competence to reduce heat stress can benefit from a combined treatment of 100 mM SA and 5 mM $CaCl_2$ (Lin et al. 2021).

The most important gaseous pollutant is tropospheric ozone (O_3) which has a wide range of deleterious effects on vegetation at both functional and genetic levels and is now anticipated to rise in many parts of the world owing to changing climatic conditions and human activities. In this regard, Marchica et al. studied the response of *Salvia officinalis* to O_3 stress at a molecular level to check the tolerance ability of the plant and identified the genetic and biochemical mechanisms of sage to cope with stress. The identification of regulatory genes WRKYs sheds light to further research on the production of O_3-resistant plants by biotechnological approaches (Marchica et al. 2019).

13.9.2 Recommendations for Future Research and Utilization

With the growing demand for high-quality organically reared sage, more research is required to harmonize organic cultivation processes, spanning seed priming to cost-effective extraction and processing procedures. Despite the fact that it's a really promising and widely available herb for pharmaceutical enterprises, extraction and separation procedures for verifying the herb's functionality are still needed. One of the main obstacles to sage cultivation is the fungal attack. Current breeding studies are mainly focused on essential oil quality and quantity. It is necessary to have a look at breeding for fungal resistance for effective cultivation. As of now, very little research has been done on the breeding for enhanced production of antioxidants like rosmarinic acid and carnosol in *Salvia officinalis.* Since both the compounds are found effective in treating metabolic and other lifestyle diseases, further research is needed to produce new cultivars with high antioxidants.

The active components of sage essential oil, thujone, and camphor are known to be neurotoxic. So, breeding programs should give more focus on the production of sage plants having essential oil with a reduced amount of thujones. So that it will be helpful for the implementation of sage extracts in neuroprotective drugs for diseases like Alzheimer's.

The growing interest in replacing synthetic antioxidants with natural antioxidants with similar efficacy could be advantageous for both improving food quality and having a favorable impact on the treatment of many human diseases. But, the different physiologically active chemicals in sage that are responsible for its pharmacological activities and the level of cytotoxicity are yet unknown, therefore medication dosage studies in the future are still a possibility. Apart from that, studies provide the promising effects of sage extracts on cognitive function and neural activity. But the use of the sage leaves among children under the age of 18 is not suggested due to a shortage of safety studies on children and teenagers. Since a major part of brain development takes place in childhood, the future research on the safety of administration of sage in kids will be helpful for enhancing their cognitive function.

Appendix Genetic Resources

Cultivar name	Characteristic feature	Location	Reference
Foggia	High alpha thujone, lowest beta-thujone	Austria	(Grausgruber-Gröger et al. 2012)
AT11	Less thujone content	Austria	(Grausgruber-Gröger et al. 2012)
Bona	High essential oil yeild	Poland	(Luwańska et al. 2017)
Newe-Ya'ar No. 4	Hybrid, medium essential oil production	Israel	(Dudai et al. 1999)
Extrakta	High yield 1.5–2% of essential oil	Germany	(Tibaldi et al. 2022)
De Resniresti	High yield	Romania	(Schmiderer and Novak 2020)
Dessislava	High yield 0.06–0.28% of essential oil	Bulgaria	(Schmiderer and Novak 2020)
Primoska	Heavy metal tolerant	Yugoslavia	(Masarovicova et al. 2004)
Krajova	Rich in hydroxycinnamic acids (3.06- 3.52%) and rosmarinic acid (0.76–1.65%)	Czech Republic	(Schmiderer and Novak 2020; Tekeľová et al. 2015)
Kew Gold	High yield of essential oil (0.95 mL/100 g)	Hungary	(Andrea Boszormenyi et al. 2009)
Regula	High leaf yield and essential oil content	Switzerland	(Schmiderer and Novak 2020)
IPPO	High leaf yield/ hector, 1.5% of essential oil content	Italy	(De Mastro et al 2006)
Woodcote Farm	Rarely flowers, hardy	North America	(Tucker and Maciarello 1990)
Tricolor	Marginally hardy and flowers rarely	North America	(Tucker and Maciarello 1990)
Berggarten	Large leaves and hardy and few flowers	North America	(Tucker and Maciarello 1990)
Icterina	Leaves edged with yellow	North America	(Tucker and Maciarello 1990)
Purpurascens	Hardy, few flowers	North America	(Tucker and Maciarello 1990)

(continued)

(continued)

Cultivar name	Characteristic feature	Location	Reference
Albiflora	Hardy, more flowers	North America	(Tucker and Maciarello 1990)
Rubriflora	Hardy, few flowers	North America	(Tucker and Maciarello 1990)
Compacta	Hardy, more flowers	North America	(Tucker and Maciarello 1990)

References

Abdani Nasiri A, Mortazaeinezhad F, Taheri R (2018) Seed germination of medicinal sage is affected by gibberellic acid, magnetic field and laser irradiation. Electromagn Biol Med 37(1):50–56

Abu-Darwish MS, Cabral C, Ferreira IV, Gonçalves MJ, Cavaleiro C, Cruz MT, Al-Bdour TH, Salgueiro L (2013) Essential oil of common sage (*Salvia officinalis* L.) from Jordan: assessment of safety in mammalian cells and its antifungal and anti-inflammatory potential. Biomed Res Int 2013:538940

Aghaei Z, Talebi M, Rahimmalek M (2017) Assessment of genetic diversity within and among sage (Salvia) species using SRAP markers. Plant Genet Resour 15(3):279–282

Ali M, Li P, She G, Chen D, Wan X, Zhao J (2017) Transcriptome and metabolite analyses reveal the complex metabolic genes involved in volatile terpenoid biosynthesis in garden sage (*Salvia officinalis*). Sci Rep 7(1):16074

Ali M, Nishawy E, Ramadan WA, Ewas M, Rizk MS, Sief-Eldein AG, El-Zayat MA, Hassan AH, Guo M, Hu GW, Wang S (2022) Molecular characterization of a Novel NAD+-dependent farnesol dehydrogenase SoFLDH gene involved in sesquiterpenoid synthases from *Salvia officinalis*. PLoS One 17(6):e0269045

Ali M, Abdelkawy AM, Darwish DBE, Alatawi HA, Alshehri D, Al-Amrah H, Soudy FA (2023) Changes in metabolite profiling and expression levels of key genes involved in the terpenoid biosynthesis pathway in garden sage (*Salvia officinalis*) under the effect of hydrazine hydrate. Metabolites 13(7):807

Arslan Ö (2022) Boron toxicity impacts on photosystem II photochemical efficiency of sage. Bot Serb 46(1):113–119

Aslani Z, Hassani A, Abdollahi Mandoulakani B, Barin M, Maleki R (2023) Effect of drought stress and inoculation treatments on nutrient uptake, essential oil and expression of genes related to monoterpenes in sage (*Salvia officinalis*). Sci Hortic 309:111610

Babaeekhou L, Ghane M, Abbas-Mohammadi M (2021) In silico targeting SARS-CoV-2 spike protein and main protease by biochemical compounds. Biologia 76(11):3547–3565

Bahadirli NP, Ayanoğlu F (2019) Preliminary results: assessment of new salvia chemotypes for herbal tea industry by hybridization. Sci Papers Ser A Agron 62(1):501–506

Bahadirli NP, Ayanoglu F (2021) Genetic diversity of Salvia species from Turkey assessed by microsatellite markers. J Appl Res Med Aromat Plants 20(1):100281

Bahadoran S, Teymouri Y, Hassanpour H, Mohebbi A, Akbari MR (2022) Effect of sage (*Salvia officinalis* L.) extract in antioxidant status and intestinal morphology of pulmonary hypertensive chickens. Vet Med Sci. https://doi.org/10.1002/vms3.804

Barberini S, Savona M, Raffi D, Leonardi M, Pistelli L, Stochmal A, Vainstein A, Pistelli L, Ruffoni B (2013) Molecular cloning of SoHPPR encoding a hydroxyphenylpyruvate reductase, and its expression in cell suspension cultures of *Salvia officinalis*. Plant Cell Tiss Org Cult 114(1):131–138

Ben Farhat M, Jordán MJ, Chaouch-Hamada R, Landoulsi A, Sotomayor JA (2016) Phenophase effects on sage (*Salvia officinalis* L.) yield and composition of essential oil. J Appl Res Med Aromat Plants 3:87–93

Ben Khedher MR, Hammami M, Arch JRS, Hislop DC, Eze D, Wargent ET, Kępczyńska MA, Zaibi MS (2018) Preventive effects of *Salvia officinalis* leaf extract on insulin resistance and inflammation in a model of high fat diet-induced obesity in mice that responds to rosiglitazone. PeerJ 6:e4166

Bharadwaj R, Thashanamoorthi G, Demiwal P, Sircar D, Ramalingam S (2023) SmERF6 promotes the expression of terpenoid pathway in *Salvia officinalis* and improves the production of high value abietane diterpenes, carnosol and carnosic acid. Plant Cell Physiol 17:pcaf007

Boszormenyi A, Hethelyi E, Farkas A, Horvath G, Papp N, Lemberkovics E, Szoke E (2009) Chemical and Genetic Relationships among Sage (*Salvia officinalis* L.) Cultivars and Judean Sage (*Salvia judaica* Boiss.). J Agric Food Chem 57(11):4663–4667

Bradley McEwen (2020) Sage advice for "cognitive improvement." J Aust Tradit-Med Soc. 26(1):6–8. https://doi.org/10.3316/informit.069855027873444

Çakir E, Bağdat RB, Katircioğlu YZ, Maden S (2017) Occurrence of root rot caused by *Phytophthora cryptogea* on Common Sage (*Salvia officinalis*) in Turkey. J Agric Sci Technol 7(6). https://doi.org/10.17265/2161-6256/2017.06.005

Cao MT, Wu JJ, Wang RH, Xu L, Qi ZC, Wei YK (2020) The complete chloroplast genome of Russian sage *Salvia yangii* B. T. Drew (Lamiaceae). Mitochondrial DNA B Resour 5(3):2590–2591

Ceylan F, Uncu AO, Soyturk Patat A, Uncu AT (2024) Whole-genome resequencing identifies exonic single-nucleotide variations in terpenoid biosynthesis genes of the medicinal and aromatic plant common sage (*Salvia officinalis* L.). Genet Resour Crop Evol. https://doi.org/10.1007/s10722-024-01900-z

Chen EG, Tsai K-L, Chung H-H, Chen J-T (2018) Chromosome doubling-enhanced biomass and dihydrotanshinone I production in *Salvia miltiorrhiza*, a traditional Chinese medicinal plant. Molecules 23(12). https://doi.org/10.3390/molecules23123106

Craft JD, Satyal P, Setzer WN (2017) The chemotaxonomy of common sage (*Salvia officinalis*) based on the volatile Constituents. Medicines (Basel) 4(3). https://doi.org/10.3390/medicines4030047

D'Aquila P, Paparazzo E, Crudo M, Bonacci S, Procopio A, Passarino G, Bellizzi D (2022) Antibacterial activity and epigenetic remodeling of essential oils from Calabrian aromatic plants. Nutrients14(2). https://doi.org/10.3390/nu14020391

Dadras H, Hayatbakhsh MR, Golpour A (2020) Dietary administration of common sage (*Salvia officinalis*) and coneflower (*Echinacea angustifolia*) extracts affects growth, blood parameters and immune responses of Beluga, *Huso huso*. Turkish J Fish Aquat Sci 20(5). https://doi.org/10.4194/1303-2712-v20_5_05

Dastanpoor N, Fahimi H, Shariati M, Davazdahemami S, Hashemi SM (2013) Effects of hydropriming on seed germination and seedling growth in sage (*Salvia officinalis* L.). Afr J Biotechnol 12(11):1223–1228

De Mastro G, Aiello N, Scartezzini F, Vender C, Brunetti G (2006) Herbage yield and essential oil quality of three cultivars of Sage (*Salvia officinalis* L.) grown in two Italian environments. Acta Hort 723:232–237

Deja A (2021) Technology for the production of somatic seeds useful for the storage of valuable genotypes of *Salvia officinalis*. Biol Life Sci Forum 11(1):43

Delfine S, Formisano C, Sirignano C, Rigano D (2024) Effect of harvest time and low water availability during balsamic period on *Salvia officinalis* L. biomass development, ecophysiological traits and essential oil yield and composition. Biochem Syst Ecol J114:104825

Dinel A-L, Lucas C, Guillemet D, Layé S, Pallet V, Joffre C (2020) Chronic supplementation with a mix of *Salvia officinalis* and *Salvia lavandulaefolia* improves Morris water maze learning in normal adult C57Bl/6J Mice. Nutrients 12(6). https://doi.org/10.3390/nu12061777

Drienovská I, Kolanović D, Chánique A, Sieber V, Hofer M, Kourist R (2020) Molecular cloning and functional characterization of two highly stereoselective borneol dehydrogenases from *Salvia officinalis* L. Phytochemistry 172:112227

Dubey P (2017) How to Grow Sage (*Salvia Officinalis*)? In: The Modern Vedic. https://themodernvedic.com/herbs-natural-foods/grow-sage-salvia-officinalis/

Dudai N, Lewinsohn E, Larkov O, Katzir I, Ravid U, Chaimovitsh D, Sa'ad D, Putievsky E (1999) Dynamics of yield components and essential oil production in a commercial hybrid sage (*Salvia officinalis x Salvia fruticosa cv. Newe Ya'ar no. 4*). J Agric Food Chem 47(10):4341–4345

Dudai N (2012) Domestication and breeding of wild medicinal and aromatic plants—thirty years of experience in Israel. Acta Hortic 175–183

Ejtahed RS, Radjabian T, Hoseini Tafreshi SA (2015) Expression analysis of phenylalanine ammonia lyase gene and rosmarinic acid production in *Salvia officinalis* and *Salvia virgata* shoots under salicylic acid elicitation. Appl Biochem Biotechnol 176(7):1846–1858

Elchaghaby MA, Abd El-Kader SF, Aly MM (2022) Bioactive composition and antibacterial activity of three herbal extracts (lemongrass, sage, and guava leaf) against oral bacteria: an in vitro study. J Oral Biosci 64(1):114–119

Etminan A, Pour-Aboughadareh A, Noori A, Ahmadi-Rad A, Shooshtari L, Mahdavian Z, Yousefiazar-Khanian M (2018) Genetic relationships and diversity among wild Salvia accessions revealed by ISSR and SCoT markers. Biotechnol Biotechnol Equip 32(3):610–617

Fern K (2014) *Salvia officinalis*. In: Useful tropical plants. https://tropical.theferns.info/viewtropical.php?id=Salvia+officinalis. Accessed 12 Jul 2022

Frankham R, Ballou SEJ, Briscoe DA, Ballou JD (2002) Introduction to conservation genetics. Cambridge University Press

Gao C, Wu C, Zhang Q, Zhao X, Wu M, Chen R, Zhao Y, Li Z (2020) Characterization of chloroplast genomes from two Salvia medicinal plants and gene transfer among their mitochondrial and chloroplast genomes. Front Genet 11:574962

Georgiana UTA, Din A, Manolescu DS, Avram S (2022) Sage extract: effect of particle size on TPC and antioxidant activity. Bioinformatic Approaches Romanian J Biophys 32(2):87–102

Ghorbani A, Esmaeilizadeh M (2017) Pharmacological properties of *Salvia officinalis* and its components. Afr J Tradit Complement Altern Med 7(4):433–440

Ghorbanpour M, Hatami M (2014) Biopriming of *Salvia officinalis* seed with growth-promoting rhizobacteria affects invigoration and germination indices. J Biol Enviorn Sci 8(22):29–36

Grausgruber-Gröger S, Schmiderer C, Steinborn R, Novak J (2012) Seasonal influence on gene expression of monoterpene synthases in *Salvia officinalis* (Lamiaceae). J Plant Physiol 169(4):353–359

Grzegorczyk I, Wysokińska H (2011) A protocol for synthetic seeds from *Salvia officinalis* L. shoot tips. Acta Biol Crac Ser Bot 53(1). https://doi.org/10.2478/v10182-011-0011-6

Gudeva LK, Trajkova F, Mihajlov L, Troicki J (2017) Influence of different auxins on rooting of Rosemary, Sage and Elderberry. Annu Res Rev Biol 1–8

Hamidpour M, Hamidpour R, Hamidpour S, Shahlari M (2014) Chemistry, pharmacology, and medicinal property of sage (Salvia) to prevent and cure illnesses such as obesity, diabetes, depression, dementia, lupus, autism, heart disease, and cancer. Afr J Tradit Complement Altern Med 4(2):82–88

Herraiz-Peñalver D, Ortiz de Elguea-Culebras G, Sánchez-Vioque R, Santana Méridas O (2015) Identification of a hybrid species of sage (*Salvia officinalis* L. *x S. lavandulifolia subsp. lavandulifolia*) through the study of the essential oil. J Essent Oil Res 27(5):363–372

Iannotti M (2022) How to grow sage. In: The spruce. https://www.thespruce.com/growing-sage-1402599. Accessed 4 Mar 2022

Ioja-Boldura O-M, Radu F, Popescu S, Borozan A (2010) Regeneration, micropropagation, callus cultures and somatic embryogenesis of common sage (*Salvia officinalis* L.). Bull Univ Agric Sci Vet Med Cluj-Napoca Hortic 67(1):308–313

Jafari S, Daneshvar M, Salmi M, Jalal-Abadi A (2017) Indirect organogenesis and plant regeneration in common sage (*Salvia officinalis* L.): an important medicinal plant of Iran. Mod Appl Sci 11(5):22

Jakovljević M, Jokić S, Molnar M, Jašić M, Babić J, Jukić H, Banjari I (2019) Bioactive profile of various *Salvia officinalis* L. preparations. Plants 8(3). https://doi.org/10.3390/plants8030055

Jažo Z, Glumac M, Paštar V, Bektić S, Radan M, Carev I (2023) Chemical composition and biological activity of *Salvia officinalis* L. Essential Oil. Plants 12(9)

Jeanne (2016) Growing sage, or *Salvia officinalis*. In: Home garden joy. https://homegardenjoy.com/site/2016/05/growing-sage-salvia-officinalis.html

El Jery A, Hasan M, Rashid MM, Al Mesfer MK, Danish M, Rebah FB (2020) Phytochemical characterization, and antioxidant and antimicrobial activities of essential oil from leaves of the common sage L. from Abha, Saudi Arabia. Asian Biomed 14(6):261–270

Jha UC, Bhat JS, Patil BS, Hossain F, Barh D (2015) Functional genomics: applications in plant science. PlantOmics: Omics Plant Sci 65–111

Jiang G-L (2013) Molecular markers and marker-assisted breeding in plants. Plant Breed Lab Fields 3:45–83

Ke C, Guan W, Bu S, Li X, Deng Y, Wei Z, Wu W, Zheng Y (2019) Determination of absorption dose in chemical mutagenesis in plants. PLoS One 14(1):e0210596

Kintzios S, Nikolaou A, Skoula M (1999) Somatic embryogenesis and in vitro rosmarinic acid accumulation in *Salvia officinalis* and *S. fruticosa* leaf callus cultures. Plant Cell Rep 18(6):462–466

Koch L, Lunau K, Wester P (2017) To be on the safe site—Ungroomed spots on the bee's body and their importance for pollination. PLoS ONE 12(9):e0182522

Liber Z, Židovec V, Bogdanović S, Radosavljević I, Pruša M, Filipović M, Han Dovedan I, Jug-Dujaković M, Šatović Z (2014) Genetic diversity of dalmatian sage (*Salvia officinalis* L.) as assessed by RAPD markers. Agric Conspec Sci 79(2):77–84

Lin K-H, Lin T-Y, Wu C-W, Chang Y-S (2021) Protective effects of salicylic acid and calcium chloride on sage plants (*Salvia officinalis* L. and *Salvia elegans* Vahl) under high-temperature stress. Plants 10(10). https://doi.org/10.3390/plants10102110

Luwańska A, Wielgus K, Seidler-Łożykowska, Lipiński D, Słomski R (2017) Evaluation of *Agrobacterium tumefaciens* usefulness for the transformation of sage (*Salvia officinalis* L.). Transgenesis Second Metablsm 153–176

Mader E, Lohwasser U, Börner A, Novak J (2010) Population structures of genebank accessions of *Salvia officinalis* L. (Lamiaceae) revealed by high resolution melting analysis. Biochem Syst Ecol 38(2):178–186

Marchica A, Lorenzini G, Papini R, Bernardi R, Nali C, Pellegrini E (2019) Signalling molecules responsive to ozone-induced oxidative stress in *Salvia officinalis*. Sci Total Environ 657:568–576

Margetts G, Kleidonas S, Zaibi NS, Zaibi MS, Edwards KD (2022) Evidence for anti-inflammatory effects and modulation of neurotransmitter metabolism by *Salvia officinalis* L. BMC Complement Med Ther 22(1):131. https://doi.org/10.1186/s12906-022-03605-1

Martini N (2022) Sage. J Prim Health Care 14(1):89–90

Masarovicova E, Kralova K, Stresko V (2004) Comparative study of uptake, accumulation and some effects of cadmium in two cultivars of *Salvia officinalis* L. Chemia i Inżynieria Ekologiczna 11:209–214

Mehta D (2012) *Salvia officinalis* Linn: relevance to modern research drive. Inventi Impact: Planta Activa 2012(4):203–207

Meyer RS, DuVal AE, Jensen HR (2012) Patterns and processes in crop domestication: a historical review and quantitative analysis of 203 global food crops. New Phytol 196(1):29–48

Miraj S, Kiani S (2016) A review study of therapeutic effects of *Salvia officinalis* L. Pharm Lett 8(6):299–303

Mohamed MA, Aly MK, Ahmed ET, Abd El-latif SA (2019) Effect of plant growth regulators on organogenesis of *Salvia officinalis* l. Plants. Minia J Agric Res Dev 39(3):401–14. https://doi.org/10.1111/j.1365-2621.1980.tb04086.x

Mohammed HA, Eldeeb HM, Khan RA, Al-Omar MS, Mohammed SA, Sajid MS, Aly MS, Ahmad AM, Abdellatif AA, Eid SY, El-Readi MZ (2021) Sage, *Salvia officinalis* L., constituents, hepatoprotective activity, and cytotoxicity evaluations of the essential oils obtained from fresh and differently timed dried herbs: a comparative analysis. Molecules 26(19). https://doi.org/10.3390/molecules26195757

Moradi M, Ghavami V, Niazi A, Seraj Shirvan F, Rasa S (2023) The effect of *Salvia officinalis* on hot flashes in postmenopausal women: a systematic review and meta-analysis. Int J Community Based Nurs Midwifery 11(3):169–178

Mossi AJ, Cansian RL, Paroul N, Toniazzo G, Oliveira JV, Pierozan MK, Pauletti G, Rota L, Santos AC, Serafini LA (2011) Morphological characterization and agronomical parameters of different species of Salvia sp. (Lamiaceae). Braz J Biol 71(1):121–129

Mourabiti F, Derdak R, El Amrani A, Momen G, Timinouni M, Soukri A, El Khalfi B, Zouheir Y (2024) The antimicrobial effectiveness of *Rosmarinus officinalis, Lavandula angustifolia,* and *Salvia officinalis* essential oils against *Klebsiella pneumoniae* and *Pseudomonas aeruginosa* in vitro and in silico. S Afr J Bot. 168:112–23

Nadeem MA, Nawaz MA, Shahid MQ, Doğan Y, Comertpay G, Yıldız M, Hatipoğlu R, Ahmad F, Alsaleh A, Labhane N, Özkan H (2018) DNA molecular markers in plant breeding: current status and recent advancements in genomic selection and genome editing. Biotechnol Biotechnol Equip 32(2):261–285

Nait Bachir Y, Sahraoui N, Cheurfa Z, Medjkane M, Ziane AH (2022) Formulation of a natural nanosystem based on β- cyclodextrin/arginine/xanthan to increase the antifungal activity of *Salvia officinalis* essential oil from Algeria (Bejaïa, Kalaa n'Ath Abas). J Res Pharm 26(3):581–597

Nicola S, Fontana E, Hoeberechts J, Saglietti D (2003) Rooting products and cutting timing on sage (*Salvia officinalis* L.) propagation. In: III WOCMAP congress on medicinal and aromatic plants-volume 2: conservation, cultivation and sustainable use of medicinal and 676, pp 135–41

Nicola S, Fontana E, Hoeberechts J, Saglietti D (2005) Rooting products and cutting timing on sage (*Salvia officinalis* L.) propagation. Acta Hortic 676(2):135–141

Papa S, Bacu A (2016) RFLP analysis of cpDNA of *Salvia officinalis* L. of Northern Albania can serve to elucidate genetic diversity among close natural populations. Eur J Biotechnol Genet Eng 3(1):65–74

Papafotiou M, Martini AN, Papanikolaou E, Stylias EG, Kalantzis A (2021) Hybrids development between Greek salvia species and their drought resistance evaluation along with *Salvia fruticosa*, under the attapulgite-amended substrate. Agronomy 11(12):2401

Pavić V, Jakovljević M, Molnar M, Jokić S (2019) Extraction of carnosic acid and carnosol from Sage (*Salvia officinalis* L.) leaves by supercritical fluid extraction and their antioxidant and antibacterial activity. Plants 8(1). https://doi.org/10.3390/plants8010016

Putievsky E, Ravid U, Diwan-Rinzler N, Zohary D (1990) Genetic affinities and essential oil composition of *Salvia officinalis* L., *S. fruticosa* Mill., *S. tomentosa* Mill. and their hybrids. Flavour Fragr J 5(2):121–123

Radosavljević I, Jakse J, Javornik B, Satovic Z, Liber Z (2011) New microsatellite markers for *Salvia officinalis* (Lamiaceae) and cross-amplification in closely related species. Am J Bot 98(11):e316–318

Radosavljević I, Satovic Z, Jakse J, Javornik B, Greguraš D, Jug-Dujaković M, Liber Z (2012) Development of new microsatellite markers for *Salvia officinalis* L. and its potential use in conservation-genetic studies of narrow endemic *Salvia brachyodon* Vandas. Int J Mol Sci 13(9):12082–12093

Radosavljević I, Bogdanović S, Celep F, Filipović M, Satovic Z, Surina B, Liber Z (2019) Morphological, genetic and epigenetic aspects of homoploid hybridization between *Salvia officinalis* L. and *Salvia fruticosa* Mill. Sci Rep 9(1):3276

Ravenscroft (2022) D *Salvia officinalis*—Sage Growing Guide. Gardeners HQ. https://www.gardenershq.com/Salvia-officinalis.php

Salomón R, Furones MD, Reyes López FE, Tort L, Firmino JP, Quintela JC, Pinilla-Rosas JM, Vallejos-Vidal E, Gisbert E (2022) Phytogenics from Sage and Lemon verbena promote growth, systemic immunity and disease resistance in Atlantic Salmon (*Salmo salar*). *Front. Mar. Sci.* 9:828497. https://doi.org/10.3389/fmars.2022.828497

Santos-Gomes PC, Seabra RM, Andrade PB, Fernandes-Ferreira M (2002) Phenolic antioxidant compounds produced by in vitro shoots of sage (*Salvia officinalis* L.). Plant Sci 162(6):981–987

Sarrou E, Ganopoulos I, Xanthopoulou A, Masuero D, Martens S, Madesis P, Mavromatis A (2017) Genetic diversity and metabolic profile of *Salvia officinalis* populations: implications for advanced breeding strategies. Planta 246(2):201–215

Schmiderer C, Steinborn R, Novak J (2023) Monoterpene synthases of three closely related sage species (*Salvia officinalis, S. fruticosa* and *S. pomifera*, Lamiaceae). Plant Physiol Biochem 196:318–327

Schmiderer C, Novak J (2020) *Salvia officinalis* L. and *Salvia fruticosa* Mill. Dalmatian and three-lobed sage. Med Aromat Stimul Plants (12):523–537. https://doi.org/10.1007/978-3-030-38792-1_16

Sellami IH, Rebey IB, Sriti J et al (2012) Drying sage (*Salvia officinalis* L.) plants and its effects on content, chemical composition, and radical scavenging activity of the essential oil. Food Bioprocess Technol 5(8):2978–2989

Sharifi-Rad M, Ozcelik B, Altın G, Daşkaya-Dikmen C, Martorell M, Ramírez-Alarcón K, Alarcón-Zapata P, Morais-Braga MF, Carneiro JN, Leal AL, Coutinho HD (2018) *Salvia spp.* plants-from farm to food applications and phytopharmacotherapy. Trends Food Sci Technol 80:242–263

Sharma Y, Fagan J, Schaefer J, Yashaswini Sharma C (2019) Ethnobotany, phytochemistry, cultivation and medicinal properties of Garden sage (*Salvia officinalis* L.). J Pharmacogn Phytochem 8:3139–3148

Sharma Y, Fagan J, Schaefer J (2020) In vitro screening for acetylcholinesterase inhibition and antioxidant potential in different extracts of sage (*Salvia officinalis* L.) and Rosemary (*Rosmarinus officinalis* L.). J Biol Act Prod Nat 10(1):59–69

Sharma A (2017) *Salvia officinalis* (Sage plant) benefits, information and uses. Medicinal plants and their uses. https://www.medicinalplantsanduses.com/salvia-officinalis-sage-herb-benefits

Sokolik OP, Prozorova GO (2022) Current research opportunities for potential phytotherapeutic agents for the treatment of pathologies of the female reproductive system. Eur J Clin Exp Med 20(1):109–116

Stojanović D, Aleksić JM, Jančić I, Jančić R (2015) A Mediterranean medicinal plant in the continental Balkans: a plastid DNA-based phylogeographic survey of *Salvia officinalis* (Lamiaceae) and its conservation implications. Willdenowia 45(1):103–118

Sunar S, Korkmaz M, SiĞmaz B, AĞar G (2020) Determination of the genetic relationships among salvia species by RAPD and ISSR analyses. Turk J Pharm Sci 17(5):480–485

Swamy MK, Sinniah UR, Ghasemzadeh A (2018) Anticancer potential of rosmarinic acid and its improved production through biotechnological interventions and functional genomics. Appl Microbiol Biotechnol 102(18):7775–7793

Szpyrka E, Słowik-Borowiec M (2019) Consumer health risk to pesticide residues in *Salvia officinalis* L. and its infusions. J Environ Sci Health B 54(1):14–19

Tavan M, Sarikhani H, Mirjalili MH, Rigano MM, Azizi A (2021) Triterpenic and phenolic acids production changed in *Salvia officinalis* via in vitro and in vivo polyploidization: a consequence of altered genes expression. Phytochemistry 189:112803

Tawfik AA, Mohamed MF (2006) Shoot differentiation and plant regeneration from thidiazuron-induced callus of *Salvia officinalis*. Acta Hortic (723):309–314

Tawfik AA, Mohamed MF (2007) Regeneration of salvia (*Salvia officinalis* L.) via induction of meristematic callus. In Vitr Cell Dev Biol-Plant 43(1):21–27

Tekeľová D, Tóth J, Czigle S, Koutsoulas A (2015) Hydroxycinnamic derivatives content in plant organs linked to harvest time of *Salvia officinalis* L. cv. *'Krajová'*. Eur Pharm J 62(1):40–46

Tibaldi G, Hazrati S, Hosseini SJ, Ertani A, Bulgari R, Nicola S (2022) Cultivation techniques and drying process can affect the inflorescence essential oil composition of three selections of *Salvia officinalis*. Ind Crops Prod 183:114923

Trivedi N, Mishra A, Kumar D (2022) Herbal medicine is the way of potential therapeutic option for the treatment of COVID-19: recent updates. J Complement Med Res 13(1):27

Tucker AO, Maciarello MJ (1990) Essential Oils of Cultivars of Dalmatian Sage (*Salvia officinalis* L.). J Essent Oil Res 2(3):139–144

Tursun AO, Sipahioglu HM, Telci I (2021) Genetic relationships and diversity within cultivated accessions of *Salvia officinalis* L. in Turkey. Plant Biotechnol Rep 15(5):663–672

Venskutonis PR (1997) Effect of drying on the volatile constituents of thyme (*Thymus vulgaris* L.) and sage (*Salvia officinalis* L.). Food Chem 59(2):219–227

Verma RS, Padalia RC, Chauhan A (2015) Harvesting season and plant part dependent variations in the essential oil composition of *Salvia officinalis* L. grown in Northern IndiaJ Herb Med 5(3):165–171

Walker JB, Sytsma KJ, Treutlein J, Wink M (2004) Salvia (Lamiaceae) is not monophyletic: implications for the systematics, radiation, and ecological specializations of Salvia and tribe Mentheae. Am J Bot 91(7):1115–1125

Wang L-C, Wei W-H, Zhang X-W, Liu D, Zeng KW, Tu PF (2018) An integrated proteomics and bioinformatics approach reveals the anti-inflammatory mechanism of carnosic acid. Front Pharmacol 9:370

Wang M, Zhao H-X, Wang L, Wang T, Yang RW, Wang XL, Zhou YH, Ding CB, Zhang L (2013) Potential use of DNA barcoding for the identification of Salvia based on cpDNA and nrDNA sequences. Gene 528(2):206–215

Warbletoncouncil (2022) *Salvia officinalis*: characteristics, habitat, distribution, properties. In: Warbleton council. https://warbletoncouncil.org/salvia-officinalis-8741. Accessed 3 Mar 2022

Wilfried D, Nina CDG, Silvia B (2021) Effectiveness of Menosan® *Salvia officinalis* in the treatment of a wide spectrum of menopausal complaints. A double-blind, randomized, placebo-controlled, clinical trial. Heliyon 7(2):e05910

Will M, Claßen-Bockhoff R (2014) Why Africa matters: evolution of Old World Salvia (Lamiaceae) in Africa. Ann Bot 114(1):61–83

Yang H, Chen H, Ni Y, Li J, Cai Y, Wang J, Liu C (2023) Mitochondrial genome sequence of Salvia officinalis (Lamiales: Lamiaceae) suggests diverse genome structures in cogeneric species and finds the stop gain of genes through rna editing events. Int J Mol Sci 24(6).

Zeidabadi A, Jafari M, Emamghoreishi M, Dabbaghmanesh MH, Akbarzadeh M (2022) The Efficacy of *Salvia officinalis* in on promotion of sexual function in postmenopausal women: a double-blind randomized controlled clinical trial. Bull Pharm Sci 45(2):823–831. https://doi.org/10.21608/BFSA.2022.131398.1332

Zimowska B (2008) Fungi threatening the cultivation of sage (*Salvia officinalis* L.) in south-eastern Poland. Herba Pol. 54:15–24

Zutic I, Putievsky E, Dudai N (2004) Influence of harvest dynamics and cut height on yield components of Sage (*Salvia officinalis* L.). J Herbs Spices Med Plants 10(4):49–61

Chapter 14
Innovative Plant Breeding Strategies for Nutrient-Rich, Climate-Smart Sesame (*Sesamum Indicum* L.)

Sarita Pandey, Md Kamaruzzaman, Abhishek Anand, and Sushil K. Kothari

Abstract The Crop sesame is commonly known by several names, including sesame, gingelly, and til, *Sesamum indicum* L. could be Considered one of the earliest oilseeds to have been utilized by humans. For centuries, it has thrived, predominantly across Asia and Africa, owing to its rich protein content and valuable edible oil. Its cultivation spans diverse environments, from semi-arid tropical regions to subtropical and temperate zones. In countries like India, sesame holds particular importance for small-scale and marginal farmers, serving as a vital source of income amidst challenging agricultural landscapes. It is grown for its seeds, which have 18–25% protein and very high-quality oil content of 38–54%. The exceptional gene pool generated by sesame's wide range of characteristics, diverse range of environmental adaptability, and great diversity must be employed to produce desirable cultivars for sesame farmers. The lack of cultivars suitable for machine harvesting and sesame's low yield are the major obstacles to its expansion. Therefore, present and future projects for sesame genetic improvement should incorporate features that increase yield and quality, local adaptability, and machine harvesting ability. This can be accomplished by combining traditional breeding techniques with genetic and genomic tools like breeding assisted by genomics. As a result, the chapter's goal is to describe sesame's breeding progress, genetic improvement opportunities, and constraints.

Keywords Sesame · *til* · Antioxidant · Bioactive components · Sesamol · Genetic variation · Micronutrients · Genomics · Production challenges

S. Pandey (✉) · M. Kamaruzzaman · A. Anand · S. K. Kothari
The Neotia University, Kolkata, West Bengal 743368, India
e-mail: saritakumari.pandey@tnu.in

M. Kamaruzzaman
e-mail: md.kamaruzzaman@tnu.in

A. Anand
e-mail: abhishek.anand@tnu.in

S. K. Kothari
e-mail: Sushil.Kothari@tnu.in

J. M. Al-Khayri et al. (eds.), *Biodiversity and Genetic Improvement of Herbs and Spices*, Advances in Plant Breeding Strategies 8, https://doi.org/10.1007/978-3-031-81838-7_14

14.1 Background

The oldest oilseed crop known for its premium quality high oil content and antioxidant qualities is sesame (*Sesamum indicum* L., 2n = 26). Because of its resistance to heat and dry conditions, it is currently grown in 80 countries. Its history of cultivation dates back to the Indian subcontinent's Harappa Valley between 3050 and 3500 BC (Bedigian and Harlan 1986). With only a small amount of the wild species, *S. radiatum*, being used, *S. indicum* is the only species under the genus *Sesamum* that is commonly grown (Zhang et al. 2011). As per Kobayashi's 1991 Index Kewensis and Supplements, 29 out of the 36 species of *Sesamum* are native to Africa, indicating that Africa is the place where sesame originated (Sharma et al. 2014). Nonetheless, Asia also has a wide variety of cultivated sesame varieties. The origin of sesame can be traced back to the northwestern region of South Asia, particularly during the period of the Harappan culture, according to Fuller's (2003) archaeological evidence of burnt sesame seeds. Fertile seeds can be produced by crossing the species *S. malabaricum Burm.* with *S. indicum* in India (Bedigian 2010).

14.1.1 Botanical Descriptions of Sesame

The tap root system of the upright annual herb sesame consists of several fibrous roots with root hairs in addition to the primary tap root. Being a eudicot, it is a erect, tall plant with a comparatively short life cycle that has two cotyledons in its seed. A wide variety of leaf forms, including lanceolate, linear, elliptic, ovate, and narrowly cordate kinds, can be found in the sesame germplasm. The leaf arrangement can be opposite, alternating, ternate, or mixed (Miao et al. 2021). Composed of a pedicel, calyx, corolla, stamen, and pistil, the flower is bisexual, tubular, bell-shaped, and has a long lip. The diverse advantages of the bioactive components of sesame seeds have been brought to light by recent study, reviving interest in this comparatively unexplored crop (Anilakumar et al. 2010). Sesame's semi-domesticated status is presumably a result of artificial selection and adaptability to different growing environments, which accounts for some of its genetic and phenotypic variability (Bedigian and Harlan 1986; Wei et al. 2015). While insect-mediated cross-pollination is common, with rates ranging from 0 to 68%, sesame is largely a self-pollinating plant (Pathirana 1994) (Fig. 14.1).

14.1.2 World Sesame Production

The impact of global warming on agriculture is profound, posing significant challenges (Fig. 14.2). These changes threaten crop yields, soil health, and water availability, which in turn jeopardize global food security. However, sesame, with its

Fig. 14.1 A flower of crop sesame. *Source* Author-Sarita Pandey

high production in hot and dry climatic regions, can be an important crop in this challenging climatic condition. Approximately 70% of sesame production occurs in regions situated between latitudes 30° S and 43° N. These nations are largely found in the hot and dry regions of Africa, Asia, Central America, and Latin America. Africa is a key place for the production of sesame. India, Myanmar, and China are the primary producers in Asia. The top 10 producers of sesame now are all developing nations from the continents of Asia and Africa. Sesame production has increased during the past ten years due to the rising globalisation of agriculture (Figs. 14.3a, b). Sesame production increased globally, going from 1,419,988 tonnes in 1961 to 6,803,824 tonnes in 2020. (Fig. 14.4).

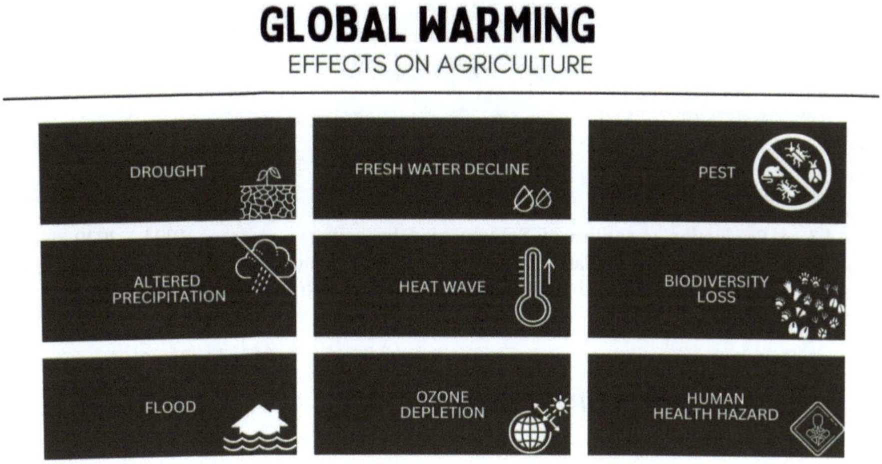

Fig. 14.2 Effect of global warming on agriculture. *Source* Author Sarita Pandey

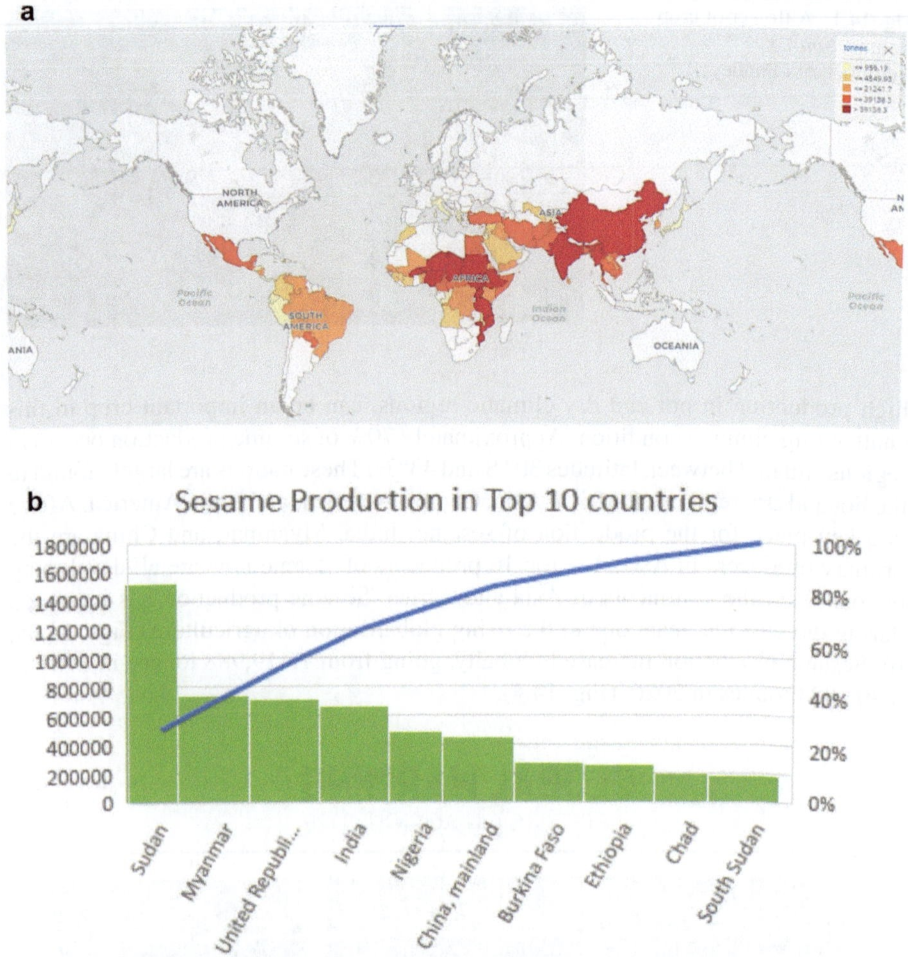

Fig. 14.3 **a** Production quantities of Sesame seed by country—Average 1994—2020 source: https://www.fao.org/faostat/en/#data/QCL/visualize, as retrieve on 06th June 2022. **b** Production quantities of Sesame seed in top 10 countries in 2020, as retrieve on 06th June 2022. *Source* Author-Sarita Pandey

14.1.3 Nutritional Importance of Sesame

Sesame is a well-known source of dietary protein because it has a high concentration of sulfur-containing amino acids, specifically methionine. Proteins make up roughly 15–25% of the dry weight of sesame seeds, or 30–50% of the mass of the defatted sesame cake. Of the total proteins contained in sesame, 60–70% and 15–25%, respectively, are made up of two primary storage proteins: 11S globulin and 2S albumin. The abundance of these inferred storage proteins in cysteine and methionine is mostly due to two 2S albumin isoforms and one 11S globulin isoform, according

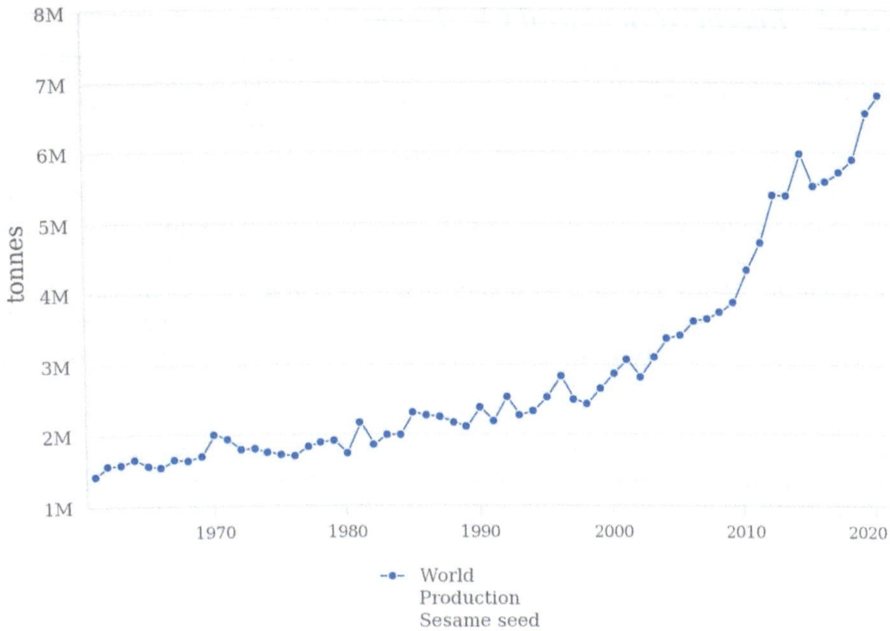

Fig. 14.4 Trend of sesame production in the world from 1961 to 2020, as retrieve on 06th June 2022. *Source* Author-Sarita Pandey

to the amino acid makeup of these proteins. Sesame oil is primarily composed of triacylglycerols (> 95%).

14.2 Modern Farming Methods and Challenges

14.2.1 Modern Farming Practices

The optimal situations for growing sesame are heavy clay soils with good drainage, raised beds, and minimal irrigation. It cannot survive salinity but likes moderately fertile, pH 5–8, slightly acidic to alkaline soils. As a principal crop, sesame is typically planted without any issues after cotton, corn, sorghum, peanuts, etc. Given the sesame seed's small size, proper field preparation is a need for a good crop stand. The ideal row spacing is between 15" and 40". Johnson grass, wild cucumber, mint, sunflower, kochia, and ground cherry should be kept to a minimum in fields. Fields planted with balanced NPK fertilisers produce the finest sesame harvests. The sesame will receive the best water from a heavy pre-irrigation.

14.2.2 Recent Agricultural Challenges

In general, challenges in sesame production revolve around issues such as the difficulty in providing inputs and financing to out-growers, as well as the limited capacity and productivity of smallholder farmers and cooperatives, compounded by environmental factors. Specifically, the prevalence of droughts or excessive rainfall leading to waterlogging poses significant threats, often resulting in crop failures and reduced productivity. Sesame cultivation predominantly occurs in rain-fed environments, where climatic variability significantly impacts output and productivity. Moreover, sesame-growing regions contend with the presence of approximately 98 types of weeds spanning 31 families, including grasses, sedges, bindweeds, and broadleaf varieties. Small-scale farmers, who constitute a majority of sesame growers, struggle with the absence of modern planting, harvesting, and threshing equipment due to financial constraints. Consequently, the lack of contemporary technologies hampers efficient farming practices and contributes to suboptimal yields. Additionally, inadequate facilities and poor fertilizer responses further impede the growth and profitability of sesame crops, exacerbating the existing challenges in cultivation.

14.2.3 Genetic Improvement Objectives

The main focus of research in sesame cultivation lies in genetic improvement to combat a range of stresses. The important stresses are pests like gall midge, webworm, seed bug, green vegetable bug, African bollworm, and crickets, which pose significant threats to sesame crops. Additionally, diseases such as bacterial blight, leaf spot, and phyllody, which cause deformations in leaves and flowers, as well as wilting, are major concerns for sesame farmers. Alongside pest and disease management, another crucial research objective is the development of non-shattering varieties of sesame, which would enhance harvesting efficiency and reduce post-harvest losses, thereby addressing a key challenge in sesame cultivation.

14.3 Germplasm Diversity

Joshi (1961) traces the roots of genetic exploration into the morphological and agronomic aspects of sesame back to the 1930s. The outcomes of traditional genetic examinations of these traits form a robust basis for delving into molecular genetic analyses of sesame (Mei et al. 2017). Presently, a multitude of sesame agronomic characteristics are undergoing extensive molecular genetic investigations. To understand the genetic underpinnings of these traits and propel advancements in molecular genetics and breeding techniques for sesame, further research into gene interaction types through various crosses, targeting both quantitative and qualitative traits, remains imperative (Hossain et al. 2010).

14.3.1 Classification and Species Relationship

Sesamum, a genus comprising 40 species, predominantly thrives in Africa and the India-Sri Lanka region. However, due to limited cytogenetic data, the chromosomal numbers are only known for approximately one-third of these species. Notably, *Sesamum indicum* shares a chromosome count of 2n = 26 with *S. capense Burn, S. malabaricum Nar., and S. schenkii Aschers*. Conversely, *S. laciniatum* boasts a somatic chromosome number of 2n = 28, while *S. angolase* and *S. prostratum* each exhibit 2n = 32. The highest somatic number, 2n = 64, is observed in S. occidentale and *S. radiatum Schum & Thonn*. Among Indian species, *S. mulayanum,* closely resembling *S. indicum*, showcases beneficial traits such as resistance to phyllody and wilt. While true wild species of *Sesamum* in tropical Africa and India produce viable F_1 seeds, F_1 hybrids are typically sterile, except for *schenkii-indicum* hybrids, which display some end-season fertility. An array of regional cultivars of S. indicum exists, although it is often regarded as a single cultivated species categorized by seed color into white sesame (*S. indicum*) and diverse sesame (*S. orientale*) (Andargie et al. 2021). Moreover, based on seed color and inheritance, it is speculated that sesame transitioned from symmetrical to asymmetrical forms and from *S. capense* to *S. indicum*, following the sequence of black, brown, yellow, and white-seeded types (He et al. 1994).

14.3.2 Sesame Germplasm

Presently, about 20,000 conserved accessions of sesame germplasm are dispersed worldwide. Most of the seed materials for sesame are stored in South Korea, China, India, and African countries (Fig. 14.5). Of these, 6658 accessions have been conserved by the National Bureau of Plant Genetic Resources (NBPGR) in India, 7698 accessions have been kept by South Korea, and over 6000 accessions have been preserved by China (Gong et al. 2016). A number of wild species were also protected by the USA, Japan, Venezuela, Kenya, Senegal, Tanzania, and several other African countries. Global sesame breeding efforts are hampered by the relatively small genetic base, even with a plethora of germplasm supplies.

Fig. 14.5 Research institutes involved in sesame research program. *Source* Author-Sarita Pandey

14.3.3 Strategies for Conserving Genetic Resources for Sesame

The strategic sources of genetic diversity used to select the necessary traits for current and next genetic enhancement efforts are sesame genetic resources. In sesame improvement research projects, numerous local and international gene banks gather and manage genetically varied sesame germplasm resources (Fig. 14.6).

Only 5% of the genetic resources for sesame are preserved in the United States; the majority, around 95%, are saved in Asia (Teklu et al. 2022). Figure 14.6 shows major institutes devoted to the preservation of sesame germplasm. A total of 27,283 sesame genetic resources have been maintained by these organizations in the US, China, South Korea, India, and the United States (Teklu et al. 2022). Notably, three of the primary collections of sesame germplasm in the world are located in India, China, and South Korea.

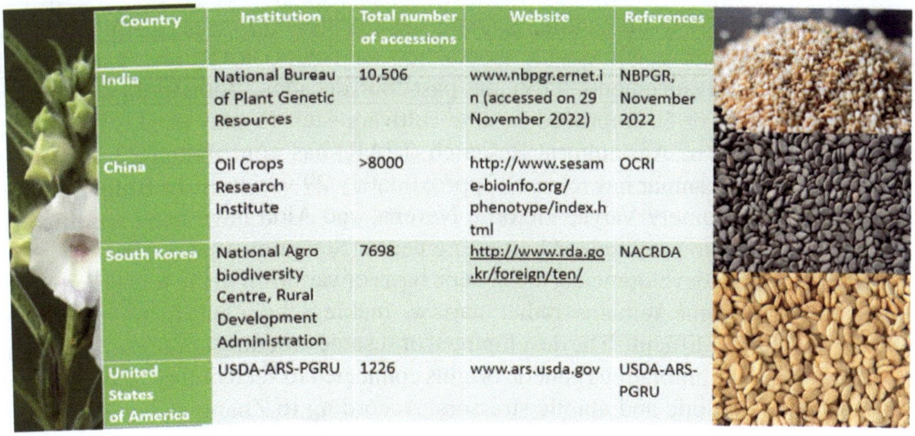

Country	Institution	Total number of accessions	Website	References
India	National Bureau of Plant Genetic Resources	10,506	www.nbpgr.ernet.in (accessed on 29 November 2022)	NBPGR, November 2022
China	Oil Crops Research Institute	>8000	http://www.sesame-bioinfo.org/phenotype/index.html	OCRI
South Korea	National Agro biodiversity Centre, Rural Development Administration	7698	http://www.rda.go.kr/foreign/ten/	NACRDA
United States of America	USDA-ARS-PGRU	1226	www.ars.usda.gov	USDA-ARS-PGRU

Fig. 14.6 The major sesame gene banks. *Source* Author- Sarita Pandey

14.4 Traditional Breeding in Sesame

14.4.1 Improvement Strategies

Sesame breeding initiatives focus on improving seed output, oil content and quality, non-shattering, synchronized maturity, and resistance to biotic and abiotic stressors as their primary improvement tactics. The majority of breeding initiatives have up to this point mostly used conventional breeding techniques to characterize the germplasm. Breeding procedures including integrated phenotyping and genotyping have been implemented recently.

14.4.2 Traditional Breeding Methodologies

Lately, there have been notable advancements in sesame research and development, mainly through the utilization of traditional breeding techniques like pure line and mass selection, hybridization, and mutation breeding. These methodologies have yielded approximately 200 improved sesame varieties globally renowned for their high yields, superior oil quality and quantity, early maturity, and resilience to several biotic stresses. Traditional breeding practices have played a pivotal role in fostering new genetic variations crucial for enhancing sesame cultivation. Understanding the available genetic diversity, its heritability, and its correlation with desirable traits is essential for effectively utilizing germplasm resources. Presently, genetic and genomic techniques, including genomics-assisted breeding and genome editing, are revolutionizing sesame improvement efforts. Tools such as molecular markers, marker-assisted breeding, genome sequencing, haplotype mapping, association mapping, and genome-wide association studies (GWAS) are instrumental in driving these advancements. Over the past four decades, both India and China have introduced over 50 improved sesame cultivars each (Yadav et al. 2022). The Ethiopian Institute of Agricultural Research (EIAR) has contributed 32 improved varieties, while Myanmar has released approximately 29 varieties. In Bulgaria, four distinct varieties namely Valya, Victoria, Nevena, and Aida have been specifically developed to cater to mechanized harvesting needs (Stamatov et al. 2018).

Even with these developments, the genetic basis of variation in the few germplasm accessions of sesame remains rather narrow, making the identification of elite breeding resources difficult. The development of sesame breeding strategies is further complicated by the ambiguous genetic origins connected to seed output, seed quality, and resilience to biotic and abiotic stressors. According to Zhang et al. (2013a, b), current biological approaches must be employed to create new breeding lines and sequence the sesame genome in order to address these issues and improve sesame attributes.

14.4.3 Heterosis Breeding in Sesame

B. Pal published the first report on sesame hybridization in 1945. Sesame has a strong potential for heterosis; one report indicates that it has a 252% hybrid vigour capacity (Zhang et al. 2021). The first light-insensitive male sterile line, Sipsms304, was created in 2009 by Chinese sesame scientists using ethyl methane sulfonate (EMS) mutagenesis (Zhang et al. 2016). For enhancement of sesame varieties, hybridization breeding combined through pedigree selection is a popular and efficient technique. Controlled crossings could be used to introduce the desired and broad diversity of traits into new kinds. Two or more paternal lines could have contributed to the cross combinations' development. The primary traditional method forbreeding new sesame varieties is by hybridization with pedigree selection. Out of the 145 types that were created between 1950 and 2013, 78 (50.6%) were created through hybridization breeding (Fig. 14.7). The remaining types were created through GMS hybrids, system selection breeding, mutation breeding, and other methods.

14.5 Molecular Breeding

14.5.1 Genome Project of Sesame

Due to the Sesame Genome Project's achievement, superior genetic resources and traits from the wild *Sesamum* species can now be used in sesame breeding. In addition to producing the first interspecific single nucleotide polymorphism (SNP) genetic

Database	Utility	Reference
http://www.ocri genomics.org/Sin base/index.html. (Sin base)	Genomics and phenotypes, etc.	Wang L. et.al. 2014
http://202.127.18.228/SesameHapMap/ (Sesame Hap Map)	Genome-wide SNP	Wei, X et.al. 2015
http://www.ncgr.ac.cn/SesameFG (SesameFG)	Genomics, phenotypes, transcriptomics	Wei, X. et.al. 2017
http://www.sesame bioinfo.org/Sisat Base/ (Sisat Base)	Genome-wide SSR	-
http://www.sesamegenome.org(The Sesame Genome Project)	Genomics	Zhang Y. et.al. 2013
http://www.sesame-bioinfo.org/phenotype/index.html (Sesame Germplasm ResourceInformation Database)	Phenotype	-
http://www.ncbi.nlm.nih.gov/genome/?term=sesame(NCBI)	Versatile	-
http://www.ocri-genomics.org/ocsESTdb/index.html (OCSESTdb)	Genomics	Ke, et.al. 2015
http://www.ocri-genomics.org/PTGBase/index.html (PTG Base)	Tandem duplication, evolution	Yu J. et.al. 2015
http://www.sesame-bioinfo.org/PMDBase (PMD Base)	Marker Information	Yu J. et. al. 2016

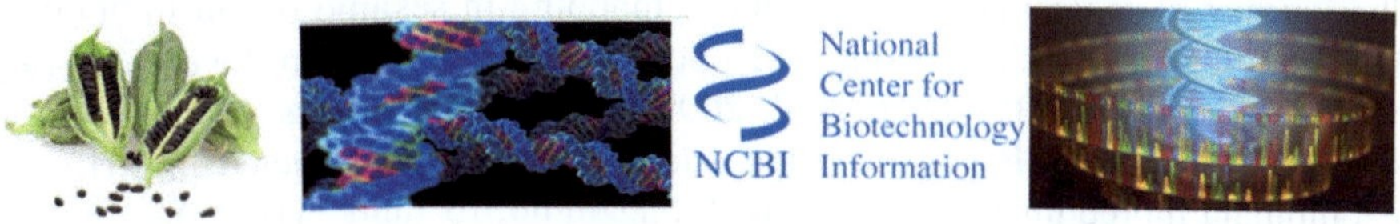

Fig. 14.7 Online genomic resources for sesame. *Source* Author Sarita Pandey

maps in sesame, the Sesame Genome Working Group generated the first interspecific hybrid populations for *S. calycinum* and *S. angustifolium.* The Sesame Genome Project (SGP) was introduced in 2010, and the Sesame Genome Working Group (SGWG) was founded by the Henan Sesame Research Centre, Henan Academy of Agricultural Sciences, China, in 2009. There were 23,713 gene models in all that were produced, with an average gene length of 1.2 kb. Additionally, utilizing novel sequencing and assembly methods, the genome of *S. indicum* has been rebuilt.

14.5.2 *Mapping and Breeding of Sesame at the Molecular Level*

Sesame stands out as an orphan crop species in plant genome research due to its ancient phylogenetic location and distant evolutionary relationship with cereals and other economically significant crops. Its self-pollination tendency and narrow genetic diversity present unique challenges in breeding (Uncu et al. 2016a, b). However, recent advancements in genetics and genomics, fueled by next-generation sequencing technologies, have propelled sesame research forward (Yu et al. 2019). Through the elucidation of crucial candidate genes, QTLs, and molecular markers associated with essential agronomic traits, breeders gain insights into sesame's genetic makeup (Zhang et al. 2019). These findings not only pave the way for molecular breeding but also underscore the importance of discovering additional functional genes and markers to enhance breeding efficiency. The widespread adoption of high-throughput next-generation sequencing has revolutionized the study of sesame molecular genetics, facilitating the identification and utilization of specific molecular markers in diversity studies (Zhang et al. 2016). Notably, the creation of SisatBase by

Dossa et al. (2017), an online database housing polymorphic SSRs, further streamlines the application of these markers in sesame research (http://www.sesame-bioinfo.org/SisatBase/)

14.5.3 Molecular Genetic Linkage Maps in Sesame

The journey of molecular genetic linkage mapping in sesame began in 2009 with the creation of the first map using an F_2 population derived from the cross between "COI1134" and "RXBS" (Wei et al. 2009). Subsequent efforts saw the development of additional SSR genetic maps, such as the one constructed in 2012 using recombinant inbred lines (RILs) from "Zhongzhi no. 13" and "Yiyangbai" cultivars (Zhang et al. 2014). Wang et al. (2017) further advanced this field by building a third SSR genetic map employing a larger RIL population and hundreds of established SSR markers. As technology progressed, reduced representation genome sequencing emerged as a powerful tool for creating SNP genetic maps in sesame, as demonstrated by Du et al. (2019). Today, the creation of genetic maps, particularly ultra-dense SNP maps, is facilitated by techniques like RAD-seq, SLAF-seq, and GBS (genotyping by sequencing) (Liang et al. 2020; Miao et al. 2020; Zhang et al. 2014). Additionally, the complete sequencing of the chloroplast (cp) genomes of S. indicum cv. Ansanggae and S. indicum cv. Yuzhi 11 has provided valuable insights into species-specific evolutionary traits, setting the stage for designed breeding in sesame in the post-genomics era (Fig. 14.8).

14.6 Exploring Tissue Culture and Genetic Modification in Sesame

Plant tissue culture and genetic alteration are the two primary techniques for agricultural genetic improvement. Sesame faces significant challenges in achieving high efficiency in in vitro callus formation and differentiation due to its unique properties. In the past few decades, numerous effective occurrences of gene transformation and plant regeneration in sesame have been reported. One of the crops that has proven to be the most difficult to induce and differentiate calluses in vitro is sesame (Zhang et al. 2019). Numerous researchers have carried out technical optimization studies over the last 40 years to determine the most efficient methods for tissue culture, plantlet regeneration, and genetic transformation in sesame (Chowdhury et al. 2017; Zhang et al. 2019).

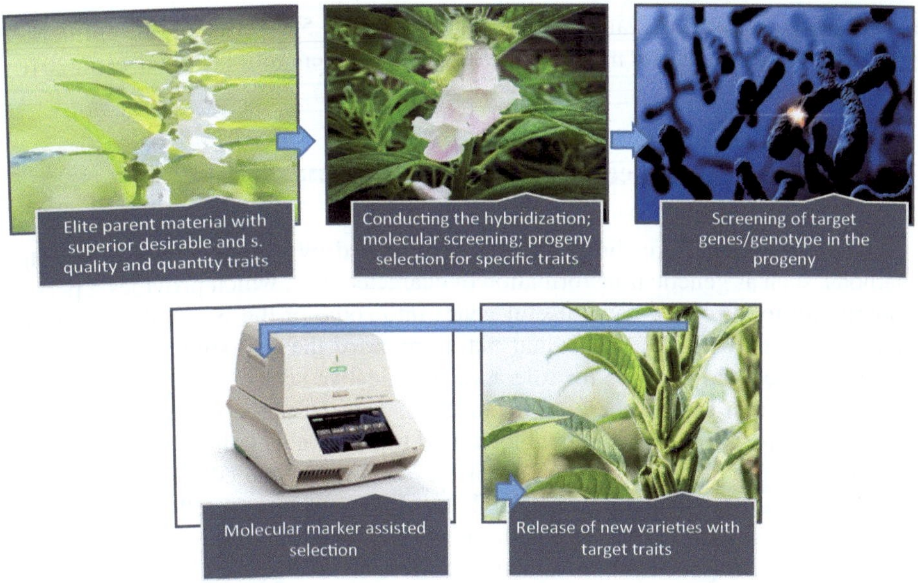

Fig. 14.8 The future of tailored breeding in sesame during the post-genomics age. *Source* Author-Sarita Pandey

14.6.1 Harnessing Tissue Culture Methods for Crossing Across Species in Sesame

Interspecific hybridization poses a formidable challenge in sesame breeding, primarily due to the low frequency of hybrids and substantial incompatibility between cultivated and wild *Sesamum* species (Yang et al. 2017). Remarkably, only hybrids between *S. malabaricum* and cultivated sesame exhibit high reproductive rates (Nimmakayala et al. 2011). To overcome these barriers, techniques such as embryo rescue and tissue culture are indispensable for creating hybrids involving species like *S. radiatum, S. schinzianum,, S. indicum* and *S. alatum* (Mukherji 1947). Establishing efficient regeneration protocols is pivotal for enhancing crop improvement through genetic transformation, yet sesame presents formidable regeneration challenges (Were 2006). Despite these hurdles, significant strides have been made over the past two decades towards achieving in vitro regeneration of sesame. Successful regeneration of *S. indicum* has been accomplished through direct organogenesis from hypocotyl and cotyledon tissues (Al-Shafeay et al. 2011; Were 2006; Yadav et al. 2010), as well as through indirect organogenesis from hypocotyl (Kwon et al. 1993; Wadeyar and Lokesha 2011) and leaf-derived callus (Raja and Jayabalan 2010). Additionally, somatic embryogenesis has been reported from cotyledon and hypocotyl-derived callus (Ram et al. 1990; Xu et al. 1997; Baskaran and Jayabalan 2006). Micropropagation techniques using shoot tip and nodal cultures have also been successfully implemented (Abdellatef et al. 2010). However, the reported success rate of plant regeneration varies widely, ranging from 4.5 to 88%, largely influenced by the genotype-dependent nature of in vitro sesame regeneration (Yadav et al. 2010).

14.7 Genetic Engineering and Gene Editing

Sesame conventional breeding can be supplemented by a variety of cutting-edge methods, such as genetic transformation of characteristics, which provides a perfect opportunity to introduce some useful genes into some of the best cultivars of the plant (Yadav et al. 2010). Chowdhury et al. released the first study on transgenic sesame's functional analysis for tolerance to oxidative stresses, salt, drought, and the pathogen that causes charcoal rot in 2017. Targeted gene modification, often known as genome editing, is a technique used to create novel allelic variants in the genomes of crop plants (David and Repkova 2017). Genome editing technologies based on clustered regularly interspaced short palindromic repeats (CRISPR), such as CRISPR/Cas9, CRISPR/Cpf1, base-editing systems, and prime editing systems, have shown promise for genetic improvement initiatives involving plants like sesame (Zhu et al. 2020). However, there aren't many investigations on CRISPR-based genome editing technologies in sesame due to their unique genomes and susceptibility to genetic manipulation (He et al. 2021).

14.8 Digitalisation in Sesame Breeding

There are several important plant breeding data management software solutions available, each with unique features. The Breeding Management System (BMS) offers comprehensive data management tools for trial design, data collection, and analysis, integrating genetic and genomic data to support multi-location trials and collaborative breeding programs (https://www.integratedbreeding.net/breeding-management-system). Field Book is a mobile app designed for efficient data collection in the field, featuring customizable data collection forms and synchronization with desktop databases (https://www.fieldbookapp.com/). KDDart provides a centralized database for plant breeding data, with tools for data collection, storage, and analysis, scalable for large breeding programs (http://www.diversityarrays.com/software). Agrobase is a comprehensive suite for plant breeding and research data management, offering tools for trial management, statistical analysis, and reporting, with GIS integration for spatial analysis (https://www.agronomix.com/) (Fig. 14.9). Seedlink is a web-based platform that includes modules for pedigree management, trial data collection, and real-time data access, facilitating collaborative work (https://www.seedlinks.com/). CyVerse, previously known as iPlant Collaborative, provides cyberinfrastructure for life sciences data management, with tools for data storage, analysis, and visualization in a collaborative environment (https://cyverse.org/). Flapjack is graphical genotyping software designed for visualizing genotype and pedigree data,

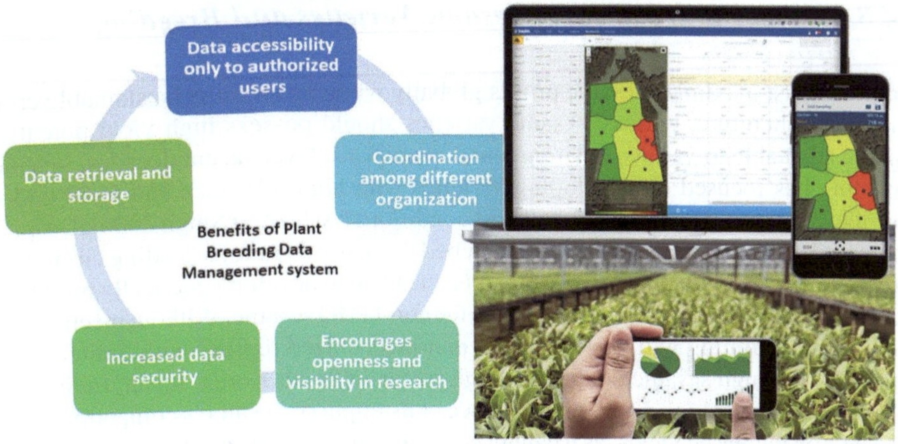

Fig. 14.9 Benefits of plant breeding management system. *Source* Author- Sarita Pandey

integrating seamlessly with breeding databases (https://ics.hutton.ac.uk/flapjack/). These solutions cater to the diverse needs of plant breeders, enhancing data collection, management, analysis, and visualization processes.

14.8.1 Scope of Precision Agriculture

Precision agriculture can significantly accelerate sesame breeding by leveraging advanced technologies to enhance various stages of the breeding process. For instance, drones equipped with multispectral cameras can monitor sesame fields, collecting high-resolution data on plant health, soil conditions, and environmental factors. This data helps breeders identify the best-performing plants under diverse conditions, expediting selection. Additionally, soil moisture sensors and precision irrigation systems ensure optimal water management, crucial for sesame's growth and yield. By integrating genomic technologies, breeders can rapidly analyze the genetic makeup of large sesame populations and correlate specific traits with field performance, such as drought tolerance or pest resistance. Automated phenotyping platforms further streamline the process by continuously monitoring plant growth and development, enabling quicker and more accurate selections. An example is a breeding program aimed at enhancing drought tolerance in sesame, where precision agriculture tools help identify and propagate the most resilient varieties, significantly reducing the breeding cycle's time.

14.8.2 The Future of Ideal Sesame Varieties and Breeding

Farmers must maximize sesame profits globally while maintaining sustainable crop rotations. Therefore, optimal sesame varieties should possess high yield potential and be adaptable to various environmental conditions. Over recent decades, sesame breeding has focused on developing varieties with high tolerance and yield under low temperatures, waterlogging, and pest conditions (Ashri and Singh 2006). Despite increased sesame yields, they haven't matched those of other crops, leading farmers to prefer more profitable alternatives. Breeders hesitate to develop genetically modified (GM) sesame due to consumer preference for non-GMO sesame. Without mechanization, sesame farming will be confined to non-mechanized regions with limited crop options. Although sesame commands high prices in South Korea, Japan, India, and China, excessively high prices could reduce consumption. Sesame's competitiveness depends on increased yields, as it cannot compete with other crops otherwise. Sesame seeds, rich in protein, oil, and antioxidants, are versatile and beneficial for health, a fact supported by tradition and science. The future of sesame lies with agronomists and breeders, who can leverage genomic technologies to expedite breeding processes. Despite being behind major crops in genomics, sesame's "genome time" is now. Future advancements will likely involve entrepreneurs creating businesses to modify traits affordably. Breeders need to identify target genomes for scientists to assemble. For example, a cultivar resistant to phyllody will be discovered in Turkey, and its causal genes identified. Traditional breeding is time-consuming, typically taking 7 to 10 years to transition from cross to variety. Multiple crosses are usually required to transfer a trait to an acceptable genotype, with a low probability of success in initial generations. Genomic selection will facilitate easier identification of plants with optimal traits from large populations in the future.

14.9 Conclusion and Prospects

In conclusion, the exploration of innovative plant breeding strategies for nutrient-rich, climate-smart sesame (*Sesamum indicum* L.) presents a promising avenue for addressing the challenges posed by evolving agricultural landscapes and changing climatic conditions. Through a combination of traditional breeding methods and cutting-edge genomic tools, significant progress has been made in un-raveling the genetic basis of key agronomic traits in sesame, such as yield, drought tolerance, and nutrient content. The elucidation of genomic resources, including molecular markers and genetic maps, has facilitated the development of improved cultivars with enhanced nutritional profiles and resilience to environmental stressors. The prospects for sesame breeding are bright, with opportunities to further optimize breeding strategies and harness the full potential of this versatile crop. Continued advancements in genomics, such as the application of genome editing technologies, will enable precise trait manipulation and accelerate the development of tailor-made cultivars suited to specific agroecological regions and consumer preferences. Moreover, collaborative efforts between researchers, breeders, policymakers, and farmers will be essential for ensuring the widespread adoption of innovative breeding practices and the sustainable cultivation of sesame.

References

Abdellatef E, Ahmed MM, Daffalla HM, Khalafalla MM (2010) Enhancement of adventitious shoot regeneration in sesame (*Sesamum indicum* L.) cultivar promo KY using ethylene inhibitors. J Phytol 2(1)

Anilakumar KR, Pal A, Khanum F, Bawa AS (2010) Nutritional, medicinal and industrial uses of sesame (*Sesamum indicum* L.) seeds—an overview. Agric Conspec Sci 75:77–685

Andargie M, Vinas M, Rathgeb A, Möller E, Karlovsky P (2021) Lignans of sesame (*Sesamum indicum* L.): a comprehensive review. Molecules 26(4):883

Al-Shafeay AF, Ibrahim AS, Nesiem MR, Tawfik MS (2011) Establishment of regeneration and transformation system in Egyptian sesame (*Sesamum indicum* L.) cv Sohag 1. GM crops 2(3):182–192

Ashri A, Singh RJ (2006) Sesame (*Sesamum indicum* L.). In: Singh RJ (ed) Genetic resources, chromosome engineering, and crop improvement. 4:231–289

Baskaran P, Jayabalan N (2006) In vitro mass propagation and diverse callus orientation on *Sesamum indicum* L.-an important oil plant. J Agric Technol 2(2):259–269

Bedigian D (2010) Cultivated sesame, and wild relatives in the genus *Sesamum*. CRC Press, Boca Raton, FL, pp 33–77

Bedigian D, Harlan JR (1986) Evidence for cultivation of sesame in the ancient world. Econ Bot 40(2):137–154

Chowdhury S, Basu A, Kundu S (2017) Overexpression of a new osmotinlike protein gene (SindOLP) confers tolerance against biotic and abiotic stresses in sesame. Front Plant Sci 8:410

David V, Repkova J (2017) Application of next-generation sequencing in plant breeding. Czech J Genet Plant Breed 53:89–96

Dossa K, Diouf D, Wang L, Wei X, Zhang Y, Niang M, Cisse N (2017) The emerging oilseed crop *Sesamum indicum* enters the "Omics" era. Front Plant Sci 8:1154

Du H, Zhang H, Wei L, Li C, Duan Y, Wang H (2019) A high-density genetic map constructed using specific length amplified fragment (SLAF) sequencing and QTL mapping of seed-related traits in sesame (Sesamum indicum L.). BMC Plant Biol 19:1–20

Fuller DQ (2003) Further evidence on the prehistory of sesame. Asian Agri-History 7(2):127–137

Gong HH, Zhao FT, Pei W, Meng QH (2016) Advances in sesame (*Seamum indicum* L.) germlasm resources and molecular biology research. J Plant Genet Resour 17(3):517–522

He F, He L, Hou L, Liu J (1994) Karyotype as related to phylogenies in sesame. J Southwest Univ Nat Sci Edn 6:573–576

He J, Zhang K, Tang M, Zhou W, Chen L, Chen Z (2021) CRISPR-based genome editing technology and its applications in oil crops. Oil Crop Sci 6:105–113

Hossain S, Ford R, Mcneil D, Pittock C, Panozzo JF (2010) Inheritance of seed size in chickpea (*Cicer arietinum* L.) and identification of QTL based on 100-seed weight and seed size index. Aust J Crop Sci 4:126–135

Joshi AB (1961) Sesamum. Indian central oilseed committee Hyderabad, India, p 109

Kwon TH, Abe T, Sasahara T (1993) Efficient callus induction and plant regeneration in Sesamum species. Plant Tissue Cult Lett 10(3):260–266

Kobayashi T (1991) Cytogenetics of Sesame (*Sesamum indicum*). Dev Plant Genet Breed 2:581–592

Liang J, Ye Y, Yan X, Yan T, Rao Y et al (2020) Construction of a high density genetic map and QTL mapping of PEG-induced drought tolerance at seedling stage in sesame using whole genome re-sequencing. Res Sq. https://doi.org/10.21203/rs.3.rs-50720/v2

Mei H, Liu Y, Du Z, Wu K, Cui CQ et al (2017) High-density genetic map construction and gene mapping of basal branching habit and number of flowers per leaf axil in sesame. Front Plant Sci 8:636

Miao H, Li C, Duan Y, Wei L, Ju M, Zhang H (2020) Identification of a Sidwf1 gene controlling short internode length trait in the sesame dwarf mutant dw607. Theor Appl Genet 133(1):73–86

Miao H, Langham DR, Zhang H (2021) Botanical descriptions of sesame. Sesame Genome 19–57

Mukherji S (1947) Relation of total soluble solids in the cell sap of Sesamum species to the degree of susceptibility and resistance to Antigastra (Lepidoptera–Pyralidæ) attack. Nature 160(4055):95–96. https://doi.org/10.1038/160095a0

Nimmakayala P, Perumal R, Mulpuri S, Reddy UK (2011) Sesamum. In: Kole C (ed) Wild corp relatives: genomic and breeding resources oilseeds. Springer, Berlin, pp 261–273

Pathirana R (1994) Natural cross-pollination in sesame (*Sesamum indicum* L.). Plant Breed 112:167–170

Raja A, Jayabalan N (2010) Callus induction and plantlet regeneration from leaf explant of sesame (*Sesamum indicum L.* Cv. SVPR-1). J Swamy Bot Club 27:93–98

Ram R, Catlin D, Romero J, Cowley C, (1990) Sesame: new approaches for crop improvement. In: Janick J, Simon JE (eds) Advances in new crops. Timber Press, Portland, pp 225–228

Sharma E, Shah TI, Khan F (2014) A review enlightening genetic divergence in *Sesamum indicum* based on morphological and molecular studies. Intl J Agri Crop Sci 7(1):1–9

Stamatov S, Velcheva N, Deshev M (2018) Introduced sesame accessions as donors of useful qualities for breeding of Introduced sesame accessions as donors of useful qualities for breeding of mechanized harvesting cultivars. Bulg J Agric Sci 24:820–824

Teklu DH, Shimelis H, Abady S (2022) Genetic Improvement in Sesame (*Sesamum indicum* L.): progress and outlook: a review. Agronomy 12(9):2144

Uncu AO, Frary A, Karlovsky P, Doganlar S (2016) High-throughput single nucleotide polymorphism (SNP) identification and mapping in the sesame (*Sesamum indicum* L.) genome with genotyping by sequencing (GBS) analysis. Mol Breed 36(12):1–12

Uncu AO, Frary A, Karlovsky P, Doganlar S (2016) High-throughput single nucleotide polymorphism (SNP) identification and mapping in the sesame (*Sesamum indicum* L.) genome with genotyping by sequencing (GBS) analysis. Mol Breed 36(12):173

Wadeyar BS, Lokesha R (2011) Studies on high frequency shoot regeneration in sesame (*Sesamum indicum* L.). Plant Tissue Cult Biotechol 21:45–52

Wang H, Zhang H, Ma Q, Wei L, Ju M et al (2017) Optimization of EMS mutagenesis condition and screening of mutants in sesame. J Henan Agric Sci 46(1):36–41

Wei LB, Zhang HY, Zheng YZ, Miao HM, Zhang TZ, Guo WZ (2009) A genetic linkage map construction for sesame (*Sesamum indicum* L.). Genes Genomics 31(2):199–208

Wei X, Liu K, Zhang Y, Feng Q, Wang L et al (2015) Genetic discovery for oil production and quality in sesame. Nat Commun 6:1–10

Were BAI (2006) Genetic Improvement of oil quality in Sesame *(Sesamum indicum* L.*)*: assembling tools. Department of crop science, Swedish university of agricultural sciences

Xu ZQ, Jia JF, Hu ZD (1997) Somatic embryogenesis in *Sesamum indicum* L cv. Nigrum J Plant Physiol 150:755–758

Yadav S, Shukha HS, Ram RA (2010) Studies on foliar of NAA, GA_3, boric acid and $Ca(NO_3)_2$ on fruit retention, growth, application yield and quality of aonla (*Emblica officinalis* Gaertn.) cv. Banarasi Hort J 23:64–67

Yadav R, Kalia S, Rangan P, Pradheep K, Rao GP, Kaur V, Siddique KH (2022) Current research trends and prospects for yield and quality improvement in sesame, an important oilseed crop. Front Plant Sci 13:863521

Yang M, Liu H, Zhou T, Qu H, Yang Y et al (2017) Production and identification of F1 interspecific hybrid between *Sesamum indicum* and wild relative *S. indicatum*. Sci Agri Sin 50(10):1763–1771

Yu J, Golicz AA, Lu K, Dossa K, Zhang Y, Chen J, Zhang X (2019) Insight into the evolution and functional characteristics of the pan-genome assembly from sesame landraces and modern cultivars. Plant Biotechnol J 17(5):881–892
Zhang H, Miao H, Ju M (2019) Potential for adaptation to climate change through genomic breeding in sesame. Genomic designing of climate-smart oilseed crops. Springer, Cham, pp 371–440
Zhang H, Miao H, Li C, Wei L, Duan Y et al (2016) Ultra-dense SNP genetic map construction and identification of *SiDt* gene controlling the determinate growth habit in *Sesamum indicum* L. Sci Rep 6:31556
Zhang H, Miao H, Wang, L, Qu L, Liu H, Wang Q, Yue M (2013) Genome sequencing of the important oilseed crop (*Sesamum indicum* L). Genome Biol 14(1):1–9
Zhang Y, Wang L, Xin H, Li D, Ma C et al (2013) Construction of a high-density genetic map for sesame based on large scale marker development by specific length amplified fragment (SLAF) sequencing. BMC Plant Biol 13:141
Zhang YX, Jian SUN, Zhang XR, Wang LH, Zhuo CHE (2011) Analysis on genetic diversity and genetic basis of the main sesame cultivars released in China. Agric Sci China 10(4):509–518
Zhang YX, Wang LH, Li DH, Gao Y, Lv HX et al (2014) Mapping of sesame waterlogging tolerance QTL and identification of excellent waterlogging tolerant germplasm. China Agri Sci 47(3):422–430
Zhang H, Langham DR, Zhao Y, Khalafalla A, Miao H (2021) Traditional breeding in sesame. In: The Sesame Genome, pp 145–158
Zhu H, Li C, Gao C (2020) Applications of CRISPR–Cas in agriculture and plant biotechnology. Nat Rev Mol Cell Biol 21:661–677

Yu J, Golicz AA, Lu K, Dossa K, Zhang Y, Chen J, Wang L, You J, Fan D, Edwards D, Zhang X (2019) Insight into the evolution and functional characteristics of the pan-genome assembly from sesame landraces and modern cultivars. Plant Biotechnol J 17(5):881–892
Zhang H, Miao H, Ju M (2019) Potential for adaptation to climate change through genomic breeding in sesame. Genomic designing of climate-smart oilseed crops. Springer, Cham, pp 371–440
Zhang H, Miao H, Li C, Wei L, Duan Y, Ma Q et al (2016) Ultra-dense SNP genetic map construction and identification of SiDt gene controlling the determinate growth habit in *Sesamum indicum* L. Sci Rep 6:31556
Zhang H, Miao H, Wang L, Qu L, Liu H, Wang Q, Yue M (2013) Genome sequencing of the important oilseed crop *Sesamum indicum* L. Genome Biol 14(1):401
Zhang Y, Wang L, Xin H, Li D, Ma C et al (2013) Construction of a high-density genetic map for sesame based on large scale marker development by specific length amplified fragment (SLAF) sequencing. BMC Plant Biol 13:141
Zhang YX, Sun J, Zhang XR, Wang HL, Zhao CH (2011) Analysis on genetic diversity and genetic basis of the main sesame cultivars released in China. Agric Sci China 10(4):509–518
Zhang YX, Wang LH, Li DH, Gao YA, Lu HX et al (2014) Mapping of sesame waterlogging tolerance QTL and identification of excellent waterlogging tolerant germplasm. China Agr Sci 47(3):422–430
Zhang H, Langham DR, Zhao Y, Khalafalla AM, Miao H (2021) Traditional breeding in sesame. In: The Sesame Genome. pp 145–158
Zhu H, Li C, Gao C (2020) Applications of CRISPR–Cas in agriculture and plant biotechnology. Nat Rev Mol Cell Biol 21(11):661–677

Chapter 15
Genetic Diversity and Breeding of Thyme (*Thymus* Spp.)

Zsuzsanna Pluhár and Zsuzsanna György

Abstract *Thymus* species are aromatic perennial herbaceous plants or subshrubs belonging to the *Lamiaceae* family. Health benefits and aromatic properties of thyme known from the ancient era, however, the spectrum of various industrial and domestic utilization areas expanded considerably. Several thyme species applied as resources of dried drugs, spices, essential oils, extracts and oleoresins. Almost all of the therapeutic effects of the extracts and essential oils applied internally are in connection with the respiratory system, since it is a good expectorant having antispasmodic and antiseptic effects as well. According to the literature reported on factors influencing composition and biological activity of thyme essential oil, we can conclude that genetic features, environmental conditions, growing as well as processing technologies have all significant effect on the final product and quality parameters. Concerning genetic background, the presence of different chemotypes with distinct chief compounds in the essential oil influence principally the areas of utilization. The main constituent of commercial thyme oil is predominantly thymol with the highest therapeutical relevance attributed to *Thymus* species. Therefore, breeding efforts made in order to develop varieties with high essential oil content as well as elevated thymol percentage. In this chapter, we present an overview of the origin, distribution, taxonomic position, genetic diversity and conservation, cultivation practices and recent developments on biotechnology and molecular biology with their application for crop improvement in association with conventional breeding methods of thyme.

Keywords Essential oil · Breeding · Medicinal herb · *Thymus* · Thyme · Horticulture

---

Z. Pluhár (✉)
Department of Medicinal and Aromatic Plants, Institute of Horticultural Science, Hungarian University of Agriculture and Life Sciences, 29-43 Villányi Street, Budapest 1118, Hungary
e-mail: Pluhar.Zsuzsanna@uni-mate.hu

Z. György
Group of Horticultural Plant Genetics, Department of Plant Biotechnology, Institute of Genetics and Biotechnology, Hungarian University of Agriculture and Life Sciences, 29-43 Villányi Street, Budapest 1118, Hungary
e-mail: Benyone.Gyorgy.Zsuzsanna@uni-mate.hu

J. M. Al-Khayri et al. (eds.), *Biodiversity and Genetic Improvement of Herbs and Spices*, Advances in Plant Breeding Strategies 8, https://doi.org/10.1007/978-3-031-81838-7_15

15.1 Introduction

15.1.1 Botanical Classification and Morphology

Thyme species classified in the genus *Thymus* belonging to the tribe *Menthae*, subfamily *Nepetoideae* and family *Lamiaceae*, involves diverse essential oil bearing taxa with high intraspecific variability (Stahl-Biskup 2002). The most important species are *Thymus vulgaris* L. (garden thyme), *T. zygis* L. (red/Spanish thyme), *T. serpyllum* L. (wild thyme), and *T. capitatus* (L.) Hoffmanns. & Link syn. *Thymbra capitata* (L.) Cav. (Spanish oregano) (Fig. 15.1).

Thymus vulgaris L. (garden thyme) and *T. zygis* (Spanish thyme) native to the West-Mediterranean region, belong to the Section VI. *Thymus*, within the genus *Thymus* (Morales 2002). Wild thyme taxa with Eurasian distribution, involved in the Section *Serpyllum* of 120 chamaephyte (Ch) species were divided into 6 subsections and possess the highest chromosomal variability within the genus (Jalas 1972),

Fig. 15.1 *Thymus* species of economic importance: **a** *Thymus vulgaris* L.- Garden thyme **b** *Thymus zygis* L. **c** *Thymus serpyllum* L. **d** *T. capitatus* (L.) Hoffmanns. and Link (syn. *Thymbra capitata* (L.) Cav.)

where *Thymus serpyllum* L. *s.l.* (Wild/Creeping thyme) is involved in the subsection *Serpyllum*.

Thymus species are aromatic perennial herbaceous plants or subshrubs with chamaephyte (Ch) life-form (Heeger 1989). They can be grouped according to plant habit, where the first type comprises thyme species of little subshrubs with upright shoots, woody at the base, usually below 50 cm (e.g. *T. vulgaris*). On the other hand, there are very small (<15 cm) creeping thymes with smooth herbaceous shoots and rooting twigs, mostly included in the section *Serpyllum* (e.g. *T. serpyllum* L.).

Thymus plants have quadrangular stems, the young being hirsute. The hairs can cover either all four faces of the stem (holotrichous) or only two faces alternating in each internode (allelotrichous). They can also be found only on the four ribs of stems (goniotrichous). The leaves can be flat and more or less wide, or with revolute margins and almost acicular. All intermediates seem to be possible. The tector hairs if exist in *Thymus* are always simple, but rarely single-celled. Leaves are very frequently ciliate at the margins, either at the whole margin or only at the base or on the petiole.

The flowers grow more or less in clusters in the nodes. Species with shorter internodes has globose or capituliform inflorescences with flower bracts (Morales 2002). The most widely cultivated thyme species, garden thyme (*Thymus vulgaris*) is a woody perennial subshrub (Ch) with fibrous root system and a height of 20–50 cm. The aerial parts of the small bush include numerous upright, densely branching stems, woody at the base and soft at the upper part, where the length of the quadrangular shoots is 10–25 cm. Leaf size is 3–5 mm × 1.5 mm, leaf lamina are more or less hairy, opposite, oval, elliptic or lanceolate, greyish green, leathery with leaf edges rolled downwards (Heeger 1989). Both surfaces are covered by essential oil producing glandular hairs (Fig. 5.2a). The spike inflorescence consists of whorls, where the flower color can be pink or white. The blooming period is due late spring or early summer, depending on the geographical site. Fruit development results in 4 little, spherical, dark brown fecund schizocarp acorns (nutlets) developing covered by sepals.

The glandular trichomes are very important structures responsible for external volatile oil secretion. Two types of glandular trichomes exist on thyme plants since early seedling age: capitate glands with the upper cells full of volatile oils and big globose (peltate) 4–12 celled glands (Fig. 5.2b), typical of *Lamiaceae*, with one basal and one stalk cell, either (Morales 1989; Yamaura et al. 1992).

Taxonomically *Thymus* is a very complex genus, because of the polymorphism of a number of species and the absence of intrageneric incompatibility. Interspecific and introgressive hybridization between related species is a very common phenomenon and the main reason of variation. The resulting hybrids often have intermediate or mixed morphological characteristics (Hernández et al. 1987; Marin 1996). However, the greatest biodiversity of *Thymus* was reported from the Iberian Peninsula and from Turkey (Salgueiro et al. 1997; Baser et al. 1996; Pérez-Sánchez et al. 2008).

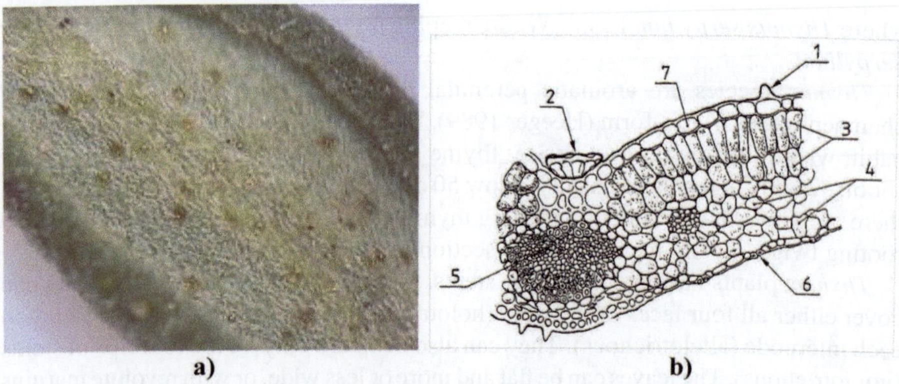

Fig. 15.2 **Peltate glandular trichomes of *Lamiaceae* type and covering hairs** on the lower leaf surface of *Thymus vulgaris* under stereo microscope (**a**) and **graphic cross section of a thyme leaf** based on a light microscopic view (**b**). *Legends* 1 Conical hair; 2 Sessile peltate glandular trichome of *Lamiaceae* type; 3 Palisade parenchyma; 4 Spongy parenchyma; 5 Collateral vascular bundle; 6 Stoma; 7 Epidermis covered by cuticule. *Source* (a) Photo: Pluhár et al. 2007; (b) Hungarian. Standard: Anonymous 1987).

15.1.2 Origin and Distribution

Thymus species are widely distributed in the Old World: Europe, Asia and North Africa, with the highest number of species in the Mediterranean region, especially in the West Mediterranean (Iberia and NW-Africa) (Fig. 15.3). Two gene centers of the genus *Thymus* are in Turkey (36 species, 50% endemic) and in the Iberian Peninsula (24 endemic of 35 species) (Morales 2002). The rate of endemic *Thymus* species is also considerable in NW-Africa (12 of 15 species), while thymes taxa are also abundant in Greece (18 species) and in Iran (17 species), either (Morales 1997; Karaman et al. 2001).

Thymes are heliophylous plants, a fact which reflects the ecology of the genus. *Thymus* plants frequently live on rocks or stones in mountain areas where the soils are well drained. However, different *Thymus* species require various substrata: *T. carnosus* lives on sand dunes near the sea, while *T. lacaitae* on gypsaceous soils, and *T. vulgaris* usually on calcareous soils.

Thyme species are highly resistant, which allow them to live under extreme climatic conditions concerning temperature and water supply. They do not avoid either cold or aridness. The high production of volatile oils can also be an adaptive characteristic for dry climate, because the volatile substances evaporate and produce a saturated atmosphere around the plant, which decrease the loss of water. Especially some species of the section *Serpyllum* can live in very cold climate, like *T. glacialis* in Siberia or *T. praecox* in Greenland.

From an ecological point of view, we can find the following correlation: bushy, woody, and erect plants are widely distributed in dry climates, whereas more fresh and humid climates usually provide habitats for plants with flat leaves, which are

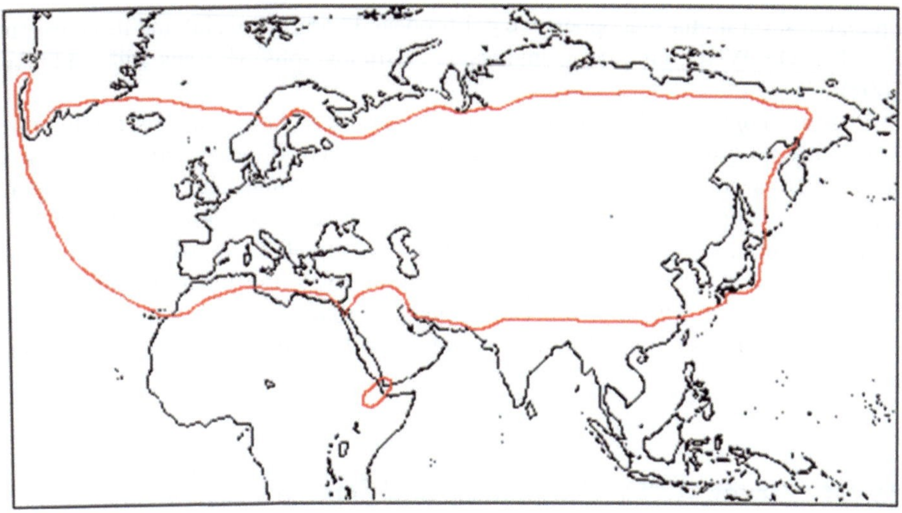

Fig. 15.3 Geographical distribution of the genus *Thymus*. *Source* Morales (1997)

woody only at the base. The latter usually are herbaceous with creeping or lying stems. Such species mostly belong to the sections *Hyphodromi* and *Serpyllum*. The production of volatile oils in this group is probably lower than in the first one (Morales 2002).

Thymus vulgaris L. is a Mediterranean species, native to Portugal, Spain, Italy and France on calcium containing soils. Moreover, it can be found in natural habitats of Middle-Asia, Algeria and Tunisia. Garden thyme is native to dry dwarf shrublands (macchia/garrigue/tomillares/frigana) close to the seashore of the Mediterranean Sea, on maritime limestone hillsides.

15.1.3 Health Benefits and Economic Importance

Thyme has outstanding health benefits due to a wide range of biological activities reported, such as carminative, antispasmodic, antitussive, expectorant, bactericidal, anthelmintic and adstringent. The medicinal and non-medicinal uses of different *Thymus* species are well-documented (Zarzuelo and Crespo 2002). Traditionally, it has been used for diseases of the upper respiratory tract, dyspepsia and chronic gastritis (ESCOP Monograph 2002). The essential oil is still considered as one of the most important active agents of *Thymus* species, having great significance from medicinal and food industrial point of views as well. The main monoterpene constituent of commercial thyme oil is predominantly thymol (together with carvacrol up to 70 percent). Fresh and dried flowering shoots of *Thymus vulgaris* L. (garden thyme) applied as spices for culinary purposes. Its drug and essential oil are official in several Pharmacopoeias (e.g. European Pharmacopoeia) and monographs (e.g. ESCOP, WHO, German Commission E Monographs) (Barnes and Anderson 2002).

Pharmacological studies support some of the traditional medicinal uses, which are principally attributed to the essential oil and flavonoid constituents. Other component such as caffeic acid, rosmarinic acid, oleanolic acid, saponins, and tannins also play important role in the medicinal treatment of different diseases (Vila 2002).

The estimated world production of thyme essential oil is 10 tons/year with the approximate value of 1.5 M US$/year. Main exporting countries are France, Germany, Poland, Hungary, the Netherlands, Serbia and USA, while the main importing countries include France, Germany, Switzerland, UK and USA (Anonymous 2017).

15.1.4 Drugs, Active Compounds and Standardization

Thymi herba (dried, crumbled leaves and flowers) and *Thymi typo thymolo aetheroleum* (hydrodistilled essential oil of thymol chemotype) are official drugs listed in the European Pharmacopoeia 10th Ed. (Anonymous 2020b). In the case of *Thymi typo thymolo aetheroleum* compounds limits are included in the specific standard of Ph. Eur. (Table 15.1, Fig. 15.4). The most frequent compounds of the oil are the phenolic monoterpene constituents (thymol and carvacrol) as well as p-cymene, γ-terpinene, borneol, linalool, and 1,8-cineol.

Dried aerial parts of *Thymus vulgaris* L. and/or *T. zygis* L. are utilized as raw materials for primary processing of *Thymi herba*. According to the pharmacopoeial specifications, *Thymi herba* must contain at least 12 mL/kg (1.2 mL/100 g) essential oil, calculated on the dry weight (DW) basis, where the relative percentage of the phenolic monoterpene constituents (thymol and carvacrol) is required to reach 40%. Flavonoids (apigenin, luteolin, thymonin, *Thymus*in, quercetin, etc.), phenolic acids (rosmarinic acid, caffeic acid, chlorogenic acid), triterpenic acids (ursolic acid, oleanolic acid), bitter compounds and resins are also detected in the drug (Stahl-Biksup 1991).

Dried flowering shoots of wild thyme (*Serpylli herba*) is also listed in the Pharmacopoeia Europaea, with a specification for the essential oil (EO) level (min. 3 mL/kg) (Anonymous 2016). In addition, bitter compounds, tannins, flavonoids and phenolic acids have also been determined in the dried aerial parts of wild thyme (Mills and Hutchins 2014; Boros et al. 2010). Essential oil (*Serpylli aetheroleum*) produced by distillation of collected flowering shoots of various native species, occurring in different parts of Europe with two gene centres found in Turkey and in the Iberian Peninsula (Morales 2002). Monographs (Mills and Hutchins 2014) refer to the collected plant material as *Thymus serpyllum* s.l. (sensu lato: in a broader sense), so the main terpenoid compound of the EO can also be different (p-cymene, carvacrol, thymol, linalool or borneol) in *Serpylli herba*. However, it has to be considered that

Table 15.1 Relative percentages (%) of compounds in thyme essential oil determined by GC/MS*, with limits specified in the Pharmacopoeia Europaea 10th and in the ISO 19817:2017 standard

Reference	Ph. Eur. 10 (2020)	ISO 19817:2017 (2017)
Product name	*Thymi typo thymolo aetheroleum*	Thymol type essential oil of thyme
Source species accepted	*Thymus vulgaris* L. & *T. zygis* L	*Thymus vulgaris* L. & *T. zygis* L
Essential oil compound	Relative percentage limits (%)	Relative percentage limits (%)
α-Thujene	0.2–1.5	0.5–1.5
α-pinene	–	0.5–2.5
β-Myrcene	1.0–3.0	1.0–2.8
α-Terpinene	0.9–2.6	0.9–2.6
p-Cymene	14.0–28.0	14.0–28.0
γ-Terpinene	4.0–12.0	4.0–11.0
Linalool	1.5–6.5	0.5–6.5
Terpinene-4-ol	0.1–2.5	0.1–2.5
Carvacrol methyl ether	0.05–1.5	–
Thymol	37.0–55.0	37.0–55.0
Carvacrol	0.5–5.5	0.5–5.5
β-Caryophyllene	–	0.5–2.0

GC/MS*: determination of relative percentages of the essential oil done by gas chromatography/mass spectrometry.
Source Anonymous 2017; Anonymous 2020b.

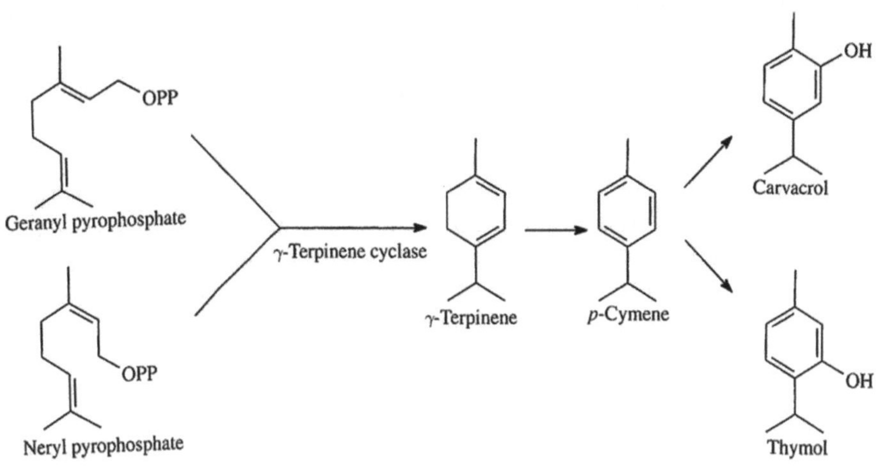

Fig. 15.4 Final steps of the biosynthesis of monoterpene phenols (thymol and carvacrol) of thyme essential oil. *Source* Stahl-Biskup (2002)

the application of the drug in phytotherapy is primarily attributed to the phenolic monoterpene molecules: thymol and/or carvacrol.

In addition, quality requirements for dried thyme (*Thymus vulgaris* L.) leaves in the rubbed form, are also included in the ISO Standard no. 6754:1996 (Anonymous 1996). From industrial aspect, the essential oil content of the dried herb is an important factor contributing to the flavor intensity. Whole thyme leaves should contain a minimum of 0.5% essential oil, which equals 5 ml/kg dried herb, and ground thyme should contain at least 0.2% essential oil to meet the requirements.

ISO 19817:2017 standard specifies characteristics of the thymol type essential oil of thyme (*Thymus vulgaris* L. and *Thymus zygis* L.), in order to facilitate the assessment of its quality (Anonymous 2017). Restrictions for certain major and minor compounds are determined, where only slight differences shown with reference to the Ph. Eur. standard on thymol type thyme oil (Anonymous 2020b) (Table 15.1; Fig. 15.4).

15.1.5 Health Benefits and Uses

Health benefits of thyme had already been discovered in the ancient Egypt (Zarzuelo and Crespo 2002). Applied internally almost all of its therapeutic effects are in connection with the respiratory system, since it is a good expectorant having anti-spasmodic and antiseptic effects as well (Bardeau 1973; Poletti 1979). Garden thyme can also be used against problems of the digestive system; it is useful in the cases of flatulence and gastric ulcer, too (Perrot and Paris 1971; Furlenmeier 1984). In the European folk medicine, it has been used against skin inflammations, ulcers, rheumatic complaints and neuralgia (Schaunberg and Paris 1977; Furlenmeimer 1984).

Recent findings verified that essential oil and polyphenol rich extracts of *Thymus* species show significant antioxidant (radical scavenging and lipid peroxidation inhibitor) potential as well as antibacterial and antifungal activity against various foodborne and human pathogens, respectively. Moreover, cytotoxic activity of *thyme* extract and antitumor potential of the essential oil were also demonstrated (Kulisic et al. 2006; Maksimovic et al. 2008; Jovanovic et al. 2022). Thus, *Thymus* species provide high-quality raw materials, which are suitable in various formulations in self-medication as well as in the pharmaceutical, food, cosmetic and chemical industries (Mills and Hutchins 2014; Jovanovic et al. 2022).

The possible therapeutic effects of the essential oil and its main compounds have been thoroughly analysed in the past decades. According to the literature data, the phenolic monoterpene compounds (thymol, carvacrol) have strong antibacterial and antifungal effect showing good efficiency against *Candida* sp. (Panizzi et al. 1993; Zambonelli et al. 1996; Sacchetti et al. 2005; Braga et al. 2010); their anti-inflammatory (Fachini-Queiroz et al. 2012), insecticidal (Szczepanic et al. 2012) and antispasmodic (Begrow et al. 2010) activities have also been described. However, there are no clinical studies available for thyme oil to support a well-established use (WEU) indication in accordance with Article 10a of Directive 2001/83/EC. According to the market overview and literature, thyme oil for oral use, cutaneous use and as a bath additive, fulfils the criteria of medicinal use throughout a period of at least 30 years, including at least 15 years within the EU/EEA. According to the Directive 2004/24/EC, thyme oil is accepted as a traditional herbal medicinal product (THMP) for oral use as an expectorant in cough associated with cold. Cutaneous use as bath additive, exclusively based upon long-standing application (Anonymous 2020a).

Significant antioxidant effect of the whole essential oil and its main compounds – thymol and carvacrol – has been analysed several times. Because of the established strong antibacterial, antifungal and antioxidant effect of the essential oil it can also be used as a natural food preservative (Nowak et al. 2012). In the research work of Ruberto and Baratta (2000) 98 pure essential oil compounds were tested as potential free radical scavengers or lipid-peroxidation inhibitors, among them several can be found in the essential oil of garden thyme. The authors emphasized that, besides thymol and carvacrol, in particular α- and γ-terpinene also had a comparable activity to that of α-tocopherol, used as reference. From this point of view, sesquiterpenes have significantly lower antioxidant activity. These compounds (β-caryophyllene, α-humulene) have rather different effect; many literature data are referring to β-caryophyllene, concerning its potential anti-inflammatory (Fernandes et al. 2007; Horváth et al. 2012), antispasmodic (Leonhardt et al. 2010), and anxiolytic-like effects (Moreira-Galdino et al. 2012). The anti-inflammatory effect of α-humulene was also investigated (Fernandes et al. 2007). With relevance to other sesquiterpene compounds of garden thyme essential oil – δ-cadinene, spathulenol, γ-muurolene – individual activity has not been investigated yet, however, several literature data are available referring to essential oils, rich in these sesquiterpenes, showing antifungal, insecticidal (García et al. 2007) and anti-inflammatory effects (Tung et al. 2008). Summarizing the literature data, the essential oil composition of garden thyme may influence significantly its possible therapeutic or food industrial applicability (Pluhár et al. 2016).

Chemotype patterns as well as methods of extraction highly influence the results of antioxidant capacity and therapeutical applications. Among natural antioxidants, essential oils containing high amount of thymol or carvacrol (in *Thymus, Origanum, Thymbra, Satureja*) were reported to possess the highest antioxidant activity (Peltoketo et al. 2000). Rosmarinic acid accounted for 22–55% of the antioxidant effect of the ethanolic extract, while the essential oils with high proportion of thymol showed high antioxidant activity (Chizzola et al. 2008). Fecka et al. (2006) reported that aqueous extracts of *Serpylli herba* and *Thymi herba*, containing salvianolic acid and rosmarinic acids as main compounds, showed significant antioxidant effect. They assumed that the activity of these compounds may be synergistic with those of the essential oil constituents.

15.2 Current Cultivation

15.2.1 Collection Versus Cultivation

Concerning raw material production, flowering shoots of various native species, occurring in different parts of Europe are collected from natural habitats for centuries. Beside garden thyme (*Thymus vulgaris* L.), further thyme species (*T. capitatus, T. mastichina, T. caespititius, T. sepyllum, T. pannonicus,* etc.) attract regional or broader interest, where flowering shoots are collected from wild habitats and used for various therapeutic purposes or applied as ornamental plants in gardens (Figueiredo et al. 2008; Maksimovic et al. 2008; Trindade et al. 2008).

Formerly, thyme drug collected in natural habitats in Spain supplying the 90% of the world demand. Nowadays, facing to the increasing interest for thyme products, the required volume is not likely to be ensured by natural populations. Moreover, habitats of thyme have been intensively exploited in the Iberian Peninsula by destructive gathering methods in traditional source areas, where the insufficient rainfall also cause serious problems. Additionally, the interests of the pharmaceutical/food industries do not focus on all chemotypes available in nature, but only few, namely thymol-, carvacrol-, and linalool-types.

Therefore, an increase in the demand of thyme of cultivated origin must be expected with standardized composition and yield of the essential oils and with uniform organoleptic properties of the leaves. Actually, *T. vulgaris* is cultivated in almost all of the countries with moderate climate, mainly in France, Italy, Poland, Germany, Hungary, North Africa, Mid-West and North America (Rumińska and Ożarowski 1990). The organic cultivation of thyme is also increasing, what is especially a relevant trend considering the demand of food and natural cosmetic industry (Carron et al. 2010).

15.2.2 Current Cultivation Practices

T. vulgaris is characterized by wide ecological tolerance, as it grows properly at the yearly average temperature of 7–25 °C, 400–2800 mm precipitation per year. Xerophyte species, which requires full sunshine and warmth in cultivation. Concerning soil type, calciferous, loose or medium hard soils are required with good texture, porosity, appropriate water capacity, enriched by nutrients. Deep situated fields should be avoided because of the frost damages of snowless winter periods. Wet fields cause slow development of plants and low level of essential oil. Well-drained light (sandy) and medium (loamy) soils, are suitable for thyme, it can also be grown in soils poor in nutrients. Common thyme field is preferably established on land previously cultivated with cereals or legumes. Bovine manure constitutes the basic fertilizer, while usual fertilisation include 80 nitrogen, 80 phosphorus and 100 potassium units (Rey and Sáez 2002).

a) b)

Fig. 15.5 Propagation material of *Thymus vulgaris*: **a** nutlets, **b** rooted cuttings. *Photos* Pluhár 2008)

In general, the length of the culture in the case of the woody perennial *Thymus vulgaris* is 4–6 years, depending on the edapho-climatic conditions of the growing area. Thyme fields are established out of crop rotation and can be applied both in conventional and in organic cultivation systems (Najar et al. 2021). Thyme propagated from seeds (nutlets), stem cuttings or by layering however, dividing plants at their roots may also help to increase plantation (Fig. 15.5). Seeds are generally sown in spring in a plastic tunnel or in greenhouses to a depth of 2 mm into seed trays as the seeds are quite small with the weight of 0.25–0.29 g/1000 seeds (Fig. 15.5). Seeds germinate in about two weeks in a seed tray and transplants are ready after about 6–8 weeks. Later on, seedlings transplanted to the field after the late spring frosts or grown further in a greenhouse. The plants can generally withstand winter frost, except for the N-European ares where *T. vulgaris* is cultivated as an annual plant (Rey and Sáez 2002; Galambosi et al. 2002).

The recommended spacing for plants is 25–30 cm apart within rows and a row spacing of 40–60 cm. Transplants, cuttings and seedlings can be suitable in beds approximately 1.2 m wide, with three rows on each bed. Ideally, beds designed for the machinery used. Plant density can be adapted according to the expected size of mature plants and moisture availability (Bencze and Pluhár 2013). Direct sowing is advantageous in organic cultivation of thyme in mountain areas, where more mechanized growing established. Sowing in stripes applied where rows close together alternate with wide inter-row spaces, which resulted in fine shoots, better drug quality, higher yields and essential oil content, if compared with the seedling growing technology (Carron et al. 2010).

Weeds must be removed 4–5 times during the season with selective herbicides in conventional or mechanically in organic cultures. Regular watering after the spring harvest guarantees the second harvest in autumn. Concerning pests, one may expect to find cicadas or spider mites, which may appear in hot summers but cause only little damages. In other periods, aphids, powdery mildew or rust may appear but they do not cause serious damage in thyme fields.

For essential oil production, thyme harvested once per annuum, when flowering begins during late spring. Nevertheless, in certain conditions two cuts per year are possible, yielding altogether 2.5–5.0 t/ha dried drug. Stems and leaves harvested just as flowering begins when cultivated for dried products, by cutting the entire plant back to about 10 to 15 cm above the ground. The second harvest performed early autumn on the branches bearing only leaves. For fresh produce, only the tips of the branches are harvested so the plants are strong enough to produce young shoots again (Rey and Sáez 2002).

To ensure the quality of products, by reducing loss of flavor through vaporization of essential oils and to maintain green color, thyme dried at temperatures lower than 40 °C. Natural drying and artificial driers used in the drying process. Sun drying results in poor quality of essential oil. After drying, the leaves separated further from the stems, sieved and graded. Essential oil is distilled using fresh harvested flowering shoots (Venskutonis 2002).

15.2.3 Current Agricultural Challenges

The drug yield and quality of *Thymus* spp. is highly dependent on genetic, environmental and agricultural factors. It is important to emphasize that growing practices properly developed with respect to the environmental conditions of the cultivation area as well as to the chemotype of the variety. On the other hand, one should recognize the chemotypes formed in natural environments in a certain location, and to know which environmental factors may cause variations between plants of the same species. This could allow the best places for cultivation and get optimal conditions for obtaining essential oils with a desired composition. It is also particularly interesting to mark the importance of harvest time in the composition, because it induces changes in volatile oil composition (Keefover-Ring et al. 2009; Pluhár et al. 2017). The quality of thyme products is influenced by the proportion of leaves (50% <) and the essential oil content. The former is the highest before flowering, while the latter in full bloom phase, so the grower have to decide on the time of harvest according to the final product of thyme cultivation (Rey and Sáez 2002). In the study of Kamondy et al. (2005) on selected lines of *T. vulgaris*, the essential oil content and the percentages of compounds depended primarily on the time of cutting (spring or autumn) and possibly influenced by the pre-harvest wheather conditions either. As a general rule, higher essential oil content and thymol % were detected at spring cutting (1.329 ml/100 g and 57.06%) than at autumn harvest (1.035 ml/100 g and 32.44%). Proportion of p-cymene has changed conversely: it was definitely higher in September (25.55%), than in May (12.84%). These results were in accordance with Hudaib et al. (2002). It could be concluded at phenolic chemotypes that cooler and more humid autumn season did not promote the thymol and carvacrol biosynthesis, which resulted in higher levels of precursor molecules (p-cymene and γ-terpinene) (Pluhár et al. 2013).

However, in the case of a 1,8-cineol chemotype of Spanish wild populations of *T. vulgaris,* Jordán et al. (2009) pointed out that the most advantageous harvesting period is in the mid-vegetative stage, when the maximum relative proportion of 1,8-cineole, borneol, camphor and monoterpenic hydrocarbons can be determined. In contrast, terpenyl acetate, α-terpineol and linalool, associated with fresh aroma, were mostly concentrated from full bloom to advanced fruit formation. Concerning the experimental clones of terpineol, linalool and geraniol chemotypes, no significant seasonal changes were found in major essential oil compounds by Pluhár et al. (2013), thus they can be harvested with high expected percentage of the dominant compounds either in spring or in autumn seasons. On the contrary, the related minor compounds of geraniol and α-terpinyl acetate went under significant seasonal changes between spring and autumn: geranyl acetate decreased by 1/3 from 13–21% to 4–7%, while α-terpineol dropped by ½ from 19–21% to 11%, respectively. However, linalyl acetate derivative of the major compound of linalool chemotype remained at constant level. The age of the thyme plantation also affects the essential oil properties. Hudaib et al. (2002) have found that throughout the entire vegetation cycle, young (2-year-old) plants provided a markedly higher essential oil yield compared to the older (5-year-old) ones, with the highest thymol ratio as well.

In addition, the effect of weather conditions on non-volatile phenolic compounds and on total antioxidant capacity were proven by Sárosi and Bernáth (2008) in the case of *T. vulgaris,* where the warm, sunny and dry growing season was favourable.

Baranauskienė et al. (2003) showed that different doses of nitrogen fertilizers have no significant effect on the essential oil content and composition in *Thymus vulgaris* plantation. However, the use of certain amounts of nitrogen fertilizers resulted in higher biomass production and, consequently, in elevated essential oil yields referring to unit area. Khazaie et al. (2008) have found in a semi-arid region of Iran that th effect of irrigation is more expressed in the second year of cultivation on the total biomass and oil production of thyme. In a pot experiment on different *Thymus vulgaris* taxa, Pluhár et al. (2015a, b) found higher levels of rosmarinic acid, total phenolics and antioxidant activity under induced drought stress conditions (40% soil water capacity), than at sufficient level (70% SWC) of water supply. However, negative effects of water deficiency were observed when measuring the amounts of essential oils and flavonoids. Concerning the physiological background of the above mentioned phenomenon, Tátrai et al. (2016) reported that 2-year-old *Thymus x citriodorus* plants under PEG induced drought stress conditions showed a morphological drought avoidance mechanism by maintaining the root system development through shoot fresh weight reduction in order to promote root absorption capacity and sustain plant growth (Fig. 15.6). Furthermore, thyme plants showed a physiological protection mechanism of tissue dehydration through stomatal closure. This resulted in balancing the plant water loss through transpiration and root water uptake, efficiently alleviating the unfavorable environmental condition. At the same time, carvacrol more significantly accumulated under high drought stress (+31.7%) as compared to the essential oils of control plants.

Fig. 15.6 Effects of PEG levels (0, C; 2%, D1 and 4%, D2) on shoot and root fresh weight (FW) and root area of *Thymus citriodorus*. Values are mean ± SD of nine replicates. *Source* Tátrai et al. 2016

We can conclude that physiological plant responses observed at *Thymus* species are generally promising, with respect to the predicted extremities and climate change tendencies. However, complementary irrigation is necessary in cultivation to maintain higher yields and essential oil production in persistent arid periods.

15.3 Germplasm Diversity and Conservation

Until nowadays, almost 270 essential oil components were indentified in thyme essential oil, where the ratio of monoterpenes is generally high, up to 90%. The main compounds – thymol and carvacrol – have phenolic characteristics (Stahl-Biskup 2002). Between 1960 and 1989, several experiments were carried out by examining the influence of biotic and abiotic factors on the essential oil diversity (Stahl-Biskup

1991). In general, the genus *Thymus* represents high infraspecific morphological variability and chemical polymorphism caused by environmental factors and genetic variation due to frequent hybridization and sexual dimorphism (gynodioecy) (Stahl-Biskup 1991, 2002; Morales 2002; Marin et al. 2003; Dajić-Stevanović et al. 2008). Polychemism of the genus *Thymus* is a consequence of the dynamic evolution, not only preserving the species but also provides a territorial favour in the process of adaptation to the continuously changing environment. Essential oil chemotypes of thyme species are considered as chemical races (Hegnauer 1978) as their populations are geographical distinct with inheritable chemical properties (Stahl-Biskup 2002).

15.3.1 Germplasm Diversity

The studies on the polymorphism of the genus *Thymus* started with the publications of Granger and Passet (1971, 1973), who reported 6 chemotypes for *T. vulgaris* after studying several populations and a great number of individuals in the south of France. Each of the six chemotypes, geraniol (G), α-terpineol (A), *tr*-sabinen hydrate or thujanol-4 (U), linalool (L), carvacrol (C) and thymol (T), is named after the dominant monoterpene in the essential oil of the plant. Having different molecular structure, the six monoterpene can be divided into two main groups: phenolic (thymol and carvacrol) and non-phenolic (the other four monoterpenes) ones. The presence of the dominant monoterpene in *T. vulgaris* is controlled by an epistatic series of five loci with the following sequence: G → A → U → L → C → T (Vernet et al. 1986). In addition to the main terpenes, various chemotypes have other predictable components, G, A and L plants can have considerable amounts of their primary monoterpene's acetate (geranyl, α-terpinyl and linalyl acetate, respectively); U plants often contain terpinen-4-ol, mycren-8-ol and linalool in addition to the main monoterpenes, *cis*-sabinene hydrate, while C and T plants have substantial levels of γ-terpinene and p-cymene. The latter are precursors in the biosynthesis of carvacrol and thymol Table 15.2.

For geraniol, linalool and α-terpineol chemotypes, the dominant monoterpenes occurs at levels up to 90–95% of the oil. The thuyanol chemotype contains up to 60–65% of a combination of thuyanol-4 and terpinen-4-ol (with some linalool and myrcenol-8). The carvacrol chemotype contains up to 85% carvacrol, whereas the thymol chemotype contains no more than 65% thymol. There has been no statistical comparison of monoterpenes variation among chemotypes or sites for each chemotype, despite the fact that the different chemotypes show marked spatial segregation in relation to a complex climatic-edaphic gradient. The two phenolic chemotypes confined to low-elevation areas close to the Mediterranean Sea, and the four non-phenolics are more abundant in areas with colder winters and deeper soils (Granger and Passet 1973; Thompson 2002). In Spanish *Thymus vulgaris* populations, the 1,8-cineole chemotype is the most frequent, thus *T. zygis* is the main resource of thymol type essential oil there (Cases et al. 2009). Essential oil composition of

Table 15.2 Relative percentages (%) of compounds in *T. vulgaris* essential oils of different origin indicating distinct chemotypes

Chemotype	Linalool	Carvacrol	Thymol	Linalool /thymol	Thymol	Thymol	Thymol	Carvacrol
Origin	France, Rhone region			France Saint-Etienne	Turkey, Icel region	Albania	Jemen	Iran, Babamaan Hills
Reference no	1			2	3	4	5	6
Component	%	%	%	%	%	%	%	%
α-thujene	–	–	–	–	2.8	–	–	
α-pinene	–	–	–	–	2.9	–	2.9	–
myrcene	–	–	–	0.7	3.5	–	2.5	–
α-terpinene	–	–	–	–	2.7	–	–	0.1–8.8
γ-terpinene	0.1–0.7	5.0–8.7	7.2–14.1	4.1	14.1	4.2–27.6	–	–
p-cymene	2.3–14.8	5.4–11.7	9.7–9.9	3.2	9.9	7.8–43.8	18.4	1.0–9.3
lInalool	**63.4–81.8**	3.7–8.0	3.9–5.1	**41.6**	3.9	–	0.6	6.9–31.8
Terpinen-4-ol	0.0–2.6	0.4–0.7	0.7	0.2	0.3	–	–	–
tHymol	1.9–2.4	3.2–14.1	**40.1–52.5**	**36.4**	**46.2**	**21.4–60.1**	**51.3**	–
cArvacrol	0.8–1.0	**30.9–57.2**	4.1–5.0	2.1	2.4	1.2–3.0	2.0	**27.2–73.1**
β-caryophyllene	3.4	2.8–3.8	2.7–4.2	1.0	1.6	1.3–3.1	4.3	0.5–1.7

Sources 1 Thompson et al. 2003; 2 Mewes et al. 2008; 3 Özcan and Chalchat 2004; 4 Asllani and Toska, 2003; 5 Al Maqtari et al. 2011; 6 Nezhadali and Sharifi 2010

different *T. vulgaris* chemotypes detected in various geographical locations presented in Table 15.2.

In several studies, different *Thymus* species reported as the sources of flavonoids (Corticchiato et al. 1995; Merghem et al. 1995; Miura et al. 2002; Marin et al. 2003; Benkiniouar et al. 2007). Polyphenol pattern in *Thymus* taxa possess physiological and chemical stability, and have wide structural variability. Several studies were oriented to use the unique composition of flavonoids in *Thymus* species as an important chemotaxonomic marker for the distinction of species (Vila 2002). For pharmacological purposes, it is also important to screen, analyze and identify those substances. Several studies, focused on flavonoid investigations in *Thymus*, confirmed the occurrence of phenolic acids (e.g. ferulic acid, caffeic acid and its derivatives p-coumaric, rosmarinic, chlorogenic acids), flavanols (e.g. quercetin, rutin), flavanols (e.g. catechin and epicatechin), flavanones (naringenin, eriodictyol, hesperetin and dihydroquercetin), and flavones (e.g. apigenin and its glucosides, luteolin and its glucosides) (Jordán et al. 2005; Vila 2002). A study conducted in Hungary on native *Thymus* taxa of sect. *Serpyllum* showed that the flavonoid profile depended mainly on genetic feature of *Thymus* species, when they planted on the same locality (Budapest, Hungary) (Boros et al. 2010; Pluhár et al. 2015a, b) (Table 15.3).

15.3.2 Genetic Resources Conservation Approaches

In Europe, at least 2000 Medicinal and Aromatic Plants (MAP) taxa are used commercially, of which two–thirds (1200–1300 species) are native to Europe. Thyme collection is the most significant in Spain an in Turkey, where various taxa are concerned. In Spain, each *Thymus* species were formerly subjected to wild crafting where altogether 75 million plants were exploited yearly resulting in a volume of 1500–1700 t/year (Blanco and Breaux 1997). In North-West Turkey, beside the highly variable *Thymus* taxa, several further species (*Origanum, Thymbra, Satureja* és *Corydothymus*) are gathered together to achieve 3000–5000 t/year drug, which contributes to the diverse quality and destruction of the habitats, either (Özhatay et al. 1997). Conservation of valuable thyme genetic resources is becoming more important than ever because of the climatic changes with extreme weather conditions and rapid industrial growth. Introduction of valuable *Thymus* taxa into cultivation is another way to preserve natural populations.

A proper understanding of interspecific and intraspecific genetic variation as well as distribution of genetic diversity helps germplasm curators and breeders to devise better conservation strategies and efficient utilization of the valuable thyme genetic resources. The international organization ECPGR (European Cooperative Programme for Plant Genetic Resources) involves the Medicinal and Aromatic Plants Working Group (https://www.ecpgr.cgiar.org/working-groups/medicinal-and-aromatic-plants), which focuses on gathering information about gene bank accessions of Europaean partner institutions. The countries involved in the Medicinal and Aromatic Plants Working Group (MAP WG), have reported previously a total of 23

Table 15.3 Content of polyphenolic compounds on dry-weight basis in the dried samples of *Thymus* species native to Hungary, collected in the experimental station (Budapest 2009)

Thymus species	Origin of accessions	Catechin	Caffeic acid	Chlorogenic acid	Epicatechin	p-Coumaric acid	Dihydroquercetin	Ferulic acid	Rutin
		[$\mu g\ g^{-1}$]	[$\mu g\ g^{-1}$]	[$\mu g\ g^{-1}$]	[$ng\ g^{-1}$]	[$\mu g\ g^{-1}$]	[$\mu g\ g^{-1}$]	[$\mu g\ g^{-1}$]	[$\mu g\ g^{-1}$]
Thymus glabrescens Willd	Csesznek 1	1.26	49.07	2.28	115.05	1.15	**39.87**	4.18	3.18
	Csesznek 2	**5.55**	51.59	15.21	55.92	0.70	4.97	5.89	2.38
	Szentbékkálla 1	1.52	51.63	**310.57**	ND	**12.57**	11.20	8.43	21.99
	Szentbékkálla 2	**6.37**	**59.44**	11.72	171.67	1.21	19.97	4.03	0.63
Thymus pannonicus All	Cegléd-bercel 1	ND	48.76	26.32	37.20	0.84	**35.91**	4.40	ND
	Cegléd-bercel 2	ND	52.03	107.97	53.25	0.74	21.62	4.87	5.66
	Cegléd-bercel 4	1.75	43.10	**207.51**	60.49	0.81	**30.35**	5.15	31.15
	Cegléd-bercel 6	**7.76**	45.94	73.89	104.00	1.95	28.83	4.27	13.77
	Buda Hills 2	2.56	50.78	38.04	24.13	1.83	6.49	**11.04**	**67.58**
	Buda Hills 3	3.79	50.41	14.12	57.13	1.07	**48.06**	3.42	1.80
	Buda Hills 4	5.71	52.63	**206.18**	ND	0.77	**40.93**	2.70	29.87
	Ada	1.08	41.41	22.17	ND	0.60	28.51	3.89	3.63
Thymus praecox Opiz	Pilis Hills 2	1.84	35.33	2.56	**5249.81**	3.07	2.67	**11.08**	8.77
	Buda Hills 1	ND	43.95	**224.11**	ND	3.56	3.09	7.09	**171.00**
	Buda Hills 3	2.99	50.78	38.04	24.13	1.83	6.49	**11.04**	**67.58**
Thymus pulegioides L	Börzsöny 2	ND	54.21	**250.28**	240.81	2.49	3.67	**12.50**	**67.73**
	Börzsöny 3	ND	44.82	132.54	200.73	4.99	11.27	9.46	21.29
	Börzsöny 4	ND	**63.97**	60.29	63.83	2.00	13.28	4.75	7.64
	Regéc	1.99	**59.95**	45.97	0.31	0.76	15.60	3.56	20.01
Thymus serpyllum L	Fenyőfő	0.06	**61.08**	26.98	132.26	0.94	2.70	2.10	6.70

(continued)

Table 15.3 (continued)

Thymus species	Origin of accessions	Apigenin-7-glucoside	Rosmarinic acid	Eriodictyol	Quercetin	Naringenin	Hesperetin	Apigenin
		[μg g^{-1}]	[μg g^{-1}]	[μg g^{-1}]	[μg g^{-1}]	[μg g^{-1}]	[ng g^{-1}]	[μg mg^{-1}]
Thymus glabrescens Willd	Csesznek 1	0.15	**1078.01**	6.45	1.88	9.57	ND	26.06
	Csesznek 2	0.19	**1062.76**	1.26	0.46	4.18	ND	9.50
	Szentbékkálla 1	**23.41**	**1131.30**	1.87	0.95	2.79	ND	30.18
	Szentbékkálla 2	ND	**1075.62**	5.01	1.75	13.44	ND	64.83
Thymus pannonicus All	Cegléd-bercel 1	ND	**1273.59**	**12.61**	1.78	**20.56**	ND	35.96
	Cegléd-bercel 2	ND	**1118.33**	6.20	2.05	7.88	ND	33.24
	Cegléd-bercel 4	0.11	**1170.48**	9.62	2.03	**23.60**	ND	58.79
	Cegléd-bercel 6	0.12	**1090.89**	**11.59**	1.95	15.26	ND	54.23
	Buda Hills 2	ND	948.51	1.54	1.19	4.59	ND	44.86
	Buda Hills 3	ND	**1060.57**	**14.53**	2.67	16.24	ND	65.56
	Buda Hills 4	ND	**1050.52**	**13.71**	3.36	18.37	ND	**120.69**
	Ada	ND	**1048.95**	6.17	1.40	8.54	ND	45.57
Thymus praecox Opiz	Pilis Hills 2	0.26	948.81	0.76	1.26	0.96	24.7	12.63
	Buda Hills 1	5.50	83.49	1.14	4.45	3.63	ND	11.78
	Buda Hills 3	0.04	948.51	1.54	**9.61**	7.56	**210.2**	28.38
Thymus pulegioides L	Börzsöny 2	ND	948.26	0.73	3.46	2.76	46.0	8.34
	Börzsöny 3	0.10	886.09	1.88	2.56	4.08	**169.9**	25.48
	Börzsöny 4	ND	**1042.77**	3.01	1.29	**24.64**	ND	20.42
	Regéc	0.19	**1148.74**	5.54	2.59	9.53	ND	**104.54**
Thymus serpyllum L	Fenyőfő	ND	**1121.31**	0.48	1.29	3.08	92.2	30.42

ND—not detected.

480 accessions preserved in ex situ conditions and from those, 4624 accessions from the 10 priority species including *T. vulgaris* and *T. serpyllum*, chosen as models by the MAP WG. For these taxa, species-specific descriptors were proposed and are available on the WG webpage. Specific lists of descriptor for *Thymus vulgaris* and *T. serpyllum* established by the ECPGR MAP WG serve as a basis of description and distinction of new variety candidates using the most important morphological and chemical traits.

Draft descriptors of *Thymus vulgaris* are available as follows: https://www.ecpgr.cgiar.org/fileadmin/templates/ecpgr.org/upload/NW_and_WG_UPLOADS/MAP_Descriptors/Thymus_vulgaris_DRAFT_DESCRIPTOR_LIST_FINAL.pdf

Draft descriptors of *Thymus serpyllum* are available as follows: https://www.ecpgr.cgiar.org/fileadmin/templates/ecpgr.org/upload/NW_and_WG_UPLOADS/MAP_Descriptors/Thymus_serpyllum_DRAFT_DESCRIPTOR_LIST_FINAL.pdf

15.3.2.1 In Situ Conservation

Formerly, thyme drug collected in natural habitats in Spain supplied the 90% of the world market needs. A continuous and increasing demand for thyme products is unlikely to be ensured by natural populations, which are threatened by destructive gathering methods and insufficient rainfall in traditional source areas. Furthermore, indigenous populations of numerous endemic thyme species occur all over the Eurasian habitats, in particular in the Iberian Peninsula and in Turkey. Wild thyme species are to be preserved and maintained in their original habitats by in situ conservation, where habitat conservation and establishment of biosphere reserves or national parks are the most important forms of in situ conservation. Unregulated use of natural resources and overcollection were quite abundant which required restrictions in some regions of Andalusia and Valencia Province (Blanco and Breaux 1997).

15.3.2.2 Ex Situ Conservation

Nowadays several research institutes are engaged in the exploitation of genetic resources of various *Thymus* species, selection of excellent accessions and use them in various breeding programmes. Cultivars created by breeders provide more homogenous biological basis for thyme growers. Valuable genetic resources found in any steps of breeding, are generally maintained by breeders in gene bank collections and other ex situ reservation forms in experimental stations, research institutes, nurseries or small scale farms.Ex situ field gene banks of thyme are developed through planting of stem cuttings to conserve them in clonal lines (Carlen et al. 2010; György et al. 2020).

15.3.3 *Cytogenetics*

In the genus *Thymus*, so far diploid, tetraploid and hexaploid chromosome numbers have been identified: 2n = 28, 30, 56, 58, 60 or 90 corresponding to diploid, tetraploid or hexaploid levels (Mártonfi and Mártonfiová 1996, Morales 2002; Elena-Rossello 1981). Auto- or aneuploid forms originate from the basic chromosome number of x = 14 or 15, however this is considered as secondary basic chromosome number, originating from x = 7. In the case of T. vulgaris 2n = 28, 30, 48, 56, 58 has been observed (Elena-Rossello 1981; Morales 2002; Chromosome Counts Database accessed 10.08.2022).

15.4 Traditional Breeding

15.4.1 *Improvement Strategies*

The processing industries need homogenous high quality and standardised raw material of thyme, especially from thymol-, carvacrol-, and linalool-types. Therefore, an increase in the demand of thyme of cultivated origin must be expected with standardized composition and yield of the essential oils and with uniform organoleptic properties of the leaves. In order to meet the requirements of the criteria of Ph. Eur. or ISO standards, breeding programmes were initiated on *Thymus* species in research institutes of several European countries and in North America to develop high performace varieties (Rey 1992; Letchamo et al. 1999; Rey et al. 2004; Carlen et al. 2010). Major strategies for improving thyme include well-defined set of activities related to the evaluation of drug yield and quality, including studies on floral biology and pollination, collection and evaluation of germplasm, selection of suitable parents, hybridization between selected parents, collection of seed and raising of seedlings in the nursery, and transplantation to progeny plots for initial screening and selection. F1 hybrid breeding as well as clonal selection can also be appropriate methods used, depending on the taxa, or the main goal of the breeding programme.

In general, the main aims of *Thymus* breeding include developing highly productive cultivars of uniform habit and flowering time with high drug and essential oil yields meet the criteria of standards. Regarding morphology, compact plant habit with densely branching upright flowering shoots including high proportion of leaves and flowers are required. Moreover, adaptability to changing environmental conditions concerning extremities of temperature and water supply are also to be considered. Varieties suitable to alternative cultivation technologies (organic farming, by direct sowing, to marginal areas, etc.) are preferred. In those Nordic areas, where growing of *T. vulgaris* in perennial culture is impossible, introduction of the most winter hardy variety'Deutscher Winter' by changing the culture length from perennial to annual was the solution for thyme growing (Galambosi et al. 2002). However, an other strategy was applied in Lithuania: instead of *T. vulgaris*, introduction and clonal selection of native thyme species (*T. serpyllum, T. pulegioides*) was a proper breeding tool to grow *Thymus* species of different major compounds (thymol, carvacrol, linalool, geraniol) in the essential oil (Loziené 2009). Similar initiatives on the introduction of an indigenous thyme species (*T. pannonicus*) into cultivation were reported from Serbia and Hungary as well (Pluhár et al. 2015a, b; Arsenijević et al. 2019).

15.4.2 Traditional Breeding Methodologies

The floral biology of garden thyme has been described in details by Assuad and Valdeyron (1975). As a result of the phenomenon of gynodioecy of *T. vulgaris*, small female flowers or larger hermaphrodite flowers can be found on separate plants within a population (Fig. 15.7). The diversification of active substances is partly the result of gender variability. Nuclear and cytoplasmic genes generate jointly the gender of the individual. The frequency of female plants is usually higher in a population than the hermaphrodite ones and is periodically changing (*T. vulgaris*: 5–95%, *T. serpyllum*: 30–95%). The female plants produce seeds of better germination capacity (Thompson 2002). To maintain the vigour of the species, cross fertilization between male fertile (MF, hermaphrodite) and male sterile (MS, female) plats, enhanced with protandry (development of anthers before pistils) is more frequently found in natural populations than autofertilization among MF individuals. The use of male-sterility for breeding and production of new varieties was first proposed by Rey (1990) (Rey and Sáez 2002). The visually observed heterogeneity of the gynodioecious thyme populations caused partly by the the different size, shape and colour of the flowers of male fertile and male sterile plants. Nevertheless, Pank and Krüger (2003) found not an inevitable correlation between this phenomenon and heterogeneity of other traits, when comparing F1 hybrid ('Varico 1') and selected varieties (e.g.'Deutscher Winter') of thyme. It was also proven that'Varico 1' was characterised by the highest level of homogeneity, which confirm that hybrid cultivar breeding is an adequate approach to improve homogeneity of thyme cultivars.

Breeding programmes are based on studies on morphological, production biological and chemical variabilities of the original populations established by seeds of different origin, among experimental conditions. In the case of the perennial *Thymus* species, not only the first but also the further years are important, when plant size and habit, leafiness, flower characteristics, fertility, winter hardiness, ecological tolerance are evaluated. After several years, individuals with promising traits are then selected and propagated by using halfwood cuttings (cloning) to maintain the high performance of the mother plant. The clones of the parent plants are investigated later in larger scale under field conditions and selected further if necessary to increase the high overall performance and homogeneity, either. After several years of improvement, selected lines or clones can be acknowledged by national authorities or by international organizations as varieties or maintained further by the breeders and growers associations (Anonymous 2008). Vegetative or generative propagation materials are

Fig. 15.7 Male fertile (MF) (**a**) and male steril (MS) (**b**) flower types of *Thymus vulgaris*. *Photos* Pluhár 2010

provided by the variety owners or authorized companies, depending on the type of cultivars. Rare chemotypes are usually maintained and propagated by cuttings for small scale production to ensure the special volatile composition. Thymol chemotypes are propagated by using F1 hybrid seeds ('Varico') or by seeds originating from standard varieties selected (e.g.'Deutscher Winter','Sloneczko','French Summer', etc.) (Anonymous 2008; Carlen et al. 2010).

15.4.3 *Cultivar Characterization and Phylogeny*

Breeding programmes conducted in Switzerland in France, Germany, Poland, Hungary, Romania, Bulgaria, Czech Republic and in Canada, resulted in different types of cultivars: selected standard and clone varieties or F1 hybrids, respectively (Rey et al. 1993; Rey et al. 2004; Anonymous 2008; Carlen et al. 2010; Pluhár et al. 2012).

In most of the countries (France, Switzerland, Italy or Spain), breeding activities started by the evaluation of *Thymus* species in natural habitats in and pre-selected populations of desirable traits were later introduced into cultivation. At the same time, parallel with the domestication, efforts made by the breeders to improve further the selected genotypes to create new varieties. The German common thyme is another good example for thyme breeding, where the empiric selection procedure, managed by the farmers resulted in the 'Deutscher Winter' variety, widely grown in several

European countries formerly. Its frost resistance allows to cultivate this variety even in northern countries, e.g. in the Netherlands in Finland or in Canada, at least in annual culture. 'Deutscher Winter' variety has greyish green dense foliage and bloom later than the other known varieties. The bush is usually large and densely branching with upright shoots, which is heterogenous in plant habit and flower morphology. High yielding cultivar with high thymol percentage but varying essential oil level resulting frequently in substandard drug quality. 'Deutscher Winter' served as a basis of further variety breeding: improved new cultivars and F1 hybrid varieties ('Varico 1–3') created by using the genetic basis of this old variety.

Heine et al. (2001) studied *T. vulgaris* varieties of different origin, involved in common cultivation practices with 3 harvests per year in Germany, between 1998 and 2000. The best 3 years' total yields were found at'Krajovy' (Czech) and 'Deutscher Winter' (Germany) cultivars, while the highest percentage of the leaf and flower fraction was detected at 'Sloneczko' (Poland) and 'De Dolj' (Romania). The growing year influenced the yields but the other features were determined genetically. There were significant fluctuations in the essential oil content ranging 1.47–2.40 ml/100 g in 1999, while between 1.69 and 2.49 ml/100 g in 2000. Thymol percentage however showed lower degree of variability (1999: 64.7–69.2%; 2000: 54.8–61.7%). 'Varico 1' and 'Varico 2' F1 hybrid cultivars showed the highest essential oil content (2.77–3.14 ml/100 g). Dudas and Böhme (2004) confirmed that'Krajovy','De Dolj' and'Deutscher Winter' were the most suitable for cultivation in Germany with the maximum yields achieved in the third year, while the best drug quality with the highest essential oil content and thymol % was detected in the second vegetation year. Nonetheless,'Varico 1' and'Varico 2' kept the outstanding essential oil producing ability till the 5th year of cultivation. The essential oil yield calculated for unit area (means: 32.9–48.6 l/ha) was the highest in the second season. Galambosi et al. (2002) proved the advantage of'Deutscher Winter' to'Varico 1' for cultivation in Finland, as the former produce 5–23% higher yields in annual cultivation. Further varieties were selected in Bugaria ('Slava') and in Canada ('Laval-1','Laval-2') (Stoeva et al. 2001; Letchamo et al. 1999).

The German Variety List reported about six frost tolerant thyme varieties available of thymol chemotype in 2002: De Dolj, Deutscher Winter, Krajovy, Sloneczko, Varico 1 and Varico 2 (Anonymous 2002). In a series of variety trials, significant differences were found among varieties, concerning the proportion of leaves (medium high: 55% to very high: 66%) and leaf drug yields (medium–low: 3.0 t/ha to high: 4.4 t/ha). The essential oil yields changed between medium–low (1.5 ml/100 g) to high (2.4 ml/100 g), while thymol percentages in a range of medium (57%) to high (67%) values.

There is a lack of further data on cultivars available from seed supplier companies ('French Summer','Standard Winter', etc.). Kosakowska et al. (2020) reported about the high level of heterogeneity in clones of'Standard Winter' cultivar, concerning morphological traits, yields, essential oil properties as well as phenolic compounds (Table 15.4). The authors emphasized that beside essential oil, the evaluation of phenoloid molecules of thyme accessions is also very important, where they found rather variable data on the spectrum and amount of flavonoids and phenolic acids.

Table 15.4 Information on selected garden thyme (*Thymus vulgaris*) cultivars

Cultivar name (available from)	Origin	Blooming period	Leaf ratio	Drug yield	Active compounds	
					Essential oil content	Thymol %
De Dolj	Romania	Early	Medium to high	**High**	Medium	Medium
Deutscher Winter	Germany	Late	Medium to high	**High**	Medium to high	Medium to high
Krayový (1952)	Czech Republic	Early to medium	Medium to high	**High**	Medium to high	Medium to high s
Sloneczko (1989)	Poland	Early to medium	**High to very high**	Medium	Medium to high	High
Standard Winter	Germany, selection	Early to medium	Medium to high	Medium	2.10–4.38	37.00–54.60
Pnnon Timol (2015)	Hungary, selection	Medium	Medium to high	Medium	1.46–2.47	67.70–72.90

Source Anonymous 2002; Kosakowska 2020, Pluhár et al. 2012

Varieties are to be examind in each region and those with the highest performance and adaptability are recommended for cultivation. Pluhár et al. (2010, 2016, 2017) reported about *T. vulgaris* selection, where the best clone (TV17) was recognized as the first thyme variety in Hungary in 2015 with the name of'Pannon Timol' (Fig. 15.8). The variety of thymol chemotype has superior traits if compared with 'Deutscher Winter': compact habit, proper regeneration ability, medium blooming period, 3.5 t/ha total drug yield of three years' cultivation, appropriate essential oil content (1.46–2.47 ml/100 g) and high thymol percentage (62.7–72.9%).

15.4.4 Traditional Breeding Limitations

Thyme breeding has generally focused on improvement of therapeutically important traits, such as essential oil content and thymol percentage together with yield components (dry mass, high leaf ratio, essential oil yield, etc.) and those morphological features, which may enhance their expression. *T. vulgaris* is an extremely variable species, in terms of morphological traits, chemical composition and represents sexual dimorphism (gynodioecy) controlled by nuclear and cytoplasmic genes. High variability gives almost unlimited possibilities to breeding programmes using genetic bases of natural populations, germplasm collections and contemporary cultivars in growing. However, when considering recent requirements of standardization of herbal products, high intraspecific diversity with wide variety of factors influencing drug quality can be rather disadvantageous. Several authors reported that selected thyme cultivars subjected to out-crossing, inbreeding depression and heterozygosity.

Fig. 15.8 A *'Pannon Timol'* variety of *Thymus* vulgaris in the second vegetation year. *Photo* Pluhár 2020

The other important traits that have been targeted by the thyme breeders are adaptability, tolerance to abiotic stresses like drought and cold, higher regeneration ability between the periods of harvests, better drug quality, and suitability to various areas of utilization (biopesticides, new formulations of cosmetics, etc.) in the future.

15.4.5 Role of Biotechnology

Biotechnology has limited significance in the breeding of thyme. There are only a few examples when biotechnological tools have been applied in course of a breeding process. Navratilova et al. (2021) established an efficient system for micropropagation and in vitro polyploidization of diploid T. vulgaris using oryzalin. The obtained tetraploid plants differed in the relative proportions of almost all of the individual terpenes, content of several important compounds (thymol, γ-terpinene, 4-cymene) increased, hence these plants are excellent new breeding materials. Another area of breeding where biotechnology might play a role is the breeding for resistance, especially for various abiotic stresses. Zia et al. (2010) studied the possibility of in vitro selection for NaCl tolerance of T. vulgaris.

15.5 Molecular Breeding

15.5.1 Molecular Marker-Assisted Breeding

There are only a few studies using molecular markers for studying *Thymus* species mostly for studying the inter- and intraspecific genetic diversity. Most of the publications are based on random molecular markers which do not require any sequence information like random amplified polymorphic DNA (RAPD), amplified fragment length analysis (AFLP) or inter simple sequence repeats (ISSR). Many studies parallel to the molecular marker analysis performed also chemical analysis of the essential oil and looked for correlation between the two. Echeverrigaray et al. (2001) assessed the genetic variability of six commercial cultivars of thyme (*Thymus vulgaris*) along with analysing the essential oil composition. All of the cultivars exhibited characteristic RAPD patterns that allowed their identification. Based on both the RAPD patterns and the chemical composition the cultivars were divided into two clusters, which coincided. However, Trindade et al. (2008) while using RAPD to reveal genetic diversity of *Thymus caespitius* found no straight correlation with the results of the chemical analysis. Similarly, no correlation was found by Lima et al. (2009) between ISSR and chemical analysis. On the other hand, when the two molecular methods – RAPD and ISSR – was combined by Trindade et al. (2009) the robustness of the data clearly increased and clustering according to the geographic origin and chemical profile was achieved. Still they conclude that these markers cannot be used to elucidate the chemical polymorphism in this species. Smolik et al. (2009) used ISSR for distinguishing seven *Thymus* species, but only *T. vulgaris* and *T. serpyllum* generated specific DNA profiles. Intergenic Spacer (IGS) between 5 s RNA subunits genes were also assessed for these species, but no polymorphism was revealed. Pluhár et al. (2012) also used RAPD to study four *Thymus* species native to Hungary. High genetic diversity was found, however neither interspecific nor on interpopulational level clustering was found. Sostaric et al. (2012) assessed genetic relation of seven *Thymus* species from Serbia with AFLP. Most of the genetic diversity was attributed to differences among individuals within populations. Species differentiation was significant in all cases, except for *T. marschallianus, T. pannonicus,* which according to Jalas (1972) are synonym to each other.

As in the case of other plants, the need for the use of sequence-specific molecular markers arose in case of *Thymus* species as well. However, as no whole genome sequence is available Karaca et al (2013) analysed ESTs (Expressed Sequence Tag) from *Salvia* and *Stenogyne* species for identifying microsatellite motifs. A total of 75 microsatellite amplifying primer pairs were tested on 12 species from the *Lamiaceae* family, including *Thymus serpyllum*. Among the tested primer pairs 52 were found to be cross-genera transferable. Later Karaca et al. (2015) tested these primer pairs on several *Thymus* species and found that 23 primer pairs were useful in the genetic characterization of *Thymus* genera as well. Applying these SSR markers greater genetic diversity was observed at the interspecific than at the intraspecific level. It was shown that hybridization is common in natural populations of the genera with two or more species living together. These 23 SSR primer pairs are the one and only published for studying *Thymus* genera, but there are no other publications with either these or other SSR markers about the genus. Recently ISSR and RAPD marker based results were published. Sarfaraz et al. (2020) assessed the genetic diversity of 11 *Thymus* species based on nine ISSR primers, which showed high efficiency for distinguishing among the different species. György et al. (2020) used 12 ISSR primers to characterize *T. vulgaris* chemotypes. Even though no chemotype specific band was found, distinguishing the five studied chemotypes was possible. Accessions belonging to the thymol chemotype showed the most distinct separation. Kim et al. (2020) used 16 RAPD primers for distinguishing a native Korean species, *T. quinquecostatus* from *T. vulgaris.* All studied cultivars belonging to the two species showed characteristic RAPD patterns that allowed their identification. Nevertheless, no significant relationship was found between the essential oil composition and RAPD data. The most recent publication is based on SRAP (Sequence Related Amplified Polymorphism) marker, which is also a random molecular marker targeting the coding regions of the genome. Sarfaraz et al. (2021) assessed the intra- and interspecies genetic variability of 11 Iranian *Thymus* species with eight SRAP primer combinations. High polymorphism was found also with this marker. Distinction of some species was unequivocal, while others clustered together revealing high admixture. As a conclusion, analyzing chemical and genetic relationship in the genus *Thymus* is possible with more or less success. It seems that studying a single species results more likely in correlation between chemical and genetic polymorphism even by using random molecular markers. The high intra- and interspecies genetic variability along with the other features beside essential oil composition are so diverse that without sequence specific markers it is less likely to find correlation between chemo- and genotype.

15.5.2 Functional Genomics

Functional genetic studies of *T. vulgaris* are concentrated around three main topics: the genetic determination of chemotypes, identification of genes involved in terpenoid biosynthesis, and the determination of gynodioecy.

Based on crossing experiments Vernet et al. (1986) demonstrated that the nature of chemotype is controlled by a series of loci, each locus having a dominant and a recessive allele among which epistatic interactions exist. Later, Thompson et al (2003) suggested, that instead of epistasis, the genetic control of monoterpene production may be a consequence of gene dosage effects along a chain of loci. Recently, Bataillon et al. (2022) showed that SNPs located in genes of the monoterpene pathway explain a large fraction of chemotype variation. Variation in three terpene-synthase loci explains the variation in chemotype identity, with one single locus accounting for as much as 78% of this. A model combining these three loci was proposed to explain virtually all chemotype variations.

As next generation sequencing (NGS) is getting more and more feasible the study of any biological sample or phenomenon has shifted towards nucleotide sequence based approach. In the last years, several studies have been published also related to *T. vulgaris* using NGS technology.

Soorni et al. (2019) performed transcriptome sequencing of five *Thymus* species to mine putative genes involved in thymol and carvacrol biosynthesis. Transcriptom sequencing is a relatively new and powerful approach for identifying gene sequences in organisms without available genome sequences. In course of the study, 11 genes were identified as being involved in thymol- and other monoterpene biosynthesis. Three unigenes (germacrene D synthase, γ-terpinene synthase, and a member of the cytochrome P450 71 family) were identified as probably responsible for terminal steps of thymol biosynthesis. All putative terpene synthases (TPSs) were assigned to the Mono-TPSb subfamily, supporting the existing view that *Thymus* TPSs belong to a distinct clade. Based on the assumption that the differences in contig number reflects differences in transcript abundance the MEP biosynthesis way seems to be the major pathway to monoterpene thymol production over the MVE pathway.

Mollion et al. (2018) also used NGS technology for locus targeted genotyping aiming to distinguish between two scenarios how gynodioecy is maintained. After transcriptome sequencing the amount and patterns of nucleotide diversity in the nuclear and cytoplasmic genome was contracted. Plentiful diversity was found at the nuclear level but reduced genetic diversity and an excess of rare polymorphisms in the cytoplasmic genome compared to the nuclear genome. Their finding suggests that cytoplasmic male sterility (CMS) genes recurrently enter the populations at low frequency via mutation or migration and go to fixation unimpeded (successive sweeps) resulting in recent fixation of CMS lineages in the studied populations.

15.6 Tissue Culture Applications

Plant tissue culture offers independence from climatic and geographical factors, elimination of pest and diseases, shorter production cycles and novel products not found in nature. *In vitro* tissue culture is established for several purposes in case of aromatic- and medicinal plants. Micropropagation can be used to rapidly multiply plants with desirable traits and create healthy, disease-free material without seasonal constraints. Somaclonal variability inherent in tissue culture provides an excellent source of genetic variability enabling selection for different purposes. On the other hand, tissue culture might also be used for preserving the existing wide range of genetic diversity on minimal media ensuring slow growth or via cryopreservation. Tissue culture has been widely used for breeding purpose as well, especially in selection for stress tolerance. Finally, establishing cell cultures for the purpose of secondary metabolite production is the most important aim of tissue culture medicinal plant species.

In recent years, numerous papers have been published about micropropagation of various species of the thyme genus. Most papers concern the micropropagation of *T. vulgaris*. but other *Thymus* species are also in the focus. One of the first papers is about micropropagation of *T. piperella* (Sáez et al. 1994). Ozudogru et al. (2011) published an efficient micropropagation protocol applicable both to young and mature explants of *T. vulgaris*, and tested it also on another *Thymus* species. Shabnum and Wagay (2011) also published about *in vitro* micropropagation of severel *Thymus* species, including *T. vulgaris*. Aicha et al. (2013) established a protocol for the micropropagation of *T. satureioides*. El-Banna (2017) has induced callus growth from shoot tip explants and then regenerated shoot growth. Affonso et al. (2009) studied the effect of different growth hormones on the growth of *in vitro* plants and their effect on the essential oil production. No qualitative changes were observed in the volatile profile of thyme plants. Kulpa et al. (2018) established a micropropagation protocol for *T. vulgaris* 'Słoneczko' and evaluated the potential of micropropagated plants for essential oil production with industrial application. They concluded that plants derived from *in vitro* cultures may be an excellent source of standardised essential oils. Both El Ansari et al. (2019) and Tevfik and Yegerova (2020) focused on optimizing the composition of media at the different stages of micropropagation. Several combinations of different growth hormones were tested for each step of micropropagation (introduction of explants in vitro, shoot multiplication, rooting, adaptaion). However, El Ansari et al. (2019) found the combinations of Kin + IAA or NAA, DPU + IBA and Ad + IBA the most efficient, while Tevfik and Yegerova (2020) found Kinetin alone the best for multiplication and IBA or IAA for rooting.

Callus induction from various *Thymus* species was the subject of several publications in the last decade as well. Abdullah et al. (2012) initiated callus from different organs of *T. vulgaris* and they also achieved a one-step regeneration. Khodakarami et al. (2013) studied the effect of plant growth hormones on callus induction of *T. vulgaris* and *T. daenensis*. Pawelec et al. (2017) also investigated the callus initiation and maintenance from *T. vulgaris*. Bakhtiar et al. (2016) induced callus growth in *T. persicus* and plant regeneration was also achieved. Several research groups not just initiated callus cultures from *Thymus* species, but also studied the content of essential oil and other secondary metabolites of the callus tissue and the potential of industrial scale production. Mosavat et al. (2019) studied the potential of applying ZnO nanoparticles to callus culture of several *Thymus* species to increase callus proliferation and the production of volatile compounds, which approach was found to be a simple and cost-effective method to be applied at industrial scale. Dyshlyuk et al. (2021) analysed *T. vulgaris* callus for identifying geroprotector substances. Beside thymol, which is the main volatile oil component, also apigenin, gallic-, chlorogenic-, and caffeicacid demonstrated geroprotective properties. Darwish et al. (2020) following the rosmarinic acid production in *in vitro* plants and callus and studying the expression of some genes from the rosmarinic acid biosynthesis pathway concluded that rosmarinic acid production and the expression of the studied genes increased in callus tissue, compared to the in vitro plants.

15.7 Hybridization

15.7.1 Conventional Hybridization

Although the wide range of variability is the prerequisite of a successful breeding program, the expected result of it is a new cultivar with high level of homogeneity. As the standard varieties bred generally by selection possess different levels of heterogeneity, hybrid cultivar breeding was suggested in the case of the heterozigous *T. vulgaris* (Pank 2010). Based on the gynodioecy, the male sterility provide controlled fertilization of selected parental lines to the breeder without the laborious flower emasculation as needed generally for crossing. Hybrid cultivar breeding programme requires homozygous parental lines: a male sterile (MS) mother line and a male fertile (MF) pollinator line with high performance and high combining ability. A so-called maintainer line is also necessary to conserve the MS mother line for further crossings. The F1 hybrid seeds produced by the parental lines every year and are used for cultivation purposes. Hybrid varieties have three main advantages: their performance exceeds those of the parental lines due to the hybrid vigour caused by heterosis, they are quite homogenous and protect the breeders'rights because they segregate and become heterozygous in the next generations. The disadvantage is that maintenace costs of parental lines and of seed production are higher if compared by selected varieties (Pank 2010). If the goal is a fertile, seed producing hybrid variety, crossing experiments are conducted by using MS-MF parents with isolation. Only the seeds of male sterile parents are collected, then the progenies are evaluated. Only the parents of the best hybrids were applied for F1 hybrid seed production for growers. It was found that'Varico 1' and'Varico 2' F1 hybrid varieties were always superior to the other accessions or parents, originated from clones or polycrosses of selected clones, either (Rey et al. 1993; Rey and Sáez 2002).

15.7.2 Hybrid Cultivars

In France continuous selection activity on different chemotypes is coordinated by ITEIPMAI, where thyme grower associates are also involved in variety trials, by testing the productivity and drug quality of newly bred lines. New F1 hybrid cultivars with high performance presented in 2008 belong to 3 distinct chemotypes, where the variety names reflect to the origin:'Thymlia' (thymol type),'Carvalia' (carvacrol type),'Linalia' (linalool type) (Anonymous 2008) (Table 15.5).

Researchers of the Agroscope Changins-Wadenswil ACW (Conthey, Switzerland) carried out a breeding programme to optimize the yield and quality of thyme varieties, introducing new accessions to the selection of MS and MF parental lines to a new F1 hybrid cultivar, which was presented in 2010 under the name of'Varico 3' (Carlen et al. 2010). This homogenous hybrid variety of thymol chemotype was suggested for cultivation in Central Europe and Switzerland, characterized by high

Table 15.5 Information on garden thyme (*Thymus vulgaris*) cultivars developed recently

Cultivar name, (available from)	Origin	Leaf ratio, %	Drug yield, t/ha	Active compounds				Essential oil yield, l/ha (sum of 3 years)
				Essential oil content, mL/100 g DW	Thymol %	Carvacrol %	Linalool %	
Deutscher Winter (Control)	Germany, (selected population)	67.00	3.31	2.90	**56.00**	3.00	2.00	166.00
Varico 3 (2010)	Switzerland (F1 hybrid)	64.00	3.86	4.90	**65.00**	3.00	2.00	332.00
Linalia (2008)	France (F1 hybrid)	67.00	2.66	4.90	11.00	23.00	**48.00**	236.00
Thymlia (2008)	France (F1 hybrid)	64.00	3.43	6.20	**30.00**	**31.00**	5.00	370.0
Carvalia (2008)	France (F1 hybrid)	66.00	2.97	5.10	**22.00**	**41.00**	6.00	279.00

Source Anonymous 2008; Carlen et al. 2010,Vouillamoz et al. 2011

essential oil content (> 4.9%), dry matter production proven to be superior to other cultivars studied (Deutscher Winter, Linalia, Thymlia, Carvalia) (Vouillamoz et al. 2011) (Table 15.5).

15.8 Conclusion and Prospects

Essential oil and extracts of thyme species have long history and relevance in therapy, as healing agents of expectorant and spasmolytic preparations used in the European traditional medicine for centuries (Mills and Hutchins 2014). Thyme (*Thymus vulgaris* L.) was known as a polymorph medicinal plant sourced exclusively from the wild habitats, but nowadays it is widely cultivated in almost all of the countries with moderate climate. An increase in utilization and consumption of the herb and the essential oil is predicted in the case of certain chemotypes (thymol-, carvacrol-, and linalool-types), where standardized composition of the essential oils and uniform organoleptic properties are expected by pharmaceutical, perfumery or food industries. Considering the great economic significance of thyme, breeding efforts were made in order to develop new cultivars providing high yield and excellent drug quality. Comprehensive research activities, studying the genetic bases of diversity resulted in several findings, supporting the relevance of hybrid variety breeding. Since *T. vulgaris* is a highly heterozygous subshrub with 4–6 years' culture, developing new cultivars through conventional breeding techniques is laborious and time consuming. Nonetheless, several selected and F1 hybrid varieties suitable to different climatic conditions have been developed and are being used at a commercial scale. However, most of these thyme cultivars/varieties have been developed from only one species with the exclusion a great number of taxa with specific traits such as stress resistance. Thus, it is recommended to develop new methods to utilize further species. High chemical polymorphism implies a high intraspecific genetic diversity which is only partly clarified by molecular biological studies on *Thymus* species (Li et al. 2009; Lima et al. 2009; Rahimmalek et al. 2009; Pluhár et al. 2012; Rustaiee et al. 2013).

15.8.1 Current Research Initiatives to Combat Global Climate Change

Drought is one of the greatest limitation to crop production and will be increasingly important due to the recognized effects of the global climate change. The growing practices of *T. vulgaris* introduced in the past decades will likely to be modified by the climate changes, the global warming and the higher incidence of extremities. One can predict that medicinal plant growers should be aware of higher probability of extreme weather conditions, which may principally influence the success of production. Tolerance against extremities, such as drought, frost or hot periods, will possibly be the main goal of breeding programs, even in the case of thyme which is highly resistant to aridity and possess by wide ecological flexibility in terms of environmental conditions, either. At the same time, cultivation technologies have to be developed accordinly. In the case of the perennial *Thymus* species it can be emphasized, as the field establishment needs high initial costs invested in the propagation material. In order to ensure the approproiate raw material supply from thyme production in the future, efforts made to understand the physiological plant responses to biotic and abiotic stress factors as well as to develop new or improved varieties suitable for wider cultivation. According to predictions and recent scientific findings, it is suggested to growers to install irrigation facilities with the possibility of supplementary fertilization in thyme growing to ensure proper plant development, high drug yields and essential levels (Tátrai et al. 2016; Németh et al. 2016).

15.8.2 Recommendations for Future Research

In spite of their economic importance, there is not enough information on the genetic relationship concerning *Thymus* species (Echeverrigaray et al. 2001; Sunar et al. 2009). However, molecular biological data may provide a solution to the taxonomic problems both in intraspecific level and among the thyme species. The use of DNA marker techniques was proven to be important in the case of polymorph species of scarce reliable taxonomic characters (Karaca et al. 2008; Ince et al. 2009). György et al. (2020) mentioned that identifying chemotype-specific molecular markers would promote breeding and selection of thymes cultivars dedicated for specific purposes according to their main volatile oil component. Hence, further studies are needed to find chemotype specific markers for *Thymus vulgaris* L. Furthermore, genetic engineering should be introduced to facilitate development of new thyme cultivars capable of producing qualitatively superior products.

On the other hand, small scale cultivation after introduction of native thyme species adapted to the local environment can also be recommended, especially in organic farming.

References

Abdullah N, Taghrid NA, Zena YK (2012) The plant callus of thyme *Thymus vulgaris* L. Coll Basic Educ Res J 11(4):623–636

Affonso VR, Bizzo HR, Lage CLS, Sato A (2009) Influence of growth regulators in biomass production and volatile profile of in vitro plantlets of *Thymus vulgaris* L. J Agricult Food Chem 57(14):6392–6395

Aicha N, Rachida TC, El Abdelmalek M (2013) Micropropagation of *Thymus satureioides* Coss. an endangered medicinal plant of Morocco. J Agricult Technol 9(2):487–501

Al Maqtari MA, Alghalibi SM, Alhamzy EH (2011) Chemical composition and antimicrobial activity of essential oil of Thymus vulgaris from Yemen. Turk J Biochem 36:342–349

Anonymous (1987) Hungarian. Standard no. MSZ 19878–87 *Thymi herba*. Hungarian Standards Institute, 1987 (in Hungarian)

Anonymous (1996) ISO Standard 6754:1996: Dried thyme (*Thymus* vulgaris L.), 2nd ed last revised and updated in 2021. ISO, Geneva, Schwitzerland

Anonymous (2002) Thymian. In: Beschreibende Sortenliste, Arznei-und Gewürzpflanzen. Deutscher Landwirtschaftsverlag GmbH, Hannover, pp 164–169

Anonymous (2008) Rapport d'Activité 2008. ITEIPMAI (L'Organisme Français de Recherche pour le Dévéloppement des Plantes à Parfum, Médicinales et Aromatiques. pp 12–13

Anonymous (2016) European Pharmacopoeia 9.0. *Serpylli herba*. 9th ed. Council of Europe, Strasbourg, France, pp 1559–1560

Anonymous (2017) ISO Standard 19817:2017: Essential oil of thyme [*Thymus* vulgaris L. and *Thymus* zygis L.], thymol type. ISO, Geneva, Schwitzerland

Anonymous (2020a) Assessment report on *Thymus* vulgaris L., *Thymus* zygis L., aetheroleum. European Medicines Agency, HMPC, 07 Aug 2020

Anonymous. (2020b) *Thymi typo thymolo aetheroeum* (Thyme essential oil of thymol type). Pharmacopoeia Europaea 10th ed no. 01/2012:1374

Arsenijević J, Drobac M, Šoštarić I, Jevđović R, Živković J, Ražić S, Moravčević Đ, Maksimović Z (2019) Comparison of essential oils and hydromethanol extracts of cultivated and wild growing *Thymus* pannonicus all. Ind Crops Prod 130:162–169. https://doi.org/10.1016/j.indcrop.2018.12.055

Asllani U, Toska V (2003) Chemical composition of Albanian thyme oil (*Thymus* vulgaris L.). J Ess Oil Res 15(3):165–167

Assuad W, Valdeyron G (1975) Remarques sur la biologie du thym, *Thymus* vulgaris L. Bull Soc Bot Fr 122:21–34

Bakhtiar Z, Mirjalili MH, Sonboli A (2016) In vitro callus induction and micropropagation of *Thymus* persicus (Lamiaceae), an endangered medicinal plant. Crop Breed Appl Biotech 16:48–54

Baranauskienė R, Venskutonis PR, Viskelis P, Dambrauskienė E (2003) Influence of nitrogen fetilizers on the yield and composition of thyme (*Thymus vulgaris*). J Agric Food Chem 51:7751–7758

Bardeau F (1973) Le pharmacie de bon Dieu. Stock, Paris, pp 279–281

Barnes J, Anderson LA (2002) Phillipson JD Herbal Medicines. Pharmaceutical Press, Chicago, London

Baser K, Kirimer N, Ermin N, Özek T (1996) Composition of essential oils from three varieties of *Thymus* praecox Opiz growing in Turkey. J Ess Oil Res 9:319–321

Bataillon T, Gauthier P, Villesen P, Santoni S, Thompson JD, Ehlers BK (2022) From genotype to phenotype: Genetic redundancy and the maintenance of an adaptive polymorphism in the context of high gene flow. Evol Lett 6(2):189–202

Begrow F, Engelbertz J, Feistel B, Lehnfeld R, Bauer K, Verspohl EJ (2010) Impact of thymol in thyme extracts on their antispasmodic action and ciliary clearance. Planta Med 76:311–318

Bencze É, Zs P (2013) *Thymus* vulgaris L. In: Bernáth J (ed) Vadon termő és termesztett gyógynövények (Wild-growing and cultivated medicinal plants). Mezőgazda Publisher, Budapest, pp 476–479 (in Hungarian)

Benkiniouar R, Rhouati S, Touil A et al (2007) Flavonoids from Thymus algeriensis. Chem Nat Compd 43:321–322. https://doi.org/10.1007/s10600-007-0118-x

Blanco E, Breaux J (1997) Results of the study of commercialization, exploitation and conservation of medicinal and aromatic plants in Spain. Unpublished report for Traffic Europe

Boros B, Jakabová S, Dörnyei Á, Gy H, Zs P, Kilár F, Felinger A (2010) Determination of polyphenolic compounds by liquid chromatography-mass spectrometry in *Thymus* species. J Chromatogr A 1217:7972–7980

Braga PC, Dal Sasso M, Culici M, Spallino A (2010) Inhibitory activity of thymol on native and mature Gardnerella vaginalis biofilms: in vtro study. Arzneimittelforsch 60(11):675–681

Carlen C, Schaller M, Carron CA, Vouillamoz J, Baroffio CA (2010) The new *Thymus vulgaris* L. hybrid cultivar 'Varico 3' compared to five established cultivars from Germany, France and Switzerland. Acta Hort 860:161–166

Carron CA, Baroffio C, Vouillamoz J (2010) Incidence du semmis ou de la plantation sur le rendement et la qualité de la sauge, du thym et de la guimauve. Rev Sui Viticult Arboricult Horticult 42(4):236–244

Cases A, Pérez B, Navarrete P, Mora E, Pena B, Peluzzo A, Calvo R, de Ron S, Varela F (2009) Variability in the chemical composition of wild *Thymus vulgaris* L. Acta Hort 826:159–165

Chizzola R, Michitsch H, Franz C (2008) Antioxidative properties of *Thymus vulgaris* leaves: Comparison of different extracts and essential oil chemotypes. J Agric Food Chem 56:6897–6904

Chromosome Counts Database. http://ccdb.tau.ac.il/Angiosperms/Lamiaceae/Thymus/Thymus%20vulgaris%20L. Accessed on 10 Aug 2022

Corticchiato M, Bernardini A, Costa J, Bayet C, Saunois A, Voirin B (1995) Free flavonoid aglycones from *Thymus herba barona* and its monoterpenoid chemotypes. Phytochem 40:115–120

Dajić-Stevanović Z, Šoštarić I, Marin PD, Stojanović D, Ristić M (2008) Population variability in *Thymus glabrescens*, Willd. from Serbia: morphology, anatomy and essential oil composition. Arch Biol Sci Belgr 60:475–483

Darwish HY, Abdelmigid H, Albogami S, Alotaibi S, El-Deen AN, Alnefaie A (2020) Induction of biosynthetic genes related to rosmarinic acid in plant callus culture and antiproliferative activity against breast cancer cell line. Pakistan J Biol Sci 23(8):1025–1036

Dudas S, Böhme M (2004) Stabilität der quantitativen und qualitativen merkmale bei Thymian (*Thymus* vulgaris L.) im fünfjährigen Sortenversuch. Fachtagung für Arznei-und Gewürzpflanzen, Jena, pp 123–125

Dyshlyuk LS, Fedorova AM, Loseva AI, Eremeeva NI (2021) Callus cultures of *Thymus* vulgaris and Trifolium pratense as a source of geroprotectors. Food Process Tech Technol 51(2)

ESCOP Monographs: thymi herba. In: the scientific foundation for herbal medicinal products. Thieme, New York

Echeverrigaray S, Agostini G, Atti-Serfini L, Paroul N, Pauletti GF, Dos Santos ACA (2001) Correlation between the chemical and genetic relationships among commercial thyme cultivars. J Agric Food Chem 49:4220–4223

ECPGR/MAP Working Group website. https://www.ecpgr.cgiar.org/working-groups/medicinal-and-aromatic-plants. Accessed on 31 Aug 2022

ECPGR/MAP WG. Draft descriptors of *Thymus vulgaris*. https://www.ecpgr.cgiar.org/fileadmin/templates/ecpgr.org/upload/NW_and_WG_UPLOADS/MAP_Descriptors/Thymus_vulgaris_DRAFT_DESCRIPTOR_LIST_FINAL.pdf. Accessed on 31 Aug 2022

ECPGR/MAP WG. Draft descriptors of *Thymus serpyllum*. https://www.ecpgr.cgiar.org/fileadmin/templates/ecpgr.org/upload/NW_and_WG_UPLOADS/MAP_Descriptors/Thymus_serpyllum_DRAFT_DESCRIPTOR_LIST_FINAL.pdf. Accessed on 31 Aug 2022

El Ansari ZN, El Mihyaoui A, Boussaoudi I, Benkaddour R, Hamdoun O, Tahiri H, Lamarti A (2019) Effect of macronutrients, cytokinins and auxins, on in vitro organogenesis of *Thymus* vulgaris L. American J Plant Sci 10(09):1482

El-Banna HY (2017) Indirect micropropagation of *Thymus* vulgaris plant. J Plant Prod 8(11):1241–1246

Elena-Roselló JA (1981) Cytotaxonomic and evolutionary studies in *Thymus* (*Labiatae*); relationships of the members of section *Thymus* Jalas. Anales Jardin Botanico De Madrid 38(1):51–59

Fachini-Queiroz FC, Kummer R, Estevão-Silva CF, Carvalho MDDB, Cunha JM, Grespan R, Bersani-Amado CA, Cuman RKN (2012) Effects of thymol and carvacrol, constituents of *Thymus vulgaris* L. essential oil, on the inflammatory response. Evidence-based Complement. Altern Med. Article no. 657026. https://doi.org/10.1155/2012/657026

Fecka I, Cisowski W, Sroka Z, Kowalczyk A (2006) The presence of polyphenolic compounds in some volatile oil containing plants and their biological activities. 37th International symposium on essential oils, Grasse, France, 10–13 Sept 2006. Abstracts, p 66
Fernandes ES, Passos GF, Medeiros RM, da Cunha F, Ferreira J, Campos MM, Pianowski LF, Calixto JB (2007) Anti-inflammatory effects of compounds alpha-humulene and (-)-*trans*-caryophyllene isolated from the essential oil of *Cordia verbenacea.* Eur J Pharmacol 569:228–236
Figueiredo AC, Barroso JG, Pedro LG, Salgueiro L, Miguel MG, Faleiro ML (2008) Portuguese *Thymbra* and *Thymus* species volatiles: chemical composition and biological activities. Current Pharm des 14:3120–3140
Furlenmeimer M (1984) Plantas curativas y sus propiedades medicinales. Schwittez, Zug, p 168
Galambosi B, Zs G, Pessala R, Hupila I, Aflatumi A, Repćak M, Svoboda PK (2002) Yield and quality of selected herb cultivars in Finland. Acta Hort 578:139–149
García M, Gonzalez-Coloma A, Donadel OJ, Ardanaz CE, Tonn CE, Sosa ME (2007) Insecticidal effects of *Flourensia oolepis* Balke (Asteraceae) essential oil. Biochem System Ecol 35:181–187
Granger R, Passet J (1971) Types chimique (chémotypes) de l'espèce *Thymus vulgaris* L. C R Acad Sci Paris 273:2350–2353
Granger R, Passet J (1973) *Thymus vulgaris* L. spontane de France: races chimiques et chemotaxonomie. Phytochem 12:1683–1691
György Z, Incze N, Pluhár Z (2020) Differentiating *Thymus* vulgaris chemotpyes with ISSR molecular markers. Biochem Syst Ecol 92:104118. https://doi.org/10.1016/j.bse.2020.104118
Heeger EF (1989) Handbuch des Arznei- und Gewürzpflanzenbaues (Drogengewinnung). VEB Deutscher Landwirtschaftsverlag, Berlin, pp 673–677
Hegnauer R (1978) Die systematische Bedeutung der ätherischen Öle (Chemotaxonomie der äterischen Öle), Dragoco Rep, pp 204–230
Heine H, Eger H, Krüger H (2001) Results of variety trials with thyme. In: ISHS world conference of medicinal and aromatic plants, Budapest 8–11 July 2001. Abstracts, p 233
Hernández LM, Tomás-Barberán FA, Tomás-Lorente F (1987) A chemotaxonomic study of free flavone aglycones from some Iberian *Thymus* species. Biochem System Ecol 15:61–67
Horváth B, Mukhopadhyay P, Kechrid M, Patel V, Tanchian G, Wink DA, Gertsch J, Pacher P (2012) *β*-caryophyllene ameliorates cisplatin-induced nephrotoxicity in a cannabinoid 2 receptor-dependent manner. Free Rad Biol Med 52:1325–1333
Hudaib M, Speroni E, Di Pietra AM, Cavrini V (2002) GC/MS evaluation of thyme (*Thymus vulgaris* L.) oil composition and variations during the vegetation cycle. J Pharm Biomed Anal 29:691–700
Ince AG, Elmasulu S, Cinar A, Karaca M, Onus AN, Turgut K (2009) Comparison of DNA marker techniques for *Lamiaceae.* Acta Hort 826:437–445
Jalas J (1972) *Thymus.* In: Tutin TG, Heywood VH, Burges NA, Valentine DH, Walters SM, Webb DA (Eds) Flora Europaea, vol 3. Cambridge University Press, UK, pp 172–182
Jordán MJ, Martínez RM, Goodner KL, Baldwin EA, Sotomayor JA (2005) Seasonal variation of Thymus hyemalis Lange and Spanish Thymus vulgaris L. essential oils composition. Ind Crops Prod 24(3):253–263
Jordán MJ, Martínez RM, Martínez C, Monino I, Sotomayor JA (2009) Seasonal variation of *Thymus hyemalis* Lange and Spanish *Thymus vulgaris* L. essential oils composition. Ind Crops Prod 29:145
Jovanović A, Balanč B, Petrović P, Djordjevic V (2022) Pharmacological potential of *Thymus* serpyllum L. (wild thyme) extracts and essential oil: a review. J Engin Process Manag 13:32–41
Kamondy L, Zs P, Ferenczy A, Sz S (2005) Kerti kakukkfű (*Thymus vulgaris* L.) törzsek produkcióbiológiai értékelése (Production biological evaluation of garden thyme (*Thymus vulgaris* L.) lines). Kertgazdaság (Horticulture) 37:197–207
Karaca M, Ince AG, Turgul S, Turgut K, Onus AN (2008) PCR-RFLP and DAMD-PCR genotyping for *Salvia* species. J Sci Food Agric 88:2508–2516
Karaca M, Ince AG, Aydin A, Ay ST (2013) Cross-genera transferable e-microsatellite markers for 12 genera of the Lamiaceae family. J Sci Food Agricult 93(8):1869–1879

Karaca M, Ince AG, Aydin A, Elmasulu SY, Turgut K (2015) Microsatellites for genetic and taxonomic research on thyme (*Thymus* L.). Turkish J Biol 39(1):147–159

Karaman S, Digrak M, Ravid U, Ilcim A (2001) Antibacterial and antifungal activity of the essential oils of *Thymus* revolutus Celak from Turkey. J Ethnopharmacol 76:183–186

Keefover-Ring K, Thompson JD, Linhart BY (2009) Beyond six scents: defining a seventh *Thymus* vulgaris chemotype new to southern France by ethanol extraction. Flav Fragr J 24(3):117–122

Khazaie HR, Nadjafi F, Bannayan M (2008) Effect of irrigation and plant density on herbage biomass and oil production of thyme (*Thymus vulgaris*) and hyssop (Hyssopus officinalis). Ind Crops Prod 27:315–321

Khodakarami F, Maleki ZB, Azimkhani R (2013) The effect of plant regulators on callus induction of thyme (*Thymus* vulgaris L. and *Thymus* daenensis)

Kim M, Moon JC, Kim S, Sowndhararajan K (2020) Morphological, chemical, and genetic characteristics of korean native thyme bak-ri-hyang (*Thymus quinquecostatus* Celak.). Antibiotics 9(6):289

Kosakowsk, O, Baczek K, Przybyl JL, Pawelczak A, Rolewska K, Weglarz Z (2020) Morphological and chemical traits as quality determinants of common thyme (*Thymus vulgaris* L.), on the example of 'Standard Winter' cultivar. Agronomy 10(6):909

Kulišić T, Dragović-Uzelac V, Miloš M (2006) Antioxidant activity of aqueous tea infusions prepared from oregano, thyme and wild thyme. Food Tech Biotech 44:485–492

Kulpa D, Wesołowska A, Jadczak P (2018) Micropropagation and composition of essentials oils in garden thyme (*Thymus vulgaris* L.). Not Bot Horti Agrobot Cluj-Napoca 46(2):525–532

Leonhardt V, Leal-Cardoso JH, Lahlou S, Albuqeurque AA, Porto RS, Celedonio NR, Oliveira AC, Pereira RF, Silva LP, Garcia-Teofilo TM, Silva AP, Magalhaes PJ, Duarte GP, Coelho-de-Souza AN (2010) Antispasmodic effects of essential oil of *Pterodon polygalaeflorus* and its main constituent β-caryophyllene on rat isolated ileum. Fundam Clin Pharmacol 24:749–758

Letchamo W, Gosselin A, Hoelzl J, Marquard R (1999) The selection of *Thymus vulgaris* cultivars to grow in Canada. J Essent Oil Res 11(3):337–342. https://doi.org/10.1080/10412905.1999.9701149

Li M, Lei Y, Yani W (2009) ISSR analysis of genetic relationships between seven varieties of thyme. J Shanghai Jiaotong Univ-Agric Sci 27(3):219–222

Lima AS, Trindade H, Figueiredo AC, Barroso JG, Pedro LG (2009) Molecular and volatiles characterization of Portuguese *Thymus caespititius* chemotypes. In: IV International symposium on breeding research on medicinal and aromatic plants-ISBMAP 2009, Ljubljana, Slovenia, 17–21 June 2009, pp 81–85

Loziené K (2009) Selection of fecund and chemically valuable clones of thyme (*Thymus*) species growing wild in Lithuania. Ind Crops Prod 29:502–508

Maksimović Z, Milenković M, Vučićević D, Ristić M (2008) Chemical composition and antimicrobial activity of *Thymus* pannonicus, All. (Lamiaceae) essential oil. Cent Eur J Biol 3:149–154

Marin DP (1996) A chemotaxonomic study of vacuolar flavonoids from some Balkan Micromeria species (Lamiaceae). Arch Biol Sci (Belgrade) 48:49–54

Marin DP, Grayer JR, Kite CG, Matevski V (2003) External leaf flavonoids of *Thymus* species from Macedonia. Biochem Syst Ecol 31:1291–1307

Mártonfi P, Mártonfiová L (1996) Thymus chromosome numbers from Carpathians and Pannonia. Thaiszia 6:25–38

Merghem R, Jay M, Viricel MR, Bayet C, Voirin B (1995) Five 8-C benzylated flavonoids from T. hirtus (Labiatae). Phytochem 38:637–640

Mewes S, Krüger H, Pank F (2008) Physiological, morphological, chemical and genomic diversities of different origins of thyme. Gen Resour Crop Evol 55(8):1303–1311. https://doi.org/10.1007/s10722-008-9329-7

Mills S, Hutchins R (2014) ESCOP Monographs: Serpylli herba—Wild Thyme, 3rd ed (online series). European Scientific Cooperative on Phytotherapy (ESCOP), Exeter, UK, p 14

Miura KH, Kikuzaki N, Nakatani J (2002) Antioxidant activity of chemical components from sage (Salvia officinalis L.) and thyme (*Thymus* vulgaris L.) measured by the oil stability index method. Agric Food Chem 50(7):1845–1851

Mollion M, Ehlers BK, Figuet E, Santoni S, Lenormand T, Maurice S, Galtier N, Bataillon T. (2018) Patterns of genome-wide nucleotide diversity in the gynodioecious plant *Thymus* vulgaris are compatible with recent sweeps of cytoplasmic genes. Genome Biol Evol 10(1):239–248

Moreira Galdino P, Nascimento MVM, Ferreira Florentino I, Campos Lino R, Oluwagbamigbe Fajemiroye J, Abdallah Chaibub B, Realino de Paula J, Monteiro de Lima TC, Costa EA (2012) The anxiolytic-like effect of an essential oil derived from Spiranthera odoratissima A. St. Hil. leaves and its major component, β-caryophyllene, in male mice. Prog Neuro-Psychopharm Biol Psychiat 38:276–284

Morales R (1989) EL género *Thymus* L. en al región mediterránea occidental (*Lamiaceae*). Biocosme Mésogéen 6:205–211

Morales R (2002) The history, botany and taxonomy of the Genus *Thymus*. In: Stahl-Biskup E, Sáez F (eds) Thyme: the genus *Thymus*. Taylor and Francis Inc., London, UK, pp 1–43

Mosavat N, Golkar P, Yousefifard M, Javed R (2019) Modulation of callus growth and secondary metabolites in different *Thymus* species and Zataria multiflora micropropagated under ZnO nanoparticles stress. Biotechnol Appl Biochem 66(3):316–322

Nowak A, Kalemba D, Piotrowska M, Czyzowska A (2012) The effects of thyme (*Thymus vulgaris*) and rosemary (*Rosmarinus officinalis*) essential oils on *Brochothrix thermosphacta* and on the shelf life of beef packaged in high-oxygen modified atmosphere. Food Microbiol 32:212–216

Najar B, Pistelli L, Ferri B, Angelini LG, Tavarini S (2021) Crop yield and essential oil composition of two *Thymus vulgaris* chemotypes along three years of organic cultivation in a hilly area of Central Italy. Molecules 26:5109

Navrátilová B, Švécarová M, Bednář J, Ondřej V (2021) In vitro polyploidization of *Thymus vulgaris* L. and its effect on composition of essential oils. Agronomy 11:596

Németh É, Szabó K, Zs P, Radácsi P, Inotai K (2016) Changes in biomass and essential oil profile of four Lamiaceae species due to different soil water levels. J Essent Oil Res 28(5):391–399. https://doi.org/10.1080/10412905.2016.1176606

Nezhadali A, Sharifi H (2010) Study of complex formation constants for some cations with O-Phenylenediamine in binary systems using square wave polarography technique. Engineering 2:1026–1030. https://doi.org/10.4236/eng.2010.212129

Ozudogru EA, Kaya E, Kirdok E, Issever-Ozturk S (2011) In vitro propagation from young and mature explants of thyme (*Thymus vulgaris* and *T. longicaulis*) resulting in genetically stable shoots. In Vitr Cell Dev Biol-Plant 47(2):309–320

Özhatay N, Koyuncu, M, Atay, S, Byfield, AJ (1997) The wild medicinal plant trade in Turkey. Dogal Hayati Koruma Dernegi, Istambul, Tirkey and Fauna and Flora International, Cambridge, UK. Unpublished report

Özcan M, Chalchat J-C (2004) Aroma profile of *Thymus vulgaris* L. growing wild in Turkey. Bulg J Plant Phisiol 30(3–4):68–73

Panizzi L, Flamini G, Cioni PL, Morelli I (1993) Composition and antimicrobial properties of essential oils of four Mediterranean Lamiaceae. J Ethnopharmacol 39:167–170

Pank F, Krüger H (2003) Sources of variability of thyme populations (*Thymus vulgaris* L.) and conclusions for breeding. Z Arzn Gew Pfl 8(3):117–124

Pank F (2010) Conventional breeding of medicinal and aromatic plants- fundamentals and examples. Acta Hort 860:135–146

Pawelec K, Kulpa D, Siwek H (2017) Callus Culture of common thyme (*Thymus vulgaris* L.). World Sci News 74:94–105

Peltoketo A, Dorman HJD, Yrjönen T, Sumanen J, Laakso I, Vuorela H, Hiltunen R (2000) Antioxidant properties of volatile oil and aqueous fractions of selected medicinal plants. Phytomed Suppl II 75

Pérez-Sánchez R, Ubera JL, Lafont F, Gálvez C (2008) Composition and variability of the essential oil in *Thymus* zygis from southern Spain. J Essent Oil Res 20:192–200

Perrot E, Paris R (1971) Les plantes medicinales. Presses Universitaires de France, Paris, p 233
Pluhár Z, Héthelyi É, Kutta G, Kamondy L (2007) Evaluation of environmental factors influencing essential oil quality of Thymus pannonicus All. (Hungarian thyme) and of Thymus praecox Opiz (Creeping thyme). J Herbs Spices Med Plants 13(1):23–43
Pluhár Z, Simkó H, Kovács K, Vida S, György Z, Sárosi S (2010) Essential oil composition of selected *Thymus vulgaris* L. clones belonging to five chemotypes. In: 41th International symposium on essential oils. 5–8 Sept 2010. Abstracts, Wroclaw, Poland, p 80
Pluhár Z, Kocsis M, Kuczmog A, Csete S, Simkó H, Sárosi S, Molnár P, Horváth G (2012) Essential oil composition and preliminary molecular study of four Hungarian *Thymus* species. Acta Biol Hung 63(1):81–96
Pluhár Z, Tátrai ZA, Ziaja Z, Radácsi P, Simkó H, Sárosi S (2013) Seasonal variation of the essential oil at different *Thymus vulgaris* L. chemotypes. In: 44th International symposium on essential oil, 8–12 Sept 2013. Abstracts, Budapest, Hungary, p 48
Pluhár Z, Simkó H, Sárosi S, Boros B, Dörnyei Á, Felinger A, Horváth Gy (2015) Determination of essential oil and polyphenolic compounds in *Thymus* species. In: II International symposium on horticulture in Europe 1099, vol 2, pp 661–670
Pluhár Z, Radácsi P, Szabó D, Malekzadekh M, Inotai K, György Z, Szabó K, Németh-Zámboriné É (2015) Effect of water supply on the accumulation levels of phenolic compounds, essential oil properties and antioxidant activity of *Thymus vulgaris*. Planta Med 81(16)—PW184
Pluhár Z, Szabó D, Sárosi S (2016) Effects of different factors influencing the essential oil properties of *Thymus* vulgaris L. Plant Sci Today 3:312–326. https://doi.org/10.14719/pst.2016.3.3.241
Pluhár Z, Szabó D, Seidler-Lozykowska K, Inotai K, Tavaszi-Sárosi S, Zámboriné Németh É (2017) Changes in essential oil properties of different *Thymus* vulgaris varieties influenced by developmental phases, growing locations and seasons. Nat Volatiles Essential Oils 4(3):61
Poletti A (1979) Plantas y Flores Medicinales. Instituto Parramon, Barcelona, pp 103–104
Rahimmalek M, Bahreininejad B, Khorrami M, Tabatabaei BES (2009) Genetic variability and geographic differentiation in *Thymus daenensis* subsp. *daenensis*, an endangered medicinal plant, as revealed by inter simple sequence repeat (ISSR) markers. Biochem Genet 47(11–12):831–842
Rey C (1990) *Thymus vulgaris* L. du Val d'Aoste (Italie): un écotype intéressant pour les zones marginales. Rev Suisse Viticult Arboric Horticult 22:313–324
Rey C (1992) Recherche et production des plantes medicinales et aromatiques en Suisse: Situation actuelle. In: Verlet V (ed) 3emes Rencontres Techniques et Economiques. Plantes Aromatiques et Medicinales. Nyons, 2-4 Decembre, 1991. CFPPA de Nyons, pp 116–121
Rey C, Palevitch D, Putievsky E (1993) Selection of thyme (*Thymus vulgaris*, L.). Acta Hort 344:404–410. https://doi.org/10.17660/ActaHortic.1993.344.46
Rey C, Sáez F (2002) Field culture, in vitro culture and selection of *Thymus*. In Stahl-Biskup E, Sáez F (ed) Thyme. The genus *Thymus*, 1st ed. Taylor & Francis, London, UK—New York, USA, pp 177–196
Rey C, Carron CA, Cottagnoud A, Schweizer N, Bruttin B, Carlen C (2004) Nouveaux hybrides de thym vulgaire. Rev Suisse Viticult Arboric Hortic 36(5):297–301
Ruberto G, Baratta MT (2000) Antioxidant activity of selected essential oil components in two lipid model systems. Food Chem 69:167–174
Rumińska A, Ożarowski A (1990) Leksykon roślin leczniczych. Państwowe Wydawnictwo Rolnicze i Leśne, Warszawa, p 565
Rustaiee AR, Yavari A, Nazeri V, Shokrpour M, Sefidkon F, Rasouli M (2013) Genetic diversity and chemical polymorphism of some *Thymus* species. Chem Biodiv 10:1088–1098
Sacchetti G, Maietti S, Muzzoli M, Scaglianti M, Manfredini S, Radice M, Bruni R (2005) Comparative evaluation of 11 essential oils of different origin as functional antioxidants, antiradicals and antimicrobials in foods. Food Chem 90:621–632
Sáez F, Sánchez P, Piqueras A (1994) Micropropagation of *Thymus* piperella. Plant Cell Tissue Organ Cult 39(3):269–272

Salgueiro LR, Procnca da Cunha P, Tomas X, Canigueral S, Adzet T, Vila R (1997) The essential oil of *Thymus villosus* L. ssp. *villosus* and its chemical polymorphism. Flav Fragr J 12:117–122

Sarfaraz D, Rahimmalek M, Saeidi G, Sabzalian MR (2020) Genetic relations among and within wild and cultivated *Thymus* species based on morphological and molecular markers. 3 Biotech, 10(7): 1–13

Sarfaraz D, Rahimmalek M, Saeidi G (2021) Polyphenolic and molecular variation in *Thymus* species using HPLC and SRAP analyses. Sci Rep 11(1):1–14

Sárosi S, Bernáth J (2008) The effect of weather conditions on the essential oil and total phenol content of different *Thymus vulgaris* L. cultivars. In: 39th International symposium on Essential Oils. Quedlinburg, Germany 7–10 Sept. Book of Abstracts, p 155

Schaunberg P, Paris F (1977) Guida de las plantas medicinales. Omega, Barcelona p 316–317

Shabnum S, Wagay MG (2011) Micropropagation of different species of *Thymus*. J Res Dev 11:71–80

Smolik M, Jadczak D, Korzeniewska S (2009) Assessment of morphological and genetic variability in some *Thymus* accessions using molecular markers. Notulae Botanicae Horti Agrobotanici Cluj-Napoca 37(2):234–240

Soorni A, Borna T, Alemardan A, Chakrabarti M, Hunt AG, Bombarely A (2019) Transcriptome landscape variation in the genus *Thymus*. Genes (Basel) 10(8):620

Sostaric I, Liber Z, Grdisa M, Marin PD, Stevanovic ZD, Satovic Z (2012) Genetic diversity and relationships among species of the genus *Thymus* L. (section *Serpyllum*). Flora-Morphol Distrib Funct Ecol Plants 207(9):654–661

Stahl-Biskup E (1991) The chemical composition of *Thymus* oils: a review of the Literature 1960–1989. J Essent Oil Res 3:61–82

Stahl-Biskup, E (2002) Essential oil chemistry of the genus *Thymus*–a global view. In: Stahl-Biskup E, Sáez F (Eds) Thyme. The genus *Thymus*, 1st ed. Taylor & Francis, London, UK-New York, USA, pp 75–143

Stoeva T, Dobos A, Máthé I, Bosseva Y (2001) Productivity and composition of essential oils of Bulgarian thyme varieties. Rasteniev'dni Nauki 38(7–10):359–361

Szczepanik M, Zawitowska B, Szumny A (2012) Insecticidal activities of *Thymus vulgaris* essential oil and its components (tyhmol and carvacrol) against larvae of lesser mealworm, *Alphitobius diaperinus* Panzer (Coleoptera: Tenebrionidae). Allelopathy J 30:129–142

Sunar S, Aksakal O, Yildrim N, Guleray A, Gulluce M, Sahin F (2009) Genetic diversity and relationship detected by FAME and RAPD analysis among *Thymus* species growing in eastern Anatolia region of Turkey. Romanian Biotech Letter 14:4313–4318

Tátrai ZA, Sanoubar R, Pluhár Z, Mancarella S, Orsini F (2016) Gianquinto G (2016) Morphological and Physiological Plant responses to drought stress in *Thymus citriodorus*. Intern J Agronomy 10:1–8. https://doi.org/10.1155/2016/4165750

Tevfik AS, Yegorova NA (2020) Clonal Micropropagation of *Thymus vulgaris* L. In Vitro. In Vitr Cell Dev Biol-Anim 56(S1):63–64

Thompson JD (2002) Population structure and spatial dynamics of genetic plymorphism in thyme. In: Thyme. The Genus *Thymus*. Stahl-Biskup E, Sáez F (Eds) Taylor & Francis, London, UK-New York, USA, pp 44–48

Thompson DJ, Chalcat J-C, Michet A, Linhart BJ, Ehlers B (2003) Qualitative and quantitative variation in monoterpene co-occurence and composition in the essential oil of *Thymus vulgaris* chemotypes. J Chem Ecol 29(4):859–880

Trindade H, Costa MM, Lima SB, Pedro LG, Figueiredo AC, Barroso JG (2008) Genetic diversity and chemical polymorphism of *Thymus* caespititius from Pico, Sa˜o Jorge and Terceira islands (Azores). Biochem Syst Ecol 36:790–797

Trindade H, Costa MM, Lima SB, Pedro LG, Figueiredo AC, Barroso JG (2009) A combined approach using RAPD, ISSR and volatile analysis for the characterization of *Thymus caespititius* from Flores, Corvo and Graciosa islands (Azores, Portugal). Biochem Syst Ecol 37(5):670–677

Tung YT, Chua MT, Wang SY, Chang ST (2008) Anti-inflammation activities of essential oil and its constituents from indigenous cinnamon (*Cinnamomum osmophloeum*) twigs. Biores Technol 99:3908–3913

Venskutonis PR (2002) Thyme—processing of raw plant material. In: Stahl-Biskup E, Sáez F (eds) Thyme. The Genus *Thymus*. Taylor & Francis, London, UK-New York, USA, pp 224–234

Vernet P, Gouyon PH, Valdeyron G (1986) Genetic control of the oil content in *Thymus vulgaris* L.: a case of polymorphism in a biosynthetic chain. Genetica 69:227–231

Vila R (2002) Flavonoids and further polyphenols in the genus *Thymus*. In: Stahl-Biskup E, Sáez F (eds) Thyme. The Genus *Thymus*. Taylor & Francis, London, UK-New York, USA, pp 144–176.

Vouillamoz J, Schaller M, Rossinelli M, Carron C-A, Carlen C (2011) 'Varico 3', nouvel hybrid de thym (*Thymus* vulgaris L.) pour la production en Suisse. Rev Suisse Viticult Arboric Hortic 43(6):370–376

Yamaura T, Tanaka S, Tabata M (1992) Localization of the Biosinthesis and Accumulation of Monoterpenoids in Glandular Trichomes of Thyme. Planta Med 58:153–158

Zambonelli A, Zechini D'Aulerio A, Bianchi A, Albasini A (1996) Effects of essential oil on phytopathogenic fungi in vitro. J Phytopathol 144:491–494

Zarzuelo A, Crespo E (2002) The medicinal and non-medicinal uses of thyme. In: Stahl-Biskup E, Sáez F (eds) Thyme. The Genus *Thymus*. Taylor & Francis, London, UK-New York, USA, pp 263–292

Zia A, Rezanejad F, Safarnejad A (2010) In vitro selection for NaCl tolerance in *Thymus* vulgaris L. J Cell Mol Res 2(2):86–92

Chapter 16
Plant Breeding Advances in Turmeric (*Curcuma longa* L.)

Madhavan Parthasarathy, Adithya Kodagu Anand, K. R. Rakesh Rao, Rashmi Ramakrishnan, Wudali Narasimha Sudheer, and Praveen Nagella

Abstract *Curcuma longa* L., known as the 'golden spice of India' works as medicine as well as a daily spice in Indian households to date. The current cultivation practice includes preparation of land, choosing unaffected rhizome, fungicide treatment, sowing, frequent maintenance, fertilization and lastly cultivation. Breeding curcuma species is quite challenging focusing on water scarcity and disease-resistant varieties. Either conventional techniques like clonal selection and polyploid breeding, or molecular techniques like bioinformatics and molecular marker-assisted breeding, or tissue culture through organogenesis, can be used to improve the situation. All the methods are done to yield a higher amount of efficient turmeric rhizomes. Currently, there are approximately 20 improved varieties of turmeric. Tissue culture methodologies are currently used for improving the varieties with consistent success and cultivation. It includes direct and indirect organogenesis, preparation of synthetic encapsulated seeds for disease prevention. Other methods that are being tested out are mutational breeding using physical and chemical mutagens, and hybridization of better varieties to yield efficient varieties. Even though genetic engineering efforts are made to improve crop sustainability in the upcoming environment, getting approval for cultivation takes time. Varieties not only should be accepted by people but also should be adaptable to the global climate change that is currently occurring. Thus, further research is going on for improving the breeding techniques of turmeric with less usage of chemical fertilizers, higher disease resistance and independence of stimulants for curcumin production. This review focuses on the advancement of current breeding practices of turmeric to date.

M. Parthasarathy · R. Ramakrishnan · W. N. Sudheer · P. Nagella (✉)
Department of Life Sciences, School of Sciences, Christ University, Bengaluru 560029, India
e-mail: praveen.n@christuniversity.in

R. Ramakrishnan
e-mail: rashmir@res.christuniversity.in

W. N. Sudheer
e-mail: wudali.sudheer@res.christuniversity.in

A. K. Anand · K. R. Rakesh Rao
Department of Studies in Biotechnology, School of Sciences, University of Mysore, Manasagangotri, Mysore, Karnataka, India

J. M. Al-Khayri et al. (eds.), *Biodiversity and Genetic Improvement of Herbs and Spices*, Advances in Plant Breeding Strategies 8, https://doi.org/10.1007/978-3-031-81838-7_16

Keywords *Curcuma longa* · Biotechnology · Mutation breeding · Bioinformatics · Cultivation

16.1 Introduction

16.1.1 Botanical Classification and Distribution

Curcuma longa L. is regarded as the 'golden spice of India', and is commonly called turmeric. It has been in use since ancient India as an ayurvedic medicine till now in households. This perennial plant features large, oblong leaves on short stems. It also produces ovate, pyriform, or oblong rhizomes that are frequently branched and have a brownish-yellow color (Rathaur et al. 2012). With 78% of the world's production, India is both the largest producer and consumer of turmeric. Under a broad genus *Curcuma*, turmeric which is scientifically known as *Curcuma longa*, is one of the species of *Curcuma* which includes *Curcuma amada* (Mango Ginger), *Curcuma angustifolia* (Wild Arrowroot), *Curcuma aromatica* (Wild Turmeric). The common household spice used for cooking is *Curcuma longa L* while *Curcuma aromatica* is used in cosmetics. The species *C. longa* was observed to have high fungicidal and antimicrobial properties compared to other species of Curcuma even though not much variation was observed between the species (Singh et al. 2002). Wild turmeric is common in the Himalayas and Western ghats flourishing in the early springtime while common turmeric is produced in majority of the southern Asian countries like China, and India in an extensive manner due to its high adaptability to the environment. Inside India, Telangana stands first in turmeric production with 3 lakh tonnes per year followed by Maharashtra with 2 lakh tonnes. Turmeric doesn't grow in winter conditions and remains in a dormant stage till spring (Mishra et al. 2018).

16.1.2 Economic and Health Benefits

Along with its usage as a food colouring and dying agent in the textile industry, turmeric is also used in medical industries with its vast known antimicrobial and healing properties. Even though having high medicinal properties, the poor solubility, and lower intestinal absorption with faster metabolic degradation have led to poor usage of turmeric extensively. Research is undergoing to improve these challenges such as the preparation of microemulsions of turmeric for better intestinal permeability. The rhizome is even used for extraction of essential oils which have both medicinal and cosmetic properties. The oil content can be improved by using the neem cake, Azotobacter, PBS etc. at a particular amount ultimately improving the plant growth. So, improving the organic nutrients supplementation for the cultivation improves the yield helping in farmer's growth (Bondre et al. 2019). The turmeric powder was observed to reduce the symptoms of arthritis upon regular intake for 4 months. The main constituent that is reported in the plant body is curcumin which is found abundant in turmeric. Turmeric was observed to have very low toxicity that even if consumed at 8 g/day, there were only minor consequences like gastrointestinal distress in humans. Care must be taken and further control should be done on the use of curcumin in the medicinal field as this can inhibit or enhance drug efficacy. Still, high-quality clinical trials involving large samples and longer duration are required to come to a conclusion on the effect of turmeric on human health (Singletary 2020).

16.2 Current Cultivation Practices and Challenges

16.2.1 Current Cultivation Practices

The cultivation practices are prepared such that even a farmer with no sophisticated equipment can follow the practices to have a higher and healthy yield of turmeric (Mirjanaik and Vishwanath 2020). The cultivation practices include procuring the correct variety suitable for that specific region. The turmeric has to be chosen according to the condition of the particular region. For example, use of IISR-Alleppey Supreme variety if the turmeric growing region gets rainfall such as in Kerala. Care has to be taken that turmeric cannot withstand the alkalinity and stagnancy of water. Modern practices can include checking the pH of the soil before cultivation which should be between 4 and 7 for optimum cultivation (Mirjanaik and Vishwanath 2020). Preparation of land by ploughing it nearly 4 times loosens the soil and helps in easier sowing of the rhizomes of turmeric. The beds have to be prepared with at least 1–2 m width and 15–20 m height in an equidistant manner of 50 cm apart from one another. These are known as partial shade-tolerant plants, so the rhizomes are planted under trees such as mango, guava etc. to utilize higher nutrition content present in the rich soil (Jayanthi and Vaideke 2015). The healthy rhizome is chosen such that it is not affected by any disease. These rhizomes are soaked in fungicides and pesticides (not exceeding 0.5%) such as Mancozeb for a minimum of 30 min before sowing it. Pre-sowing it in vermicompost helps in faster seedling formation in 30 days where it usually takes 2–3 months in the traditional method (Mirjanaik and Vishwanath 2020). The greater the diameter of the rhizome, the higher will be the number of seedlings arising. Seed rate is another factor to consider which means the number of seeds that have to be sown at a particular area, which in this case, can vary between 1500 and 2000 rhizomes per hectare area. The seeds can be stored in a well-ventilated room till sowing. The seeds planted at a wide spacing were observed to yield tall plants with higher yields ((Mirjanaik and Vishwanath 2020). The best season to sow is just before summer like April to May in regions like Karnataka and Tamil Nadu. Irrigation is provided upon observing the moisture content which can vary between 7–10 days for 20–25 cycles (Jayanthi and Vaideke 2015). The use of farm yard manure such as cow dung along with inorganic nitrogen and azospirillum of 10 kg/ha has proven to increase the yield of turmeric. The addition of other bacteria such as phosphobacteria can also be used to increase the yield depending on the cost (Mirjanaik and Vishwanath 2020). Harvesting time is when the leaves turn yellow and the stem starts to dry out. The rhizome is dug up into clumps, and the fingers are taken off of the mother rhizome and processed further (Jayanthi and Vaideke 2015).

16.2.2 Current Agricultural Challenges

In India, the green revolution began because of the country's growing population and its increased food needs. It has produced greater yields, decreased rates of poverty, improved infrastructure, more food availability, and cheaper food costs. With better farming techniques, the green revolution even increased the yield of spices (Balkrishna et al. 2021; Selvan et al. 2021). But there are still a few challenges that are faced by the farmers. A few of the main challenges faced by the cultivars are a lack of skilled labours, climate change, high labour costs, diseases to the spices and lack of financial support from the government (Hershon and Gnanamoni 2018). Due to the higher population, a large amount of land is exploited which includes woodland and forest land leading to water and wind erosion. The six main factors contributing to land degradation in the area are poverty, insecure land tenure, poor land use, deforestation, inadequate land management practices, and a lack of available land as a result of population growth. Soil quality declines with extensive farming. The application of chemicals to improve productivity decreases the soil's organic matter leading to lesser soil quality. The organic matter improves the microbial community which enriches the soil by its activity. The use of chemicals leads to acidification of soil leading to its deterioration and lesser biodiversity (Hershon and Gnanamoni 2018).

A primary issue that needs to be addressed is the lack of water. For example, 6000 m^3 of water is required to cultivate 1 hectare of turmeric. Due to changes in climate and low rainfall in a few areas, production has lessened. In a few instances, due to excess water usage leading to water logging, salination of water occurs lessening the soil quality (Selvan et al. 2021). The status of agriculture in society also plays a minor role and happens to be a challenge. Young generations prefer jobs over farming due to lesser income from this. There is a flow of people from villages to cities but not vice versa leading to an increase in urbanization leading to slums formation. The socioeconomic status of a farmer leads to compelled sales of turmeric at a lower price or sale to intermediates who gain profit from it but not the farmers (Balkrishna et al. 2021). Control of the disease is also an important challenge. In order to control the disease spreading, a few patches of grown crops have to be burnt which leads to yield loss. If the disease is ignored or observed lately, it will lead to crop loss by rhizome rot. The use of pesticides and fungicides can reduce this but also increase the cost of cultivation which ultimately falls on farmers. The application of biological agents, like Trichoderma, can lessen the rot disease; however, one of the main disadvantages is that farmers may not be well-versed in this field (Janani et al. 2021).

16.2.3 Genetic Improvement Objectives

Turmeric is a triploid species that are grown by vegetative propagation of the root rhizomes. Hybridization is ineffective due to this leading to limitation of genetic improvement to genotype selection, clonal selection and mutation breeding. Hybridization is possible in a few conditions leading to the production of recombinant varieties (Brijesh and Ajjappala 2023). Biotechnological approaches have helped in producing higher yield, highly tolerant plants which the wild type cannot survive normally. One such example can be given on producing salt-tolerant turmeric plants. Few varieties are observed to be sensitive for salinity that the cell function goes into arrest until salinity decreases. Few cultivars gain resistance by forming vacuoles of salts and disassociating them and releasing them into the environment while few breeding techniques undergo random mutation and selection of resistant ones (Bandi et al. 2022).

Several transgenic methods are tried even now to suppress this saline stress. Biotechnological techniques are necessary for gene transfer and gene improvement of desired traits because vegetative propagation is the method of reproduction. Prevention of diseases can be achieved by plant tissue culture technique known as micropropagation in which the rhizome is either converted to shoots and roots or transformed into a callus, an undifferentiated mass of cells from which many plants can be grown, using synthetic or semi-synthetic media before acclimatizing in the natural soil environment (Bandi et al. 2022). Bioreactor-based liquid culture was found to be a cheaper, highly effective technique but care must be taken while doing the technique for degradation of callus. In order to control somaclonal variation, molecular markers are used for checking the clonal fidelity and restricting the use of growth hormones can be done. Many improved varieties are produced mostly using *Agrobacterium tumefaciens*-mediated transformation (He and Gang 2014). Particle bombardment is another technique which can be followed for the establishment of stable genetic transformation. According to Shirgurkar et al. (2006), gold nanoparticles can be impounded with a plasmid that harbors the desired gene and directed towards the cell's nucleus.

16.3 Germplasm Biodiversity and Cultivation

16.3.1 Germplasm Diversity

A species' ability to adapt to changes in its environment and maintain long-term survival depends on the representation of genetic variation in the population. Evaluating plant populations' genetic diversity is crucial to developing sensible and long-lasting strategies for using genetic resources. Breeders of plants and biotechnologists use this genetic material as the foundation upon which to build new cultivars or varieties. The method of reproduction is an important factor in determining the genetic makeup of populations because it affects how genes are transmitted across time and space. Clonal plants produce a genetically identical individual called a genet, which is able to produce more independent ramets. Clonal plants are typically expected to show low genetic diversity and limited gene flow between populations as a result of this distinctive life history trait, though this is not the only outcome that can occur (Sasikumar 2005).

Today, India boasts about twenty improved varieties of turmeric thanks to techniques like mutation breeding, open-pollinated progeny selection (true turmeric seedlings), and germplasm/clonal selection. A limited number of studies have focused on the genetic fidelity of micropropagated plants and isozyme-based characterization; nevertheless, morphological characterization of curcuma species has been attempted, and molecular characterization is still in its infancy. Furthermore, Singh et al. (2016) studied the genus from the angles of anatomical characterization and biochemical profiling. Among the turmeric varieties, 'Prabha' and 'Prathibha' are notable as they are the first-ever varieties developed through open-pollinated progeny selection. Most other improved turmeric varieties in India have been derived through germplasm or clonal selection, except for Co-1, BSR-1, and BSR2, which are the result of mutant selection. Climate change and habitat degradation are threatening to wipe out the distinctive features of the local germplasm, even in spite of the Indian Supreme Court's efforts to outlaw deforestation and the stringent enforcement of laws by local authorities to protect the island's ecological integrity. The island's ecosystem, including its genetic resources, is predicted to be impacted by the effects of climate change, which include changes in rainfall patterns, increased atmospheric CO concentration, and an increase in average air temperature of 1.4–5.8 °C (Sasikumar 2005).

16.3.2 Cultivars Characterization and Phylogeny

The initial and crucial step in describing and classifying any crop species is morphological characterization. It is important in identifying and choosing the desired features in crop plants. Anu et al. (2020) conducted a study that aimed to describe the morphological and anatomical characteristics of curcuma species and turmeric varieties. But there has not been much research done on molecular characterization; the majority of studies have concentrated on isozyme polymorphism and species identification using trnK and 18S rRNA genes (Aswathi et al. 2023). Although the agro-morphology of a few improved varieties of turmeric has been thoroughly studied in diverse environments, because of genotype × environment interactions, many of these traits show instability and intra-varietal variation in different environments. As a result, they are insufficient markers that cannot accurately capture the genetic diversity. On the other hand, markers based on DNA sequences are comparatively stable and insensitive to changes in their surroundings. These markers are dependent on DNA sequence polymorphisms. Consequently, molecular characterization would improve our comprehension of the observed agro-morphological diversity in enhanced cultivars while also offering perceptions into the true genetic diversity between them (Anu et al. 2020).

Aswathi et al. (2023) state that characterization entails the thorough gathering of data on a variety of topics, such as distribution, habitat, flowering time, floral characteristics, quantitative features of floral parts, and qualitative and quantitative features of above- and below-ground characteristics. For the same species or parts of the plant, different essential oil profiles have been presented by multiple authors. It is interesting to note that different publications by the same authors have reported different values for essential oil profiles within the same species or plant parts. The disparity could potentially result from various factors such as the specimens' incorrect taxonomic identification, plant varieties, maturity variations, and analytical techniques used (Anu et al. 2020). An investigation into isozyme polymorphism in a *C. longa* germplasm collection was conducted by Shamina et al. (1998). Among the 15 accessions under study, a number of enzymes showed notable polymorphism, including acid phosphatase, superoxide dismutase, esterase, polyphenol oxidase, peroxidase, and catalase (Shamina et al. 1998). Although turmeric is primarily propagated vegetatively, Aswathi et al. (2023) noted that the observed variability in isozyme patterns indicates the influence of both natural selection and intentional human selection over time in the development of locally adapted cultivars.

16.3.3 Genetic Resource Conservation Approaches

Genetic conservation of turmeric is essential for preserving genetic diversity and ensuring the long-term survival and sustainable use of this valuable crop. Turmeric genetic conservation involves the systematic collection, preservation, and management of diverse turmeric germplasm. There are three types of genetic conservation. They are in-situ, ex-situ and on-farm conservation. "In-situ conservation" refers to the preservation of turmeric genetic resources in their natural settings, such as protected areas, botanical gardens, and national parks. It focuses on maintaining the species within its original ecological context. The in-situ method of cultivation is provided in Sect. 2.1. Turmeric can be cultivated in its natural habitat while preserving conventional agricultural methods thanks to in situ cultivation. It contributes to the maintenance of local biodiversity and cultural heritage associated with turmeric farming.

The term "ex-situ conservation" describes the management and preservation of turmeric's genetic resources away from their native environment. This approach involves maintaining living plants, seeds, or other plant parts in controlled environments such as gene banks, botanical gardens, or field collections. Tissue culture is the best ex-situ conservation method. It involves surface sterilization with standard disinfectants like $HgCl_2$ followed by the inoculation on plant growth regulators supplemented MS media for disease-free turmeric plants from healthy rhizomes and ultimately helps in the conservation of parent genes (Sinchana et al. 2020). On-farm conservation encourages farmers to continue growing and conserving traditional turmeric landraces in their fields. This strategy encourages the sustainable use of turmeric genetic resources while acknowledging the importance of farmers as guardians of genetic diversity. The use of different manures such as neem cake, NPK fertilizers, and Azotobacter for improving soil fertility and enhancing turmeric growth helps the farmer in maintaining the same genetic factors throughout the cultivars (Bondre et al. 2019). International cooperation and collaboration among researchers, gene banks, and organizations are important for sharing germplasm, knowledge, and best practices in turmeric conservation. Global collaboration in the conservation of genetic resources is facilitated by platforms like the International Treaty on Plant Genetic Resources for Food and Agriculture.

16.3.4 Cytogenetics

Cytogenetics is the study of chromosomes, their structure, function, and behaviour within cells. In the case of turmeric, cytogenetic studies have provided insights into its chromosomal characteristics and genetic makeup. These studies aid in breeding programs, genetic improvement, and the conservation of turmeric genetic resources. *C. longa* is a polyploid with $x = 7$ and a nonaploid. Prior research reveals a significant variance in the number of somatic chromosomes in *C. longa* over a period of more than 70 years. The chromosome number varies from 46 to 96 which shows the large diversity in the intra-species level. The small size and large number of chromosomes in Zingiberaceae plants make karyomorphological studies particularly challenging. So, full-fledged field research all over India will give a statistical number on the number of chromosomes of the turmeric and may even inform the reason behind this variation inside the species itself (Bhadra and Bandyopadhyay 2015).

16.4 Traditional Breeding

Generally speaking, the only techniques utilized to enhance turmeric crops were clonal selection, induced mutation, and subsequent selection. Increasing production, achieving a high curing percentage, and increasing curcumin content were the main objectives of crop improvement. Pujari et al. created the high-yielding variety Krishna (7.2 t/ha) for widespread cultivation in Maharashtra (Pujari et al. 1986).

16.4.1 Improvement Strategies

The first step in traditional breeding is to identify superior genotypes of turmeric that have desirable qualities like high yield, resistance to disease, and quality traits. Based on visual inspection and evaluation of plants in the field, this selection takes into account various attributes such as leaf characteristics, disease resistance, rhizome size and shape, and plant vigor. Mass selection is the process of choosing individuals or groups of plants from a population that have superior features and employing their offspring as the following generation. This approach enables the gradual accumulation of desired features in the population. To ensure that the desired features are stable and maintained, strict selection based on numerous generations must be used (Ayer 2017). Turmeric is typically propagated vegetatively using rhizomes. Clonal propagation ensures that the desirable traits identified in selected individuals are preserved and passed on to the next generation. By carefully selecting and propagating superior rhizomes, breeders can maintain genetic integrity and improve the overall quality of the crop (Ayer 2017). Farmers must be involved in the breeding process when cultivating traditional turmeric in order to guarantee the creation of types that satisfy their particular requirements and preferences. The performance of chosen genotypes and varieties is evaluated thoroughly throughout the breeding process through field trials. Monitoring features including production, disease resistance, rhizome quality, and adaptability to various agro-climatic situations are required for this. For choosing superior individuals and identifying potential kinds for release and further multiplication, field trials offer useful information (Bandi et al. 2022).

16.4.2 Traditional Breeding Methodologies

16.4.2.1 Clonal Selection

In the breeding of turmeric, clonal selection is a crucial method that entails identifying and cultivating exceptional individual plants with desired characteristics (Ayer 2017). It is a crucial method for improving varietal quality. Most of the enhanced varieties that have been made public thus far are clonal or germplasm selections based on comparative yield evaluation trials that were conducted specifically for each site to find turmeric types with the highest potential for yield curcumin content and curing percentage (Devi et al. 2019). Due to the conventional clonal selection methods, lengthy development time and modest progress, it is less significant for improving the quality of turmeric when compared to other molecular methods (Ayer 2017).

16.4.2.2 Artificial Selection

Another crucial method in the breeding of turmeric plants is artificial selection, commonly referred to as selective breeding. It entails purposeful breeding manipulation and direction to promote the inheritance of desired traits (Anandaraj et al. 2014). The precise qualities that breeders want to improve or add in turmeric plants are chosen. These qualities could include increased yield, superior flavour, increased curcumin content, improved disease resistance, or particular morphological traits. To combine the required features, the chosen parent plants are cross-pollinated in a regulated manner. Pollen is moved from one parent plant's anthers to another's stigma during the process. The turmeric varieties produced by the cross-pollination are analyzed to determine how well they perform and whether they display the desired features. Artificial selection is often an iterative process that involves numerous breeding cycles. To improve and strengthen the desired features, the selected superior plants are subsequently crossed with other superior individuals or with their parents during each cycle (Bahadur et al. 2016). The performance of the selected turmeric lines or varieties produced through artificial selection is evaluated in a variety of environmental settings through field tests. Once successful, these enhanced lines or varieties are made available to farmers for commercial production. By concentrating on the inheritance of desired features, artificial selection in turmeric breeding enables breeders to hasten the generation of turmeric varieties with specific properties. It makes it possible to specifically enhance turmeric plants to accommodate different agronomic, financial, and consumer demands (Ayer 2017).

16.4.2.3 Recurrent Selection

Recurrent selection entails a cyclical process of selecting, crossing, and reselecting plants from the population (Gupta et al. 2015). The initial population is assessed for desirable features like yield, disease resistance, or quality attributes. The first parental plants for the following breeding cycle are chosen from plants that exhibit the desired qualities. Cross-pollination between the chosen parent plants results in plants that possess a medley of their properties. To guarantee certain crossings and avoid unintended pollination, controlled pollination techniques like bagging or hand pollen transfer may be used. The offspring from the carefully regulated cross-pollination are raised and tested for the desired qualities. Depending on the evaluation, field tests, lab tests, or sensory evaluations to assess the effectiveness and caliber of the offspring. Superior descendants with the desired traits are picked out of the progeny based on the evaluation. These plants are chosen to be the parents of the next generation because they display traits that are better than those of the parental population. The future generation is subsequently raised by the chosen superior plants. Over several breeding cycles, the process of controlled cross-pollination, progeny assessment, and selection is repeated. Each cycle seeks to enhance the desired features and choose the best candidates. The enhanced features get more reliable and consistent within the population as recurrent selection goes on (Ayer 2017). Over numerous mating cycles, recurrent selection enables the ongoing enhancement and improvement of particular features in turmeric populations. Breeders can gradually improve the genetic makeup and general performance of turmeric cultivars by periodically choosing and crossing individuals with desired features. This method is especially helpful for enhancing complicated features that are regulated by numerous genes or for fine-tuning the expression of particular traits to accommodate market or customer preferences (Verma et al. 2015).

16.4.2.4 Polyploidy Breeding

Polyploidy is the presence of many sets of chromosomes in a cell or organism. The possible benefits of polyploidy in plant breeding include greater vigor, larger-sized plants, altered morphology, and altered gene expression (Ayer 2017). Polyploidy can occur naturally or be purposefully created. Although polyploidy breeding has not been thoroughly investigated in turmeric, it has been carried out in certain other plant species.

Polyploidy induction: Turmeric can be made polyploid by using chemicals (such colchicine or oryzalin) or by using tissue culture procedures. These techniques interfere with the natural process of cell division, which leads to chromosome duplication and the development of polyploid plants (Verma et al. 2015). The performance and desirable features of the polyploid progeny obtained from the chosen individuals are evaluated. Comparative analyses between polyploid and diploid individuals can be performed to see if the polyploids have any favourable traits. For subsequent breeding efforts, polyploid individuals with desired features might be employed as parent plants. It is possible to use self-pollination or controlled cross-pollination to combine and stabilize the desired features in succeeding generations. This entails choosing and breeding offspring that regularly exhibit the desired qualities. The agronomic performance of the chosen polyploid lines or varieties, including yield, quality, and adaptation to various settings, is assessed in the field. If successful, the enhanced polyploid lines can be made commercially available for farmers to cultivate (Bandi et al. 2022). It is significant to mention that turmeric polyploidy breeding may present various difficulties and restrictions. Turmeric can be difficult to induce into polyploidy, polyploids may have lower fertility or sterility, and polyploids must be carefully chosen and assessed to find those with favorable features. Furthermore, the inheritance patterns and trait stability of polyploids might be complicated, necessitating more study and comprehension. More investigation is required to determine the potential advantages and drawbacks of polyploidy breeding, particularly in the case of turmeric, and to create improved polyploid varieties with superior features for commercial use.

16.4.2.5 Synthetic Breeding or Polycross Breeding

Turmeric is frequently bred via polycross breeding to boost genetic diversity and the likelihood of beneficial trait combinations. Multiple parent plants are simultaneously cross-pollinated, allowing for the exchange and recombination of genetic material among a greater population of people (Ayer 2017). A wide range of parent plants with desirable features are chosen as the genetic sources. These parent plants might have diverse traits including a high yield, resistance to disease, or distinctive flavor profiles. The parent plants are carefully chosen, and they are grown near together so that they can naturally cross-pollinate. Pollen is transferred across the several parent plants with the help of pollinators like wind or insects. The flowers are allowed to mature into fruits or capsules that carry the seeds after cross-pollination. These seeds come from the several parent plants that have been used in polycross breeding. These seeds contain the genetic diversity derived from the several parent plants. Elite clones are chosen from the identified superior individuals for subsequent breeding. To maintain and uniformize the desired features, these elite clones are often multiplied vegetatively using techniques like rhizome division or tissue culture (Anindita et al. 2020).

16.4.3 Traditional Breeding Limitations

Traditional breeding techniques have been used successfully to improve turmeric, but they do have some drawbacks. The process of creating superior turmeric variants is slow because each breeding cycle can take several years. This prolonged time scale might slow down the introduction of new cultivars with particular features. In addition, complex qualities controlled by several genes can be difficult to treat when there is limited genetic variation (Ahmar et al. 2020). Field evaluations are a key component of traditional breeding techniques that are used to evaluate turmeric plant performance. However, because environmental factors can affect how characteristics are expressed, it can be difficult to determine a plant's true potential in various situations. Variations in the environment might hide or obfuscate the expression of desired features, resulting in inconsistent selection results. The main method of reproduction for turmeric is cross-pollination, which necessitates particular circumstances for successful hybridization. The reliance on sexual reproduction can restrict the ability to pass along desired qualities because the progeny may develop a variety of phenotypes as a result of genetic recombination. A significant amount of time and work may be required for controlled pollination and feature selection (Kaiser et al. 2020). In traditional breeding, the selection of superior plants is based on phenotypic evaluations, which rate observable qualities. However, it is difficult to precisely predict the inheritance of particular features because phenotypes can be modified by a variety of genetic and environmental factors. Because of this lack of accuracy and predictability, getting intended results may be more unpredictable (Kaiser et al. 2020).

16.4.4 Role of Biotechnology

Biotechnology is essential to the breeding of turmeric because it provides instruments and techniques that speed up trait improvement, boost genetic diversity, and allow the development of improved varieties. With the use of genetic transformation techniques, biotechnology enables the introduction of particular genes into plants that produce turmeric. This permits turmeric to acquire advantageous features, such as disease resistance, herbicide tolerance, and higher curcumin content. The introduction of novel features that can be challenging to acquire through conventional breeding techniques can be sped up by genetic transformation (Singh et al. 2020). Molecular markers are used in marker-assisted selection (MAS) to identify and select plants that possess specific genes or DNA regions associated with desired traits. With the aid of MAS, breeders can more effectively screen vast populations of turmeric plants, resulting in the early selection of individuals with the required features. This technique allows for the specific breeding of turmeric varieties with improved characteristics, such as disease resistance or higher yield potential (Ashraf et al. 2012). Genomic selection makes use of cutting-edge genomic and statistical tools to forecast turmeric plant performance based on DNA profiles. Genomic selection allows for the early selection of superior individuals by estimating the breeding value of plants by assessing a large number of genetic markers dispersed across the genome. By choosing plants with advantageous genetic backgrounds for several qualities, this method improves the efficacy and accuracy of turmeric breeding (Newell and Jannink 2014). Using small plant tissue samples, tissue culture techniques enable sterile fast reproduction of turmeric plants. It makes it easier to grow a lot of genetically similar plants, enabling the clonal multiplication of elite clones or better people. Additionally, tissue culture provides the chance to save and grow priceless plants that could be challenging to grow using conventional techniques, which improves the effectiveness and consistency of breeding programs for turmeric (Hussain et al. 2012). In order to identify and diagnose diseases and pests, techniques like polymerase chain reaction (PCR) and DNA sequencing can be used. This enables prompt management methods and the selection of disease-resistant or tolerant plants in breeding programs. The genetic diversity of turmeric germplasm collections can be maintained and preserved using biotechnology approaches like cryopreservation and in vitro conservation. The process of cryopreservation enables the long-term storage of plant material in liquid nitrogen at extremely low temperatures, preserving priceless turmeric genetic resources for upcoming breeding initiatives (Taşkın et al. 2013).

16.5 Molecular Breeding

Molecular breeding, commonly referred to as molecular marker-assisted breeding involves identification of beneficial features in plants and choosing them at the molecular level using genetic markers. These markers are linked to particular genes or genomic areas that regulate crucial qualities like disease resistance, potential yield, or quality attributes. Instead of relying simply on time-consuming and expensive phenotypic evaluations, breeders can effectively screen huge populations of plants and choose those with the necessary features by employing molecular markers (Ayer 2017). Inter-simple sequence repeat (ISSR) and random amplified polymorphic DNA (RAPD) are the most often used DNA-based molecular marker techniques to assess the genetic diversity in turmeric genotypes. According to the polymorphism utilizing both RAPD and ISSR markers, there is a 62% association between genetic similarity and place of origin (Singh et al. 2012). The selection of turmeric plants with high curcumin levels in the lab is supported by the discovery of a positive association between curcumin concentrations in seedlings and mature rhizomes (Nadgauda and Mascarenhas 1986). It is crucial to take into account the associations of features for which selection is to be made when choosing parental lines, as this shows that the growth and yield of turmeric have an impact on phytoconstituents. The selected potential genotypes may be used as parental materials to generate cultivars and varieties with improved quality attributes (Singh et al. 2012). Similarly, somaclonal variations can be detected in vitro using biochemical markers like protein, isozymes, essential oil concentration, etc. Protein over- and under-expression can be utilized to separate variations for selection (Pikulthong et al. 2016). It will be easier to gather the variation among genotypes of turmeric for elite genotype advancement in breeding programs with the use of molecular markers (Sasaki et al. 2002).

16.5.1 Marker Assisted Selection

This is a unique and indirect selection strategy in which the marker is selected not by phenotypic significance but rather by the genome-level trait of interest. These molecular markers are invariably connected to linkage maps and genomics. Finding the ideal marker for the desired trait or any marker linked to a QTL would increase the chances of developing cultivars with the required traits. This method has been integrated with next-generation sequencing methods and molecular biology methods like SNPs, AFLPs, In-DELs, and SSRs. Marker-assisted breeding used to be labor- and time-intensive (Singh et al. 2018). MAB encourages the selection of suitable parents with desirable features and permits early selection of offspring to encourage tolerance for various pressures. Through accurate QTL mapping, this technique can also enhance the genetic properties of turmeric cultivars, enabling the development of salinity tolerance (Bhandari et al. 2019).

16.5.2 Functional Genomics

By revealing information about the activities and interactions of the genes in the turmeric genome, functional genomics plays a crucial role in turmeric breeding. To comprehend how genes contribute to particular traits or biological processes, it entails the study of gene expression, protein function, and regulatory mechanisms (Maghuly et al. 2021). Using functional genomics methods, such as RNA sequencing (RNA-seq), it is possible to analyze turmeric's gene expression patterns in great detail. Breeders can find genes that are differentially expressed and linked to crucial features, including disease resistance or curcumin production, by comparing the transcriptomes of several turmeric varieties or under different environments. The activities of particular genes in turmeric can be clarified utilizing functional genomics techniques like gene knockout or gene silencing using tools like CRISPR/Cas9 or RNA interference (RNAi) (Liu et al. 2021). Studies on functional genomics can reveal the networks and regulatory components that govern the expression of genes in turmeric. Finding the transcription factors, enhancers, promoters, and other regulatory regions involved in the control of significant genes is a part of this. Knowing the regulatory processes makes it easier to control gene expression in turmeric breeding to modulate desired features. Integrating functional genomics data with other omics data, such as proteomics, metabolomics, transcriptomics and genomics, can help develop a systems-level understanding of the biology of turmeric (Pazhamala et al. 2021).

16.5.3 Bioinformatics

Bioinformatics plays a crucial role in the turmeric breeding process by providing techniques and resources to analyze and interpret genetic information. Bioinformatics can be used to locate and annotate genetic markers found in the turmeric genome, such as simple sequence repeats (SSRs) and single nucleotide polymorphisms (SNPs). The study of genetic diversity and molecular breeding depends heavily on these markers. Utilizing bioinformatics methods makes it easier to create molecular markers linked to crucial turmeric characteristics. These markers can be utilized in MAS, which quickens the breeding process by having breeders choose plants based on particular genetic markers connected to desired qualities (Mu et al. 2022). Breeders can easily introduce features from other species to turmeric by identifying conserved areas, genes, and processes associated with key traits through genome comparison. To understand the patterns of gene expression in various tissues or under various circumstances, bioinformatics techniques are used to evaluate transcriptomic data, such as that obtained through RNA sequencing (RNA-seq). This knowledge makes it easier to discover genes linked to crucial agronomic features as well as those involved in particular biological processes (Schmutzer et al. 2017). Based on similarities to known genes in other animals, bioinformatics tools like sequence homology and protein structure prediction can help deduce the function of turmeric genes (Tan et al. 2022). Turmeric breeders can more efficiently create superior varieties with desired qualities by utilizing the bioinformatics tools and resources to acquire insights into the genetics and biology of turmeric, speed up the breeding process, and analyze the results (Vassilev et al. 2005) (Table 16.1).

Table 16.1 Bioinformatics tools and websites that can be used in Turmeric plant research (Mu et al. 2022)

Database name	URL	Note
SMART	http://smart.embl-heidelberg.de/	A comprehensive toolkit for interactive big biological data analysis
FED	http://www.hi-tom.net/FED	Exogenous component detection platform for genome editing
MAFFT	https://mafft.cbrc.jp/alignment/server/	Online tool for matching sequences
Protter	http://wlab.ethz.ch/protter/start/	An online tool for mapping protein structures
STAG-CNS	https://github.com/srbehera11/stag-cns	A tool for finding sequentially conserved non-coding sequences across any number of species
BAR	http://www.bar.utoronto.ca/welcome.htm	Platform for plant biology analysis tools
CRISPR-P	http://crispr.hzau.edu.cn/CRISPR2/	A better toolkit for CRISPR/Cas9 genome editing in plants
Plant MetGenMAP	http://bioinfo.bti.cornell.edu/cgi-bin/MetGenMAP/home.cgi	Web-based resources for thorough analysis and incorporation of metabolite and gene expression variations within the framework of metabolic pathways
iTAK	http://itak.feilab.net/cgi-bin/itak/index.cgi	Programs for recognizing and categorizing protein kinases and plant transcription factors
PlantPAN	http://plantpan2.itps.ncku.edu.tw/	Techniques to identify plant transcription factor binding sites
SnpHub	http://guoweilong.github.io/SnpHub/	A centralized web server architecture for analyzing vast amounts of genomic variation data
EvolView	https://www.evolgenius.info/evolview/	Web-based resources for system tree management, annotation, and visualization

16.6 Tissue Culture Applications

16.6.1 Regeneration Studies

According to Loyola-Vargas and Ochoa-Alejo (2018), plant tissue techniques can undoubtedly be effective supplemental tools for plant breeding and genetic improvement programs. A significant chance exists for regaining somaclonal variants or mutants with particular agronomic or industrial traits that can be expressed at the cell or plant level. The genetic variability found in callus tissue and cell cultures can be caused by genetic or epigenetic changes (Larkin and Scowcroft 1981). The average production and quality of turmeric are poor, despite the fact that India is one of the top producers in the world and that there are very few high yielding varieties of this crop. The main problems are the low concentration of curcumin and essential oils in the cultivars that are currently available, the slow rate of multiplication, the absence of the required high yielding genotype, and loss from disease during cultivation and storage (Ayer 2017). Therefore, it is crucial to create turmeric genotypes with improved drug yielding potential, including higher curcumin, higher oleoresin, and higher rhizome yield, as well as improved quality and quantity of essential oil. This will help to establish turmeric cultivation as an industry. Plant biotechnology has gained importance in the modern era because clonal selection has limited ability to improve turmeric. A number of techniques have already been used to improve turmeric as a crop, such as in vitro pollination for the purpose of breeding disease-resistant plants, seedling selection for the purpose of advancing clonal selections, micropropagation for the generation of disease-resistant plant material, and the production of microrhizomes for the purpose of germplasm exchange and conservation. By cultivating tissues in vitro in aseptic conditions, the technique mimics the ambient conditions found in vivo and encourages growth and development without endangering the structural and functional integrity of the plant. The commercial and exchange value of the intended plant type makes the use of tissue culture techniques for plant regeneration in the fields of horticulture and agriculture concerning. In vitro, axillary or adventitious buds give rise to new plants during the process of organogenesis that grow into shoots and eventually become adventitious roots. The process by which plant cells proliferate from a callus mass is known as indirect organogenesis (Haque et al. 2022).

16.6.1.1 Direct Organogenesis

Direct organogenesis is the process of growing buds or shoots straight out of tissue, without the need for a callus stage. Salvi et al. (2000) produced a plant shoot from the immature *C. longa* inflorescence. From young inflorescences immature floral buds were isolated and cultured on MS basal medium. The obtained shoots could be successfully planted, moved to the field, and raised till maturity (Salvi et al. 2000). When cultured on MS medium supplemented with different combinations of auxins and cytokinins, turmeric rhizome buds demonstrated effective regeneration. It was recorded that 1 mg/L of NAA in combination with 2 mg/L BAP and 0.1 mg/L NAA in combination with 1 mg/L KN showed the highest number of shoots per explant (Sunitibala et al. 2001). Thidiazuron showed increased efficiency for micropropagation of turmeric. When the terminal buds of turmeric are divided longitudinally and used as explants, they are cultured on a growth medium containing 18.17 μM and precultured on 72.64 μM of thidiazuron, resulting in the highest frequency of shoot multiplication (Prathanturarug et al. 2003, 2005). Axillary buds from the rhizome of turmeric were cultured on MS medium supplemented with 9% of sucrose, 12 μM of BA and 0.3 μM of NAA and resulted in increase efficiency for the induction of micro rhizomes and showed good proliferation efficiency when transferred to soil (Islam et al. 2004). Multiple shoots were obtained from sterilized sprouted buds of turmeric rhizome when cultured on MS medium supplemented with 2 mg/L BAP. Further best rooting efficiency of shoots were recorded on half strength MS medium supplemented with IBA (Rahman et al. 2004). Rhizome buds of turmeric when cultured on MS medium supplemented with 4 mg/L BAP along with 1 mg/L NAA showed increased proliferation of shoots and rooting. Rooted plantlets were proliferated best in field conditions (Nasirujjaman et al. 2005). Linsmaier and Skoog's medium supplemented with 3 mg/L BAP showed best ability for initiation and multiplication of shoots when vegetative bud was used as an explant (Roopadarshini 2010). The effect of light intensity and plant growth regulator in the growth medium for micro rhizome production in turmeric was studied. It was observed that low light intensity was favourable and kinetin at increased concentrations played an important role in micro rhizome formation (Hashemy et al. 2009). Sprouted shoot tip explants of turmeric Salem variety were cultured on MS medium supplemented with 2 mg/ L BAP along with 0.2 mg/L IAA showed good response for shoot induction and proliferation, whereas highest rooting was observed in 2 mg/L IBA supplemented half-strength MS medium (Sinchana et al. 2020). Similar to this, another study on the Salem variety revealed that rooting was achieved on medium supplemented with 4.9 μM IBA, but shoot slice explant produced the greatest number of shoots when cultured on MS medium supplemented with 13.32 μM BAP along with 5.37 μM NAA (Pittampalli et al. 2022).

16.6.1.2 Indirect Organogenesis

Plant cells develop from a callus mass through a process known as indirect organogenesis (Haque et al. 2022). An indirect organogenesis method for *C. longa* in vitro propagation was developed by Zapata et al. (2003). The vegetative shoots of the rhizomes grown aseptically on MS culture medium supplemented with 1.5 mg/L 2,4-D and 0.2 mg/L BAP to induce dedifferentiation and the production of calluses. Callus culture was established from shoots tip explants of turmeric when cultured on MS medium supplemented with 2,4-D at 2.5–3 mg/L concentration. Subsequent culturing of callus on MS medium supplemented with higher concentrations of sucrose and 0.1 mg/L of NAA and 1 mg/L of KN showed induction of rhizomes

(Sunitibala et al. 2001). Turmeric's capacity for callus induction and plant regeneration, in conjunction with auxins and cytokinins, was examined. It was observed that 2 mg/L NAA along with polyamines i.e., putrescine, spermine, and spermidine in 1:1:1 ratio at 5 or 10 mmol/L concentration showed successful establishment of callus, whereas BAP supplemented medium with polyamines at 10 mmol/L concentration showed best efficiency for regeneration through indirect organogenesis (Viu et al. 2009). White compact callus was induced from vegetative bud explant when cultured on Linsmaier and Skoog's medium supplemented with 3 mg/L 2,4-D, and subsequent culturing of callus on 3.5 mg/k BAP fortified medium showed best ability for the proliferation of shoots (Roopadarshini 2010).

16.6.2 Synthetic Seeds

According to Magray et al. (2017), synthetic seeds are defined as artificially encapsulated somatic embryos, shoot buds, cell aggregates, or any other tissue that can be sown as a seed and that has the ability to germinate in vitro or ex vitro and that keeps this ability even after storage. Synthetic seeds offer many benefits over conventional in vitro culture techniques, including being easier to handle, plant, and transport, as well as being reasonably priced to produce. According to Shajahan et al. (2019), alginate encapsulation-based synthetic seed production technology is regarded as an effective option for elite genotype propagation and germplasm storage.

Shoot buds of turmeric plants were encapsulated with hydrogels like sodium alginate and carboxymethyl cellulose made out of Linsmaier and Skoog medium supplemented with 4 mg/L BAP. Synthetic seeds made of 4% sodium alginate were found to exhibit the best matrix and complexing agent when dipped in 50 mM calcium chloride and incubated for 30 min. The seeds were then stored at 4 °C to preserve viability (Gayatri et al. 2005). Similarly, somatic embryos of turmeric plants were encapsulated using 4% sodium alginate and 2 M glycerol and 0.4 M sucrose. Synthetic seeds were formed using 0.1 M calcium chloride. Upon incubation for 20 min followed by storage for 6–8 months the viability of seeds were checked and showed regeneration efficiency when cultured on MS medium supplemented with 1 mg/L BAP along with 0.5 mg/l NAA (Nirmal Babu et al. 2016).

16.7 Genetic Engineering and Gene Editing

The essential spice crop plant turmeric, which has the sterile triploid genetic makeup of $2n = 3x = 63$, rarely flowers. Practically no seeds have been set. The plant spreads vegetatively through the rhizomes that are underground (Shirgurkar et al. 2006). Because this vegetative growth is vulnerable to pathogen accumulation and transmission, amplification of especially beneficial stocks is a time-consuming process (He and Gang 2014). Enhancing this crop through traditional breeding is extremely difficult. Nevertheless, these selection techniques are limited to mutants produced during vegetative propagation. Attractive genotypes with higher yield, curcumin content, disease and insect/pest resistance, adaptability, etc., can be selected for the purpose of creating new cultivars. Today, genetic engineering is being used to produce crops with desired qualities, and it is especially interesting to employ in species with low genetic variety, like turmeric. Plants can develop more quickly when biotechnological techniques are applied, such as gene transfer for pest and disease resistance. However, a method for in vitro regeneration must be available in order for genetic transformation to be successful (Shirgurkar et al. 2006).

Novel techniques combining tissue culture and genetic engineering techniques are needed to support breeding programs and effective transformation approaches for determining gene function and improving physiological features for this species. According to He and Gang (2014), the only tissue culture techniques applied to this species thus far have been in vitro micropropagation and plant regeneration from callus culture.

16.7.1 Methodologies

These days, transgenic techniques are a widely used and useful tool for enhancing plant phenotypes and carrying out functional analyses of genes implicated in different physiological, biochemical, and molecular mechanisms of metabolic pathways. Particle bombardment and Agrobacterium-mediated transformation are the two most widely used techniques for producing transgenic plants. Agrobacterium is considered to be better than biolistics because it can introduce larger regions of DNA with little to no DNA rearrangement and with fewer copies of inserted transgenes at greater efficiencies and at a lower cost. As of right now, no trustworthy technique exists that describes an Agrobacterium-mediated strategy for this species, despite the description of a particle bombardment-based strategy (Shirgurkar et al. 2006). He and Gang (2014) also talked about the development of two transient expression systems in turmeric, one based on leaves and the other on calluses, both of which were mediated by *Agrobacterium tumefaciens*. Furthermore, He and Gang (2014) established an efficient method for somatic embryogenesis and regeneration.

16.7.1.1 Particle Bombardment Method

Shirgurkar et al. (2006) reported the successful regrowth of stabilized transformed plants by particle bombardment of the callus and selection for herbicide resistance to glufosinate. Particle bombardment was found to be an effective technique for the stable transformation of the turmeric plant. Cultures developed from shoots were exposed to gold particles coated with plasmid pAHC25, which contained the bar and gusA genes, both driven by the maize ubiquitin promoter. Glufosinate-containing media were used to choose transformants. Transgenic lines were generated from 50%

of the bombardment calluses on the selection medium. Transgenic shoots multiplied, recovered, and stabilized transformed plantlets from these. Using the polymerase chain reaction (PCR) and histochemical GUS assay, the steady transformation was verified. Transformed plantlets showed resistance to glufosinate. Of the turmeric calluses that are bombarded on selection medium, 50% develop transgenic calluses. Different selection mediums frequently had five shoots per callus. When the shoots are placed in a medium for multiplication, they multiply by a factor of 4–8 each month. These results show how well callus culture works as a biolistic transformation mechanism for turmeric (Shirgurkar et al. 2006).

16.7.1.2 Agrobacterium Mediated Gene Transformation

According to Pikulthong et al. (2016), *Curcuma longa* L., *Curcuma mangga* Valeton, and Zijp somatic embryos were used to create a whole in vitro regeneration system, which was subsequently used to modify the genes in these plants. Utilizing Murashige and Skoog (MS) medium supplemented with 2.0 mg/L N6-benzyladenine (BA), they developed a somatic embryo from the buds of *Curcuma longa* L. and *Curcuma mangga* Valeton & Zijp varieties in order to produce multiple shoots. Following that, the shoots were cultivated on MS media supplemented with different concentrations of two plant growth regulators—naphthaleneacetic acid (NAA) and 2,4-dichlorophenoxyacetic acid (2,4-D)—or both. The medium containing both auxins—5 mg/L 2,4-D and 5 mg/L NAA—was found to be the most successful in inducing somatic embryos after 4 weeks of culture. The somatic embryos were then placed in media with certain growth regulators to encourage the production of shootlets and rooting. Two genes involved in the synthesis of curcuminoid compounds in turmeric, diketide CoA synthase (DCS) and curcumin synthase (CURS), were also cloned and inserted into the two species through Agrobacterium-mediated transformation. The presence of the target and marker genes in the transformed somatic embryos was verified using the polymerase chain reaction assay. The modified somatic embryos could live for four weeks after being cultured. According to Pikulthong et al. (2016), this is the first report of a complete in vitro regeneration system that can be used for gene editing in *C. longa* and *C. mangga* somatic embryos. Genetic engineering can be used to create plants with desired traits; this technique is also helpful for enhancing species with low genetic diversity (Shirgurkar et al. 2006). Currently, Shirgurkar et al. (2006) and He and Gang (2014) have published the fewest papers on gene transfer in turmeric. Thus, a successful plant transformation protocol is needed for the genetic modification of turmeric.

16.8 Mutation Breeding

According to Oladosu et al. (2016), mutagenesis is the process by which abrupt, heritable changes in an organism's genetic composition occur. These changes are not caused by genetic segregation or genetic recombination, but rather by external chemical, physical, or biological factors that can include a range of chemicals, radiation, enzymes, etc. Mutational breeding produces mutants with desired traits on the plant materials used in cultivation, which can then be used to crossbreed with other cultivars (Bandi et al. 2022). Three forms of mutagenesis are used in mutation breeding. These are called "induced mutagenesis," where mutations are brought about by chemical mutagen treatment or radiation exposure (gamma rays, X-rays, ion beams, etc.); Site-Three types of mutagenesis are used in mutation breeding. These include induced mutagenesis, which is brought on by DNA insertions, insertion mutagenesis, which is brought on by DNA insertions brought on by genetic transformation and insertion of T-DNA or activation of transposable elements, and site-directed mutagenesis, which is the process of causing a mutation at a specific location in a DNA molecule. Two types of mutagenesis exist: directed mutagenesis, which involves introducing a mutation at a specific location within a DNA molecule, and insertion mutagenesis, which results from DNA insertions caused by genetic transformation, T-DNA insertion, or transposable element activation (Bado et al. 2015). The first step in plant breeding is to identify genotypes with the desired genes among the existing varieties; if one does not already exist, one must be created. Mutations are the primary source of variety in nature, and without them, plant breeding would not be possible. The primary goal of mutation-based breeding in this context is to produce and improve well-adapted plant varieties by modifying one or two essential characteristics to increase yield or quality (Pathirana 2011).

Two primary results of various advancements in the field of crop improvement through mutation breeding are new genetic stocks possessing improved characters or a higher capacity to combine traits, and improved varieties designed for commercial cultivation. These attributes could be increased output, enhanced nutritional value, resistance to pests and illnesses, early maturity, tolerance to salt, drought, etc. (Bado et al. 2015). The development of novel cultivars has been the primary objective of mutation breeding; however, the genetic stocks produced can be employed in plant breeding for a number of purposes, including as a parent in hybrid breeding programs or as a donor parent in traditional breeding programs. Besides these, the mapping of genes is a very specific objective of mutation research itself. These days, a large amount of research in molecular genetics and genomics uses the technique of induced mutagenesis to identify genes by reducing their phenotypic expression (Oladosu et al. 2016).

16.8.1 Types of Mutagens Used in Mutation Breeding

Mutations are the cause of genetic variability in many crops, both naturally occurring and artificially created. In order to create artificial mutations, a variety of physical and chemical mutagens are used. Physical mutagens include all forms of radiation, including X-rays, gamma rays, neutrons, beta particles, alpha particles, proton or deuterons, UV rays, etc. Chemical mutagenesis (Acriflavine, acridine orange, ethidium bromide, etc.) is caused by a number of substances, including alkylating agents (ethyl methanesulfonate (EMS), methyl methanesulfonate, nitrogen mustard, etc.), base analogues (bromo uracil (5BU), aminopurine (2AP), tetramethyl urea (TMU), and acridine dyes (Chaudhary et al. 2019). Another type of mutagenesis technique is the use of biological agents such as T-DNA, retrotransposons, and transposons. By causing target mutations, genome editing techniques like clustered regularly interspaced short palindromic repeats (CRISPR) and transcription activator-like effector nucleases (TALENS) can be used as an alternative mutagen (Agarwal and Kumar 2021). These mutagens are used to build resistance in a range of sensitive crops, increasing production and quality features against harmful infections. For crop development, a range of mutagens are available, and each one is essential to the effect that they have—positive or negative—on crops. Many of these compounds cause chromosome damage in plants by generating oxygen-derived radicals, which can be generated spontaneously or deliberately following the induction of mutagens (Kolakar et al. 2018).

16.8.2 Steps in Mutation Breeding

The primary methods for using induced mutations are based firmly in genetics and are relatively simple (Kolakar et al. 2018). The parent variety's seeds or tissue, referred to as M0, are exposed to radiation or other treatments in the first phase. Chromosome rearrangements or gene alterations brought on by these treatments may affect a particular character in the M1 generation. Plants of the M1 generation are homozygous but heterozygous, and only dominant mutations have been found. Seeds from the M1 generation are utilized to produce the M2 generation in order to choose plants with the desired traits. M2 seeds are planted in test plots and subjected to various screening techniques. The chosen M2 material is subsequently grown on to become M3 generation. Screening for quantitatively inherited traits, homozygosity, and field performance stability is done on the M3 generation. Following location testing and field multiplication, these promising homozygotic mutants can be released as mutant varieties and used as breeding stock in cross breeding programmes. They can also multiply in the field in later generations M6–8 (Agarwal and Kumar 2021).

16.8.3 Mutation Breeding in Turmeric

In comparison to untreated cultivars, the addition of ethyl methane sulfonate (EMS) boosted the treated cultivars ' resistance to salinity. Additionally, the cultivars treated with EMS showed a considerable improvement in the symptoms brought on by salinity stress. Therefore, the use of EMS may enable turmeric to display saline tolerance comparable to other crops such as rice, barley, and wheat (Bandi et al. 2022). Devi and Chezhiyan looked into how gamma radiation affected turmeric crops, Salem Local -G 1 (CL144), Alleppey finger turmeric -G 2 (CL146), and PTS 43 -G 3 (CL147) were the three turmeric genotypes used in the experiment. They were each given a control dose of 1.0, 1.5, 2.0, 2.5, 3.0, 3.5, and 4.0 kR of gamma radiation. The results showed that the plants matured earlier and had better yield per plant and curing percentage at 2.0 and 2.5 kR. The yield and cure % were negatively impacted by greater gamma doses, and these higher doses also delayed maturity. G 1 (CL144), one of the genotypes used, responded well to gamma radiation. The architecture of the turmeric rhizome and how growth period (maturity) affects rhizome growth and development are also covered in the study. According to the study, gamma irradiation can be used to increase turmeric yield and cure %, but the best dose must be chosen based on the genotype and other variables (Devi and Chezhiyan 2006).

16.9 Hybridization

A key factor in crop improvement is hybridization. It is evident that gene flow between diverging taxa produces diverse phenotypes thanks to the advent of numerous genomic methods and genomic analyses with the evolution of breeding and biological sciences (Goswami and Kuchay 2023). Unlike other common crops, turmeric cultivars are difficult to improve through hybridization. Clonal selection using tissue culture is essentially the only method to improve cultivar quality because the majority of turmeric cultivars are unable to yield true seeds. Scientists used explants such as vegetative buds, fragments of rhizome, or callus formed from buds and leaves, cultured in vitro using different culture media. Different explant and culture media combinations could result in various clone products comparable to how callus produced from buds might renew shoot, while alternative tissue sources do not produce the same results and these two are based on distinct culture conditions, callus formed from leaf could regenerate new plants. The main selection objectives are high rhizome production, high curcumin, essential oil, and oleoresin contents, as well as resistance to specific diseases like rhizome rot and leaf blotch (Nair 2019). According to past performance, high fresh rhizome yields could reach 35 t/ha on average, with certain varieties reaching over 50 t/ha in some regions. Normal dry rhizome production is less than 10 t/ha. A good performance would be seven or eight tonnes per hectare. To create some excellent kinds, several breeders employed mutation breeding techniques. Ionizing radiation like X-rays and chemical mutagens like EMS are possible mechanisms. The optimum method is micropropagation using in vitro micro rhizomes, particularly for turmeric as it cannot generate seeds. N-(Phenylmethyl)-9H-purin-6-amine (BA), 1-naphthylacetic acid (NAA), and ancymidol are provided with MS, the most popular medium. Additionally, using this technique would result in novel hybrid types, complete in vitro hybridization and pollination, and regenerate plantlets (Shi 2020).

16.9.1 Somatic Cell Hybridization

The process of combining the genetic material of two distinct plant species or varieties is known as somatic cell hybridization. In the case of turmeric, this method has produced new cultivars with desired characteristics. The process of somatic hybridization involves a number of steps, including the source of the protoplasts, their isolation and plating, plant regeneration, the fusion of the protoplasts, selection processes, and the identification and characterization of the resulting somatic hybrid plants (Sink et al. 1992). In turmeric, somatic cell hybridization has proven effective in creating new cultivars with enhanced yield, resistance to disease, and other desired characteristics. One example of the successful use of somatic cell hybridization in turmeric is the development of a new variety called Pratibha. This variety was created by fusing protoplasts from two different varieties of turmeric, Roma and Suguna. It was discovered that Pratibha outperformed both Roma and Suguna in terms of yield and exhibited superior resistance to diseases like rhizome rot and leaf spot (Ravindran et al. 2007). Another example is the development of a new variety called Rajendra Sonia. This variety was created by fusing protoplasts from two different species of turmeric, *Curcuma longa* and *Curcuma aromatica*. Rajendra Sonia was found to have a higher yield than both parent species, as well as improved resistance to diseases such as leaf spot and rhizome rot. In addition to improving yield and disease resistance, somatic cell hybridization has also been used to introduce other desirable traits into turmeric. For example, a new variety called Pragati was developed by fusing protoplasts from two different varieties of turmeric, Suvarna and Roma. Pragati was found to have a higher level of curcumin, the compound responsible for turmeric's yellow color and health benefits, than both parent varieties (Sahoo and Chand 2010). Somatic cell hybridization has been found as a useful technique for creating new varieties of turmeric with improved yield, disease resistance, and other desirable traits. The success of this technique in producing new varieties such as Pratibha, Rajendra Sonia, and Pragati demonstrates its potential for use in turmeric breeding programs.

16.9.2 *Hybrid Cultivars*

Turmeric is a major spice crop in India and a native of Asia and India. It belongs to the Zingiberaceae family. After undergoing specific post-harvest/processing steps like clearing, curing, drying, polishing, and powdering, this commercial crop is planted for its edible underground stems, or tuberous rhizomes, which are used for the flavor and color of culinary dishes (Choudhary and Rahi 2018). In the study by Nair (2019), intervarietal crosses of turmeric including three commercial varieties and one first generation seedling with high curcumin content and a morphological marker led to fruit setting in only one combination. One fruit with four seeds was produced by the cross of the commercial cultivar Suranjana with the high curcumin seedling (SLP 389/1). Three of these seeds germinated, and by 5–8 months after seeding, hybrids had taken root. Hybrids' rhizomes were multiplied, and the plants produced from them were evaluated for morphology, chromosomal number, yield, and breeding behaviour using pot culture tests. The morphological characteristics that were noted were plant height, leaf lamina length, leaf lamina breadth, petiole length, collar girth of the main tiller, flowering status, and fruit set on self-pollination. Compared to their parents, hybrid plants varied in size. Hybrid-1 had a yield of 812.50 g/pot, which was higher than its pollen father, "Suranjana," but on par with it. The yield of Hybrid-2 and Hybrid-3 was comparable to that of the SLP-389/1 female parent (Nair 2019; Sahoo and Chand 2010). The commercial cultivars employed in the study were triploids with $2n = 63$, according to chromosome number analysis of the genotypes used for the crossover study, but the first-generation seedling SLP 389/1 had $2n = 78$.The hybrids produced by the cross of SLP 389/1 ($2n = 78$) and Suranjana ($2n = 63$) displayed variance in two chromosome numbers and one chromosome that was identical to the female parent. Hybrid-1 had $2n = 82$ chromosomes, Hybrid-2 had $2n = 80$ chromosomes, and Hybrid-3 had $2n = 78$ chromosomes. Hybrid-2 had the highest fruit set on self-pollination, Hybrid-1 came in second, and Hybrid-3 had the lowest fruit set. While Hybrid-3 had yellowish green emerging shoots similar to the pollen parent Suranjana, Hybrid-1 and Hybrid-2 inherited the morphological marker, the red colour of emerging shoots comparable to the mother parent. The scope of investigating recombination breeding through guided crossings in turmeric was established (Nair 2019) (Table 16.2).

16.10 Conclusion and Prospects

16.10.1 *An Overview of Current Status*

Turmeric's pharmacological activities are attributed to its secondary metabolites, including curcuminoids, oleoresin, and essential oil. Traditional clonal selection has limitations, leading to biotechnological approaches like hybridization, protoplast fusion, somaclonal variations, and transgenic development. Molecular techniques

Table 16.2 Different cultivars of turmeric (Anandaraj et al. 2014)

Varieties	Source	Characteristics
Megha turmeric	ICAR Research Complex for NEH Region, Shillong, Meghalaya	Strong rhizomes and a high curcumin content make it ideal for North East Bengal and the hill regions of North West India. Dry recovery 16.37%
Sobha	College of Horticulture, Vellanikkara of Kerala Agricultural University, Kerala	Large mother rhizome with closer inter nodes and medium-sized bold fingers. The rhizomes ' inner core is a deep orange color. Additional tertiary rhizomes. Average yield of 5.74 t/ha. Maturity 240–270 days
Sona	Kerala Agricultural University, Thrissur, Kerala	Medium-sized, bold orange-yellow rhizomes without tertiary fingers Field resistant to leaf blotch A mean yield of 7.05 t/ha Ideal for Kerala's central regions
Roma	Orissa University of Agriculture and Technology, Pottangi	Grows well in Tamilnadu, Andhra Pradesh, and Orissa. 6.0% curcumin 19.8% dry recovery
Varna	Kerala Agricultural University, Thrissur, Kerala	Medium-sized, bold orange-yellow rhizomes with tertiary fingers. Field-tolerant of blotches on leaves. An average yield of 6.37 t/ha
IISR Alleppey Supreme	Indian Institute of Spices Research, Kozhikode, Kerala	Suitable for cultivation all over India.210-day crop duration, 5.55% curcumin, 16.0% oleoresin , and 19.0% dry recovery
IISR—Suguna	Indian Institute of Spices Research, Kozhikode, Kerala	Resilient to the short-lived rhizome rot disease; 30–40 tonnes of fresh rhizomes/ha
IISR Pragati	Indian Institute of Spices Research, Kozhikode, Kerala	Appropriate for regions where turmeric is grown but has severe irrigation issues. Exceptionally productive variety, yielding 38 t/ha on average (fresh rhizomes). Consistent and high curcumin variety (5.02%) in all places. Somewhat impervious to nematode infestation by root knots
IISR Prabha	Indian Institute of Spices Research, Kozhikode, Kerala	A rhizome with a reddish-yellow color that is high yielding and of good quality A mean yield of 7.30 t/ha

(continued)

Table 16.2 (continued)

Varieties	Source	Characteristics
Kanthi	College of Horticulture, Vellanikkara of Kerala Agricultural University, Kerala	Big mother rhizome with medium bold fingers and closer inter nodes Average yield of 7.34 t/ha Maturity 240–270 days
IISR Prathiba	Indian Institute of Spices Research, Kozhikode, Kerala	A reddish-yellow-colored rhizome that is high yielding and of good quality. free of illnesses in the fields of farmers. An enhancement of value-added variety. Crop Life: 225 Days. 7.23 tons of fresh rhizomes per hectare of land
IISR—Sudarsana	Indian Institute of Spices Research, Kozhikode, Kerala	Resistant to the short-term rhizome rot disease 30–40 tonnes fresh rhizome/ha
IISR Kedaram	Indian Institute of Spices Research, Kozhikode, Kerala	Suitable for cultivation all over India.. Curcumin (5.5%), oleoresin (13.6%), maturity (210 days), and drought (18.9%)
Rajendra Sonia	Rajendra Agricultural University, Dholi	Immune to the diseases that cause leaf spot and leaf blotch. 200–210 days after sowing is harvesting. Potential yield: 400–450 q/ha. Curcumin levels range from 7.5 to 8%
Suranjana	Uttar Banga Krishi Viswavidyalaya, Pundibari, Cooch Behar	5.7% curcumin, 10.9% oleoresin, 4.1% essential oil, 21.2% dry recovery, resistant to rhizome scales, tolerant of leaf blotch and rhizome rot, and somewhat resistant to shoot borer, duration 235 days
Rasmi	Orissa University of Agriculture and Technology, Pottangi	6.4% curcumin, 13.4% oleoresin, 4.4% essential oil, and 23.0% dry recovery over a 240-day period
Narendra Turmeric-1	Narendra Deva University of Agriculture and Technology, Kumarganj, Faizabad	High curcumin content and broad adaptability
Duggriala red (Cl. 317)	Dr. YSR Horticultural University, AICRPS, Jagtial, Andhra Pradesh	Strong, long, and dark orange in color, the rhizomes of this high-yielding variety are plump and long Dry recovery 23.5%

like marker-assisted selection, molecular sequencing, gene identification, cloning, and profiling enhance breeding efforts, offering faster and cost-effective results. Gene expression analysis, transcriptome sequencing, marker-assisted selection, and quantitative traits statistics provide valuable knowledge for advancing elite turmeric genotypes and improving phyto-constituents in turmeric varieties.

Genetic engineering and biotechnology have revolutionized turmeric breeding programs by improving genotypes with improved phyto-constituents, improving turmeric quality and medicinal properties. These advancements can overcome salinity stress through methods like double haploid production, marker-assisted breeding, wild relatives, mutation breeding, transgenic approaches, and candidate gene selection. Climate-smart turmeric varieties can be developed by manipulating genes related to disease resistance, pest resistance, and abiotic stress tolerance. These molecular approaches improve nutritional quality, rhizome size, curcumin content, and essential oil constituents in turmeric. Understanding climate change challenges allows genetic manipulations to improve stress tolerance, reduce yield losses, and benefit farmers while maintaining soil health.

16.10.2 Current Research Initiatives to Combat Global Climate Change

Plant breeding is crucial for developing crop varieties that can withstand global climate change challenges. Researchers are focusing on identifying and incorporating genetic traits that confer heat and drought tolerance, flood and submergence tolerance, saltwater intrusion and soil alkalinity, and enhanced resistance to pathogens and pests. Integrated pest management strategies are being developed to mitigate the impact of these factors. Plants play a vital role in carbon sequestration, and researchers are exploring ways to enhance crop varieties' capacity to capture and store carbon. Research is being conducted to identify and breed crops with increased biomass production and improved carbon allocation.

Climate change also affects crop nutritional quality, potentially impacting human health. Plant breeders are working to develop climate-resilient crop varieties that retain or enhance their nutritional quality under changing environmental conditions, selecting traits that confer flexibility in phenology, stress tolerance, and yield stability across different environments.

16.10.3 Recommendations for Future Research and Utilization

In many of the nations where turmeric is grown, it is an essential part of traditional and indigenous medical practices. Turmeric has been used for more than 4,000 years, but its potential uses have not yet been completely explored. Studies conducted over the past few decades have clearly shown that curcumin may have greater antioxidant capacity than alpha-tocopherol. Such a feature is frequently used as a chemopreventive medication for a range of malignancies including colon, breast, prostrate, esophageal, lung, and oral cancers as well as an inhibitor of atherosclerosis, proliferation of viruses, and bacteria. Demethoxycurcumin and bisdemethoxycurcumin, two more curcuminoids molecules, have also been discovered to have cytotoxic, antioxidant, and anti-inflammatory effects on cancer cell lines. The pharmaceutical industry, particularly as a source of medication, greatly benefits from the contents of such medicinally significant chemicals. Curcumin is an interesting chemical that will advance the development of innovative therapies as researchers gain a better knowledge of its therapeutic potential (Prasath et al. 2018).

One of the problems for the future in the current situation is the sustainable production of turmeric. The widespread agricultural practices of using chemical fertilizers to increase curcuminoids and sesquiterpenoids is intended to disrupt the functional characteristics and organizational structures of microbial communities. Applications of PGPR not only lessen the damaging effects of synthetic chemicals on plants and soil, but also encourage a healthy environment for plant–microbe interactions. Research on the relationships between plants and microbes is still going strong. Complex microbiomes may be studied using advancements in high-throughput technology like next-generation sequencing, which will enable greater sample quantities and promote deeper analysis of microbial populations. For investigating, locating, and describing the contributions of genetic and metabolic components involved in the interactions between host plants and endophytes, the emerging "omics" techniques are useful tools. For instance, metagenome sequencing has shown favourable impacts on the primary metabolites of plants and identified critical functions necessary for bacterial endophytes to survive inside their host plants. The study of noncultivated groups thriving in combination with plants will be made possible by coupled cultivation-independent and better cultivation methods (Kumar et al. 2017).

The development of prospective new cultivars with increased yields, curcumin concentration, and disease resistance can take advantage of biotechnology and recent advances in plant genomes. It is necessary to develop and implement methods for boosting turmeric productivity through more efficient resource usage to produce better yields while fostering the sustainability of farming systems and moving away from subsistence farming toward market-oriented agriculture (Prasath et al. 2018). Thus, knowledge based on transcriptome sequencing, marker assisted selection, gene expression analysis, and quantitative characteristics statistics can be used in breeding programmes to improve turmeric genotypes' phyto-constituents and advance the elite genotypes (Ayer 2017). Additionally, there is a chance to create innovative value-added goods by emphasizing turmeric's therapeutic benefits instead of or in addition to conventional ones. Such value-added programs can increase the income of small and marginal farmers, whose livelihoods depend on this crop.

In the future, precision breeding and the application of cutting-edge bioinformatics tools to interpret complex genetic traits will be key components of turmeric breeding, as these methods are continuously refined. Developing cultivars with improved curcumin content or resistance to newly emerging diseases will require a critical integration of omics approaches with both conventional and contemporary breeding techniques. To fully realize the benefits of these advancements and guarantee that turmeric stays a fundamental component of both conventional medicine and contemporary agriculture, cooperation among scientists, breeders, and farmers will be essential.

References

Agrawal L, Kumar M (2021) Improvement in ornamental, medicinal, and aromatic plants through induced mutation. J Appl Biol Biotech 9:162–169

Ahmar S, Gill RA, Jung K-H, Faheem A, Qasim MU, Mubeen M, Zhou W (2020) Conventional and molecular techniques from simple breeding to speed breeding in crop plants: recent advances and future outlook. Int J Mol Sci 21. https://doi.org/10.3390/ijms21072590

Anandaraj M, Prasath D, Kandiannan K et al (2014) Genotype by environment interaction effects on yield and curcumin in turmeric (*Curcuma longa* L.). Ind Crops Prod 53:358–364

Anindita PA, Putri TK, Ustari D, Maulana H, Rachmadi M, Concibido V, Suganda T, Karuniawan A (2020) Dataset of agromorphological traits in early population of turmeric (*Curcuma longa* L.) local accessions from Indonesia. Data Brief 33:106552

Anu S, Navas M, Dan M (2020) Morpho-anatomical characterisation of the rhizomes of ten species of *Curcuma* L. (Zingiberaceae) from south India. J Spices Aromat Crop 38–47

Ashraf M, Akram NA, Mehboob-Ur-Rahman FMR (2012) Marker-assisted selection in plant breeding for salinity tolerance. Methods Mol Biol 913:305–333

Aswathi AP, Raghav SB, Prasath D (2023) Assessment of genetic variation in turmeric (*Curcuma longa* L.) varieties based on morphological and molecular characterization. Genet Resour Crop Evol 70:147–158

Ayer DK (2017) Breeding for quality improvement in turmeric (*Curcuma longa* L.): a review. Adv Plants Agric Res 6. https://doi.org/10.15406/apar.2017.06.00238

Bado S, Forster BP, Nielen S et al (2015) Plant mutation breeding: current progress and future assessment. Plant Breed Rev 39:23–88

Bahadur V, Yeshudas V, Meena OP (2016) Nature and magnitude of genetic variability and diversity analysis of Indian turmeric accessions using agro-morphological descriptors. Canad J Plant Sci 96:371–381

Balkrishna A, Phour M, Thapliyal M, Arya V (2021) Current status of Indian agriculture: problems, challenges and solution. Biol Forum–An Int J 13:361–374

Bandi AS, Kaushik P (2022) Mechanisms and approaches for salt tolerance in turmeric: a breeding perspective. OBM Genet 6:1–1

Bhadra S, Bandyopadhyay M (2015) Karyomorphological investigations on some economically important members of Zingiberaceae from Eastern India. Caryologia 68:184–192

Bhandari A, Jayaswal P, Yadav N et al (2019) Genomics-assisted backcross breeding for infusing climate resilience in high-yielding green revolution varieties of rice. Indian J Genet Plant Breed 79. https://doi.org/10.31742/ijgpb.79s.1.5

Bondre SV, Nagre PK, Kale VS, Wagh AP, Gupta NS (2019) Effect of organic nutrient management on quality, processed finger yield and economics in turmeric. Inter J Chem Stud 7:3066–3069

Brijesh H, Ajjappala B (2023) Micropropagation strategies in medicinally important turmeric (*Curcuma* sp): Current research and future challenges. J Appl Biol Biotechnol 11:1–8

Chaudhary J, Deshmukh R, Sonah H (2019) Mutagenesis approaches and their role in crop improvement. Plants 8. https://doi.org/10.3390/plants8110467

Choudhary AK, Rahi S (2018) Organic cultivation of high yielding turmeric (*Curcuma longa* L.) cultivars: a viable alternative to enhance rhizome productivity, profitability, quality and resource-use efficiency in monkey–menace areas of north-western Himalayas. Ind Crops Prod 124:495–504

Devi HU, Chezhiyan N (2006) Impact of gamma rays on turmeric crop (*Curcuma longa* L.). J Horticul Sci 1:124–128

Devi RA, Kamatyanatti M, Kumar R et al (2019) Breeding for quality improvement in major spice crops: a review. Think India J 22:629–644

Gayatri MC, Roopadarshini V, Kavyashree R, Kumar CS (2005). Encapsulation and regeneration of aseptic shoot buds of turmeric (*Curcuma longa* L.). Plant Cell Biotechnol Mol Biol 6(3–4):89–94

Goswami M, Kuchay MA (2023) Hybridization: importance, techniques and consequences. Recent Trends Agric 185

Gupta AK, Mishra R, Lal RK (2015) Genetic resources, diversity, characterization and utilization of agronomical traits in turmeric (*Curcuma longa* L.). Ind Crops Prod 77:708–712

Haque MI, Singh PK, Ghuge S, Kumar A, Rai AC, Kumar A, Modi A (2022) A general introduction to and background of plant tissue culture: past, current, and future aspects. In: Chandra Rai A, Kumar A, Modi A, Singh M (eds) Advances in plant tissue culture. Academic Press, pp 1–30

Hashemy T, Maki H, Yamada Y, Kaneko TS, Syono K (2009) Effects of light and cytokinin on in vitro micropropagation and microrhizome production in turmeric (*Curcuma longa* L.). Plant Biotechnol 26:237–242

He R, Gang DR (2014) Somatic embryogenesis and Agrobacterium-mediated transformation of turmeric (*Curcuma longa*). Plant Cell Tissue Organ Cult 116:333–342

Hershon WV, Gnanamoni SM (2018) A study on production of spices and the problems faced by the cultivators and traders. IJETER 6:63–66

Hussain A, Ahmed I, Nazir H, Ullah I (2012) Plant tissue culture: current status and opportunities. In: Recent advances in plant in vitro culture, pp 1–28

Islam MA, Kloppstech K, Jacobsen HJ (2004) Efficient procedure for in vitro microrhizome induction in *Curcuma longa* L. (Zingiberaceae)–a medicinal plant of tropical Asia. Plant Tissue Cult Biotechnol 14:123–134

Janani V, Mangai NS, Naveenkumar R (2021) Review on detection methods of turmeric disease and methodology using artificial neural network. In International conference on smart data intelligence

Jayanthi M, Vaideke A (2015) A study on cultivation practices of turmeric farmers in Erode district. Ind J Appl Res 5:100–102

Kaiser N, Douches D, Dhingra A, Glenn KC, Herzig PR, Stowe EC, Swarup S (2020) The role of conventional plant breeding in ensuring safe levels of naturally occurring toxins in food crops. Trends Food Sci Technol 100:51–66

Kolakar SS, Nadukeri S, Jakkeral SA et al (2018) Role of mutation breeding in improvement of medicinal and aromatic crops: review. J Pharmacogn Phytochem 7:425–429

Kumar A, Singh AK, Kaushik MS, Mishra SK, Raj P, Singh PK, Pandey KD (2017) Interaction of turmeric (*Curcuma longa* L.) with beneficial microbes: a review. 3 Biotech 7:357

Larkin PJ, Scowcroft WR (1981) Somaclonal variation—a novel source of variability from cell cultures for plant improvement. Theor Appl Genet 60:197–214

Liu Q, Yang F, Zhang J, Liu H, Rahman S, Islam S, Ma W, She M (2021) Application of CRISPR/Cas9 in crop quality improvement. Int J Mol Sci 22. https://doi.org/10.3390/ijms22084206

Loyola-Vargas VM, Ochoa-Alejo N (2018) An introduction to plant tissue culture: advances and perspectives. Methods Mol Biol 1815:3–13

Maghuly F, Myśków B, Till BJ (2021) Functional genomics for plant breeding. Int J Mol Sci 22. https://doi.org/10.3390/ijms222111854

Magray MM, Wani KP, Chatto MA, Ummyiah HM (2017) Synthetic seed technology. Int J Curr Microbiol Appl Sci 6:662–674

Mirjanaik RH, Vishwanath YC (2020) Advances in production technology of turmeric. J Pharmacogn Phytochem 9:1198–1203

Mishra J, Bhardwaj A, Misra K (2018) *Curcuma* sp.: the nature's souvenir for high-altitude illness. In: Misra K, Sharma P, Bhardwaj A (eds) Management of high altitude pathophysiology. Academic Press, pp 153–169

Mu H, Wang B, Yuan F (2022) Bioinformatics in plant breeding and research on disease resistance. Plants 11. https://doi.org/10.3390/plants11223118

Nadgauda RS, Mascarenhas AF (1986) A method for screening high curcumin-containing turmeric (*Curcuma longa* L.) cultivars in vitro. J Plant Physiol 124:359–364

Nair KP (2019) Turmeric (*Curcuma longa* L.) and Ginger (*Zingiber officinale* Rosc.)—World's invaluable medicinal spices: the agronomy and economy of turmeric and ginger. Springer Nature

Nasirujjaman K, Uddin MS, Zaman S, Reza MA (2005) Micropropagation of turmeric (*Curcuma longa* Linn.) through in vitro rhizome bud culture. J Biol Sci 5:490–492

Newell MA, Jannink J-L (2014) Genomic selection in plant breeding. Methods Mol Biol 1145:117–130

Nirmal Babu K, Divakaran M, Pillai GS, Sumathi V, Praveen K, Raj RP, Akshita HJ, Ravindran PN, Peter KV (2016) Protocols for in vitro propagation, conservation, synthetic seed production, microrhizome production, and molecular profiling in turmeric (*Curcuma longa* L.). Methods Mol Biol 1391:387–401

Oladosu Y, Rafii MY, Abdullah N, Hussin G, Ramli A, Rahim HA, Miah G, Usman M (2016) Principle and application of plant mutagenesis in crop improvement: a review. Biotechnol Biotechnol Equip 30:1–16

Pathirana R (2011) Plant mutation breeding in agriculture. CAB Rev Perspect Agric Vet Sci Nutr Nat Resour 2011:1–20

Pazhamala LT, Kudapa H, Weckwerth W, Millar AH, Varshney RK (2021) Systems biology for crop improvement. Plant Genome 14:e20098

Pikulthong V, Teerakathiti T, Thamchaipenet A, Peyachoknagul S (2016) Development of somatic embryos for genetic transformation in *Curcuma longa* L. and *Curcuma mangga* Valeton & Zijp. Agric Nat Resour 50:276–285

Pittampalli B, Jogam P, Thampu RK, Abbagani S, Peddaboina V (2022) High-frequency plant regeneration and genetic homogeneity assessment of regenerants by molecular markers in turmeric (*Curcuma longa* L.). In Vitr Cell Dev Biology—Plant 58:169–180

Prasath D, Kandiannan K, Leela NK et al (2018) Turmeric. Horticultural reviews. Wiley., Hoboken, NJ, USA, pp 99–184

Prathanturarug S, Soonthornchareonnon N, Chuakul W, Phaidee Y, Saralamp P (2003) High-frequency shoot multiplication in *Curcuma longa* L using thidiazuron. Plant Cell Rep 21:1054–1059

Prathanturarug S, Soonthornchareonnon N, Chuakul W, Phaidee Y, Saralamp P (2005) Rapid micropropagation of *Curcuma longa* using bud explants pre-cultured in thidiazuron-supplemented liquid medium. Plant Cell Tissue Organ Cult 80:347–351

Pujari PP, Patil RB, Sakpal RT (1986) Krishna, a high yielding variety of turmeric. Indian Cocoa, Arecanut and Spices Journal 9:65–66

Rahman MM, Amin MN, Jahan HS, Ahmed R (2004) In vitro regeneration of plantlets of *Curcuma longa* Linn. A valuable spice plant in Bangladesh. Asian J Plant Sci 3:306–309

Rathaur P, Raja W, Ramteke PW, John SA (2012) Turmeric: the golden spice of life. Int J Pharm Sci Res 3(7):1987–1994

Ravindran PN, Babu KN, Sivaraman K (eds) (2007) Turmeric: the genus *Curcuma*. CRC press

Roopadarshini V (2010) High frequency shoot multiplication and callus regeneration of turmeric. Int J Biotechnol Biochem 6:723–734

Sahoo AK, Chand PK (2010) Somatic hybridization and cybridization in *Curcuma:* production and characterization of somatic hybrids and cybrids for crop improvement and phytochemical production. Plant Cell Rep 29:1315–1326

Salvi ND, George L, Eapen S (2000) Direct regeneration of shoots from immature inflorescence cultures of turmeric. Plant Cell Tissue Organ Cult 62:235–238

Sasaki Y, Fushimi H, Cao H, Cai S-Q, Komatsu K (2002) Sequence analysis of Chinese and Japanese Curcuma drugs on the 18S rRNA gene and trnK gene and the application of amplification-refractory mutation system analysis for their authentication. Biol Pharm Bull 25:1593–1599

Sasikumar B (2005) Genetic resources of *Curcuma*: diversity, characterization and utilization. Plant Genet Resour 3:230–251

Schmutzer T, Bolger ME, Rudd S, Chen J, Gundlach H, Arend D, Oppermann M, Weise S, Lange M, Spannagl M, Usadel B (2017) Bioinformatics in the plant genomic and phenomic domain: The German contribution to resources, services and perspectives. J Biotechnol 261:37–45

Selvan SS, Wahid A, Patel A, Kumar V, Sahu P (2021) Challenges in Indian agriculture. Agric Rev. https://doi.org/10.18805/ag.R-2103

Shajahan A, Raju CS, Mehaboob VM, Aslam A (2019) Somatic embryogenesis and synthetic seed technology of *Curcuma* spp. In: Faisal M, Alatar AA (eds) Synthetic seeds: germplasm regeneration, preservation and prospects. Springer International Publishing, Cham, pp 363–375

Shamina A, Zachariah TJ, Sasikumar B, George JK (1998) Biochemical variation in turmeric (*Curcuma longa* Linn.) accessions based on isozyme polymorphism. J Hortic Sci Biotechnol 73:479–483

Shi S (2020) Assessment of turmeric (*Curcuma longa* L.) varieties for yield and curcumin content. Auburn University

Shirgurkar MV, Naik VB, von Arnold S, Nadgauda RS, Clapham D (2006) An efficient protocol for genetic transformation and shoot regeneration of turmeric (*Curcuma longa* L.) via particle bombardment. Plant Cell Rep 25:112–116

Sinchana NS, Kattimani KN, Prabhuling G, Sudesh K, Jagadeesha N (2020) Standardization of tissue culture protocol for turmeric (*Curcuma longa* L.) Cv. Salem. Int J Chem Stud 8:2721–2726

Singh G, Singh OP, Maurya S (2002) Chemical and biocidal investigations on essential oils of some Indian Curcuma species. Prog Cryst Growth Charact Mater 45:75–81

Singh S, Panda MK, Nayak S (2012) Evaluation of genetic diversity in turmeric (*Curcuma longa* L.) using RAPD and ISSR markers. Ind Crops Prod 37:284–291

Singh S, Waman AA, Bohra P, Gautam RK, Roy SD (2016) Conservation and sustainable utilization of horticultural biodiversity in tropical Andaman and Nicobar Islands, India. Gen Res Crop Evol 63:1431–45.

Singh VK, Singh BD, Kumar A, Maurya S, Krishnan SG, Vinod KK, Singh MP, Ellur RK, Bhowmick PK, Singh AK (2018) Marker-assisted introgression of Saltol QTL enhances seedling stage salt tolerance in the rice variety "Pusa Basmati 1." Int J Genomics Proteomics 2018:8319879

Singh RK, Prasad A, Muthamilarasan M, Parida SK, Prasad M (2020) Breeding and biotechnological interventions for trait improvement: status and prospects. Planta 252:54

Singletary K (2020) Turmeric: potential health benefits. Nutr Today 55:45

Sink KC, Jain RK, Chowdhury JB (1992) Somatic cell hybridization. In: Kalloo G, Chowdhury JB (eds) Distant hybridization of crop plants. Springer, Berlin Heidelberg, Berlin, Heidelberg, pp 168–198

Sunitibala H, Damayanti M, Sharma GJ (2001) In vitro propagation and rhizome formation in *Curcuma longa* Linn. Cytobios 105:71–82

Tan YC, Kumar AU, Wong YP, Ling APK (2022) Bioinformatics approaches and applications in plant biotechnology. J Genet Eng Biotechnol 20:106

Taşkın H, Baktemur G, Kurul M, Büyükalaca S (2013) Use of tissue culture techniques for producing virus-free plant in garlic and their identification through real-time PCR. ScientificWorldJournal 2013:781282

Vassilev D, Leunissen J, Atanassov A, Nenov A, Dimov G (2005) Application of bioinformatics in plant breeding. Biotechnol Biotechnol Equip 19:139–152

Verma S, Singh S, Sharma S, Tewari SK, Roy RK, Goel AK, Rana TS (2015) Assessment of genetic diversity in indigenous turmeric (*Curcuma longa*) germplasm from India using molecular markers. Physiol Mol Biol Plants 21:233–242

Viu AFM, Viu MAO, Tavares AR, Vianello F, Lima GP (2009) Endogenous and exogenous polyamines in the organogenesis in *Curcuma longa* L. Sci Hortic 121:501–504

Zapata EV, Morales GS, Lauzardo ANH, Bonfil BM, Tapia GT, Sánchez ADJ, Aparicio AJ (2003) In vitro regeneration and acclimatization of plants of turmeric (*Curcuma longa* L.) in a hydroponic system. Biotecnol Apl 20(1):25–31

Chapter 17
The Enigmatic Charm of *Curcuma inodora*: Considerations for Conservation and Improvement

Pallavi Vyas, Sneha Sabu, and Penna Suprasanna

Abstract The most significant source of curcumin is from the genus *Curcuma*, which has long been utilized in numerous traditional remedies. *Curcuma inodora*, is one of the 40 species of *Curcuma* found in India, it has remained a little-studied species despite its diverse qualities and significance in the pharmaceutical and floriculture industries. This plant species has several potential medicinal and nutritional benefits including in-vitro anti-bacterial, anti-fungal, anti-oxidant, and anti-inflammatory properties. Studies on conventional and in vitro biotechnological efforts have been made towards improvement. In this chapter, we have tried to present the recent information on the various aspects of *Curcuma inodora*, including its taxonomical diversity, therapeutic properties, breeding strategies, *in-vitro* culture techniques and prospects, which can serve as a reference for future research projects and related initiatives.

Keywords Taxonomy · Ethnobotany · Phytochemistry · Micropropagation · Future aspects · *Curcuma inodora*

P. Vyas (✉) · P. Suprasanna
Amity Institute of Biotechnology, Amity University Maharashtra, Mumbai, India
e-mail: pvyas@mum.amity.edu

P. Suprasanna
e-mail: psuprasanna@mum.amity.edu

S. Sabu
Friedrich schiller University Jena, Jena, Germany

Erfurt Research Centre for Horticultural Crops (FGK), University of Applied Sciences, Erfurt, Erfurt, Germany

S. Sabu
e-mail: sneha.sabu@fh-erfurt.de

J. M. Al-Khayri et al. (eds.), *Biodiversity and Genetic Improvement of Herbs and Spices*, Advances in Plant Breeding Strategies 8, https://doi.org/10.1007/978-3-031-81838-7_17

17.1 Introduction

More than 1200 species of ginger belong to the *Zingiberaceae* family, which is primarily found in the tropics and subtropics. Its distribution center is in the Indo-Malayan region, but it also ranges from tropical Africa to Central and South America (Kress et al. 2002). Over 250 of these gingers, specifically those from the USA, Europe, and South-East Asia, are commonly grown as ornamental plants elsewhere in the world (Mohanan 2013). Around 200 species are known in India under 21 genera, primarily in the Andaman and Nicobar Islands with over 60 ornamental species, south India, and northeast India (Sabu 2006). Only about 25 species, the bulk of which are exotics, are employed in gardens despite the fact that there are several potential ornamental species in the wild (Kumar 2013; Kumar et al. 2013). *Boesenbergia, Curcuma, Globba, Hedychium, Larsenianthus,* and *Zingiber* are significant genera that are utilized as ornamentals (Km et al. 2013)

There are 120 species in the genus *Curcuma* (*Zingiberaceae*), most of which are found in South and South-East Asia but also occur in China, Australia, and the South Pacific. India is home to 40 of these species. Melghat, located in Amaravathi district of Maharashtra has been linked to four different *Curcuma* species (Bhogaonkar and Devarkar 1999; Dhore and Joshi 1988) mainly *C. longa* L. is cultivated while *C. inodora* Blatt., *C. pseudomontana* J. Graham and *C. decipiens* Dalzell are wild (Dagawal and Bhogaonkar 2020). *C. pseudomontana* is used in traditional medicine as a cure to jaundice also to diabetes (Padal et al. 2012), against body swellings and for enhanced lactation (Ramarao et al. 2000). Fresh tubers of these are consumed as a blood purifier (Acharya et al. 2012; Lechtenberg et al. 2004). The curative phenolic chemicals known as curcuminoids have been shown to be abundant in the dried rhizome of *Curcuma longa L.* (Awika and Rooney 2004; Bhogaonkar n.d.-a). It is used in several ways in Ayurveda and traditional medicine world over. Among these, *C. inodora* is widely distributed throughout Maharashtra and is very common and abundant in Melghat (Dagawal and Bhogaonkar n.d.-a).

The rhizome of *Curcuma inodora* Blatt. (Zingiberaceae), which is used to cure a variety of ailments including constipation, psychosomatic problems, and muscular discomfort, also has anti-inflammatory, antipyretic, and wound healing qualities. (Jadhao and Bhuktar 2019). As there is not much discussions and studies carried out on *Curcuma inodora*, this chapter profoundly discuss more about the Taxonomy, origin and distribution, diversity, ethnobotanical and medicinal properties, phytochemical profiles also with its conservation and improvement.

17.2 Taxonomy

Curcuma inodora Blatt. (*Zingiberaceae*) also known as "Jangali halad". *Curcuma purpurea* is a synonym of *Curcuma inodora.* This is an Angiosperm plant belongs to family *Zingiberaceae*;

sub-family *Zingiberoideae.*

17.3 Morphological Description

Curcuma inodora is perennial herb which is tuberous in nature and grows on hill tops. Their leaves are radical, 16-31 × 14cm, broadly elliptic lanceolate. They have an inflorescence of spikes; bracts rosy green purplish. Capsules 3-valved with arillate seeds. Also, the Fls. and Frts is July - September. The precise locations are Adwadi, Anjaneri hill, Vani, Harsul in Maharashtra (Sangale and Kshirsagar n.d.). In accordance with Prabhu et al.'s (2014) variability study, the rhizome of *Curcuma inodora* is conical, pale yellow in the centre, and whitish in the periphery; the root bulbs are many, ovoid, and white inside. Tall, leafy shoot, 20–50 cm. Eight-to-ten-centimeter long pseudostem. Laminae are 15-30 in numbers, 7–12 cm, elliptic, base oblique, tip acuminate, minutely hairy along the conspicuous side veins on the upper side, bottom side glabrous; ligule 2 mm long, hyaline; leaves distichous, 3-6, petiolate; petiole 20-25 cm long. Inflorescence is either lateral or central; peduncle 5–10 cm long; spike 10–16 cm long with a definite violet coma. Large, 4-5.5 × 1.5 cm, rounded-tipped, oval or lanceolate coma bracts. Fertile bracts are 3.5–4 × 1.5 cm, are pale green with a purple band at the tip and are not recurved. Small, linear, purple, minutely pubescent, outer 1.5-0.6 cm, interior 4 2 mm, bracteoles. Flowers that are as long as the bracts, 4.5–5.5 cm. Calyx tubular, white, 1–1.2 cm long, with a small, crooked trilobe at the tip. Tube minutely pubescent. Corolla tube, about 2.5 cm long, deep purple, pubescent; lobes uneven, dorsal lobe, about 1.6 cm long, apiculate; lateral lobes, about 1-1.3 cm long, ovate-lanceolate, deep purple, glabrous. Obovate, obscurely trilobed, emarginate tip, purple labellum with a central stripe of vivid yellow. Oblong, 1.5–1.6 0.5–0.7 cm, deep purple lateral staminodes. Anthers are about 4 mm long, thecae parallel, and the filament is purple. The thecae are white with a pink base. There are two linear, 5 mm long epigynous glands. Trilocular, 2.8-3.2 2-2.2 mm in diameter, thickly pubescent ovary with numerous ovules on axile placenta. Long, filiform, pink style; bilobed stigma. Brown, hairy, globose to ovoid fruit, 1-1.2 0.9 cm (Prabhu Kumar et al. 2014). A *C.* inodora plant which was collected from Bhatan village in Raigad district of Maharashtra is shown in figure 17.1a to provide an idea about the look and appearance of the plant. Figures 17.1b, c and d show the different parts viz. inflorescence, flower, roots and rhizomes of the *C.* inodora plant respectively.

17.4 Distribution and habitat

Typically, it is thought that species without tubers are not beneficial commercially. But, the commercial use of various *Zingiberaceae* species in the United States of America was documented by Burch 1998 (Burch 1998). Khumkratok et al. (2012) and Mood and Larsen (2001) reported the ornamental value of some South East Asian species of the genus, including *C. alismatifolia, C. aurantiaca, C. cordata, C.*

Fig. 17.1 A **a**: *Curcuma inodora* plant. **b:** Inflorescence of *Curcuma inodora*. **c**: Flower of *Curcuma inodora*. **d**: Roots and rhizomes of the *C. inodora*

inodora, C. parviflora, C. roscoeana, C. alismatifolia, C. rabdota, C. bicolor, and C. glans (Khumkratok et al. 2012; Mood and Larsen 2001; Velayudhan 2015).

In India, the genus is represented by 40 species (Lechtenberg et al. 2004; Sabu 2006; Velayudhan 1996). There are 40 species in the genus that can be found in India (Lechtenberg et al. 2004; Sabu 2006; Velayudhan 1996). East, North, East, and Southern peninsular regions make up most of the members' distribution. The

Indian *Curcuma* is split into two subgenera based on morphological characteristics. *Eucurcuma* has three main sections: Stolonifera, Nontuberosa, and Tuberosa. (Velayudhan 1996) and two species of *Valeton's Paracurcuma* from 1918 (Valeton 1918). *Curcuma* species from the section Nontuberosa are abundant in the Western Ghats region, which is in the states of Kerala, Tamil Nadu, Karnataka, Goa, and Maharashtra and runs parallel to the Arabian Sea coast. (Velayudhan 2015).

Endangered to Peninsular India, it has been more frequently reported from Andhra Pradesh (Raju et al. 2006) and ranges from Maharashtra to North Karnataka (Mangaly 1993; Sabu 2006) and Gujarat (Phatak and Oza 1959). Even though there are several species with high horticultural potential, Indian horticulturists are yet to give these species significant attention. It is also widespread in plains, roadside cuts, and laterite deserts. It grows in semi-evergreen forest undergrowth at an altitude of 20–200 m above mean sea level in rocky open areas (Prabhu Kumar et al. 2014).

17.5 Ethnobotanical and medicinal uses

For thousands of years, people have used the medicinal and nutritional benefits of the *Curcuma* genus. Most information on the ethnobotanical and ethnopharmacological uses of *Curcuma* taxa come from the Asian continent (Apavatjrut et al. 1999; Baghel et al. 2013; Ibn Sinâ 2015; Li et al. 2011; Rajkumari and Sanatombi 2017; Sasikumar 2005; Velayudhan 1996). The commercially grown species of the *Curcuma* genus, *C. domestica* Valeton (often known as "turmeric"), is one of the most widely used spices in the world. *C. longa*, which has demonstrated a wide range of ethnobotanical uses and is frequently used as a condiment, food preservative, and coloring agent (Sasikumar 2005). Among the *Curcuma* species, *Curcuma inodora* is less studied. *Curcuma inodora* Blatter J, a potential ornamental ginger indigenous to peninsular India (Sabu 2006). Since the inflorescence appears before the leafy shoots even emerge from the underground rhizomes, it is frequently referred to as "hidden purple ginger" or "surprise ginger." (Prabhu Kumar et al. 2010; Sabu et al. 2011). Additionally, locals use it in traditional medicine and refer to it as "Scentless turmeric" and "hidden lily" (Jangli halad). It is being used to treat cuts with fresh rhizome paste, a potent antiseptic. Some people employ the smoke from dried rhizomes in Tantrik, Vashikarana, and Mayajal Kriyas to hypnotize their subjects (P. Y. Bhogaonkar and Devarkar 1999). The specialized prefix "inodora" refers to the rhizome's non-aromatic character (Sabu 2006). The plant is propagated by dividing rhizomes and can be cultivated as a ground cover plant and a crop of cut flowers. Its value as an ornamental cut-flower plant is increased by the variety in the bloom color and bract color (Kumar 2013). The root stock paste is applied as a cure to glandular diseases as well as for piles (Bhogaonkar n.d.-b; Bhogaonkar and Kadam 2007; Shah and Gopal 1982), psychosomatic disorders and also for constipation (Dagawal and Bhogaonkar 2021; Jagtap et al. 2006). Small pieces of *C. inodora* rhizomes steeped in a cup of water for 4-5 hours and decanted once a day for 1-2 days are used by tribal residents of Nandurbar district, Maharashtra, to treat psychosomatic illnesses, constipation, and muscle discomfort. (Patil and Bhaskar 2006). At concentrations of 200 mg/kg of body weight, the rhizome's methanol extract exhibited anti-inflammatory, antipyretic, and wound healing action. (Bharati et al. 2009). The rhizome of *Curcuma inodora* Blatt. (Zingiberaceae), which is used to cure a variety of ailments including constipation, psychosomatic problems, and muscular discomfort, also has anti-inflammatory, antipyretic, and wound healing qualities. (Jadhao and Bhuktar 2019). If exampling a promising study to justify the medicinal properties of *Curcuma inodora*, Bharathi et al.'s study looked into the scientific justifications for the traditional usage of *C. inodora* zingi for a variety of medical purposes. The dried methanol extract of the rhizomes of *Curcuma inodora* contains alkaloids, phenols, polyphenols, saponins, tannins, anthraquinones, and steroids, which may explain why the plant's methanol extract is effective against all experimentally induced laboratory models of inflammation, pyrexia, and wound healing (Bharati et al. 2009).

Recently, Nagoress et al. (2023) reported the *in-vitro* anti-bacterial, anti-fungal, anti-oxidant, and anti-inflammatory properties of *Curcuma inodora* rhizome oils. This study showed that environmental conditions have an impact on the phytochemical composition of undiscovered *C. inodora* rhizome oil. It also showed that Hydro Distilled Rhizome Essential Oil (HDREO) might be a rich source of the "most-used herbal component," or -Bisabolol, which is known for its unique identity. The antibacterial, antioxidant, and anti-inflammatory potentials of HDREO suggest that it will be helpful in the creation of the corresponding medications. This study also advised conducting a thorough *in-vivo* analysis of this species' rhizome oil for the preparation of antibacterial, antifungal, antioxidant, and anti-inflammatory medications. (Nagore et al. 2023).

Insulin and other oral hypoglycaemic/antihyperglycemic medications are the mainstays of the management of blood glucose levels that are within the normal range. There is a resurgence of interest in plant-based medications that modulate physiological effects in the prevention and treatment of diabetes because all these treatments have insufficient efficacy and are linked to unfavourable side effects. The alpha glycosidase enzyme and glucose uptake by yeast cells are both inhibited by the leaves of *Curcuma inodora*, according to a study by Vidyasabbani et al. published in 2021. This research provides a solid biochemical foundation for the management of diabetes through the inhibition of glucose uptake and stimulation of glucose absorption, both of which regulate blood glucose levels (Alluri and Satla 2017).

17.6 Phytochemical properties

The rhizome of *Curcuma inodora* Blatt (Zingiberaceae), endowed with anti-inflammatory, antipyretic, and wound-healing properties, is used to treat psychosomatic problems, constipation, and muscular pain (Jadhao and Bhuktar 2019).

According to a study conducted by Jadhao and Bhuktar (2019), a variety of bioactive substances, including 1alpha, 25-dihydroxy-26,27-dimethyl 20,21,22,22,23,23-hexadehydro-24a-homovitamin D3, Trp Gln Trp, Arachidonyl amine N-(1-methyl-2-hydroxy-2-phenyl-ethyl), Lys Ile, Arg 6R-vitamin D3 6,19-(4-phenyl-1,2,4-triazoline-3,5-dione)adduct, methylprednisolone succinate Prednicarbate, N-Palmitoyl-L-Serine, 1-(9Z,12Z,15Z-octadecatrienoyl)phosphoric Acid, Docosahexaenoyl -2- (4Z,7Z,10Z,13Z,16Z,19Z)Nona oxaoctacosan-28-ylp- (butylamino) benzoate, sn-glycerol, benzonatate (2,5,8,11,14,17,20,23,26), and (17E)Salmeterol, 1,2-ditetradecanoyl-sn- glycero-3-phospho-(1'-sn glycerol), 1,2-ditetradecanoyl-sn-glycero-3-phospho-(1'-sn glycerol), and Digitoxin are other examples. The results of the High-Resolution Mass Spectroscopy (HRMS) study revealed the presence of many compounds with varied chemical structures.

The first study on the chemical components of this species' essential oil was published by Kumar et al 2017. The hydro distillation of air-dried *Curcuma inodora* Blatt (Zingiberaceae) leaves yielded volatile oil, which was then subjected to chemical analysis using GC and GC-MS. The major constituents of the above oils were beta.-Pinene (4.09%), Caryophyllene (31.33%), alpha-caryophyllene (5.97%), Benzofuran (21.29%), gamma.-Elemene (5.18%), alpha.-Bisabolol (2.32%), beta.-Elemenone (3.09%), Ethanone (2.76%), Phytol (3.19%), 1, 6- Dagawal and Bhogaonkar 2021 gathered 12 different variations of *C. inodora, C. pseudomontana,* and *C. longa.* Upon screening of free amino acids using High Performance Thin Layer Chromatography (HPTLC), the authors found that *Curcuma inodora* accessions contained four to eleven amino acids, whereas *C. pseudomontana* and *C. longa* each had 10 amino acids. All the examined samples included glycine and proline. Each species and its variants have a unique amino acid composition, which makes it possible to use it as a distinguishing feature. The distinctness and relatedness of the species are both reflected in the amino acid HPTLC screening results (Dagawal and Bhogaonkar 2021).

The use of *Curcuma inodora* Blatt by traditional healers for a variety of diseases is supported by the presence of several bioactive components. Table 17.1 presents the information on phytochemicals in *Curcuma inodora* plant. However, isolation of specific phytochemical components and their bioactivity testing could lead to a brand-new medication (Jadhao and Bhuktar 2019).

17.7 Breeding for improvement and conservation

Breeding is essential for obtaining new varieties with desirable characteristics like disease resistance, better yield, higher nutrient content, aesthetic appearance of flower and fruits. It ensures the availability of natural resources for future generations and safeguards biodiversity. For decades, improved varieties in horticultural crops have been developed through breeding. Selectively crossbreeding the plants with desirable traits led to the development of novel variants in many ornamental and medicinal

Table 17.1 Phytochemicals in *Curcuma inodora* and its effects (Dagawal and Bhogaonkar 2021; Jadhao and Bhuktar 2019)

Name of compound	Molecular formula	Effects	Reference
Ganoderic acid	$C_{30}H_{44}O_3$	Improves nervous system, eyesight and liver functions, helps internal organs, improves memory, increases vitality, strengthens spleen function, calms the soul, improves lung function, Protects the kidneys, Improves the function of ears, joints, muscles, helps skin	(Wasser 2005)
Trp Gln Trp	$C_{27}H_{30}N_6O_5$	Melatonin and serotonin production which help to regulate apatite, sleep-wake cycle, mood and pain	(Pmc_7699789 n.d.)
N-(1-methyl-2-hydroxy-2-phenylethyl) arachidonyl amine	$C_{29}H_{43}NO_2$	unknown	https://pubchem.ncbi.nlm.nih.gov/compound/5283415
Methylprednisolone succinate	$C_{26}H_{34}O_8$	Used against arthritis, blood disorders, certain cancers, eye disorders, allergic reactions	(Al-Maini and Urowitz 2007)
Prednicarbate	$C_{27}H_{36}O_8$	Helps against itching, swelling, discomforts due to skin conditions	(15657633 n.d.)
N-Palmitoyl-L-serine phosphoric acid	$C_{19}H_{38}NO_7P$	Used against human osteosarcoma	(Jan et al. 2003)
Salmeterol	$C_{25}H_{37}NO_4$	Controls asthma, prevent bronchospasm,	(Peters and Faulds 1995)
Digoxin	$C_{41}H_64O_{13}$	Treat heart diseases	(Virgadamo et al. 2015)

(continued)

Table 17.1 (continued)

Name of compound	Molecular formula	Effects	Reference
Proline	$C_5H_9NO_2$	Wound healing, Collagen formation, Cartilage generation, skin repairing, joint repairing, healing of gut line	(Vettore et al. 2021)
Lucin	$C_6H_{13}NO_2$	Protein synthesis, tissue regeneration, metabolism	(Casperson et al. 2012)
Phenylalanine	$C_6H_{11}NO_2$	Melanin production	(van Ginkel et al. 2019)
Histidine	$C_6H_9NO_2$	Used to treat neurological disorders, atopic dermatitis, metabolic syndrome, diabetes, uraemic anaemia, ulcers, inflammatory bowel diseases, malignancies, and muscle performance	(Holeček 2020)
Arginine	$C_6H_{14}NO_2$	Immune regulation, blood pressure regulation	(McRae 2016)
Glycine	$C_2H_5NO_2$	Liver and heart health, reduce muscle loss, reduce diabetic risks, improve sleep	(Razak et al. 2017)

plants (Ahmad et al. 2020; Carlen 2016; Wang et al. 2020). Genetic variations are the foundation for all sorts of breeding approaches which are described in the paragraphs below.

17.8 Different breeding approaches

Multiple breeding approaches are available and being used in horticultural plants to develop new varieties with improved trains. Conventional breeding techniques, mutation breeding, hybridization, genetic modification, marker assisted selection (MAS) and genomic selection are some of the approaches often sued for plant improvement. Conventional breeding involves the selection and crossing between plants showing desirable traits to produce a progeny with better characteristics. In mutation breeding, mutations are induced by using different chemical or physical mutagen. Further, positive mutants (those which are showing improvement in traits) are selected and bred to create a new cultivar. Genetic modification is the insertion, deletion or modification in specific genes which leads to change in the phenotype, hence improvement in properties. This technique is more precise compared to the previous ones as here specific genes are targeted for modification. In MAS, different molecular markers like ISSR, RAPD, microsatellite (SSR), SCoT etc. are used to locate the gene which is responsible for specific phenotype in the entire genome. Thus, breeders don't have to wait till the phenotypic expression of the gene. Genomic selection uses genomic data to select a suitable candidate gene for modification (Varshney and Dubey 2009). Ayer (2017) reviewed the efforts made for quality improvement in turmeric (*C. longa*) through advanced biotechnological breeding approaches including MAS, genome selection (Ayer 2017). Recently, genetic diversity analysis was performed in 152 cultivars of *C. longa*, having high curcumin content i.e. > 1.50 % and were collected from four different states of India by using 39 SSR primers. The study identified the curcumin rich genotypes of *C. longa* which are useful for future crop improvement programs. Additionally, highly heterogenous cultivars of *C. longa* were also reported which need to be protected through cryopreservation for future generations (Gogoi et al. 2023). Genetic diversity analysis of 24 local cultivars of *Curcuma longa* from Anand in Gujrat, India by implementing SSR markers led to the identification of a potent *Curcuma* cultivar named as Kasturi (ATG-21) with superior properties in terms of fresh and dry finger rhizome yield per plant, curcumin content and rhizome oil content (Lavudya et al. 2024). Significant variation was reported in the natural population of *C. inodora* plant hence, breeding studies were not performed in large number. As the plant has ornamental value due to its beautiful inflorescence, scientists implemented mutation breeding to get multiple inflorescences on the same plant, to get color variation in the inflorescence, to get plants with more height. Varying concentration (1–3%) of three popular mutagens namely EMS, acridine and colchicine were used to induce mutations in *C. inodora*. The changes observed were in leaf size, leaf number, plant height, initiation of flowering and size of inflorescence. Interestingly, emergence of two inflorescence was observed by treating the rhizome with 1% EMS for 12 hours (Prabhukumar et al. 2015).

17.8.1 In vitro culture and micropropagation

Micropropagation is an effective technique for achieving multiple goals like conservation of endangered or near threatened plants, regeneration of positive mutants into complete plant showing improved properties, mass scale propagation by ensuring the continuous availability of miniature planting material throughout the year and development of genetically uniform plant with desirable color in case of ornamental plants (Suprasanna and Jain 2022). Development of *in-vitro* culture protocols for *C. inodora* is much needed as it is a near threatened plant with a very beautiful inflorescence due to which it has ornamental potential. Despite having tremendous medicinal value and ornamental potential, *C. inodora* plant is still underexplored due to its endemic nature and seasonal appearance. Additionally, variability in inflorescence color and pattern makes its identification very difficult. Micropropagation protocols are available for various species of genus *Curcuma* which will be a great lead for the similar study with *C. inodora*. *Paracurcuma* and *Eucurcuma* are the ornamental gingers from Thailand which are exported to other countries due to their amazingly beautiful inflorescence. To fulfil the ever-increasing demand, *in-vitro* propagation was explored to achieve the rapid shoot multiplication which may leads to the development of miniature planting material in short span of time. Young inflorescence was used as explant and significant shoot formation was observed in modified MS medium, supplemented with a combination of NAA and BA (Ruamrungsri et al. 2022). Successful regeneration (85%) of complete plant under greenhouse conditions was reported in *C. karnatakensis* through direct as well as callus mediated organogenesis. Rhizome bus was used as explant and it was cultured on MS agar medium supplemented with appropriate combinations of growth regulators (Shanthala et al. 2021). *In-vitro* plantlet regeneration studies were performed in multiple species of genus *Curcuma* by using different explants like rhizome buds, leaf base, petiole and stem in *C. longa* (Salvi et al. 2002), leaf sheath in *C. amada* (Soundar Raju et al. 2013), axillary bud explant in *C. cungustifolia* (Das Bhowmik et al. 2016)and shoot buds in *C. caesia* ((Haida et al. 2022). As vegetative propagation through rhizome is the dominant method of propagation in the genus *Curcuma*, micro-rhizome (small sized rhizome) is an effective way of germplasm conservation. Micro-rhizomes are easy to transport and regenerate therefore, very promising for long-term conservation of germplasm. Micro-rhizomes had been successfully developed in *C. longa* and *C. caesia* by lowering the sucrose concentration in the culture medium (Brijesh and Ajjappala 2022). Table 17.2 summarizes the micropropagation attempts made with *Curcuma* spp.

Table 17.2 Micropropagation in *Curcuma* sp. (Brijesh and Ajjappala 2022)

Species	Explants	Nutrient media	Inference	Reference
Curcuma longa	Rhizome buds	MS+BAP, Kinetin Riboside, Zeatin, DMAP+Adenine+Adenine Sulphate+Metatopolin (10 μM) + IAA (1 μM)	Multiplication of shoots are more efficient in liquid medium	(Salvi et al. 2002)
	Leaf sheath	MS+4.49μM 2,4-D+1.32, 2.20+2.64 μM BAP.	Regeneration of plants from somatic embryos	(Raju et al. 2006)
	Leaf, petiole, stem, and root explants embryogenic cell-suspension derived from roots	Schenk and Hildebrandt (SH)+ 0.5 mg/L TDZ+various concentrations of BAP	Production of shoots and roots	(Jie et al. 2019)
	Leaf sheath, leaf lamina, leaf base, apical buds, and lateral buds	MS+3 mg/L 2,4-D+0.8 mg/mL Kinetin	Callus induction	(Gurav et al. 2020)
Curcuma amada	Leaf sheaths	MS+2.0 mg/L 2,4-D+0.5 mg/L BAP+3% sucrose	Embryonic callus induction	(Soundar Raju et al. 2013)
Curcuma angustifolia	Axillary buds	MS+4.4–22.2 μM BAP+2.9–5.7 μM IAA+2.3–23.2 μM Kn+2.7–5.4 μM NAA+67.8-271.5 μM Adenine Sulphate	Production of multiple shoots	(Das Bhowmik et al. 2016)
Curcuma caesia	Rhizome buds	MS+3.0 μM BA+0.5 μM NAA	Production of shoots	(Shahinozzaman et al. 2013)
	Shoot buds	MS+B5+15 μM of BAP d 6 μM of IBA	Callus induction	(Haida et al. 2022)
Curcuma zedoaria	Axillary bud	MS+BAP (2.2–22.2 μM)+ Kn (2.3–23.2 μM)+ IAA (2.9–11.4 μM)+ NAA (2.7–10.2 μM NAA)+ Adenine Sulphate (33.9–203.6 μM)	Production of roots and shoots	(Jena et al. 2020)

17.9 Diversity analysis of natural or regenerated plants through molecular markers

Existence of intraspecific variations is a key factor behind the varying phytochemical profiles of medicinal plants. Analysis of genetic diversity among natural populations is a track to locate a genotype with desirable features. Similarly, to produce true-to-type plants, confirmation of clonal fidelity is an essential step. A variety of DNA-based molecular markers like ISSR, SSR, SCoT, CBDP etc. were successfully implemented in multiple *Curcuma* sp. for genetic diversity and clonal fidelity analysis. (Jena et al. 2019) implemented ISSR markers to ensure genetic uniformity in *C. aungustifolia*. ISSR markers were also utilized to confirm the genetic homogeneity of regenerated plants through *in-vitro* culture in *C. zedoaria* (Jena et al. 2019). Start Codon Targeted (SCoT)-PCR system proved to be an effective tool for analysis of genetic diversity and phylogenetic relationship among cultivars of *C. longa* from Kerala, India (Mondal and Bandyopadhyay 2024). Identification of *C. inodora* becomes very difficult due to existence of intraspecific as well as interspecific morphological variations. ISSR markers were applied to derive a phylogenetic relationship among *C. longa*, *C. pseudomontana* and 12 varients of *C. inodora* (Dagawal and Bhogaonkar n.d.-b). A total 305 bands were obtained with 5 ISSR primers out of which 299 were polymorphic. Varying environmental and geographical factors create variations in genotype as well as phenotypes. This highlights the medicinal and ornamental potential of *C. inodora.*

17.10 Conclusion and prospects

Despite of showing great genetic diversity, medicinal properties (due to the presence of wide range of phytochemicals) and aesthetic look, the *C. inodora* plant is infamous. Maharashtra is the one biodiversity hotspot for this plant so, the plant is known and used only by local tribes of Maharashtra. Due to the excessive exploitation by local population, it gained status of near threatened plant and became endangered in Peninsular India. Like other common *Curcuma* species, *C. inodora* plant also demands attention. Development of *in-vitro* propagation strategies will surely be helpful in mass propagation and conservation of this near threatened plant. Additionally, hairy root culture and mutation breeding techniques has potential to generate plantlets with higher levels of alkaloids, phenolic compounds, flavonoids, volatile oils in rhizome and leaves. Mutation breeding might potentially be used to enhance the ornamental value of *C. inodora* plant by generating variations in flower in terms of size, number, and color. Further, implementation of DNA based molecular markers is beneficial to confirm the genetic fidelity of mutants as well as *in-vitro* regenerants to produce true-to-type plants with remarkable ornamental potential. To conclude, *C. inodora* plant not only be conserved for future but also needs to be mass propagated for its medicinal and ornamental potential. This opens the new avenues for employment generation and economic growth of local tribes.

References

15657633 (n.d.)

Acharya V, Sharma V, Patra PK, Naik ML, Kanungo VK (2012) Plants used by kamar, gond and halba tribe of Dhamtari district of Chhattisgarh for relief of sickle cell disease. Recent Res Sci Technol 4(3):1–3

Ahmad R, Anjum MA, Balal RM (2020) From markers to genome based breeding in horticultural crops: an overview. Phyton 89(2):183

Alluri R, Satla S (2017) Comparative In vitro antidiabetic and antioxidant activity of Pulicaria wightinia, *Curcuma* inodora, Derris scandens leaf extracts. Res J Pharm Technol 10(6):1615–1620

Al-Maini M, Urowitz M (2007) Chapter 45—Systemic Steroids. In: Tsokos GC, Gordon C, Smolen JS (eds) Systemic Lupus Erythematosus, pp 487–497. https://www.sciencedirect.com/science/article/pii/B9780323044349500567

Apavatjrut P, Anuntalabhochai S, Sirirugsa P, Alisi C (1999) Molecular markers in the identification of some early flowering *Curcuma* L.(Zingiberaceae) species. Ann Bot 84(4):529–534

Awika JM, Rooney LW (2004) Sorghum phytochemicals and their potential impact on human health. Phytochemistry 65(9):1199–1221

Ayer DK (2017) Breeding for quality improvement in turmeric (*Curcuma longa* L.): a review. Adv Plants Agric Res 6(6):1–4

Baghel SS, Baghel RS, Sharma K, Sikarwar I (2013) Pharmacological activities of *Curcuma caesia*. Int J Green Pharm (IJGP) 7(1)

Bharati D, Sonawane SA, Kanase KG, Undale VR, Abhynkar MM, Bhosale AV (2009) Evaluation of anti-inflammatory, antipyretic and wound healing activity of *Curcuma Inodora* (Zingibaraceae). Res J Pharmacogn Phytochem 1(1):35–40

Bhogaonkar MDP (n.d.-a) HPTLC fingerprinting of phenolics profile of three *Curcuma* species

Bhogaonkar MDP (n.d.-b) HPTLC fingerprinting of phenolics profile of three *Curcuma* species

Bhogaonkar PY, Devarkar VD (1999) Additions to the Flora of Melghat. In: The directorate project tiger Melghat, Amravati, India (Technical Bulletin No. VII)

Bhogaonkar PY, Kadam VN (2007) Plant related beliefs and taboos of Banjara tribe of Umarkhed (Dist. Yavatmal MS.). Indig Knowl: Appl 97

Brijesh, H., and Ajjappala, B. (2022). Micropropagation strategies in medicinally important turmeric (*Curcuma* sp): Current research and future challenges. J Appl Biol Biotechnol

Burch D (1998) The commercial market for gingers in the US Heliconia Soc. Int Bul 9(3):1–4

Carlen C (2016) Breeding and cultivation of medicinal plants. In: Herbal medicines: development and validation of plant-derived medicines for human health. CRC Press, Boca Raton, FL, USA, pp 79–91

Casperson SL, Sheffield-Moore M, Hewlings SJ, Paddon-Jones D (2012) Leucine supplementation chronically improves muscle protein synthesis in older adults consuming the RDA for protein. Clin Nutr 31(4):512–519

Dagawal M, Bhogaonkar P (2020) HPTLC fingerprinting of phenolics profile of three *Curcuma* species. Int J Herb Med 8(2):88. www.florajournal.com

Dagawal, M., and Bhogaonkar, P. (n.d.-a). HPTLC screening of amino acid profile of curcuma species from Melghat forests Dist. Amravati Maharashtra

Dagawal M, Bhogaonkar P (n.d.-b) Infraspecific variations and Molecular characterization by using ISSR Markers in Curcuma Species

Dagawal M, Bhogaonkar P (2021) HPTLC screening of amino acid profile of *curcuma* species from Melghat Forests Dist. Amravati Maharashtra. Int J Pharm Sci Res 12(7):3750. https://doi.org/10.13040/IJPSR.0975-8232.12(7).3750-55

Das Bhowmik SS, Basu A, Sahoo L (2016) Direct shoot organogenesis from rhizomes of medicinal Zingiber Alpinia calcarata Rosc. and evaluation of genetic stability by RAPD and ISSR markers. J Crop Sci Biotechnol 19:157–165

Dhore MA, Joshi PA (1988) Flora of Melghat tiger reserve. Tech Ser 1

Gogoi A, Munda S, Paw M, Begum T, Siddiqui MH, Gaafar ARZ, Kesawat MS, Lal M (2023) Molecular genetic divergence analysis amongst high curcumin lines of Golden Crop (*Curcuma longa* L.) using SSR marker and use in trait-specific breeding. Sci Rep 13(1):19690

Gurav SS, Gurav NS, Patil AT, Duragkar NJ (2020) Effect of explant source, culture media, and growth regulators on callogenesis and expression of secondary metabolites of *Curcuma* longa. J Herbs Spices Med Plants 26(2):172–190

Haida, Z., Sinniah, U. R., Nakasha, J. J., and Hakiman, M. (2022). Shoot induction, multiplication, rooting and acclimatization of black turmeric (*Curcuma caesia* Roxb.): an important and endangered *Curcuma* species. Horticulturae 8(8):740

Holeček M (2020) Histidine in health and disease: metabolism, physiological importance, and use as a supplement. Nutrients 12(3):848

Ibn Sinâ HA (2015) Al-Qanun fi'l-Ṭibb (canon of medicine). Almaʿee, Tehran

Jadhao AS, Bhuktar AS (2019) Phytochemical profile of *Curcuma Inodora Blatt*. Rhizome extract. Think India J 22

Jagtap SD, Deokule SS, Bhosle SV (2006) Some unique ethnomedicinal uses of plants used by the Korku tribe of Amravati district of Maharashtra India. J Ethnopharmacol 107(3):463–469

Jan CR, Lu YC, Jiann BP, Chang HT, Wang JL, Chen WC, Huang JK (2003) Novel effect of N-palmitoyl-l-serine phosphoric acid on cytosolic Ca 2+ levels in human osteoblasts. Pharmacol Toxicol 93(2):71–76. https://doi.org/10.1034/j.1600-0773.2003.930203.x

Jena S, Ray A, Sahoo A, Sahoo S, Dash B, Kar B, Nayak S (2019) Rapid plant regeneration in industrially important *Curcuma* zedoaria revealing genetic and biochemical fidelity of the regenerants. 3 Biotech 10(1):17. https://doi.org/10.1007/s13205-019-2009-9

Jena S, Ray A, Sahoo A, Sahoo S, Dash B, Kar B, Nayak S (2020) Rapid plant regeneration in industrially important *Curcuma zedoaria* revealing genetic and biochemical fidelity of the regenerants. 3 Biotech 10(1):17

Jie EY, Ahn MS, Lee J, Cheon YI, Kim CY, Kim SW (2019) Establishment of a high-frequency plant regeneration system from rhizome-derived embryogenic cell-suspension cultures of *Curcuma longa* L. Plant Biotechnol Rep 13:123–129

Khumkratok S, Boongtiang K, Chutichudet P, Pramaul P (2012) Geographic distributions and ecology of ornamental *Curcuma* (Zingiberaceae) in Northeastern Thailand. Pak J Biol Sci: PJBS 15(19):929–935

Km, P., Mamiyil, S., and Thomas, V. (2013). Effect of temperature and light on the promotion of off-season flowering in Island Purple ginger, Boesenbergia siphonantha (Baker) M. Sabu et al. (Zingiberaceae)—A promising ornamental ginger from Andaman Islands, vol 17, pp 144–147

Kress WJ, Prince LM, Williams KJ (2002) The phylogeny and a new classification of the gingers (Zingiberaceae): evidence from molecular data. Am J Bot 89(10):1682–1696

Kumar KMP (2013) Effect of temperature and light on the promotion of off-season flowering in Island Purple ginger, Boesenbergia siphonantha (Baker) M. Sabu et al. (Zingiberaceae)–a promising ornamental ginger from Andaman Islands. J Hortic For Biotechnol 17(1):144–147

Kumar KMP, Sabu M, Thomas VP (2013) Effect of temperature and light on the promotion of off-season flowering in island purple ginger, Boesenbergia siphonantha (Baker) M. Sabu et al. (Zingiberaceae)-A Promising Ornamental Ginger from Andaman Islands. For Biotechnol 17(1). www.journal-hfb.usab-tm.ro

Lavudya S, Patel MP, Kumar S, Singh A, Sakure AA, Pandya M (2024) Identification of elite genotypes through morpho-biochemical and molecular characterization of turmeric (*Curcuma longa* L.). S Afr J Bot 166:161–168

Lechtenberg, M., Quandt, B., and Nahrstedt, A. (2004). Quantitative determination of curcuminoids in *Curcuma* rhizomes and rapid differentiation of *Curcuma domestica Val.* and *Curcuma xanthorrhiza Roxb.* by capillary electrophoresis. Phytochem Anal: Int J Plant Chem Biochem Tech 15(3):152–158

Li S, Yuan W, Deng G, Wang P, Yang P, Aggarwal B (2011) Chemical composition and product quality control of turmeric (*Curcuma longa L.*)

Mangaly JKL (1993) A taxonomic revision of the South Indian species of *Curcuma Linn.* (Zingiberaceae). Rheedea 3:139–171

McRae MP (2016) Therapeutic benefits of L-arginine: an umbrella review of meta-analyses. J Chiropr Med 15(3):184–189

Mohanan K (2013) Variability studies in 'Peacock Ginger', *Kaempferia elegans Wall.* (Zingiberaceae). Ann Plant Sci 2(05):138–140

Mondal S, Bandyopadhyay M (2024) Characterization of genetic diversity in *Curcuma longa L.* cultivars using start codon targeted (SCoT) markers. S Afr J Bot 168:452–461

Mood J, Larsen K (2001) New *Curcuma*s from South-east Asia. New Plantsman 8(4):207–217

Nagore P, Lokhande P, Mujawar H (2023) In-vitro Bioevaluation, pharmacokinetics and molecular docking study of unexplored Bisabolol-rich *Curcuma inodora Blatt.* essential oil from Konkan region: a biodiversity hotspot. Plant Sci Today 10(2):223–231

Padal SB, Ramakrishna H, Devender R (2012) Ethnomedicinal studies for endemic diseases by the tribes of Munchingiputtu Mandal, Visakhapatnam district, Andhra Pradesh, India. Int J Med Aromat Plants 2(3):453–459

Patil HM, Bhaskar VV (2006) Medicinal uses of plants by tribal medicine men of Nandurbar district in Maharashtra

Peters DH, Faulds D (1995) Salmeterol: an appraisal of its quality-of-life benefits and potential pharmacoeconomic positioning in asthma. PharmacoEconomics 7:562–574

Phatak VG, Oza GM (1959) Occurrence of *Curcuma inodora Blatt.* at Pavagadh (Gujarat). J Bombay Nat Hist Soc 56:368–369

pmc_7699789. (n.d.)

Prabhu Kumar KM, Sabu M, Thomas VP, Nimisha G, Mohanan KV (2010) Red ginger: a promising crop for floriculturists. Plant Hortitech 10(1):40–43

Prabhu Kumar KM, Thomas VP, Sabu M, Prasanth AV, Mohanan KV (2014) Systematic and variability studies on "hidden purple ginger", *Curcuma inodora Blatter* J. (Zingiberaceae)—an endemic promising ginger from Peninsular India. Webbia 69(1):123–130. https://doi.org/10.1080/00837792.2014.913353

Prabhukumar KM, Thomas VP, Sabu M, Prasanth AV, Mohanan KV (2015) Induced mutation in ornamental gingers (Zingiberaceae) using chemical mutagens viz. colchicine, acridine and ethyl methane sulphonate. J Hortic For Biotechnol 19:18–27

Rajkumari S, Sanatombi K (2017) Nutritional value, phytochemical composition, and biological activities of edible *Curcuma* species: a review. Int J Food Prop 20(sup3):S2668–S2687

Raju VS, Reddy CS, Ragan A (2006) *Curcuma* L. (Zingiberaceae) in Andhra Pradesh: a preliminary study. J Eco Taxon Bot 30:773

Ramarao N, Rajendran A, Henry AN (2000) Increasing the secretion of breast milk-indigenous practices in Andhra Pradesh. AncT Sci Life 19(3–4):176

Razak MA, Begum PS, Viswanath B, Rajagopal S (2017) Multifarious beneficial effect of nonessential amino acid, glycine: a review. Oxidative Med Cell Longev

Ruamrungsri S, Akarapisarn A, Teerarak M (2022) Micropropagation of young inflorescence *Curcuma* hybrid In vitro. Int J Agric Technol 2022 18(3):1355–1366

Sabu M (2006) Zingiberaceae and Costaceae of south India. Indian Association for Angiosperm Taxonomy

Sabu M, Thomas VP, Kumar KMP, Mohanan KV (2011) Package of practices of ornamental gingers. Indian Association for Agniosperm Taxonomy

Salvi ND, George L, Eapen S (2002) Micropropagation and field evaluation of micropropagated plants of turmeric. Plant Cell Tissue Organ Cult 68:143–151

Sangale MP, Kshirsagar SR (n.d.) Taxonomy and occurrence of some rare, endangered and endemic plants of Sahyadri Ranges of Nashik region. J Basic Sci 5(2):1–7

Sasikumar B (2005) Genetic resources of *Curcuma*: diversity, characterization and utilization. Plant Genet Resour 3(2):230–251

Shah, G. L., and Gopal, G. V. (1982). ethnobotanical profile of the dangies. J Econ Taxon Bot

Shahinozzaman M, Ferdous M, Faruq M, Azad M, Amin M (2013) Micropropagation of black turmeric (*Curcuma caesia* Roxb.) through in vitro culture of rhizome bud explants. J Cent Eur Agric

Shanthala AA, Dilkalal A, Umesh TG (2021) An efficient Invitro approach for direct and callus mediated regeneration of *Curcuma karnatakensis*–an endemic plant of karnataka. J Herbs Spices Med Plants 27(3):229–241. https://doi.org/10.1080/10496475.2020.1806166

Soundar Raju C, Kathiravan K, Aslam A, Shajahan A (2013) An efficient regeneration system via somatic embryogenesis in mango ginger (*Curcuma amada* Roxb.). Plant Call Tissue Organ Cult (PCTOC) 112:387–393

Suprasanna P, Jain SM (2022) Biotechnology and induced mutations in ornamental plant improvement. Acta Hortic 1334:1–12. https://doi.org/10.17660/ActaHortic.2022.1334.1

Valeton T (1918) New notes on the Zingiberaceae of Java and Malaya n archipelago. Bull Jard Bot Buitenzorg Ser 2(27):1–166

van Ginkel WG, van Reemst HE, Kienstra NS, Daly A, Rodenburg IL, MacDonald A, Burgerhof JGM, de Blaauw P, van de Krogt J, Santra S (2019) The effect of various doses of phenylalanine supplementation on blood phenylalanine and tyrosine concentrations in tyrosinemia Type 1 patients. Nutrients 11(11):2816

Varshney RK, Dubey A (2009) Novel genomic tools and modern genetic and breeding approaches for crop improvement. J Plant Biochem Biotechnol 18:127–138

Velayudhan KC (1996) The conspectus of the genus *Curcuma* in India. J Econ Tax Bot 20:375–382

Velayudhan KC (2015) Ornamental *Curcuma* species in Western Ghats of India. Indian J Plant Genet Resour 28(03):269–277

Vettore LA, Westbrook RL, Tennant DA (2021) Proline metabolism and redox; maintaining a balance in health and disease. Amino Acids 53(12):1779–1788

Virgadamo S, Charnigo R, Darrat Y, Morales G, Elayi CS (2015) Digoxin: a systematic review in atrial fibrillation, congestive heart failure and post myocardial infarction. World J Cardiol 7(11):808

Wang W, Xu J, Fang H, Li Z, Li M (2020) Advances and challenges in medicinal plant breeding. Plant Sci 298:110573

Wasser SP (2005) Reishi or ling zhi (Ganoderma lucidum). Encycl Diet Suppl 1:603–622

Correction to: Advances in Dill (*Anethum graveolens* L.) Molecular Breeding Strategies

Mai M. Wahba, Mohamed A. Abdein, Haifa A. S. Alhaithloul, Suliman M. S. Alghanem, and Maha M. Elshamy

Correction to:
Chapter 4 in: J. M. Al-Khayri et al. (eds.), *Biodiversity and Genetic Improvement of Herbs and Spices*, Advances in Plant Breeding Strategies 8, https://doi.org/10.1007/978-3-031-81838-7_4

In the original version of the book, the surname Abedin has been removed and replaced by Abdein. The book and the chapter have been updated with the changes.

The updated version of this chapter can be found at
https://doi.org/10.1007/978-3-031-81838-7_4

J. M. Al-Khayri et al. (eds.), *Biodiversity and Genetic Improvement of Herbs and Spices*, Advances in Plant Breeding Strategies 8, https://doi.org/10.1007/978-3-031-81838-7_18

Index

J. M. Al-Khayri et al. (eds.), *Biodiversity and Genetic Improvement of Herbs and Spices*, Advances in Plant Breeding Strategies 8, https://doi.org/10.1007/978-3-031-81838-7

H

I

MIX
Papier aus verantwortungsvollen Quellen
Paper from responsible sources
FSC® C105338

If you have any concerns about our products,
you can contact us on
ProductSafety@springernature.com

In case Publisher is established outside the EU,
the EU authorized representative is:
Springer Nature Customer Service Center GmbH
Europaplatz 3, 69115 Heidelberg, Germany

Printed by Libri Plureos GmbH
in Hamburg, Germany